*In the mythology of the Indians of the
Northern Pacific Coast, Raven brought light to the world,
freed the tides, and formed the lakes and rivers.
He married Fog Woman, who provided him with Salmon.
But because he was voracious, Raven lost
Salmon and ruined his world.*

CHARLES E. WARREN

*The Department of Fisheries and Wildlife
Oregon State University, Corvallis, Oregon*

IN COLLABORATION WITH
PETER DOUDOROFF

Biology and Water Pollution Control

W. B. SAUNDERS COMPANY
Philadelphia • London • Toronto

W. B. Saunders Company: West Washington Square
Philadelphia, Pa. 19105

12 Dyott Street
London, WC1A 1DB

1835 Yonge Street
Toronto 7, Ontario

Biology and Water Pollution Control

ISBN 0-7216-9120-X

Print No.:   9   8   7   6   5   4   3   2

PREFACE

This book of mine has little need for preface, for indeed it is 'all preface' from beginning to end.

D'Arcy Thompson, 1942

We must suppose that D'Arcy Thompson was serious when he thus began the preface to his monumental book, *On Growth and Form.* Though a long discourse on its subject, his book remains principally one of possibilities for the future, possibilities to stimulate and guide the studies of those who follow him. Hopefully, the book we are here beginning will also be a preface, a preface to the broader application of biology to certain of man's environmental problems. The reader must inevitably come to the conclusion that this book is more one of possibilities than of accomplishments in water pollution biology. Even though biology has contributed to the solution of some problems of environmental pollution, the possibilities of biology have barely begun to be exploited.

A reader seeking in this book answers to water pollution problems will find few. Any such answers must go beyond the realm of biology. They must be worked out through our governmental institutions. They must be based not only on sound biology but also on sound sociology and technology. Few too in this book can be the answers to some of the most important biological questions, for such is the state of biological knowledge. Here we can pose some of these questions, and here we can suggest some of the most promising ways to obtain answers.

We hope to explain to the widest possible audience the biology that is relevant to water pollution control. We want this book to be read not only by biologists but also by engineers, social scientists, economists, industrial managers, public administrators, politicians, and the interested public. All are involved in the effort that we hope will lead to wiser use of our waters. All can contribute more effectively, we believe, if they better understand the possible role of biology in water pollution control. Many biological ideas, including some of the most important ones, are difficult or elusive. Nevertheless, we think that this book can be read with understanding not only by biologists but also by others who are interested. Throughout, we have endeavored to begin with beginnings and to bring to the reader the current biological thought that we consider to be relevant to water pollution control.

Biology can aid man more than it has in his struggle for a better environment. Recognizing the differences as well as the problems resulting from man's cultural and technological development, biologists still view the biological problems of man in much the same way as they do those of other animals. This view brings some perspective and a realm of knowledge to bear on man's problems. In agriculture and medicine, we have seen the value of biology. Hopefully, we will see more use of biological thought and knowledge in the search for solutions to the problems of human population growth. The problems of environmental pollution seem sometimes to be almost hopelessly entangled with population growth.

Biological thought, investigation, and planning have contributed all too little to the search for solutions to our problems of environmental pollution. Society has encouraged relatively few of its biologists to undertake this work; such neglect can no longer be afforded. Those who began applying biology to problems of water pollution were able to exploit only a few of the powerful approaches the great traditions of biology have given us. Perhaps this book can serve as a reminder of how truly relevant to man's environmental problems are these traditions and approaches.

The length of this book derives not from its depth but from its breadth. Each chapter is a bare introduction to its topic. To some readers, the coverage of particular chapters will seem inadequate. To readers having other backgrounds, the same chapters may seem to cover too much. This difficulty was recognized when we decided to write for a wide audience, and it cannot really be resolved. We can only ask for the patience of our readers. Some readers may be dismayed that our coverage of the literature and knowledge of water pollution biology is not wider. Our aim has been to explain, for those who wish to know, the biology that is relevant to water pollution control. We have in addition attempted to present an adequate picture of the state of knowledge in this area, without giving detailed consideration to much of the literature. More than this, space would not permit. Those who would read further will find ample references to books, literature reviews, and original investigations.

Some years ago, Peter Doudoroff, my long-time friend and colleague, and I envisioned this book as one whose writing we would share equally. In the years that have followed, Pete's responsibilities made this impossible. I have been fortunate, however, to have his counsel and help in many ways, and our readers will benefit from his broad experience, which shows in many chapters. His contributions to Chapters 2, 3, 5, 13, and 14 have been particularly great, and he has reviewed in his meticulous way every word in the book. But in assuming responsibility for the writing, I also assumed responsibility for the content, not all of which may satisfy Pete or our readers.

I have enjoyed immensely the writing of this book. If, as an introduction to biology and water pollution control, it can be read with enjoyment, I will be doubly rewarded.

Corvallis, Oregon   CHARLES E. WARREN

ACKNOWLEDGMENTS

Now as this particular endeavor nears completion, I find difficulty in properly expressing my keen awareness and appreciation of the knowledge and efforts of those who have made it possible. One man alone cannot do much, either in science or in making one of science's stories available. Many well know that they have contributed substantially to the making of this book. But there are many others, most of whom have passed on, whose biological studies are its foundation.

Sometime in the fall of 1965, I was chatting with Thomas G. Scott, my department head, about notes I had been making for a book I hoped one day to write. It was his suggestion that I begin immediately. This suggestion and a sabbatical leave granted to me by Oregon State University during the 1966–67 academic year prevented this book from being later than it is. His encouragement made this task easier.

Many of the topics in this book I could not have treated with confidence without the help of individuals who know them better than do I. Their thoughtful criticisms and suggestions have made this a better book, though probably not as good a book as they may have wished. John R. Donaldson reviewed Chapters 5, 6, and 20; Raymond C. Simon, Chapter 7; and C. Ladd Prosser, Chapters 7 and 8. Hugo Krueger helped me a great deal with the section on The Problem of Toxic Substances, which is in Chapter 8. K. Radway Allen reviewed Chapters 11, 16, 17, and 19; Gerald E. Davis, Chapters 11 and 17; and Terry D. Roelofs, Chapter 16. Colin G. Beer and John D. McIntyre reviewed Chapter 12. Chapter 21 was reviewed by H. A. Hawkes and Donald C. Phillips. Robert W. Brocksen and John B. Sprague read the entire manuscript. Carl E. Bond helped me with some matters on the structure of fish. Lyle D. Calvin examined the mathematical representations I have used.

For nearly two decades now, my colleagues, the staff and students of the Pacific Cooperative Water Pollution Laboratories, have contributed greatly to knowledge of the biology relevant to water pollution control. Much of this book is theirs. While I have been writing it, Marjorie A. Jackson has done much of the office work that I should have done and has kept my schedule sufficiently free for me to write. She also found time to type the entire original draft and a good share of the final draft. Most of the final draft was typed by Velda D. Mullins.

Marthanne B. Norgren conscientiously checked all of the literature citations and prepared the bibliographic materials. In this, she had the assistance of our helpful science librarians. And she along with Dorothy K. Carson did the proofreading.

Mark E. Warren did all of the technical illustrations. To me, the hours we spent together in preparing these were one of the rewarding experiences of writing this book. William M. Reasons and his staff in the Photographic Service made our work easier. Virginia Taylor created the representation of the Indian legend of Raven and Fog Woman that appears on the cover. Edward Malin helped me with the details of this legend.

Twenty-five years ago, Roland E. Dimick introduced me to aquatic biology, and a few years later this foresighted man convinced me that I should work on problems related to water pollution. During the years that have followed, he has been my teacher, my colleague, and my very close friend. This book is his too.

Since the beginning of this endeavor, the enthusiastic and very competent assistance of the staff of W. B. Saunders Company has been all any writer could wish. Tyler Buchenau, a great biology editor, with knowledge, understanding, and patience, has shown me the way.

Finally, what can one say in appreciation of the role his family must inevitably play in such an undertaking? I think Peggy, my wife, and Mark, Daniel, and Catherine understand. And it was my father, Edward J. Warren, who introduced me to what Robert M. Hutchins has since called *The Great Conversation*.

CHARLES E. WARREN

CONTENTS

PART II   CONDITIONS OF LIFE IN THE AQUATIC ENVIRONMENT

PART III   MORPHOLOGY AND PHYSIOLOGY

PART IV ECOLOGY OF THE INDIVIDUAL ORGANISM

Chapter 23

The Roles of Water Pollution Biologists 387

PART I

Introduction

The changes that man brings about in his environment are determined largely by his necessities, his knowledge, and his values. Very early, man developed codes of behavior that worked to protect his environment and his well-being. But as technology, population, and patterns of living began their rapid change with the Industrial Revolution, early codes of behavior became inadequate to deal with the developing environmental problems of great technological, sociological, and ecological complexity. Efforts have been made to stem the environmental degradation that has followed, but only recently has general public awareness of the seriousness of these problems become great enough to offer hope that the tide may be turned. New codes of behavior deriving from new values are long overdue. These codes must allow for the necessities and a good share of the enlightened desires of man. But the problems are great, and if evolving codes are to guide man to the kind of world he should want, man's knowledge of his environment and of himself must be greatly increased.

1 Historical Background

What is all Knowledge too but recorded Experience, and a product of History; of which, therefore, Reasoning and Belief, no less than Action and Passion, are essential materials?

THOMAS CARLYLE, 1830, p. 84

The absence of romance in my history will, I fear, detract somewhat from its interest; but if it be judged useful by those inquirers who desire an exact knowledge of the past as an aid to the interpretation of the future, which in the course of human things must resemble if it does not reflect it, I shall be content.

THUCYDIDES, *c.* 431 B.C., p. 354

THE INDUSTRIAL REVOLUTION

Water pollution as we know it today—a problem afflicting large regions—began with the Industrial Revolution. When man hunted or herded, or when most of his then small numbers were tied to the land growing crops, the disposal of his wastes created no acute problem. Primitive man, first through biological evolution and then through cultural evolution, must have acquired behavior leading to practices of cleanliness. Such practices occur quite generally among other terrestrial vertebrates living in their natural environments. Mosaic law and the Zoroastrian religion provided sanitary codes undoubtedly based on health needs. Waste disposal must have created health and nuisance problems for large cities of early history, but, by today's standards, these cities were small and widely separated and their problems more local (Gray, 1940; Reynolds, 1946).

Advances in agricultural methods in the eighteenth century increased the productivity of the men growing crops and paved the way for some to leave the land and begin other occupations. Discoveries and inventions along with the use of capital during this period led to the development of early factories, where ever-greater numbers of people were employed in the growing cities. This early movement of people to the cities increased throughout the Industrial Revolution and continues today, even in the primarily agricultural countries. Population growth in the countries of the western world has increased city populations far more than those of rural areas. Improvements in agriculture, the greater productivity of the factory worker as well as the farm worker, the use of fossil fuels and electricity, and other improvements have in these countries generally permitted higher standards of living, in spite of Thomas Malthus' (1798) grim warning of food shortage, poverty, and disease.

Nevertheless, the rapid growth of cities in England in the early nineteenth century resulted in large concentrations of people in areas having no adequate community organization. In this period of rapid change, the economic condition of the factory worker was not always good, and conditions in the communities in which he lived became

3

rapidly worse. Sir Edwin Chadwick, in his survey of English cities in the 1840's, found water supply and waste disposal facilities almost indescribably foul (Finer, 1952). Contaminated water supplies led to cholera and typhoid epidemics, and John Snow implicated sewage-polluted water in the Broad-street Well cholera epidemic of 1854 in London (Wilson, 1952). Tremendous problems of providing safe water and adequate waste disposal as well as slow acceptance of the significance of the early work of bacteriologists—including the discovery of the typhoid bacillus by Eberth in 1880 (Morgan, 1952)—were impediments to correction of conditions leading to serious epidemics.

Sanitary conditions in the cities of the United States did not become quite so intolerable as in the cities of England. The Industrial Revolution came later in America, many cities were located on large waterways, and the Americans were able to benefit from many of the experiences of the English. H. L. Bowditch, Chairman of the State Board of Health of Massachusetts, and Rudolph Hering, for the National Board of Health, went to England and Europe to study waste disposal practices from 1870 to 1880. Still, cities in the United States had their own considerable problems of water supply and waste disposal. Chicago was disposing of wastes into Lake Michigan, which provided her water supply, even with epidemic typhoid prevalent. The Chicago Drainage Canal—constructed on the basis of Hering's recommendation to transfer these wastes to the Mississippi River drainage via the Illinois River—alleviated but did not solve this problem (Chase, 1965b, p. 49). This led to litigation beginning in 1900. The State of Missouri claimed that sewage from Chicago was a hazard to the residents of St. Louis who used unpurified water from the Mississippi below the confluence of the Illinois River (Fuller, 1912, pp. 246–253, 258–261; Chase, 1965c, p. 133). Other litigation concerning removal of water from Lake Michigan into another drainage basin continues to the present time (Glass, 1965).

WATER DISTRIBUTION, WASTE COLLECTION, AND WASTE TREATMENT

Systems of water supply and waste collection reached levels of advancement in the Mesopotamian Empires of Assyria and Babylonia that were not reached in modern civilization until the latter part of the nineteenth century (Gray, 1940). From water-flushed latrines in houses, laterals leading to sewers constructed of brick have been uncovered in the ruins of Eshnunna near Baghdad, ruins which have been dated about 2500 B.C. The upper levels of the ruins of Mohenjo-daro—about 2550 B.C.—of the Indus civilization contain evidence of latrine, bathing, and kitchen facilities connected by terra-cotta pipes to brick-lined pits having near their tops outlets to covered street drains. The Aegean civilization on the island of Crete—about 3000–1000 B.C.—transported water under pressure in pipes and used lateral and main sewers, parts of which are functional today. The so-called sewers of Rome had few house connections and were designed for the removal of surface and underground water, not human wastes, which were thrown into the streets from where they entered drains and passed to the Tiber. Conditions in the cities of Europe during the Middle Ages became unspeakable, apparently little thought or effort being given to providing sanitary water or waste disposal (Gray, 1940). Had Europeans been aware of the sanitary advances of early civilizations, perhaps conditions would not have been permitted to reach the extremes of the Industrial Revolution. But rapid growth of cities with little community organization still would have presented an obstacle difficult to surmount.

Sewers in the United States and Europe were originally designed for the removal of storm and other drainage water. The discharge of human wastes into sewers was forbidden by law in London until 1815, in Boston until 1833, in Paris until 1880 (Metcalf and Eddy, 1930, p. 10), and in Baltimore until the turn of the century (Chase, 1964b, p. 59). Eventually, foul conditions developing in cities led to direct introduction of these wastes into sewers. A law was passed in London in 1847 that required connection of all houses and cesspools to sewers. As late as 1922, 20,000 houses remained unconnected with the sewers in Baltimore (Metcalf and Eddy, 1930, p. 10). Introduction of wastes into storm sewers in many cities led to combined sewer systems; it was not until later that the advantages of separate sewers for high volume storm waters and for wastes became apparent. Before 1870 in the United

States, there were only two sewer systems designed according to engineering principles, Julius Adams having designed the Brooklyn system and E. S. Chesbrough the Chicago system, during the 1850's. Also about this time, the English engineer Thomas Hawksley demonstrated the feasibility of providing continuous supplies of water to houses. This and the water closet made possible a water-carriage system for disposal of wastes from the home.

Although the introduction of wastes into sewers improved living conditions around houses, it led to foul conditions in rivers and streams. Smaller streams and drainage channels were covered and eventually became parts of sewer systems. Waterways not converted to sewers became increasingly bad. This was particularly so in England, where cities were usually located on smaller rivers than in the United States. Odors arising from the Thames during 1858 and 1859 made life in London almost intolerable (Gray, 1940); conditions in some American waters were not much better, the Chicago Drainage Canal becoming known as "Bubbly Creek" and having in places a thick scum upon which people could walk safely (Fuller and McClintock, 1926, p. 5). In England, the "Nuisances Removal Act" was passed in 1855; but, in spite of the cholera epidemics, more attention appears for some time to have been given to pollutional interference with industrial and agricultural uses of water than to health effects.

Studies leading to the development of waste treatment practices were undertaken primarily to control pollutional conditions in rivers and streams. Owing to the smaller rivers in England, waste treatment received more attention there than in the United States in the latter part of the nineteenth century. The Chinese have maintained the fertility of their soils for centuries by returning to them all vegetable, animal, and human wastes. Victor Hugo, in *Les Misérables*, expressed great concern that Paris, in passing her wastes to the sea, was not similarly exploiting their value. Prince Albert in England was also concerned about such wastage of potential fertilizer. Irrigation with sewage was practiced to some extent in England, but only a few attempts were made in the United States. Land disposal of wastes never developed extensively in these countries owing to the large tracts of land needed adjacent to population centers, the sewage farms of Berlin occupying 43,000 acres in 1910 and those of Paris occupying 12,584 acres in 1924 (Chase, 1964c, p. 49). Chemical treatment with subsequent sedimentation was used to some extent in sewage treatment prior to the development of methods based on understanding of biological oxidation of organic matter, methods not requiring great tracts of land.

Chase (1964d, p. 50) has pointed out that in the late nineteenth century and early twentieth century:

Three fundamental concepts developed from the research of English, American and other scientists, namely: 1) aerobic biological conversion of unstable organic matter in solution to stable matter; 2) anaerobic biological decomposition and stabilization of organic solids to relatively inert humus, and 3) disinfection by chlorination. Upon this triumvirate of basic principles established by the scientists are based all subsequent developments in the treatment of sewage and other liquid-borne organic wastes.

Sewage treatment as it has evolved may be divided into three groups of processes. Solid materials are first separated from the liquid sewage by screening devices and by settling basins or tanks, which are sometimes used in conjunction with chemical precipitation. The removal of most of the settleable solid materials is usually termed *primary treatment*. Oxidation of dissolved or colloidal organic materials in the liquid sewage, *secondary treatment*, is brought about by the activities of bacteria and other organisms in the presence of oxygen. This may be accomplished in large oxidation ponds, in so-called trickling filters, or by the activated sludge process. Trickling filters contain beds of rock having attached biological growth through which the sewage is passed. The activated sludge process involves passing sewage through tanks in which it is aerated and kept in contact with a bacterial floc produced by the process. Solid materials resulting from primary and secondary treatment are subjected to digestion under anaerobic conditions in tanks. The end products of this digestion are gases, liquids, mineral compounds, and nondigestible organic materials generally referred to as humus, which can be of value for agricultural purposes. There has been much concern in recent years over nuisance blooms of algae caused by increased concentrations of nitrates and phosphates in lakes.

These minerals may be present in high concentrations in effluents after secondary treatment. Such undesirable effluent characteristics have led to interest in further waste treatment, sometimes termed *tertiary treatment*, sometimes *advanced treatment*.

Prior to World War I, most waste treatment research and development was concerned with sewage, but industrial wastes of many sorts were becoming increasingly a problem. Many of these wastes could be treated by the processes developed for sewage, but not others, particularly those containing toxic metals. To the present day, we are confronted with the problem of removing toxic substances from large volumes of water or waste.

CONFLICTING INTERESTS

The changes in society that were set in motion with the beginning of the Industrial Revolution inevitably led to conflicting interests in the uses of water. Conditions in the rivers of England that rendered them unsuitable for industrial and agricultural uses apparently were initially responsible for the pollution control efforts of the mid-nineteenth century. Bacteriology was in its infancy. It was only with increasing public awareness of the dangers of discharging human wastes into domestic water supplies that concern for public health became an important stimulus for pollution control. Public health has continued to be a consideration; but with the development of chlorination, waters unsuitable for other uses could be made safe for human consumption. Thus, concern for public health was reduced to the point that it has not usually been a major impetus to water pollution control.

In England as well as in the United States, earliest attention was given to problems caused by domestic sewage, but textile, paper, leather, metal, and other industries were locating along rivers, and other problems needing attention developed. By 1915, the Royal Commission on Sewage Disposal reported (Chase, 1965a, p. 58):

It has, moreover, been clearly evident to us that manufacturing processes are tending to become more complicated, chemically and bacteriologically as well as mechanically, and that in the absence of systematic regulation waste liquors are likely to become a source of increasing pollution.

This trend toward increasing complexity of industrial processes and wastes has continued to the present day. Many industries require considerable amounts of water of high quality. Where their wastes have been discharged untreated, receiving waters have often been rendered unsuitable for other uses.

Some land use practices have had undesirable effects on the waters of the United States. Improper removal of timber has led to rapid water runoff and so caused erosion, siltation of streams, low summer flows, and high stream temperatures. Poor agricultural practices have resulted in similar damages. Mining for coal and metals became important in the nineteenth century, and drainage from mines made many waters unfit for any use. Later, petroleum wastes too became a major problem.

Thus, many surface waters in the United States became unfit for domestic, agricultural, and some industrial uses. Heated effluents in raising the temperature of receiving waters reduced their value for cooling. Other effluents made some waters useless for food processing. Corrosive wastes damaged industrial equipment and shipping facilities. Minerals made some waters unsuitable for use in boilers. Many industries and communities were forced, at additional expense, to obtain their water from greater distances.

The great Atlantic Coast fisheries on anadromous and estuarine species from New England to Chesapeake Bay declined drastically after about 1880. The Atlantic salmon (*Salmo salar*) and then the American shad (*Alosa sapidissima*) all but disappeared. Poor fisheries management, dams across rivers and streams, and pollution were responsible for this. The once fabulous clam and oyster fisheries of the Atlantic Coast have in large part been rendered useless since the turn of the century. Typhoid epidemics caused by consumption of shellfish from sewage-polluted waters led public health agencies to condemn vast areas, so preventing further harvest.

The conservation movement—which received much impetus from the White House Conference called by President Theodore Roosevelt on May 13, 1908—must be credited for most of the pollution control and abatement undertaken. The National Research Council (1966, p. 71) reported that:

It is perhaps of considerable significance that neither matters of human health nor industrial demands for water of high purity have, historically, inspired public action to the degree that it has been inspired by demands for pure water for swimming, recreation, fishing, and wildlife observation.

Nevertheless, in the face of opposition by some groups, and with general public apathy, the state and federal governments have until very recently been reluctant to enact effective water pollution control legislation. With regard to commercial fisheries alone, Dr. J. L. McHugh, fishery scientist and administrator, has been quoted (Wright, 1966, p. 133):

The fishing industry of the United States, however, consists of many small more or less independent segments, few of which are of dominant importance locally. Therefore, when urban or industrial development impinges upon fishery interests, the fisheries often suffer.

There have been, of course, many reasons for inaction over the deteriorating condition of our waters; no single group or interest bears the burden of responsibility for this. The problems resulting from the development of community-industrial complexes have no easy solutions. The economic importance to communities of their industrial segments has discouraged communities from pressing for corrective measures. And, if industries have been slow to seek adequate solutions for their waste disposal problems, so also have communities failed to approve the bonds necessary for sewage treatment facilities.

Ultimately, perhaps, the problem of waste disposal is more than one of conflicting uses of water. The discharging of all sorts of highly obnoxious materials into our waters is an affront to civilized values and damaging to our culture.

LAW AND REGULATION

Many of the basic rules of water use in the United States evolved as a part of the *common law* of England during the period between the Norman Conquest and the American Revolution. Common law developed bit by bit as judges made decisions that were logical extensions from previous decisions. The common law was mainly concerned with private rights, because a large proportion of the cases coming before judges

in feudal England involved conflicts between private litigants over land rights. Generally, water came to be considered a part of the land, and with ownership of the land went certain rights, *riparian rights*, regarding streams flowing across it in known and defined channels. Under common law, the riparian owner had absolute right to use a stream for domestic purposes or livestock water, ordinary or primary uses. For extraordinary or secondary uses like irrigation or power, so long as these were incidental to the use of the land, he also had rights, but they were more limited. He had no right to export the water or to use it for purposes unconnected with the land. The riparian owner, under common law, had the right to the water of the stream in its natural state. He could bring action against those responsible for changing its quality, even though he sustained no actual damages. Nevertheless, in England it has been held that one can obtain the right to change the quality of a stream after having done so continuously for 20 years (Klein, 1962, p. 18).

Although riparian rights are not recognized in some western states, they remain an important part of the law relating to water in most parts of the United States, where they have been modified in various ways to better meet local or regional needs (Banks, 1961, p. 157). Riparian doctrine could not be successfully applied in arid western states where large quantities of water were needed for mining and irrigation. From the few streams having adequate flow, it was sometimes necessary to transfer water great distances for use at other locations. In consequence, the *appropriative doctrine* of water rights evolved in western states. Unlike riparian rights, appropriative rights were established through use and could be lost through disuse, first users obtaining first rights. In some instances, appropriative doctrine was superimposed on riparian doctrine (Banks, 1961, p. 157). Under appropriative doctrine, the traditional concepts of priority have generally been applied when applications have competed for appropriation from a limited supply: domestic and municipal uses have been given first priority and irrigation second. No appropriative rights appear ever to have been given for the use of water for waste disposal (Banks, 1961, p. 158). The administration of appropriative rights is pursuant to state laws.

As municipal and industrial pollution became a problem, the courts tended to deal with pollution cases as nuisance problems. Thus, even when changes in water quality were the result of unreasonable uses, the riparian owner had to show that damage was done to him as an individual, if he were to be successful in litigation. Handling of pollution cases became punitive after the damage was done (McKee and Wolf, 1963, p. 29). Failure of riparian owners to undertake litigation frequently permitted serious degradation of water quality as a result of municipal and industrial development. This development often assumed such importance in the economy of areas that the courts, in balancing advantages and disadvantages, tended only to grant recovery of damages to the riparian owner and refused to enjoin continued pollution (Banks, 1961, p. 158). Deplorable water conditions and neglect of the public interest finally led most states to undertake water pollution control by statutory means.

The legislatures of many states gave the state health departments statutory powers and responsibilities for pollution control and gave corollary powers and responsibilities for the protection of fish and wildlife to appropriate agencies. The primary responsibilities for control usually fell on the sanitary engineering or environmental sanitation divisions of the health departments. The power of these departments depended on the common law, specific statutes, and quasi-legal rules and regulations of the boards of health. Such a foundation for pollution control primarily by health and wildlife agencies, with no agency having overall, final, adequate, or even clear authority, proved unsatisfactory and led several states to enact specific water pollution control legislation. McKee and Wolf (1963, p. 29) have stated that:

Generally, these newer acts supplement the existing statutes by providing for a single administrative agency that is responsible for control and abatement of water pollution in the state. Provisions are often made (a) for representation of the several interests concerned in water pollution, either on an advisory board or on the control board itself, (b) for the definition and determination of pollution, usually by stream standards or classifications, (c) for the enforcement of abatement orders, with penalties for non-compliance, and (d) for hearings and appeals of parties affected by actions of the agency.

Thus, in the United States, we have slowly moved away from water pollution control mainly by court action after the damage has been done and toward establishment of guidelines that help to protect the interests of all concerned with problems resulting from municipal and industrial growth. The resolution of complex water pollution problems can more reasonably be accomplished by well-defined administrative procedures, the courts being left "to deal only with the adjudication of individual rights as they become subjects of lawsuits" (Freeman, 1963, p. 87).

Recognition of national responsibility in water pollution control came much earlier in England than in the United States and led to adoption by Parliament of the "Public Health Act" in 1875.

This Act clearly recognized for the first time that care of public health was a national responsibility and established a system of local health administration setting down amongst other things the duties of local authorities with regard to the disposal and treatment of sewage. (Klein, 1962, p. 6)

The "Rivers Pollution Prevention Act" adopted in 1876 formed the basis of all legal action connected with water pollution until as recently as 1951, not only in England and Wales but also, with modifications, in Scotland and Ireland (Klein, 1962, p. 6).

The first specific federal water pollution control legislation in the United States, the "Rivers and Harbors Act of 1899," was solely for preventing impediments to navigation (Forsythe, 1961, p. 251). The "Public Health Service Act of 1912" authorized investigations of water pollution related to disease. In 1924, Congress adopted the "Oil Pollution Act" specifically for control of oil discharges in coastal waters where these caused damage to aquatic life, harbors and docks, and recreational facilities (Forsythe, 1961, p. 252).

With the federal government assuming no general responsibility for water pollution control, conditions in the waters of the United States became increasingly bad during the first half of the twentieth century. Few state or local governments had adequate statutory authority to prevent, control, or abate water pollution, and such authority as existed was rarely comprehensive in protecting all legitimate water uses. Many of those having vested interests opposed the passage of legislation that would enable effective control at the state level. Furthermore, many of the problems were of an interstate nature. Recognition of the need for federal entry into this area of public concern was thus long overdue when the "Water Pollution Control Act of 1948" was passed. Important as was this recognition, this act did not permit development of effective action by the federal government.

Passage of the "Water Pollution Control Act of 1956" marked the beginning of effective federal action. This act provided for: development of comprehensive water pollution control programs in cooperation with states; broadening already available technical assistance to states; conducting and contracting research; a national water quality network for collecting basic data; grants to aid in construction of waste treatment plants; grants to state and interstate agencies for water pollution control programs; cooperation among federal and state agencies to control pollution from federal installations; and encouragement of interstate cooperation and adoption of uniform state laws. The act identifies public water supply, propagation of fish and wildlife, recreation, agriculture, and industry as uses that should be protected from the effects of pollution. To avoid leaving anything out, a category for other legitimate uses also was included (Forsythe, 1961). The 1961 amendments to this act broadened the scope of federal enforcement activity to include not only interstate but also navigable waters and, on request of the governor, intrastate waters.

With the "Water Quality Act of 1965," the federal government was empowered to review and accept or reject water quality standards adopted by the states. The states were required, under this act, to submit to the federal government for approval sets of water quality standards and plans for their realization. Guidelines were provided the states, and failure of the states to propose and adopt suitable programs would result in federally imposed standards. Thus, with the entry of the federal government into the setting of water quality standards, a new era in the control of water pollution in the United States began. Serious problems of water pollution in our country will no doubt require federal capabilities and prestige in their solution. But this alone will not suffice. These problems are woven into the fabric of our society and its resource utilization.

Their removal will require the expenditure of great effort and money by all levels of government, by industry, and by our citizens.

RECENT TRENDS

In recent years, water pollution has come to be recognized for what it is: a part of the critical problem of water resource management. The providing of adequate water has become perhaps our greatest natural resource problem; this has been amply documented in many places. Here, we need only point out that on a national basis we must make a relatively fixed supply meet a rapidly increasing demand by providing insofar as possible the right quantity and quality of water, when and where it is needed. This can be accomplished only if there are extensive programs of water quality control. Indeed, Congressman John A. Blatnik of Minnesota, who has fathered much of the national water pollution legislation, has said:

Is America really suffering a water shortage? The answer is 'No.' Our country is suffering from one hundred years of mismanagement, waste, devastation and neglect of its water resources. . . . But there is no shortage of water anywhere in the United States except in the traditionally arid sections of the West. . . . What we are suffering is a very real shortage of *useable water!* (Wright, 1966, p. 164)

Most of the water used for purposes other than irrigation will in the near future be reused many times before returning to the atmosphere through evaporation. Satisfactory maintenance of the quality of this water presents difficult social, economic, and technological problems.

After suspended materials have been removed from wastes and most of the dissolved organic materials oxidized, there remain in most wastes small amounts of materials that with inadequate dilution may reduce the value of receiving waters. Nutrients like nitrates and phosphates present in wastes having received only secondary treatment may lead to blooms of algae, which can reduce the value of waters for fishing and other recreational uses. Such blooms also can lead to taste and odor problems in domestic water supplies. We are faced with a problem of unknown dimensions in the occurrence in our water supplies of persistent toxic substances such as some of the pesticides. The accumulation of undesired substances in water is difficult to prevent when it is used repeatedly. Advanced levels of waste treatment will become generally necessary, but this will present both economic and technological difficulties.

The temperature of water is a quality characteristic of great importance for some of its uses. Most domestic, agricultural, and industrial uses of water increase its temperature somewhat, but large increases occur when it is used for cooling nuclear or fossil-fueled generators of electric power. Many important commercial and recreational fisheries are on species having narrow limits of temperature tolerance, and the very existence of these fisheries depends on the maintenance of suitable temperatures. Only the most careful water use management can make this possible.

Over the past 20 years, billions of dollars have been expended by the public and by industry for waste treatment facilities. Even so, many of our waters have improved but little, and not enough to satisfy most of those concerned. The reason for this, of course, is that the installation of treatment facilities has barely kept pace with further urban and industrial development. But general public concern has led to the approval of programs that will result in the expenditure in the years ahead of many more billions of dollars for improving the condition of the waters of the United States. Along with increased public concern over the quality of our waters, there has come increased awareness of the complexity of the circumstances that have led to their degradation. Many now realize that we are all responsible for the conditions that have developed, realize that pollution is not just something someone else causes. The federal government has in recent years invested heavily in research into the causes and effects of pollution and into possible solutions for the problems resulting; on a lesser scale, industry and state governments have done likewise. Regional studies — like the "Delaware Estuary Comprehensive Study" (Federal Water Pollution Control Administration, 1966a) conducted by the federal government in cooperation with state and local governments and industry — have documented the complexity of pollution problems and have suggested alternative approaches to their solution, even the costs and benefits of each approach. The Ohio River Valley Water Sanitation Commission, ORSANCO,

was formed in 1949 as an interstate agency involving eight states for the purpose of improving the condition of the Ohio River and its tributaries. With general public support, the actions of local governments and industry have resulted in important steps toward intelligent use of one of the great rivers of our country. ORSANCO was not the first interstate agency devoted solely to water pollution control; five came into being even before the "Water Pollution Act of 1948," which encouraged such compacts. Regional solutions for pollution problems will become ever more necessary with urban and industrial growth. But even regional authorities have been unable to cope with all of the problems, and the authority and resources of the federal government will be required in the solution of some. Increasing public concern over the past 20 years has caused Congress, which earlier was reluctant to act, to adopt legislation directing extensive federal involvement in a national program to better the use we make of our waters. Such tendencies toward federal involvement, which many deplore, are almost invariably the result of local or state inability or unwillingness to cope with social problems.

SELECTED REFERENCES

Banks, H. O. 1961. Priorities for water use. Pages 153–167. *In* Proceedings the National Conference on Water Pollution, 1960. U.S. Department of Health, Education, and Welfare, Washington, D.C. x + 607pp.

Chase, E. S. 1964–65. Nine decades of sanitary engineering. (An article in 9 parts.) Water Works and Wastes Engineering, April 1964–June 1965.

Forsythe, R. A. 1961. The needs and obligations of federal agencies. Pages 250–269. *In* Proceedings the National Conference on Water Pollution, 1960. U.S. Department of Health, Education, and Welfare, Washington, D.C. x + 607pp.

Freeman, P. R. 1963. Judicial expression. Pages 64–87. *In* J. E. McKee and H. W. Wolf (Editors), Water Quality Criteria. 2nd ed. California State Water Quality Control Board Publication 3–A. xiv + 548pp. + map.

Gray, H. F. 1940. Sewerage in ancient and mediaeval times. Sewage Works Journal 12:939–946.

McKee, J. E., and H. W. Wolf (Editors). 1963. Water Quality Criteria. 2nd ed. California State Water Quality Control Board Publication 3–A. xiv + 548pp. + map.

National Research Council. Committee on Pollution. 1966. Waste Management and Control. National Academy of Sciences, Washington, D.C. [xii] + 257pp. (NAS Publication 1400.)

U.S. President's Science Advisory Committee, 1965. Restoring the Quality of Our Environment. Environmental Pollution Panel Report. U.S. Government Printing Office, Washington, D.C. xii + 317pp.

2 The Meaning of Pollution

But in this men stand not usually to examine, whether the idea they and those they discourse with have in their minds be the same: but think it enough that they use the word, as they imagine, in the common acceptation of that language; in which they suppose that the idea they make it a sign of is precisely the same to which the understanding men of that country apply that name.

JOHN LOCKE, 1689, p. 253

. . . and there was confusion over the very definition of pollution, and the decision of what was to be abated.

MCKEE AND WOLF, 1963, p. 29

The word pollution is an adaptation of the Latin *pollutionem,* meaning defilement, from *polluere,* to soil or defile. It was used in its present form in France in the twelfth century. The *Oxford English Dictionary* gives its Middle and Modern English forms in examples of usages dating from the fourteenth century. These examples suggest that the various forms of this word were early most frequently used to express defilement of man, his beliefs, or his symbols by physical, moral, or spiritual contamination. Examples of usage of *pollution* or the verb form *pollute* with reference to physical contamination of terrestrial or aquatic environments are given only for the eighteenth and nineteenth centuries.

The use of the word pollution to signify contamination of water increased in the latter part of the nineteenth century, and during the twentieth century its most common usage has undoubtedly been with reference to water, soil, and air contamination. To many people, water pollution no doubt means the introduction into water of anything "dirty," regardless of the amount or effects of the material introduced. This idea of pollution does not easily include the in- troduction of small amounts of radioactive materials or toxic substances or the causing of temperature or salinity changes. As times change, man's concepts change, and words take on new meanings.

The words used to express particular ideas are determined in part by historical usage and in part by the knowledge, interests, and values of their users. Thus, the meanings given to words not only differ in time but differ between groups of people and between individuals. Common usage ultimately determines correct usage; few if any words have a single accepted definition. This makes imperfect the communication among human beings by means of language; mathematical representations appeal to many minds for this reason.

The different meanings different groups and individuals quite properly give to the same words and the resulting imperfect communication lead to difficulties in human social behavior. We frequently see this in our personal lives, although here such difficulties are rarely of great moment. But, particularly under democratic forms of government, when different groups use the same word to express different ideas in attempting to bring about pertinent social action through governmental processes, more serious difficulties may arise. Laws are enacted by legislative bodies attempting to carry out the will of the people. While considerable effort is expended by legislatures to make the wording of laws clear, many groups influence these bodies during their deliberations. These groups frequently have different interests and use the same words for slightly different ideas. In attempting to compromise and satisfy the interests of different groups, the legislators must often use key words that mean different things to different groups; and sometimes intentionally and sometimes unintentionally the meaning of these words is left vague, and the intent of the legislation remains unclear. When a governmental agency attempts to carry out a law enacted by a legislature, conflicts arise if the agency's interpretation of the law differs from the interpretations of groups affected by actions of the agency. This sometimes leads to litigation, and then the courts must determine the intended meaning of words and phrases used by the legislators in writing the law. As in personal discourse, ambiguity cannot always be avoided when laws are written. Words that have social significance and that are much used by many individuals and groups take on many shades of meaning, and these words stimulate legislation and appear in laws. Such a word is *pollution*.

It may be helpful for us to give and discuss examples of different usages of the word pollution. Persons interested in water pollution control should recognize these, for obvious reasons. Although there may be nothing linguistically incorrect in different usages, some usages are undoubtedly more useful than others in conceptualizing pollution problems so as to make possible the effective use of our technological and scientific capabilities and social institutions in resolving these problems. In concluding this chapter, we will give our definition of pollution in an attempt to make our own usage clear. But even here, we are sure we will sometimes necessarily and sometimes unconsciously be inconsistent.

To a great many people, water pollution means the introduction into natural waters of anything that to them appears to be foreign. These people may be unaware that the apparently foreign substances frequently occur naturally in these waters and enter them by processes that would not generally be considered pollution. This idea of pollution includes neither the concept of measurable change in the receiving water nor the concept of reduction in the value of that water for any use by man. Most other ideas of pollution include in varying degrees one or the other of these two concepts.

Some biologists have included in their definitions of pollution the concept of measurable change in the aquatic environment, but they have not included the concept of reduction in the value of any use of the water by man. Ruth Patrick (1953, p. 33) has defined pollution as "any thing which brings about a reduction in the diversity of aquatic life and eventually destroys the balance of life in a stream." F. P. Ide (1954, p. 87) has written that "pollution is any influence on the stream brought about by the introduction of materials to it which adversely affects the organisms living in the stream. . . ." As we will explain in Chapter 18, when plant and animal species have the opportunity to colonize a water, an assemblage of species having rather definite characteristics will develop over a period of time. A change in the quality of the water may then lead to the development of a different assemblage of species. According to these definitions, this change in the biological community would be evidence of pollution; no consideration would be given to whether or not any change had occurred in the value of the water to man. Changes in aquatic communities are not necessarily accompanied by decrease in the usefulness of waters for man. The production of valuable species of fish may be increased (Warren et al., 1964) and certainly would not always be decreased. Furthermore, small changes in the characteristics of a natural water that could make use of the water hazardous for man might not result in measurable changes in the composition of its aquatic community.

A definition of pollution that has application under common law but that does not take into account the question of damage to the usefulness of water has been given by Coulson and Forbes (1952, p. 198): "the addition of something to water which changes its natural qualities so that the riparian proprietor does not get the natural water of the stream transmitted to him." In illustrating this definition, they give examples of kinds of pollution that have been controlled by legal action: the addition of hard water to soft water; the raising of the temperature of water; and the addition of some substance that reacts with another substance already in the water and thus causes pollution, even though individually the substances are harmless. Use of the adjective *harmless* in the third example suggests that a court decision was based on some damage to the water user, not on a change in the *natural qualities* of the water by a substance that is in itself harmless. Increasing the hardness or the temperature of water could bring about court action based on a decrease in the value of the water for a user, but such changes would not always affect the usefulness of a particular water.

On reading surveys of judicial proceedings relevant to water pollution (McKee, 1952; Freeman, 1963), one comes to the conclusion that the courts have usually interpreted pollution to mean a change in the quality of water brought about by human activities leading to a decrease in the value of the water to some user. This was made exceptionally clear in a case in Colorado in 1934 (McKee, 1952, p. 100):

For the purposes of this case, the word 'pollution' means an impairment, with attendant injury, to the use of the water that plaintiffs are entitled to make. Unless the introduction of extraneous matter so unfavorably affects such use, the condition created is short of pollution. In reality, the thing forbidden is the injury. The quantity introduced is immaterial.

Although evidence that a defendant was indeed responsible for some measurable change in water quality has been important in most cases, it appears that it has not been the mere change but some damage to the usability of the water that has been the basis of most abatement orders and damage awards.

The intent to protect water uses from the effects of pollution rather than merely to prevent water quality changes is reflected in federal and many state laws. The federal "Water Pollution Control Act of 1956," as we noted in Chapter 1, lists five categories of water use to receive protection from the adverse effects of pollution. According to the Water Code, State of California (McKee and Wolf, 1963, p. 1):

because of the widespread demand and need for the full utilization of the water resources of the State for beneficial uses, it is the policy of the State that the disposal of wastes into the waters of the State shall be so regulated as to achieve highest water quality consistent with maximum benefit to the people of the state. . . .

Gorlinski (1957, p. 61), in reviewing the legal basis for water pollution control in California, has written:

Under California statutes, when the waters of the State, including both surface and underground supplies, are impaired by a discharge of sewage or industrial wastes to a degree which creates an actual hazard to the public health, a 'contamination' is said to exist. When such impairment is of a degree which does not create an actual public health hazard, but which does adversely and unreasonably affect the waters of the State for domestic, agricultural, navigational, recreational or other beneficial uses, a 'pollution' is said to exist. A third objectionable condition, 'nuisance' is said to exist when damage results to any community by reason of odors or unsightliness caused by what the law terms 'unreasonable practices' in the disposal of sewage or industrial wastes.

Thus, California law distinguishes between the meanings of *contamination, pollution,* and *nuisance.* Gorlinski goes on to explain the reasons for these distinctions:

Contamination, by the very urgency of its nature, is subject to immediate correction upon order of local or state health departments. The solution of pollution and nuisance problems on the other hand, is delegated to nine regional boards and the State Water Pollution Control Board created for this purpose. These latter problems call for balancing economic and benefit considerations, often with a resultant compromise of the interests involved.

Generally, the introduction into water of material that makes its use hazardous to public health is considered within the meaning of the word pollution. The "Drinking Water Standards" of the U.S. Public Health Service (1962, p. 2) state that:

Pollution, as used in these Standards, means the presence of any foreign substance (organic, inorganic, radiological, or biological) in water which tends to degrade its quality so as to constitute a hazard or impair the usefulness of the water.

In recent years in the United States, there has come an increased awareness of the need for man to live wisely in his environment. Attention has been directed toward the interrelationships between man and his environment; and there has been concern

over man's total environment, all its resources, all its uses, and all its effects on man.
From this point of view, pollution was defined recently in two important reports.

The Report of the Environmental Pollu-
tion Panel of the U.S. President's Science
Advisory Committee (1965, p. 1) entitled
Restoring the Quality of our Environment begins:

Environmental pollution is the unfavorable
alteration of our surroundings, wholly or largely
as a by-product of man's actions, through direct
or indirect effects of changes in energy patterns,
radiation levels, chemical and physical constitution
and abundances of organisms. These changes may
affect man directly, or through his supplies of
water and of agricultural and other biological
products, his physical objects or possessions, or
his opportunities for recreation and appreciation
of nature.

Waste Management and Control, a report by
the National Research Council Committee
on Pollution (1966, p. 3), begins:

Pollution is an undesirable change in the physical, chemical, or biological characteristics of our
air, land, and water that may or will harmfully
affect human life or that of other desirable species,
our industrial processes, living conditions, and
cultural assets; or that may or will waste or deteriorate our raw material resources.

Even these broad definitions of pollution
are concerned not just with environmental
change but with change that will have undesirable consequences for man. Man cannot
use the resources of the earth without bringing about change in his environment. But
he must not change the earth to his own det-
riment.

On the basis of all of these considerations,
we have chosen to define water pollution as
*any impairment of the suitability of water for any
of its beneficial uses, actual or potential, by man-
caused changes in the quality of the water.* There
are changes that come about in water as the
result of natural processes that render the
water unsuitable for some uses; these changes
have sometimes been termed *natural pollution* (Hynes, 1960, p. 2). Because of the
social implications of man-caused changes in
water quality, we prefer to distinguish clearly
between *pollution* and *natural pollution.*

SELECTED REFERENCES

Freeman, P. R. 1963. Judicial expression. Pages 64–87.
In J. E. McKee and H. W. Wolf (Editors), Water Qual-
ity Criteria. 2nd ed. California State Water Quality
Control Board Publication 3–A. xiv + 548pp. + map.

Gorlinski, J. S. 1957. Legal basis for water pollution control in California. Pages 61–63. *In* Waste Treatment
and Disposal Aspects to Development of California's
Pulp and Paper Resources. California State Water
Pollution Control Board Publication 17. [102] pp.

Klein, L. 1962. River Pollution. 2. Causes and Effects.
Butterworth, London. xiv + 456pp.

McKee, J. E., and H. W. Wolf (Editors). 1963. Water
Quality Criteria. 2nd ed. California State Water
Quality Control Board Publication 3–A. xiv + 548pp.
+ map.

National Research Council. Committee on Pollution.
1966. Waste Management and Control. National Academy of Sciences, Washington, D.C. [xii] + 257pp.
(NAS Publication 1400.)

U.S. President's Science Advisory Committee. 1965.
Restoring the Quality of Our Environment. Environ-
mental Pollution Panel Report. U.S. Government
Printing Office, Washington, D.C. xii + 317pp.

3 Water Quality Standards and Water Use Classification

As a basis for pollution abatement, we need to establish environmental quality standards. Such standards imply that the community is willing to bear certain costs or to enforce these costs on others in order to maintain its surroundings at a given level of quality and utility. For each pollutant the elements that must be taken into account are: its effects; technological capabilities for its control; the costs of control; and the desired uses of the resources that pollutants may affect.

U.S. PRESIDENT'S SCIENCE ADVISORY COMMITTEE, 1965, p. 15

A marked difference of opinion exists among water-pollution-control authorities, in industry, in private practice, and in public agencies, regarding the advantages and disadvantages of various types of standards and systems of classification.

MCKEE AND WOLF, 1963, p. 30

THE ORIGIN OF STANDARDS

A society must find ways of realizing insofar as possible its collective desires regarding the use of its natural resources. This is not a rapid, an orderly, or an efficient process, and few if any individuals or groups will be entirely satisfied with the ways found or the ends attained. It is the very nature of a society, which is composed of diverse elements, that the desires of some of its elements are subordinated in some degree to the society's requirements for persistence. Although in a healthy society the needs and even a good proportion of the desires of its elements must be satisfied, the full satisfaction of individuals is not always possible within the limits of the society; neither will such individual satisfaction assure the general well-being. Yet the successful society must find ways, however gropingly, of satisfying its elements while remaining a society.

With reference to water pollution, we noted in Chapter 1 that failure of individuals

suffering damages to enter into litigation through the courts allowed water quality conditions to deteriorate, and there was failure to protect the public interest. Even when private litigation was undertaken and decisions were reached, the body of law that resulted was more concerned with private rights than with public rights. In balancing advantages and disadvantages, the courts sometimes refused to enjoin continued pollution, because this might work economic hardship on a community, and one aspect of public welfare was protected if not another. Statutory law became necessary to attain better balance between private and public interests. Agencies empowered to protect public interests in water quality usually have had to show that some private or public interest was being damaged by the activities of other private or public groups, if corrective actions were to be taken. Such damages are difficult to establish, and frequently the courts have been confronted with volumes of technical data on which decisions had to be based.

15

Freeman (1963, p. 87) has written:

Such determination involves extremely technical, complicated, and scientific problems that reasonably might be thought more suitable for resolution by administrative procedures than by the trial of particular damage claims, the varying decisions of which may not always provide specific standards.

These and other considerations have led state and federal governing bodies to develop and adopt various kinds of standards for the protection of water quality and uses. Hopefully, this would simplify and improve pollution control procedures.

Thus, various kinds and systems of standards have evolved as one of the ways in which our society proposes to balance the needs and desires of individuals and groups with the public interest regarding water use. As is to be expected in a society composed of diverse elements and interests and having a federal government, these systems of standards are not always consistent, either within or between themselves. Further, there not only is disagreement as to the proper use of standards, there also is disagreement over just what a standard is or should be.

Water quality standards and water use classification can provide a conceptual and administrative framework for utilizing scientific and technological information in attaining the water use objectives of our society. Because we intend to explain in this book how we believe biology can be utilized most effectively in water pollution control, we must discuss water quality standards and water use classification at some length. The most effective utilization of biology will depend on the systems of standards and classification adopted. In developing our views on the use of biology and other sciences in pollution control, we have been forced to the conclusion that some systems of standards and classification will be more effective than others. In concluding this chapter, we will describe a system that we believe could be effective. We are appreciative of the problems inherent in the development, adoption, and conduct of regulatory procedures and of the reality that no system of standards or procedures can be as effective as might be wished. Nevertheless, to make clear our ideas on the effective use of biology in water pollution control, we must relate them to the most effective system of standards and classification we can conceive, even though this

system undoubtedly has shortcomings and may never anywhere be adopted.

STANDARDS, CRITERIA, OBJECTIVES, AND CLASSIFICATION

The terms *standard*, *criterion*, and *objective* are frequently used in connection with water pollution control. Some have used these words almost interchangeably while others have made or implied distinctions in their meanings. McKee (1960, pp. 19–20), in his paper "The Need for Water Quality Criteria," has distinguished between these terms in a manner favored by many:

The term 'standard' applies to any definite rule, principle, or measure established by authority. The fact that it has been established by authority makes a standard somewhat rigid, official, or quasi-legal; but this fact does not necessarily mean that the standard is fair, equitable, or based on sound scientific knowledge, for it may have been established somewhat arbitrarily on the basis of inadequate technical data tempered by a cautious factor of safety. Where health is involved and where scientific data are sparse, such arbitrary standards may be justified.

. . . .

The word 'objective' represents an aim or a goal toward which to strive and it may designate an ideal condition. Most certainly, however, it does not imply strict adherence nor rigid enforcement by an agency or health department. It is gaining favor among engineers on boards and commissions that strive to achieve water pollution control by persuasive methods and cooperative action. An example would be the objectives for the various reaches of the Ohio River, as promulgated by the Ohio River Valley Water Sanitation Commission.

A 'criterion' designates a means by which anything is tried in forming a correct judgment respecting it. Unlike a standard it carries no connotation of authority other than that of fairness and equity; nor does it imply an ideal condition. When scientific data are being accumulated to serve as yardsticks of water quality, without regard for legal authority, the term 'criterion' is most applicable.

The U.S. President's Science Advisory Committee (1965, p. 59), in its report *Restoring the Quality of Our Environment*, defines the term standard for purposes of discussion as:

a definite concentration of pollutant adopted locally or generally by any agency of government as a maximum with the intent of requiring compliance pursuant to its legal authority.

In recognizing a standard as a value established by legal authority, this definition is consistent with usual present usage. In specifying "a definite concentration of pollutant," this definition is probably too narrow; not all pollutional changes in water quality can be measured in units of concentration. And, for reasons we will discuss, the term standard should probably be reserved for broadly applied values; regulations adopted only locally might better be termed *requirements*.

Both the term standards and the term criteria are used in the federal "Water Quality Act of 1965" (Section 5a):

Such State criteria and plan shall thereafter be the water quality standards applicable to such interstate waters or portions thereof.

Here, standards are to be water quality criteria together with plans for their implementation adopted by the states and approved by the federal government. Legislative acts—as we pointed out in Chapter 2—frequently are not entirely clear as to the meaning of key terms. Although this may be understandable, it sometimes adds to the difficulties of governmental agencies endeavoring to carry out the intent of legislation. Thus, the Federal Water Pollution Control Administration (1966b, p. 7) in its "Guidelines for Establishing Water Quality Standards for Interstate Waters," pursuant to the Water Quality Act, included the statement that:

No standard will be approved which allows any wastes amenable to treatment or control to be discharged into any interstate water without treatment or control *regardless of the water quality criteria and water use or uses adopted.* [italics ours]

The Federal Water Pollution Control Administration, in attempting to carry out the intent of this act, appears to have added still another dimension to the meaning of standard.

Words like standard and criterion are most useful in the formulation and conduct of public policy if their meanings are clear to those responsible for these functions. Moreover, the ways in which scientists and others conceptualize and make their contributions to the public business depend on the meanings and uses of key words. To explain the role of biology in water pollution control, we have found it desirable, if not necessary, to define very carefully our usage of the expressions *water quality standard* and *water quality criterion*. We believe these definitions will be useful to others concerned with pollution problems.

Water quality standard: Any more or less permanent and widely applicable rule authoritatively establishing for regulatory purposes the limit of some unnatural alteration of water quality that is to be permitted or accepted as being compatible with some particular intended use or uses of water.

Water quality criterion: Any definite limit of variation or alteration of water quality expertly judged, on the basis of scientific data, not to have some specified, usually adverse, effect on the use of water by man or on organisms inhabiting the water.

Water quality standards can be established by a regulatory agency through selection from an array of water quality criteria, the choice depending on the use of the water to be protected and on the degree of impairment of that use that is judged acceptable as a matter of public policy. They should be distinguished from any special regulations or requirements, often temporary, intended only for the control of waste discharges into a particular body of water. Water quality standards as we have defined them are sometimes termed *receiving water standards* and sometimes *stream standards*. Water quality standards may be intended for wide application to all waters or to a large class of waters of a state, region, or river basin; or they may be intended for only geographically much more restricted application. As we have already noted, however, prescribed limits of alteration of a single body of water might better be considered *requirements* than standards.

Waters are often classified by regulatory agencies according to the actual or potential uses of the water that are to be protected, which depend largely on the natural characteristics of the waters and the history of their use. Without such classification of waters, wide application of uniform standards over a large geographic area usually is impossible. It will be helpful if we here define classification.

Water use classification: Administrative classification of waters with the avowed intention that all waters of a given class will be maintained in some degree suitable for the same beneficial use, the different degrees of protection to be provided a particular use being indicated by further classification into subclasses, and appropriate uniform standards being enforced to accomplish the intention.

In this definition, we recognize the need to classify waters not only according to the major uses to be protected but also according to the desired level of protection of each use. How particular waters are to be classified—what beneficial uses and levels of use are to be protected—is a matter that should be decided largely on the basis of informed public opinion or sentiment, expressed at public hearings and otherwise. Selection of uniform standards permitting the desired uses and levels of use is a highly technical matter best left to administrators having the counsel of scientists and engineers.

Water quality standards should be distinguished from *effluent standards*, which set specific limits on the permissible characteristics of effluents. Effluent standards sometimes require that undiluted waste waters be suitable for desired beneficial uses. Usually, however, effluent standards are based on the assumption that the waste waters, before further use, will be considerably diluted by the receiving waters. Effluent standards are commonly employed by regulatory agencies as a means of maintaining water quality standards in receiving waters. They provide a method for these agencies to apportion among several waste sources the capacity of a receiving water to dilute and assimilate wastes.

One general effluent standard receiving considerable attention in the United States deserves special comment. This standard simply prescribes that the best and most complete treatment technically and economically feasible be adopted for all waste waters. Since virtually complete removal of most substances from waste waters is usually technically feasible, the application of this standard depends on proof of economic feasibility, which is not often easy. Although favored by some fishery biologists, this standard does not itself insure adequate protection for aquatic life. Probably the strongest argument that has been advanced in its support is that the incompleteness of our knowledge of the water quality requirements of living organisms

renders receiving water quality standards unreliable. Nevertheless, since the requirements of living organisms have no relation to economic feasibility of waste treatment practices, the wide adoption of this standard could be hazardous to aquatic life. In this book we will be mainly concerned with water quality standards. Reasonable effluent standards usually will be based on water quality standards, for conditions in receiving waters and not characteristics of effluents directly affect aquatic organisms. Water uses depend on the quality of receiving waters.

Properly formulated and definite water quality standards can be an important aid in regulating waste disposal and enforcing pollution laws. In the absence of such standards, a regulatory agency—in order to successfully prosecute violators—must usually prove that damage to beneficial uses of water or to valuable aquatic life has occurred and is attributable to a particular pollutant. When standards have been adopted and given the force of law, it is only necessary to prove that changes in water quality have gone beyond the standards. This is usually much easier than proving that particular water uses have been harmed. The courts of law have usually accepted and enforced reasonable standards based on scientific data:

> The absence of any litigation or reported cases in which duly promulgated standards or requirements were held up to the searching light of judicial inquiry to determine their reasonableness indicates that the courts are uniformly anxious to have, and have readily adopted, scientifically developed criteria to assist them in their determination. (Freeman, 1963, p. 87)

The establishment of standards has been welcomed by many representatives of industry who have viewed reasonable, explicit, and stable requirements preferable to vague prohibitions as a basis for the planning and construction of costly waste treatment facilities. Reluctance on the part of some segments of industry to support the establishment of standards has been based largely on concern that standards might be too broadly applied, even where not appropriate. Inasmuch as one avowed purpose of standards is to promote uniformity of waste disposal regulation, this concern is not unreasonable. Some conservationists have opposed the establishment of water quality standards because they might be difficult to modify should they prove inadequate for protection of resource values. Nevertheless, there is coming increased reliance on standards, usually coupled with some classification of waters according to the particular uses to be protected.

THE SCIENTIFIC BASIS FOR STANDARDS AND THE SOCIAL BASIS FOR CLASSIFICATION

If water quality standards as we have defined them are to be most useful, they must be designed to provide selected levels of protection for given water uses, must be rather generally applied, and must be reasonably stable. For certain domestic, agricultural, and industrial uses of water, adequate water quality criteria based on scientific knowledge are now available to permit the selection of water quality standards that will provide the desired levels of protection. We say different levels of protection for particular uses because, taking water for human consumption as an example, we may want the water suitable directly from its source, suitable only after chlorination, or suitable only after some more extensive treatment. Water quality standards can be selected from available criteria that would provide for one of these levels of protection but not for the others. Thus, water quality standards appropriate for each level of protection can be based on scientific knowledge; but social and economic considerations must enter into selection of the level of protection to be given a use. It is through water use classification — the establishing of classes of water for major uses and subclasses for levels of protection to be given those uses — that social, economic, and technological considerations should enter into the choice of which set of previously established standards will be used in a particular instance.

Water quality criteria based upon scientific knowledge from which water quality standards can be selected for the protection of aquatic organisms are far more inadequate than those from which standards for the protection of most other water uses can be selected. Scientific determination of the water quality necessary for the survival, reproduction, and growth of aquatic species is usually more difficult than determining the water quality needs of other water uses.

And man has yet to be sufficiently interested in aquatic organisms to expend the effort necessary to determine fully their needs. It is the purpose of this book to explain how these needs of aquatic organisms can be determined and how information concerning them can best be employed in water pollution control. But even when the scientific basis for establishing standards for the protection of aquatic organisms becomes more adequate, the standards selected for application in a particular instance should be determined by the level of protection of the organisms that is desired; social and economic considerations must enter into the selection of the level of protection.

Biologists are likely to regard available scientific information on the environmental requirements of organisms as the only consideration in the choice of water quality standards to protect these organisms. Yet the water quality necessary to permit some production of a species may be very different from that which will permit its greatest possible production. Conditions ideal for the production of a particular fish species may be neither always attainable nor always essential to human use of that species. Considerations other than biological ones must somehow enter into public policy decisions concerning standards for the protection of aquatic life.

Some administrators evidently believe that, since water pollution control obviously involves social, economic, technological, and political considerations, water quality standards should be adapted to local conditions and revised from time to time in response to advances in waste treatment technology and variations in social, economic, or political climate. This view, with which we do not agree, appears in part in the federal "Guidelines" (Federal Water Pollution Control Administration, 1966b):

If it is impossible to provide for prompt improvement in water quality at the time initial standards are set, the standards should be designed to prevent any increase in pollution. (p. 5)

. . . .

It is anticipated that after the initial setting of standards, periodic review and revision will be required *to take into account changing technology of waste production and waste removal* and advances in knowledge of water quality requirements developed through research. [italics ours] (pp. 9–10)

It seems standards would be changed with changes in technology, not just when new scientific information indicates that existing standards are inadequate to provide the levels of protection for particular uses for which they were designed. Social, economic, and technological changes must be considered in water pollution regulation, but this can best be done by a system of classification. The use classification for a particular water can be changed with social changes and another previously established set of water quality standards adopted for that water. In this way, water quality standards recommended for particular uses and levels of protection need not be changed; only the ones adopted need change.

This distinction between the idea of water use classification and the idea of water quality standards is important if we are to have a workable system of water quality control. Perhaps a hypothetical example will be helpful in making clear the importance of this distinction. Let us imagine a stream being used as a source of domestic water and receiving an industrial waste containing a toxicant that cannot be removed from the water by ordinary water treatment processes. Concentrations of this toxicant higher than 5 mg/l in drinking water have been shown to be injurious to some human consumers; a maximum of 2 mg/l has been widely accepted as safe in drinking waters. Let us further assume that provision already has been made for what has generally been considered to be the best practicable waste treatment; yet concentrations of the toxicant in the drinking water sometimes reach 10 mg/l. Surely 10 mg/l should not be accepted as a reasonable standard, simply because treatment is the best feasible, if the stream is to be protected as a good source of drinking water. Either the stream should be classified as one that cannot for technological and economic reasons be so protected, or a more restrictive standard should be adopted and somehow enforced as soon as possible. The drinking water standard initially adopted for this toxicant may be later modified on the basis of new and more reliable toxicological information; but the availability or economic feasibility of waste treatment and control methods should be immaterial in establishing standards for safe drinking water. Establishing standards for other water uses should, in principle, be no different.

Finally, then, waters that are classified for a particular use — because of their natural characteristics, their historical usage, and social considerations — may still provide that use in different degrees. A particular set of water quality standards may be necessary to insure any particular degree of a use. Each set of standards in a series of sets relevant to a use can be based on scientific evidence without social considerations being taken into account. However, the set of standards selected in a particular instance should be determined by the level of protection desired for a use; in this, social and economic factors must be determining considerations. They form the social basis for water use classification. Sets of standards that will provide different levels of protection for the different uses should be selected from available water quality criteria. These form the scientific basis for water quality standards. Thus, standards need not change with social, economic, or technological change; only the choice from previously formulated sets of standards need change. Standards should change only when new scientific knowledge indicates that they do not provide the originally expected levels of protection for particular uses. This conceptualization permits stability in standards even with social, economic, or technological change. Changes in the standards themselves would be made only on a basis of accumulation of new scientific knowledge, which in most fields, unfortunately, is not so rapid as we have sometimes been led to believe.

A SYSTEM OF WATER USE CLASSIFICATION AND WATER QUALITY STANDARDS

This concludes our argument on the need for a system of water use classification and water quality standards and the need to distinguish clearly between classification and standards, an argument that, we fear, may seem to our reader involved and not entirely clear. It is our contribution to the general argument that has reached from the man on the street to the highest councils of our government. It is an important argument, because the development of this country and the proper use of its natural resources are at stake. Individuals, citizens' groups, industrialists, and governing bodies have been involved. As in most arguments, the backgrounds and desires of those involved have determined their differing positions; and, as in most arguments, differences in the ideas participants attach to key words have made much of this argument trivial, or even meaningless. We have endeavored to be clear as to the meanings we attach to important terms; but we know only too well our own weaknesses and the difficulty of reducing social and biological complexity to a few terms or paragraphs. Distinctions we have made may seem hazy to our reader, either because things are not really so distinct or because we have failed to write well.

In either event, we wish to impose on our reader's patience a bit more. From our argument, we would now like to abstract the major considerations in water use classification and water quality standards, as we see them. These considerations could serve as a basis for a workable system of water use classification and water quality standards; if not this, we hope they will at least further aid our reader in comprehending the argument that has involved so many.

When waters are classified or reclassified according to uses, the intended degree or level of protection of each approved use should be determined independently of that of any other approved use. As soon as the intended use status of a water has been determined by classification, water quality standards previously approved as being appropriate for the intended levels of protection of the various uses should be adopted, the most restrictive of these standards superseding any less restrictive ones. Classification of a particular water, according to uses and levels of protection, should be based primarily on economic, technological, political, and other social considerations. Which set of standards is most likely to provide the desired levels of protection for different uses is primarily a scientific matter.

Each use of water has its own water quality requirements which, once known, can be formulated as a set of standards. Where multiple uses are to be protected, the standards adopted must provide for all of their requirements. A broad use category such as agricultural irrigation includes distinguishable irrigation uses of water that do not have the same water quality requirements. Some crops are more sensitive to high salt concentrations than are others. Different industrial uses of water likewise have different water quality requirements. The require-

ments of different species of fish are not the same. A workable system of classification must take this into account. Moreover, for each specific kind of agricultural, industrial, or fisheries use, water quality may be suitable in various degrees and that use to some extent still be possible. Thus, a system of classification must indicate not only the particular kind of use to be protected but the level of water quality protection to be provided that use.

We can further develop the idea of different levels of protection for a particular kind of use by considering a species of fish utilized in recreational or commercial fisheries. Waters in which these fisheries are to be protected could be so classified as to provide the following levels of protection:

A. No pollutional impairment of the survival, reproduction, growth, production, or palatability of the fish; thus no reduction in the value of the fishery.

B. Pollutional impairment of the survival, reproduction, growth, production, or palatability of the fish not to be such as to greatly reduce the value of the fishery.

C. Pollutional impairment which may greatly reduce the value of the fishery but not render it valueless.

From sets of standards previously formulated on the basis of scientific evidence only, the set that would provide for the desired level of protection of the fishery could be adopted. In adopting the standards for a particular body of water, then, the use and level of protection would be determined on the basis of social considerations as would be the set of standards selected; but the standards themselves would be based only on scientific evidence. Any water classified for fishery use could also be classified for other uses—domestic, agricultural, or industrial—according to the level of protection each is to receive. From the sets of standards appropriate for the different uses, a set of standards could be adopted that is adequate to provide the desired levels of protection for all established uses of a particular water.

Water quality standards should derive directly from water quality criteria that are subject to change only when scientific advances necessitate their modification. Which standards are adopted for a particular water should depend on the uses and the levels of protection to be provided. The definition of acceptable water quality by means of standards should be comprehensive, and the standards should be explicit, not subject to widely varying interpretation. Sets of water quality standards should define limits of virtually all possible unnatural alterations of waters that can materially interfere with the water uses to be protected. Appropriate scientific tests of compliance with standards should be specified as necessary for clear definition of these limits. Whenever possible, numerical standards should be adopted.

Water quality standards that are not clearly based on the best available scientific information are likely to be unduly arbitrary and variable. A set of water quality standards should represent an estimate of the quality of water required by some use or uses of water, not merely the highest quality attainable at some time or place.

Enforcement of sets of standards delimiting only some of the possible alterations of water quality that can seriously interfere with water uses can be inequitable and is likely often to be ineffective. Vague or general standards that can be variously interpreted are not easily enforceable. For example, the mere statement that no toxic substances shall be present in concentrations injurious to aquatic life is not sufficiently explicit. The kinds of aquatic organisms and the kinds of evidence, tests, or criteria of injury to be considered should be specified. Otherwise, proving noncompliance with a standard will be difficult if not impossible, except in instances of gross violation. Neither adequate protection for valuable aquatic life nor a sound basis for the planning of costly pollution abatement measures is provided by imprecise standards.

Standards are generally developed for wide application. Yet the circumstances surrounding the uses of particular waters are often unique. Whenever reliable evidence indicates that generally recommended standards are not appropriate for particular waters and their uses, the standards should be superseded by requirements pertaining specifically to those waters and their uses.

At the present stage of the evolution of water quality control, general standards are necessary to provide reasonable protection for the uses of most waters. Limits of knowledge and manpower make it necessary to generalize. But the value of most waters demands that ultimately we use them as fully and wisely as possible. This will require us to have not only general but detailed knowledge of particular waters and their uses. On the basis of this knowledge,

requirements superior to general standards for protecting desired uses should be adopted. Under different circumstances, reliable evidence may show that requirements in some places should be more stringent and in other places less stringent than the generally applicable standards.

Effluent standards specifying the quality of waste waters to be discharged or other waste treatment requirements specifying treatment methods or facility capacity are usually essential to the enforcement of water quality standards. They should be clearly related to defined water quality objectives. Not being water quality standards, waste treatment requirements may, however, sometimes be designed to provide for future growth of population or industry and thus prevent immediate full utilization of the assimilatory capacity of receiving waters.

Various waste treatment requirements are usually necessary to enforce water quality standards and attain water quality objectives. Population and industrial growth may, however, prevent the attainment of objectives when some assimilatory capacity of receiving waters is not reserved for such growth. It may sometimes be necessary to require that industry demonstrate the capability of waste treatment beyond that immediately necessary, if water quality standards are to be maintained with future industrial development.

More than ever, our society is seeking ways of making better and wiser use of natural resources. Water pollution control is an important part of better and wiser water resources management. In the final analysis, we must do this for man, not for fish or some other animal, man now and in the future. So people's needs and desires must somehow be taken into account in the regulation of water pollution; hopefully, the desires will be not just for man now but for generations to come. And man does have some responsibility for the other organisms of the earth, not only for his own good. The social and scientific problems to be faced in providing for all are great. Systems of water use classification and water quality standards must reflect the complexity of these problems and provide rationales for their solution. Such systems must be adequate and sensitive to the needs of the people and able to incorporate all relevant knowledge. Although these systems must have the firmness necessary for recognition and use, they also must have the plasticity necessary for easy adaptation to changing needs and knowledge. We believe such systems of water use classification and water quality standards will need to incorporate in some way the main ideas we have emphasized.

SELECTED REFERENCES

Federal Water Pollution Control Administration. 1966b. Guidelines for Establishing Water Quality Standards for Interstate Waters. U.S. Department of the Interior. 12pp.

Freeman, P. R. 1963. Judicial expression. Pages 64–87. *In* J. E. McKee and H. W. Wolf (Editors), Water Quality Criteria. 2nd ed. California State Water Quality Control Board Publication 3-A. xiv + 548pp. + map.

McGauhey, P. H. 1968. Engineering Management of Water Quality. McGraw-Hill Book Company, New York. [vii] + 295pp.

McKee, J. E. 1960. The need for water quality criteria. Pages 15–26. *In* H. A. Faber and Lena J. Bryson (Editors), Proceedings of the Conference on Physiological Aspects of Water Quality. U.S. Public Health Service, Washington, D.C. [xi] + 244pp.

———, and H. W. Wolf (Editors). 1963. Water Quality Criteria. 2nd ed. California State Water Quality Control Board Publication 3-A. xiv + 548pp. + map.

4 Biological Approaches to Water Pollution Problems

The menace of pollution to our inland streams and rivers is too well known to require definition. In fact, unsightly and noisome conditions due to pollution are encountered so often that they have come to be accepted by many as the usual order of things. It is true, however, that many cases of pollution could be remedied and the streams so affected restored to an acceptable state for recreation, fishing, and general use with reasonable expense, if all parties concerned would cooperate. To obtain this cooperation, it is necessary to understand the situation and to judge it fairly. Much confusion and misunderstanding has arisen in attempts to define the extent of pollution and to place the responsibility for damage to fisheries, because of the lack of available information on the conditions to be defined.

M. M. ELLIS, 1937, pp. 365-366

Clearly, there are separate levels of interest here. The first says, 'We need to know the state of things. We'll find out how much waste we can add without damaging the stream, and we'll back off as much as we can afford from this level. In that way, we'll be good citizens and we'll keep out of trouble.'
The second says, 'That's dandy, we'll do just that because it is a sensible and direct answer to our immediate problem. But let's find out how the wastes act on living things so that we can understand as well as control. This will give us something beyond immediate experience that can be generally useful.'

CHARLES RENN, 1960, pp. 158-159

LEVELS OF BIOLOGICAL ORGANIZATION

Biological systems from the biochemical level to the level of the plant and animal community are very complex, but the more we know of these systems the more we must marvel at their beautiful organization and integration. Biologists believe that nearly all the characteristics of these systems are in some way important to the success of the organisms or the communities of which they are a part. This is to say that these characteristics have adaptive value, that they help the organism, its population, or its com- munity to adapt to environmental condi- tions. Since Charles Darwin (1859) published his monumental *The Origin of Species*, biologists have come to accept that these adaptive systems have evolved through *natural selec- tion* from among naturally occurring vari- ations. Those organisms having heritable characteristics favoring their survival and reproduction are more successful in leaving their kind than are organisms not so fa- vored; thus, they come to predominate in successive generations. In this way, highly complex, organized, and integrated bio- logical systems, both around and within us, have evolved over very long periods of time, because these systems have been most successful in meeting the needs imposed on organisms by their environments.

The systems within an animal cannot persist in any biologically important way apart from the whole integrated animal. Nor can animals in nature generally persist away from the biological communities of which they are integrated parts. Although it is sometimes useful to view biological systems as having different levels of organization, it is important to remember that these levels are integrated with each other in many ways. Nevertheless, biologists of all kinds have usually found it helpful to define the problems they study within the limits of particular levels of biological organization: subcellular, cellular, organ system, intact living organism, population, or community. Placing limits on particular biological investigations is necessary, and this is one way of doing so. But in interpreting our findings, we are usually ultimately concerned with populations or communities, not just with biochemical pathways, organ systems, or even individual animals, except mainly in the case of man. In making these interpretations, it is necessary that we take into account the restrictions that information from only a single level of organization places on our conclusions.

Water pollution biologists, like other biologists, have usually concentrated their individual efforts on one or another of the levels of biological organization. This will become apparent as we discuss the history of water pollution biology. Four of the parts of this book are entitled Morphology and Physiology, Ecology of the Individual Organism, Population Ecology, and Community Ecology. These parts correspond to four levels of organization into which biological systems could be categorized, and each part includes chapters elaborating approaches appropriate for studies of that level of biological organization. But now, before we go on to consider biological approaches to water pollution problems, perhaps it will be helpful for us to describe these four major levels of biological organization.

Systems Within the Organism

Integrated by nervous and hormonal mechanisms, the systems within an animal

operate together and give the animal its specific characteristics. Through evolution, the systems of each species have come to be in some degree unique. The systems that are characteristic of a particular species are passed from one generation to the next through its reproductive cells. A biochemical material, deoxyribose nucleic acid or DNA, present in the chromosomes of the reproductive cells, appears to control the development of the biochemical, physiological, organ, and even some behavioral characteristics of an animal in such a manner that it will closely resemble its parents. These characteristics make an animal what it is and make it possible for it to do what it does in meeting the exigencies of life. Biochemical systems are involved in all the activities we consider characteristic of a living organism: receiving stimuli from the environment, moving in relation to these, obtaining energy and materials and utilizing these for its needs, and reproducing its kind. To the extent that full understanding of these activities may be realized, information on biochemical systems will be necessary.

No clear distinction can be made between biochemical and physiological systems. Physiologists work with biochemical systems and biochemists are concerned with physiological systems. Physiology is the application of physical and chemical principles to the study of life. Physiologists have studied the functioning of the nervous, endocrine, circulatory, respiratory, digestive, and reproductive systems. The study of these systems may go back to physics and biochemistry or on to the physiology and performance of the whole organism.

Organ systems, as they have evolved, have provided for perception, nervous and endocrine control, circulation, respiration, food gathering and digestion, locomotion, and reproduction. Biologists are concerned not only with the structure but also with the function and adaptive value of particular systems. Knowledge of the structure of these systems and of their functioning under different environmental conditions is essential to understanding of the responses of the organism to factors in its environment.

The biochemical, physiological, and organ systems of an animal tend to keep it in a stable state. This homeostatic condition can often be maintained in the face of environmental change by mutual adjustments in the operation of systems within the organism. The greater the environmental change, however, the more serious may be the internal displacement; environmental change beyond some point will reduce the probable success of the organism.

Some biochemical, physiological, and morphological changes in animals may lead directly to their death or may so reduce their biological capabilities as to decrease their probability of surviving or reproducing. Other changes in organisms are adaptive and increase their probability of biological success under changed conditions. In studying the effects of environmental factors on the biochemistry, physiology, and morphology of an organism, we must attempt to distinguish between those changes that decrease the capabilities of the organism and those changes that are adaptive. In this, we must consider these changes as they influence the organism as a whole. Studies of biochemistry, physiology, and morphology can help us to understand the responses of the individual organism to changes in its environment.

The Individual Organism as a Whole

The individual organism as a whole is a remarkably integrated and adapted biological system, within the limits of its natural environment. Beyond these limits, the organism tends to lose its ability to adapt, and this may lead to its disintegration. Together the systems of the whole organism operate to favor its survival, reproduction, and growth. Behavior is a characteristic of the whole living organism; through their behavior, animals tend to place themselves favorably in their physical, biological, and social environments.

When we study the individual organism as a whole, we are concerned with its overall responses to ranges of different environmental factors. We are concerned with the effects these factors have on an animal's survival, reproduction, growth, and movement. Physiological responses like the respiration or osmoregulation of the whole organism are often investigated to obtain insight into its probable success under different conditions. To understand the whole organism, we need physiological and even biochemical information, but such information is most

valuable when considered in relation to the organism's problems of living in its natural environment.

As Huntsman (1948) so nicely stated, the responses of individual organisms to conditions in their environment ultimately determine the distribution and abundance of their populations. To understand the distribution and abundance of organisms, we must know how their individual responses are affected by environmental factors.

The Population

Perhaps we can most usefully think of a population as a group of organisms in which interbreeding is largely unrestricted. Thus, the individuals of a population will be of the same species and race. But all of the individuals of a species rarely if ever belong to the same population, for species generally have rather wide geographic distributions, and various barriers reduce the likelihood of individuals in one area interbreeding with individuals from another area. The populations of a species may occupy distinct geographic areas at all times in the life history of their individuals, or populations may mix and yet remain distinct if they breed separately either in space or time.

From a strictly biological point of view, as well as from a practical point of view, it is the population and not the individual organism that is important. The population, not the individual, represents a continuing, evolving, biological entity. It is the population, composed of reproducing and growing individuals, that continues to be of use to man or that may become a nuisance. In seeking to control in a beneficial way his environment, man manages populations, not individuals.

To the biologist, the population, being a level of organization intermediate between the individual organism and the biological community, represents a convenient unit to study. The biologist seeks to understand how the distribution and abundance of populations are controlled by environmental factors. He may obtain information on the distribution and numbers of individuals in space and in time as well as information on birth rates, death rates, movements, and production. Ultimately, however, in order to understand the distribution and abundance of a population, he needs not only information on the responses of the individual organisms to environmental factors but also information on the role of the population in its biological community.

The Biological Community

At a location having not too extreme physical and chemical characteristics, where plant and animal species have an opportunity to colonize, an assemblage of species will occur after a period of time. Complex interrelationships will develop among these species and between them and the conditions and resources of the location. After a period of time, if conditions are not too variable, this assemblage will take on rather definite characteristics. It can then be considered a biological community.

Because plants and animals generally distribute themselves or are distributed rather widely by various means, colonization by a species at a location having suitable conditions and resources is a general occurrence. In consequence, different locations having similar conditions and resources are often occupied by biological communities recognizably the same or similar. Opportunities for colonization vary, of course, and widely spaced locations having similar conditions and resources may come to be occupied by different communities. Nevertheless, if every location, regardless of conditions, were to have its own very distinctive community, the concept of the biological community would not be of general value.

A biological community, being dependent on the conditions and resources of its location, may change if these change. Such a community change is a reflection of changes in the plant and animal populations that compose the community. Because of complex interrelationships, a change in conditions or resources that causes one or more of these populations to change may cause many of the populations to change and be replaced by others. This will result in another assemblage of species distinguishable as a different community. We may be interested in communities for other reasons, but now we will mention only three: we cannot understand the success or failure of particular populations apart from their communities; communities may be charac-

teristic or indicative of particular kinds of environments; and environmental conditions and resources not only determine the nature of communities but are themselves modified by communities.

A SHORT HISTORY OF WATER POLLUTION BIOLOGY

The Development of a Field

The development of water pollution biology has now spanned about 100 years. It was born during the latter half of the nineteenth century, when a few men began to apply the biological knowledge of their time to solution of the grave problems of water pollution that had been allowed to develop. The growth of this field was uneven and slow until about 1950. Chlorination of drinking water reduced the concern for public health early in the twentieth century, and treatment of sewage made many situations less obnoxious. Without general public concern, population and industrial growth caused water quality problems of great biological, technological, and sociological complexity. By 1950, the American public had awakened to a national shame. Local, state, federal, and industrial coffers had to be more widely opened, if further rapid deterioration were not to follow. But more than money was required; knowledge too was needed. Engineering knowledge was approaching the needs of the time, but biological knowledge was not. Few biologists had been interested in water pollution, and now there was much to learn. So 1950 became a turning point: public interest and resources became available to adequately support needed biological investigation. A field in which biologists could apply their special knowledge and earn a living was opening. Many have since entered this field, and its development has been rapid. With scientific interest, financial resources, and time, the past 20 years have seen the greatest growth in our knowledge of the biology of polluted waters.

In any period of time, scientists who devote themselves to man's practical concerns attempt to apply their special knowledge in relevant ways. The knowledge each scientist can apply is limited by the state of knowledge in his field and by his personal interests and time, whether or not these are adequate for solution of the problem at hand. These things have been and always will be so; they are so in water pollution biology.

Development of bacteriology during the latter half of the nineteenth century permitted some biologists to identify organisms causing human diseases and thus to demonstrate the need for pure drinking water. This was obviously relevant to problems of waste disposal and water supply. Later, a few biologists contributed to the development of waste treatment processes, but this must be credited mainly to engineers and chemists. A few biologists began to study the biological communities of waters receiving wastes. But, until recently, there were never many; and the extent and complexity of the problems they faced were enormous. Some began to employ biological assays to determine what concentrations of toxic substances might be lethal to aquatic life. Investigations of aquatic communities and laboratory studies of toxicity became the foundation of water pollution biology. These were the attempts of a few to solve the problems of many. We now know that the efforts of these men could only be inadequate, even in their time of lesser problems. But what they did then, given the same circumstances, we would do now. Applied scientists in all periods have tried to make their work as good and as relevant as financial and human resources and the state of knowledge have permitted.

The foundation of water pollution biology was laid by a few dedicated pioneers. It was laid neither broad enough nor deep enough. It could not be, not by the efforts of a few. Biological knowledge was not fully exploited, either in breadth or depth; no few men can do this. And the public until recently did not provide for more. Now, new areas of water pollution biology are being developed, developed to the extent that this kind of biology would not be recognized by its pioneers. This is necessary, and the fine scientists who originated our field would not want it otherwise. Those of us who work in water pollution biology today should not forget how much we owe to them. May we make our work in our time as relevant to the problems of water pollution as they did their work in an earlier time.

Community Studies: Waste Treatment,
Biological Indices, and More Recent
Concerns

Biological waste treatment is the work of populations of plants and animals that together compose certain specialized biological communities. This was not soon recognized by those who began developing methods for treating sewage. Indeed, it was first believed that mechanical filtration of sewage through sand and other substrates was the only process involved. Later, in 1870, Frankland suggested the possible role of chemical oxidation (Johnson, 1914a). The original idea that purification was accomplished by filtration is still reflected in the use of terms like "trickling filter" and "biological filter." The role of bacteria in the nitrification and purification process was demonstrated in 1877 by Schloesing and Muntz (Johnson, 1914a). Some now prefer to term such facilities "bacteria beds," even though this is not fully descriptive of the kinds of organisms involved.

Though Sorby in 1883 associated the disappearance of sedimentary fecal material with the presence of certain crustacea and worms, Dunbar in 1900 appears to have first drawn attention to the importance of higher plants and animals in the functioning of bacteria beds (Johnson, 1914a). Good studies of the roles of some of these higher organisms followed (Johnson, 1914a, 1914b). Later biologists have explained the roles of many of the kinds of plants and animals composing the biological communities effecting treatment of wastes percolated through materials having organisms attached (Lloyd, 1945; Hawkes, 1961).

The activated sludge process, another important method of treating organic wastes, also depends on communities of living organisms, but it differs in some respects from the biological filtration process. The communities inhabiting biological filters are exposed periodically to the atmosphere, whereas the organisms involved in the activated sludge process are continually in a water medium in which oxygen is available. The communities of biological filters include many species of higher animals, which consume much of the biological growth produced; the communities involved in the activated sludge process are composed primarily of microscopic plants and animals, and excess biological growth needs to be continually withdrawn from the system.

A number of investigators had experimented with aeration treatment of sewage before Clark and Gage—during 1912 and 1913 at the Lawrence Experiment Station in Massachusetts—found growths of organisms to greatly enhance purification of aerated sewage (Metcalf and Eddy, 1930, p. 636). Ardern and Lockett (1914) found that leaving such biological growths in aeration tanks and drawing off only the clarified sewage greatly reduced the time necessary for oxidation of raw sewage. Since these workers first coined the expression "activated sludge" for this biological growth, the roles of bacteria and protozoa in this process have been studied by many investigators (Barker, 1949).

Stephen Forbes, who is generally credited with one of the earliest statements of the biological community concept (Forbes, 1887), became very interested in the effects of organic pollution on the biological communities of the Illinois River. He and Richardson developed, through their remarkably extensive and careful studies, one of the early systems of classifying into zones of pollution the reaches of rivers receiving organic wastes (Forbes and Richardson, 1913; Richardson, 1928). This they did on the basis of the presence or absence of various species of animals having different tolerances for conditions developing with waste decomposition. They were not the first, for Kolkwitz and Marsson (1908, 1909) had already developed their *saprobic system* of zones of organically polluted rivers. These men classified a great number of both microscopic and larger plants and animals according to conditions under which these organisms were found, and they used the presence and absence of different organisms to indicate the degree of organic pollution. From the work of these and other men, there developed the idea of *biological indicators of pollution.*

Many if not most biologists have believed that organisms living in water provide a more sensitive and reliable measure of the suitability of conditions there than do chemical and physical measurements. Thus, of the total biological effort expended on pollution problems, a very large part has been devoted to the search for biological measurements of pollutional conditions. And, because of the differing backgrounds and interests of the biologists involved, many ways of using biological organisms and communities for this purpose have been proposed. All are based on the idea that changes in physical and chemical conditions will favor some species, harm other species, and have little effect on the remainder. The appearance or disappearance of particular species or changes in the composition of communities should, then, be indicative of changes in physical and chemical conditions.

Some biologists have criticized the efforts of Kolkwitz and Marsson as well as those of Forbes and Richardson because, in the opinion of these biologists, the early workers placed too much reliance on the tolerance classification and the presence or the absence of particular species. These more recent biologists have believed the composition of a community to be more reliable as an index of environmental conditions than would be the presence or absence of particular species. This criticism is based largely on misreading of the earlier works, few biologists having relied for this purpose entirely on the presence or absence of a few species. Two more valid criticisms are that it is difficult to classify organisms according to their tolerance of different conditions and difficult to use a formal scheme of zones of organic pollution (Hynes, 1960, p. 163). Effects of toxic pollution are different from effects of organic pollution on stream communities (Carpenter, 1924; Jones, 1938), and some biologists have erred in using for streams receiving toxic materials systems of classification developed for decomposable organic wastes.

Some systems of using organisms as an index of conditions in aquatic environments have not given much regard to the kinds of organisms present and have emphasized mainly their variety and abundance (Patrick, 1949). If this is to be done, the best approaches are probably those based on information theory that yield diversity indices (Wilhm and Dorris, 1968). Some biologists have considered particular groups of organisms, such as the diatoms (Patrick et al., 1954), to be the most useful indices of environmental conditions. But most biologists have taken a broader view, though there has been a tendency for many to favor benthic animals, particularly insects. And most biologists have given due regard not only to the particular species present and to their probable tolerance of different conditions but also to the general composition of the community (Gaufin and Tarzwell, 1956; Hynes, 1960). The use of the life of waters to indicate their conditions will continue to contribute to the effort to control water pollution. But rather than dominate biological studies related to pollution as in the past, this approach should occupy a position more commensurate with the contribution it is likely to make.

There are far more important reasons for studying aquatic communities than to use biological organisms as indicators of environmental change, as useful as this may be. Biological communities typical of the different waters of the earth are important to man in their own right. When they are upset by environmental change, nuisance problems may develop, or species of value to man may no longer be successful, for all species are dependent in many ways on the persistence of their communities. About the turn of this century, changes in the plant life of Switzerland's beautiful Lake Zürich were first observed. Fortunately, biologists have studied these changes from the time they were first observed, and their cause, enrichment by domestic drainage, has been identified (Minder, 1938; Thomas, 1957). Lake Zürich has lost much of its beauty because of obnoxious growths of algae. All over the world, lakes have been similarly damaged, but studies patterned after those on Lake Zürich are providing the understanding necessary to prevent such damage to other lakes (Lackey, 1945; Hasler, 1947; Edmondson, 1968).

Plants are dependent on nutrients and solar energy, animals on plants and other animals for their food; the community is this web of life. When it is changed by pollutional conditions, some organisms, perhaps valuable species of fish, no longer find their necessities. One very important reason for studying communities is to gain understanding of how they sustain species of interest to man. Forbes (1887, republished 1925, p. 537) wrote:

If one wishes to become acquainted with the black bass, for example, he will learn but little if he limits himself to that species. He must evidently study also the species upon which it depends for its existence, and the various conditions upon which *these* depend.

The importance of understanding energy and material transfer in ecological systems was beautifully set forth by Lindeman (1942) in his classic paper "The Trophic-Dynamic Aspect of Ecology." Not all of Lindeman's ideas are useful today; we should not expect them to be. But his emphasis on energy and material transfer gave new direction to community studies. His view is important to water pollution biology because, after all, most of the wastes man introduces into waters alter their energy and material economy in one way or another.

Population Studies

Most often it is a change in the distribution and abundance of one or at most a few populations of organisms that first alarms the public over the effects of pollution. A bloom of some species of algae may lead to a taste and odor problem in a water supply, or a valuable fish population may decline. The biologist may want to study changes in the community and changes in the survival, reproduction, and growth of the species to understand changes in its distribution and abundance; but the man on the street wants only drinking water that is not bad or fishing that is again good. Studies of the population of interest and of conditions leading to changes in its abundance are a practical step, even if only an initial one, toward resolving such problems.

Populations of algae and other microorganisms often cause taste and odor problems in domestic water supplies. About 1850 in England, Hassall pointed out the value of microscopic examination as an aid in drinking water analysis (Tarzwell, 1963). In the United States, Sedgwick (1889) greatly stimulated the application of biological methods to the solution of water supply problems. Whipple, a leader in this field, published the first edition of his book *The Microscopy of Drinking Water* in 1899 (Whipple, Fair, and Whipple, 1927). In the late nineteenth and early twentieth centuries, taste and odor problems were being traced to particular species of algae and other organisms present in water supplies; Whipple reviewed many of these early studies in later editions of his book. Moore and Kellerman (1905) were the first to use copper sulfate to eradicate nuisance blooms of algae. Even though other algicides are now available, this is still the most widely used method of control. Tarzwell (1963) reviewed some of the recent efforts to find biological controls as well as better chemical controls for microorganisms causing taste and odor problems.

Many fish populations declined with the development of pollutional conditions. Klein (1962, pp. 4–5) states that population and industrial growth along the River Mersey and the River Irwell resulted in the virtual elimination of all fish life from these streams in England by the early nineteenth century. The shad (*Alosa sapidissima*) fishery of the Delaware River, much like fisheries of other rivers along the North Atlantic Coast of the United States, reached a peak in the late nineteenth century, and then it suffered a decline from which it has never recovered. The discharge of domestic and industrial wastes, poor fishery management, and the construction of dams are believed to have been responsible for this decline (Sykes and Lehman, 1957). Although pollutional conditions no doubt have resulted in the decline and even in the disappearance of valuable populations of fish and shellfish in many areas of the world, much of the evidence is circumstantial. In few studies have the causal relations been clearly demonstrated. Most investigations have documented the presence or absence of different species at different locations rather than determined how pollutional changes have affected survival, reproduction, or growth in particular populations.

Thompson (1925) made observations on populations of fish of different species in the Illinois River, where low oxygen conditions resulted from ice cover and the introduction of sewage; he attributed declines in the fisheries on these species in part to pollution. In their studies of the fishes of Champaign County, Illinois, Thompson and Hunt (1930) noted that small amounts of domestic wastes increased the productivity of streams and resulted in greater size, abundance, and variety of fish; larger amounts of wastes decreased the abundance of fish; and very large amounts virtually eliminated some species. Ellis (1937) found that an abundant and varied fauna of warm-water species of fish nearly always occurred where 5 mg/l or more of dissolved oxygen was present at the 982 stations he studied on rivers and streams in the United States; a poorer fish fauna existed where less than 4 mg/l was present. This classic study has often been cited as evidence that dissolved oxygen standards for the protection of fish should be 5 mg/l or more.

Loss of valuable oyster and clam fisheries has resulted from closure by health authorities of grounds contaminated with human wastes; it also has resulted from decreases in populations due to other kinds of wastes. Hopkins, Galtsoff, and McMillin (1931) investigated the effects of environmental conditions on the survival, growth, and reproduction of a Puget Sound oyster population that declined after the installation of a sulfite process pulp and paper mill. Galtsoff et al. (1938) similarly investigated a decline in the production of an oyster population in the York River on Chesapeake Bay, where a kraft process pulp and paper mill was located. These studies are particularly noteworthy, for, even today, studies of the survival, reproduction, and growth of populations affected by pollutional conditions are all too rare.

Introduction into the aquatic environment of chlorinated hydrocarbon pesticides and radioactive materials has greatly concerned biologists, because these dangerous materials may accumulate through the food chains of valuable species. Kerswill and Elson (1955) investigated the effects of DDT on the Miramichi River population of juvenile Atlantic salmon (*Salmo salar*), when this pesticide was used to control spruce bud worms in a forested area of Canada. Radioactive wastes introduced into White Oak Lake, Tennessee, were found by Krumholz (1956) to result in the accumulation of radioactive materials in the tissues of the fish. This apparently led to shorter life spans and lower growth rates; two species apparently were slowly disappearing.

It is the elaboration of new tissue by fish populations, their production, that makes them of value to man. Knowledge of how environmental conditions, pollutional and other, affect the production of populations is useful in their management. Two classic papers provided a conceptual framework for studies of fish production. Ricker and Foerster (1948) computed the production of juvenile sockeye salmon (*Oncorhynchus nerka*) in Cultus Lake, Canada, and showed how data on the growth rate and biomass of each age group of fish could be used for such computation. Allen (1951) calculated the production of brown trout (*Salmo trutta*) in the Horokiwi Stream of New Zealand, studied the food resources of this population, and used laboratory data on food consumption and growth of trout to estimate the amount of food the natural population consumed. Even though this is a promising approach, few have studied the effects of pollutional conditions on the production of fish populations. In Berry Creek, a controlled experimental stream, we were able to demonstrate how trout production was increased by the introduction of sucrose, which increased the production of *Sphaerotilus* and insect larvae and so the availability of trout food (Warren et al., 1964).

Sophisticated yet practical methods have been developed for studying the population dynamics of fish and other animals in nature (Ricker, 1958). These methods often make possible the estimation of birth, death, and growth rates of animals. Water pollution biologists have not exploited these methods as one day they must, if the effects of pollution on populations are to be understood. To explain how environmental conditions change the distribution and abundance of populations, it is necessary to relate these conditions to birth, death, growth, and movements.

Studies of the Individual Organism:
Toxicity Bioassays and
Environmental Requirements

Even when we have information on the birth, death, growth, and movements of animals in nature, it is not always possible to relate these to particular environmental conditions. And often we will be unable to learn much about survival, reproduction, growth, and movements in the natural environment, either because of limited investigational resources or because of the difficulty of the problem. It is nearly always helpful to turn to the laboratory to learn how particular factors affect the animals we study, and this biologists usually do. But whether we start our investigations in nature or in the laboratory, we must always remember that the simplicity we seek and can obtain in the laboratory is the Achilles' heel of the laboratory experiment. We must be careful in our extrapolations back to nature with her beau-

tiful complexity. Still, biologists need the laboratory approach and they know its proper role. Studies of the individual organism are usually laboratory studies.

Determination of the levels of environmental factors that soon lead to the death of animals is usually easier than finding levels that may decrease an animal's life span, reproduction, or growth. This and the apparent importance to animals of at least staying alive have caused biologists to emphasize studies of lethal effects, even though in nature sublethal effects are in the long run probably more important. Lethal effects of toxic substances on fish have received a large share of the attention of water pollution biologists.

Tests of the acute toxicity of various pollutants were undertaken in the late nineteenth century by a few men who were concerned that toxic materials in industrial wastes might have lethal effects on aquatic life (Penny and Adams, 1863; Weigelt, Saare, and Schwab, 1885). Many of the early studies of the toxicity to aquatic animals of salts, acids, drugs, and other substances were not necessarily directed toward the solution of water pollution problems (Garrey, 1916; Powers, 1917) — many still are not — but others were. In 1924, Kathleen Carpenter published the first of her important papers on the toxicity to fish of heavy metal ions present in waters draining from old lead and zinc mines in Wales (Carpenter, 1924, 1925, 1927, 1930). Jones (1938) was to continue and expand this work. Not only industrial wastes and mine drainage but the decomposition products of sewage are sometimes toxic to aquatic life; Longwell and Pentlow (1935) studied the toxicity of some of these to trout.

The chemical, engineering, and public health viewpoints dominated water pollution control efforts during the first half of this century. Apart from the bacteriologists, most of the few other biologists involved sought ways of making chemical and engineering studies more relevant to the problem of protecting aquatic life. The toxicity tests that Ellis (1937) and many others before him performed were intended primarily to provide a basis for interpreting chemical analyses that were more routinely done. It became increasingly apparent, however, that the toxicity of a substance in one water was often very different from its toxicity in another. Further, it became apparent that several toxic substances, when present together, can act differently than they do individually. Chemical analyses of complex wastes were found to be inadequate for predicting their toxicity.

These and other considerations led Hart, Doudoroff, and Greenbank (1945) and Doudoroff et al. (1951) to advocate more general use of toxicity tests with fish as test animals for direct evaluation of the toxicity of wastes. Because of the obvious need to make these tests practical and still reliable and reproducible, these authors also proposed appropriate standardization of methodology. Inasmuch as living organisms were used as reagents to evaluate a biologically important property of water pollutants—their potency as toxicants—such tests were termed and are now generally known as "toxicity bioassays." The value of these tests as the primary means of evaluating routinely the toxicity of effluents—not just as a means of providing information necessary for the interpretation of results of chemical analyses—quickly came to be widely recognized. In due time, the bioassay methods for evaluation of acute toxicity of water pollutants to fish recommended by Doudoroff et al. (1951) were included, with only minor modifications, as a standard method in the publication *Standard Methods for the Examination of Water and Wastewater* (American Public Health Association et al., 1960).

Aquatic organisms other than fish are also of value to man, either directly or as links in important food chains. Some of these have long been known to be even more susceptible than fish to many toxic pollutants. Aquatic animals that are smaller than fish and can be cultured in the laboratory may be more convenient test organisms. The water flea *Daphnia* was used as a sensitive test animal by Ellis (1937) and other investigators. Anderson (1944, 1946) refined methods for using this animal and thus laid a foundation for standardization of the procedures. Oyster and mussel larvae, whose abnormal development is a sensitive index of toxicity, also have recently been used in toxicity bioassays of industrial wastes entering marine or estuarine waters (Dimick and Breese, 1965; Woelke, 1965). We can expect standard methods for using organisms other than fish in toxicity bioassays to be widely adopted in the near future.

Prevention of lethal concentrations of toxic substances in aquatic environments is essential; the acute toxicity bioassay provides a means for their estimation. But as our concern for the well-being of aquatic organisms has increased, so has our desire to maintain conditions that do not decrease their longevity, reproduction, or growth. Hart, Doudoroff, and Greenbank (1945) suggested that some fraction of the concentration of a substance that is acutely toxic to fish used as test animals might be permitted in receiving waters and provide a margin of safety for the overall well-being of fish and other organisms. Here, then, we would extrapolate from acute toxicity information to estimate waste concentrations that in the long run would not be injurious to populations in receiving waters. There has been and continues to be interest in the use and refinement of "application factors" for more reliable estimation of permissible concentrations of toxic wastes on the basis of acute toxicity bioassays (Aquatic Life Advisory Committee, ORSANCO, 1955; Federal Water Pollution Control Administration, 1968).

But such use of acute toxicity bioassay information requires very great extrapolations indeed; studies of chronic effects of toxic substances are needed. Thus, some biologists have concerned themselves with the estimation of long-term lethal thresholds of toxicity (Wuhrmann, 1952; Burdick, 1957). Others, so as to derive application factors, have attempted to relate acutely toxic concentrations to the concentrations of substances that do not depress the reproduction and growth of fish (Mount and Stephan, 1967; Mount, 1968). The finding that different toxic substances often are simply additive in their effects (Herbert, Jordan, and Lloyd, 1965; Edwards and Brown, 1967) appears to have brought us closer to solution of problems resulting from the introduction of many toxic substances into the same waters.

Thus far in our short history of water pollution biology, we have devoted the most space to biological indices and bioassays; this is not at all disproportionate to the historical emphasis in our field. But bioindices and bioassays are only two approaches to biological studies of pollution problems, and toxic substances represent only one kind of factor in the aquatic environment. Studies of the effects of pollution have needed a broader biological base, one that now is rapidly developing.

The first biologists to study the environmental requirements of aquatic organisms were not particularly interested in problems of water pollution. The requirements of plants and animals, apart from the changes man causes in their environments, will always interest biologists. The contributions of these biologists, whatever their interests, are often relevant to water pollution biology. Temperature, dissolved oxygen, and salinity are factors governing the distribution and abundance of aquatic organisms. The tolerance of fish to these factors and the importance of an organism's previous experience with a factor in determining its tolerance were early explored (Packard, 1905; Sumner, 1905; Loeb and Wasteneys, 1912). Lethal levels of temperature, to take only one example, depend very much on an organism's recent temperature experience (Hathaway, 1927a; Sumner and Doudoroff, 1938). Before temperature became widely recognized as a pollution problem, it had become the most carefully studied lethal factor in the aquatic environment (Fry, Brett, and Clawson, 1942; Doudoroff, 1942, 1945; Brett, 1944, 1946; Fry, Hart, and Walker, 1946; Hart, 1947). The discharge of waste heat into receiving waters has now made temperature increase one of our most serious pollution problems; knowledge of the temperature requirements of organisms, whatever stimulated its search, is relevant to this problem. The story is no different for oxygen, salinity, and other factors that have become pollution problems.

Historically, the effects of environmental factors on the reproduction and growth of aquatic animals have received much less attention than lethal effects; only recently has this serious imbalance tended toward correction. Still, there were those who studied the effects of temperature on the development of fish (Gray, 1928) and on their growth (Hathaway, 1927b). Dilling, Healey, and Smith (1926) studied the effects of lead on the growth of a marine fish. Since 1950 and stimulated primarily by concern over pollutional conditions in the aquatic environment, studies of the effects of temperature, oxygen, and other environmental conditions on the development and growth of aquatic animals have fortunately come into their own.

In the survival, reproduction, and growth of animals, behavior, often inborn behavior, plays an important role; Whitman (1899) was among the early biologists recognizing

this. Shelford and Allee (1913) also saw the importance of behavior; they studied the avoidance reactions of fish in gradients of oxygen, carbon dioxide, and nitrogen dissolved in water. Later, Shelford (1917) was to study the responses of fish to toxic substances in wastes from gas works. The avoidance responses of fish to temperature, oxygen, and toxic substances have continued to interest biologists (Doudoroff, 1938; Jones, 1947, 1952). But territorial, reproductive, feeding, and still other behavior of animals is probably more important to their success than avoidance of unsuitable conditions, and to these kinds of behavior water pollution biologists will need to direct much more attention.

Before we leave this section on the individual organism as a whole, we must consider another important response of the intact living organism that properly belongs here: the metabolic rate of the organism, its rate of oxygen consumption. F. E. J. Fry's (1947) "Effects of the Environment on Animal Activity" is one of the truly important contributions to our understanding of the relationships between animals and their environments. Using metabolic rate as an integrated response of the whole organism, Fry conceptualized how temperature and oxygen interact to control and limit the metabolic rate and hence the scope for activity of animals. His ideas together with Ivlev's (1945) ideas on bioenergetics and growth provide a basis for describing in energy terms the effects of environmental conditions on metabolism and growth (Warren and Davis, 1967). This has broad application not only in our attempts to understand animals in their natural environments but also in our attempts to understand how they may be affected by changes in temperature, oxygen, salinity, toxicant concentrations, and even food availability.

Studies of Systems Within the Organism

It is of the nature of men, scientists and nonscientists alike, that their interests vary. Some scientists are most interested in the mechanisms of phenomena, others in the significance of phenomena. This is fortunate, for real understanding of phenomena requires knowledge of both their mechanisms and significance. It is, however, unfortunate that so few scientists are concerned with both levels of understanding; biology is the poorer for this. Understanding of the operational mechanisms of any level of biological organization can come only from study of the previous level; understanding of a level's significance can come only from studies of its succeeding level (Bartholomew, 1964).

The significance of an individual organism lies in its contribution to its population. Explanation of the effects of most environmental factors on the individual can come only from studies of systems within the individual. Man is mainly interested in the success of populations and communities. But to explain this, he must know the changes environmental factors bring about in the survival, reproduction, and growth of the individual organism. We reach another level of understanding when we can explain how environmental factors interact with systems within the organism in causing these changes in its responses. But studies of systems within the organism contribute little to water pollution biology unless it is shown how these systems influence survival, reproduction, growth, or activity; otherwise, we have not explained the significance of systems within the organism.

In recent years, biologists of many interests have become concerned about water pollution. They have wanted to contribute their special knowledge to the solution of problems man has created. Many of these biologists have found changes within organisms to be associated with changes in environments. Some of these changes are adaptive: they contribute to the continued success of organisms in environments changed by pollution. Others are not adaptive: they are steps in the disintegration of organisms. We must learn which changes are adaptive and which are destructive. Whether or not changes are adaptive depends primarily on how they influence survival and reproduction. The most useful knowledge of systems within the organism is that which explains the organism's success or failure in surviving and reproducing.

Physiologists who have studied osmoregulation of aquatic organisms have explained the mechanisms of systems of obvious adaptive value (Smith, 1930, 1932; Keys, 1931; Krogh, 1937). And those who have studied the effects of temperature and carbon dioxide on the oxygen dissociation curves of hemoglobin have been able to suggest probable adaptive functions (Krogh and Leitch, 1919; Kawamoto, 1929; Root, 1931; Black, 1940). These men most certainly did not consider themselves water pollution biologists. Yet salinity, temperature, oxygen, and carbon dioxide problems are often associated with water pollution. The work of these men was good, and it is relevant to the problems of today. They set high standards for those of us who would study systems within the organism.

But not always are the changes we observe in animals so clearly either adaptive or destructive. Perhaps we can assume that the destruction of blood cells in catfish exposed to lead acetate is a bad thing (Dawson, 1935). Still, is the destruction enough to harm the performance of this animal? Is an increase in the red blood cells of trout exposed to sublethal concentrations of potassium chromate (Schiffman and Fromm, 1959) an adaptive response favoring the continued success of this animal in a toxic environment? Other toxic substances have been shown to alter the electrophoretic patterns of the blood serums of fish (Fujiya, 1961). There are changes in the enzymes, cells, tissues, and organs of fish exposed to toxic substances (Weiss, 1961; Crandall and Goodnight, 1963). Wood (1960) has suggested histological examination as a means of determining the substances responsible for fish mortalities in polluted waters. The things men do and find, their ideas, the positions they take, all are a part of history. The studies we have just mentioned and the men who conducted them are a part of the recent history of water pollution biology. There will be more work of this sort, work on systems within the organism; it is needed to explain the mechanisms of responses of the whole organism. If it is not used to explain how environmental conditions interacting with systems within the organism affect its survival, reproduction, growth, and activity, biology in general as well as water pollution biology will be the poorer.

POLLUTION PROBLEMS AND THE BIOLOGIST

We have traced the efforts of biologists to apply their special knowledge to the solution of water pollution problems since the time scientists became interested in these problems. Water pollution problems occur when certain of man's uses of water conflict. Thus they are ultimately social and economic in nature. But because many of the characteristics and uses of waters either depend on or are influenced by aquatic communities, the solution of pollution problems frequently requires biological knowledge. It is the responsibility of water pollution biologists to provide the kinds of knowledge that are most needed by those responsible for water pollution control.

Water pollution control policies are but a part of the overall water resource management policy of a society. The state of scientific and technical knowledge influences these broad policies, and particular policies involve scientific and technical problems, but the primary responsibility for their establishment does not rest with the scientific disciplines. Water resource management policy develops in a society from public action through various social and governmental institutions and is influenced by geographic, historic, governmental, economic, and other social circumstances.

Even though development of public policy is not a special responsibility of the scientific disciplines, biologists and other scientists should help to provide the information necessary for development of sound policies and aid in making these policies effective. The activities of biologists that contribute to water pollution control can perhaps be categorized into four areas: providing understanding of the effects of water quality changes on aquatic life and resources; identifying aspects of existing water pollution control policies that do not take into account biological knowledge; developing biological approaches and methods of value in solving water pollution problems; and utilizing biological approaches and methods in water pollution control efforts. These roles of water pollution biologists we will treat only briefly now, for they will be considered in more detail in Chapter 23.

Without understanding of the effects of water quality changes on aquatic life and resources, the development of sound policies and their effective employment are virtually impossible. Pollution control policies for the protection of aquatic resources such as fisheries may be useless or even harmful if they are not based on knowledge of the effects of water quality changes. Biologists in-

volved in water pollution research must obtain this information and make it available to those responsible for pollution control.

Public policy relating to water pollution, by the very nature of the processes through which it develops, is frequently inconsistent with the state of scientific and technological knowledge. This may render policies ineffective in accomplishing public objectives. Biologists must be informed on public policy and relevant biological knowledge, for it is their responsibility to call to the attention of authorities and also the public any inconsistencies that exist.

Effective utilization of biology in water pollution control requires appropriate biological approaches. Classical approaches, based on biological principles, will usually be adapted for application. Which approaches are likely to be most effective depends to a considerable extent on what policies may be and how they are being implemented. Biologists must develop and evaluate approaches to the solution of pollution problems that will aid in making public policy effective. Because public policy is continually changing, and because scientific investigation requires considerable time, biologists should also be exploring approaches that could become valuable with changes in policy.

Needs for water pollution control measures and the effectiveness of such measures require continual evaluation. In this, biological knowledge and the application of biological approaches and methods are necessary. This responsibility falls largely on biologists employed by regulatory agencies and industrial organizations.

The roles of biologists in water pollution control will be determined by public policy, the nature of biological systems, the state of biological knowledge, and the possible applications of biology to the solution of pollution problems. Perhaps now we should examine considerations biologists take into account in determining how best to contribute to the solution of these problems.

THE APPROACHES BIOLOGISTS CHOOSE

In beginning this chapter, we explained how biological systems can be separated conceptually into levels of organization, and we suggested four possible levels: systems within the organism; the individual organism as a whole; the population; and the biological community. The biological systems that constitute each of these levels of organization are extremely complex, are not thoroughly understood, and cannot be encompassed within the scope of a single investigation. A lifetime of effort would be insufficient to elucidate the relationships within even one level of a given biological complex. Recognizing this, most biologists approach with even narrower conceptual frameworks their studies of particular levels of biological organization. The history of biology is the story of biologists who have exploited fruitful approaches to the solution of important biological problems. A fruitful approach to a problem is a conceptualization of the problem that will permit substantial progress toward its ultimate solution. Such an approach should identify relevant knowledge and suggest hypotheses and necessary experiments or observations; moreover, it should provide a system of analysis and synthesis of the findings that will make possible the most generally important conclusions. It will usually involve assumptions that permit a complex system to be viewed, at least initially, as being simpler than it is in reality. This allows the biologist to focus his attention on important aspects of a biological system; thus, hopefully, it permits him to accumulate sufficient information to reach conclusions regarding these aspects. But, as much as simplification may be necessary, there is danger in reaching conclusions regarding whole systems on the basis of their parts, and a good approach should indicate the relevance and the limits of possible conclusions.

Biologists, then, in their studies of complex biological systems, not only usually direct their attention primarily to a single level of organization, but also frequently exploit approaches that further focus their attention on some functional aspect of a level of organization. It should be apparent from our brief historical review that water pollution biologists, like other biologists, select particular approaches for their studies of complex biological systems. At the suborganismic level, some have used biochemical or histological approaches. Of biologists working at the level of the whole organism, some have studied the effects of environmental factors on the survival of organisms, others effects on reproduction, development, growth, or metabolism. Biologists interested in the effects of water quality changes on

populations have sometimes studied the distribution and abundance of these populations; at other times they have studied production. Some of those studying communities have been concerned with biological indices of pollution; others have been concerned with the effects of environmental change on energy and material transfer through communities. These are but a few of the possible approaches to studies of various levels of biological organization. These and other approaches are sometimes used independently; sometimes, and usually preferably, several approaches to the same problem are used in a complementary manner.

As we have noted, four of the parts of this book are devoted to different levels of biological organization from the suborganismic level to the community level. In chapters within each of these parts, we will describe and evaluate different approaches that have been used or suggested for studies of the effects of water quality changes on biological systems. From a strictly biological point of view, some of these approaches are more promising than others. A biologist hoping to solve a particular problem selects the approach he believes to be most likely to lead him to a solution. But a biologist investigating a pollution problem has other than strictly biological considerations to take into account in choosing an approach. The most promising biological approach to a pollution problem will in part depend on the kind of information that will be socially most useful in controlling the pollution.

The biological information that will be socially most useful in solving a particular pollution problem will depend to a considerable extent on the desires of society with regard to the water involved and on the administrative and legal procedures developed to make these desires obtainable. The meaning that a society and its governmental institutions come to adopt for the word *pollution* has bearing on the obtainability of these desires and on possible methods of obtaining them (Chapter 2). Of quite direct bearing on the choice of the best biological approach for the solution of a particular pollution problem are the adopted system of water quality standards and water use classification and the classification that is given the water involved (Chapter 3).

The biological approach likely to provide the information socially most useful in solving a pollution problem will depend on what we might term the *limits of acceptable biological change* for the polluted water. In Chapter 22, we will consider the problem of determining the limits of biological change that are socially acceptable. Now we will explain only how the limits of acceptable change adopted for waters having a particular use classification will influence the choice of the biological approach most likely to provide socially useful information.

A society may choose to maintain some of its waters in very nearly their natural or primitive state for aesthetic reasons or for scientific or limited recreational use. These waters would then be classified for such uses, and other possibly conflicting uses would be strictly controlled. Water quality standards insuring that the biological communities of these waters would remain in their natural state would need to be adopted. It would be impossible to arrive at water quality standards for these waters through studies of the physiology, survival, reproduction, growth, and behavior of the hundreds of species of organisms composing their communities. Detailed studies of the distribution, abundance, and production of the populations of these many species would be difficult if not impossible. Probably the most effective biological approach to determining the limits of water quality conditions that would not lead to changes in these communities would be studies of their species composition under different conditions, studies conducted at the biological community level of organization. Studies of similar waters, some of which have been affected by water quality changes and some of which have not, could yield information on the limits of water quality changes that permit the existence of the desired biological communities. This information could then be used in the original formulation of the necessary standards. Later, studies of the species composition of the communities in the waters being maintained under these standards could be used to evaluate the adequacy of the standards.

For another class of waters, a society may be willing to accept some change in the original biological communities but unwilling to tolerate any reduction in the capacity of these waters to produce some species of fish having commercial or recreational value. Studies of the effects of water quality changes on the ability of individuals of this species to reproduce and grow would be nec-

essary in determining suitable water quality standards. Also, studies of the production of this fish species, studies at the population level of organization, would be necessary. And limited community studies would in addition be needed, studies directed toward determination of the conditions under which the production of food organisms suitable for this fish would be adequate. But these studies would not seek conditions under which no changes in the biological communities would occur, only conditions under which no changes in the production and usefulness of the valuable fish species would occur.

There may be other waters so valuable to a society for industrial or agricultural uses that they cannot be maintained either in their natural state or in a state permitting unrestricted production of some valuable species. Yet, that society may be willing to bear the cost of maintaining these waters in a condition permitting the existence of populations of some desirable but relatively tolerant aquatic species. Perhaps the most effective biological approach for determining the water quality standards that will permit this would be to conduct studies at the level of the whole organism. Studies of the limits of water quality change permitting survival and some reproduction and growth of the desired species would be most useful. Such limits can be expected to permit the production of some of the food organisms of these species.

Algae and bacteria in lakes and streams sometimes cause nuisance problems. Conditions under which these problems will not occur can frequently be determined most effectively through studies at the suborganismic and organismic levels of organization along with limited studies in nature. Microorganisms often cause nuisance problems when substances introduced into waters can be used by these organisms in their nutrition. Identification of the kinds and levels of these substances enhancing the reproduction and growth of nuisance organisms requires both biochemical and physiological studies. Conditions in nature are, however, complex; and biochemical and physiological studies of the nuisance organism must be coupled with studies of its ecology, if suitable control measures are to be found.

In concluding this chapter on biological approaches to water pollution problems, we must emphasize that satisfactory studies at any level of biological organization are difficult and time consuming. They cannot soon if ever be made specifically on the populations and communities of organisms of all of the waters that might be subject to pollution. Moreover, as we pointed out in Chapter 3, it is much easier to enforce reasonable water quality standards than it is to prove that particular waste discharges are responsible for damage to aquatic resources. Indeed, one of the principal reasons for the development of sound standards is that this should make extensive biological and other investigations of every water receiving wastes unnecessary. Standards developed for a particular class of waters should be adequate to provide the desired protection for all waters of that class. Because physical and chemical characteristics of waters can usually be monitored much more simply than can biological ones, standards will usually be formulated in physical and chemical terms, even though they may be based on biological considerations. Once the limits of acceptable biological change have been established for waters of a particular class, it is the responsibility of biologists to exploit the most effective biological approaches in determining the set of standards that will insure these limits will not be exceeded. After a set of standards has been adopted, biological studies of some waters being maintained under these standards will be necessary to determine whether or not this set of standards provides the protection for which it was designed. The need for revision of standards will undoubtedly become apparent in this way. Thus, biological investigation will play one of its most important roles in developing and evaluating standards.

SELECTED REFERENCES

Doudoroff, P., B. G. Anderson, G. E. Burdick, P. S. Galtsoff, W. B. Hart, Ruth Patrick, E. R. Strong, E. W. Surber, and W. M. Van Horn. 1951. Bio-assay methods for the evaluation of acute toxicity of industrial wastes to fish. Sewage and Industrial Wastes 23:1381–1397.

Ellis, M. M. 1937. Detection and measurement of stream pollution. Bulletin of the U.S. Bureau of Fisheries 48(22):365–437.

Forbes, S. A. 1887. The lake as a microcosm. Bulletin of the Peoria Scientific Association. (Reprinted 1925. Bulletin of the Illinois State Natural History Survey 15:537–550.)

Fry, F. E. J. 1947. Effects of the environment on animal activity. University of Toronto Studies Biological Series 55. Ontario Fisheries Research Laboratory Publication 68. 62pp.

Hawkes, H. A. 1962. Biological aspects of river pollution. Pages 311–432. *In* L. Klein, River Pollution. 2. Causes and Effects. Butterworth, London. xiv + 456pp.

————. 1963. The Ecology of Waste Water Treatment. Pergamon Press Ltd., Oxford. ix + 203pp.

Huntsman, A. G. 1948. Method in ecology—biapocrisis. Ecology 29:30–42.

Hynes, H. B. N. 1960. The Biology of Polluted Waters. Liverpool University Press, Liverpool. xiv + 202pp.

Ivlev, V. S. 1945. Biologicheskaya produktivnost' vodoemov. Uspekhi Sovremennoi Biologii 19:98–120. (Translated by W. E. Ricker, 1966. The biological productivity of waters. Journal of the Fisheries Research Board of Canada 23:1727–1759.)

Jones, J. R. E. 1947. The reactions of *Pygosteus pungitius* L. to toxic solutions. Journal of Experimental Biology 24:110–122.

Kolkwitz, R., and M. Marsson. 1908. Ökologie der pflanzlichen Saprobien. Berichte der Deutschen Botanischen Gesellschaft 26a:505–519. (Translated 1967. Ecology of plant saprobia. Pages 47–52. *In* L. E. Keup, W. M. Ingram, and K. M. Mackenthun (Editors), Biology of Water Pollution. Federal Water Pollution Control Administration, U.S. Dept. of the Interior. iv + 290pp.)

————, and ————. 1909. Ökologie der tierischen Saprobien. Beiträge zur Lehre von der biologischen Gewässerbeurteilung. Internationale Revue der Gesamten Hydrobiologie und Hydrogeographie 2:126–152. (Translated 1967. Ecology of animal saprobia. Pages 85–95. *In* L. E. Keup, W. M. Ingram, and K. M. Mackenthun (Editors), Biology of Water Pollution. Federal Water Pollution Control Administration, U.S. Dept. of the Interior. iv + 290pp.)

Lindeman, R. L. 1942. The trophic–dynamic aspect of ecology. Ecology 23:399–418.

Patrick, Ruth. 1950. Biological measure of stream conditions. Sewage and Industrial Wastes 22:926–938.

Richardson, R. E. 1928. The bottom fauna of the middle Illinois River, 1913–1925: its distribution, abundance, valuation, and index value in the study of stream pollution. Bulletin of the Illinois State Natural History Survey 17:387–475.

Ricker, W. E., and R. E. Foerster. 1948. Computation of fish production. Bingham Oceanographic Collection Bulletin 11:173–211.

Tarzwell, C. M. 1963. Sanitational limnology. Pages 653–666. *In* D. G. Frey (Editor), Limnology in North America. University of Wisconsin Press, Madison. xvii + 734pp.

Whipple, G. C. Revised by G. M. Fair and M. C. Whipple. 1927. The Microscopy of Drinking Water. 4th ed. John Wiley & Sons, Inc., New York. xix + 586pp. + 19 plates.

PART II

Conditions of Life in the Aquatic Environment

Conditions of life in lakes, rivers, estuaries, and the sea are largely determined by the properties of water, including its physical and chemical stability and its solvent powers. These properties make all life possible. Life can exist only within a narrow range of conditions, which water tends to maintain in aquatic and terrestrial environments. Each organism that lives in water or otherwise uses it changes in some degree the energy and materials the water carries. Man is no exception. Many changes that organisms bring about in water are essential to the existence of their own and other kinds. But man has been careless in his use of water. Too many of the changes he has caused have not been essential to his existence and have endangered the existence of aquatic life. This has already destroyed much of the utility and beauty of the waters of the earth and now endangers man himself. He cannot persist in this carelessness. By careful use of water, man can maintain the conditions necessary for aquatic life and so provide for his esthetic as well as his material needs.

5 The Physical and Chemical Environment

WATER, MEDIUM OF LIFE

To prehistoric man there must have come with his earliest rational thoughts an awareness of the importance of water to all life. To read in Aristotle that Oceanus, God of the Sea, was the father of creation in ancient Greek mythology is no surprise. Nor is it a surprise that Thales, founder of Greek philosophy and science, considered water to be the origin of all things. Modern science has changed man's view of nearly everything; it has not lessened the wonderment which water occasions in everyone, scientist and nonscientist alike.

What is there about water that occasions such general wonderment? Its importance to life—instinctively known to all animals and known even more to man through learning—in part explains this. That it evokes the deepest feelings of beauty or fear, according to circumstances, cannot be disregarded. But other things are important to life; other things are beautiful or frightful. And scientists are notorious in their reduction of the important, the beautiful, and the frightful to causal explanations. From where does their sense of wonderment come? This is not to say that scientists are different in any important way from other men; they are not. Still, for the scientist the physical and chemical properties of water are perhaps the greatest source of wonderment. So far as he knows, there is no other substance in the universe so suited to making life possible. This is the main theme of Lawrence Henderson's classic *The Fitness of the Environment* (1913, pp. 130–131):

> Such are the facts which I have been able to discover regarding the fitness of water for the organism. The following properties appear to be extraordinarily, often uniquely, suited to a mechanism which must be complex, durable, and dependent upon a constant metabolism: heat capacity, heat conductivity, expansion on cooling near the freezing point, density of ice, heat of fusion, heat of vaporization, vapor tension, freezing point, solvent power, dielectric constant and ionizing power, and surface tension.
>
> In no case do the advantages which these properties confer seem to be trivial; commonly they are of the greatest moment. . . . Water, of its very nature, as it occurs automatically in the process of cosmic evolution, is fit, with a fitness no less marvelous and varied than that fitness of the organism which has been won by the process of adaptation in the course of organic evolution.

We will be considering in this chapter, however briefly, physical and chemical conditions of life in the aquatic environment as these are determined by solar radiation, the surface of the earth, and the properties of water, oxygen, carbon dioxide, and other substances. But, before we do, let us reflect momentarily on the remarkable properties of water itself.

The *specific gravity* of water having no dissolved solids is taken to be 1.00000 at a temperature of 4 C. This is the temperature at which water has its maximum density and a weight of 1.00000 gram per milliliter. Increases in the salt content of water increase its density and hence its specific gravity, sea water having a specific gravity averaging 1.02822 at 4 C. For most inland waters, changes in density resulting from changes in salt content are not great. The most important changes in density of water result from changes in temperature, for these changes have profound effects on the suitability of water as an environment. Changes in the density of water with changing tempera-

ture—the increase in density as temperature declines to 4 C and the decrease in density with further decline in temperature to 0 C—lead to important movements of water. Water movement transports materials. But these changes in density are even more important because the heaviest water, that at 4 C, tends to move to the bottom of a body of water and is at a temperature above the freezing point. Water near the freezing point of 0 C tends to be at the surface, and hence only here can bodies of water generally freeze. This prevents the waters of the earth from being turned into blocks of ice obliterating life. And ice itself has a density still less than that of water, so it floats.

Because water has a specific gravity 775 times as great as that of air at 0 C and 760 mm Hg, its buoyant effect is similarly greater. Thus, supportive structures and energy expenditure necessary to resist the force of gravity are much less for aquatic organisms than for terrestrial ones. But the high density of water does present certain problems of life: an animal such as a fish that must pump over its respiratory surfaces large amounts of water has much work to do. The *viscosity* of water, its frictional resistance to a moving body, is another problem of life in the aquatic environment. The buoyant action and viscosity of water together with its great solvent power and movement combine to make it the greatest geological force modifying the face of the earth.

The *specific heat* of water is much higher than that of most other common substances, only liquid ammonia and hydrogen having greater ones. The specific heat or heat capacity of a substance is the number of small calories necessary to raise the temperature of 1 gram of the substance 1 C. For water between 0 and 1 C, this is 1.000. Thus, to change the temperature of water much, large amounts of heat must be introduced. The temperatures of the waters of the earth are therefore remarkably stable. The very high latent heat of melting and latent heat of evaporation of water contribute to this stability. The *latent heat of melting* is the number of calories required to convert 1 gram of solid to 1 gram of liquid at the same temperature; it is approximately 80 for water. Much heat, then, must be introduced into ice before it will melt or, conversely, removed from water before it will freeze. The *latent heat of evaporation* of a substance is the number of calories necessary to convert 1 gram of liquid to vapor, approximately 536 for water. Evaporation tends to hold the temperature of water bodies constant in the face of heat introduction. Only ammonia has a latent heat of melting higher than that of water. The latent heat of evaporation of water is the highest known. So the heat properties of water temper the climate of the earth in both the aquatic and terrestrial environments and keep them remarkably stable from night to day and season to season, in a way that no other substance could. In these properties alone, water is uniquely suited as the medium of life.

The heat properties of water are not the only ones that make it so uniquely fitted for life. "As a solvent there is literally nothing to compare with water" (Henderson, 1913, p. 111). Acids, bases, and salts are all soluble to some degree in water; kinds of organic substances soluble in water are almost endless. Yet, apart from electrolytic dissociation and hydrolysis, themselves so important in geological and biological processes, solution in water commonly has little if any effect on the solutes. They can pass into and out of solution and still retain their peculiar characteristics that make possible their roles in the great cycle of life. Water brings to the aquatic organism life's essentials; through the medium of water, these enter not only the aquatic but also the terrestrial organism. Once there, only water could carry them throughout the organism in ways making the synthesis and metabolism of protoplasm possible. Here indeed, in its solvent powers, water is the ideal medium of life.

The *surface tension* of water, which may trap a hapless water flea, is importantly involved in the process of adsorption, wherever it occurs, within or without the organism. Substances dissolved in water collect on surfaces, be they geological or biological in origin. Surface tension is an important force in this. And whenever transportation by capillary action occurs, it is by the force of surface tension, which of all common liquids except mercury is greatest for water.

As with so many other of its properties, we often take for granted the wonderful transparency of pure water. Yet, without its properties of *light transmission*, water would not be a suitable environment for biological communities, which depend upon solar radi-

ation as their primary energy resource. Thus, not only in absorbing radiant energy and holding it as heat but also in transmitting it and making photosynthesis possible, water makes the aquatic environment suitable for life.

So, then, we have in water a most beneficent medium for life, life whose characteristics as we know them could depend on the properties of no other substance of which we are aware. For life, both stability and complexity of the external as well as the internal environment of the organism are essential. Stability—the tendency toward constancy—and complexity—the existence of many interacting components—are favored by the nature of water, foremost by its heat properties, its great solvent powers, and its chemical inertness. A more probable medium for the origin and maintenance of life can hardly be conceived.

PHYSICAL CONDITIONS IN THE AQUATIC ENVIRONMENT

Morphology and Substrate of Basins and Channels

The morphology and substrate of basins and channels influence if not determine the movements and conditions of water and the aquatic communities that can exist. Conversely, water to a great extent and biological communities to a lesser but still important extent determine the morphology and substrate of basins and channels. Conditions in the aquatic environment cannot be understood apart from this reciprocal relationship. Limnologists, oceanographers, and geologists have devoted their lives to its study, a study to which here we can devote only a few words.

In ways we little understand, movements of the crust of the earth must have originally formed the basins of the sea, the continents, and the mountains. Whether or not there was an Atlantis to sink into the sea, we know crustal changes continue to modify the face of the earth, above and below its waters. Volcanic activity has had and continues to have obvious effects on water basins and channels in some places. But water is perhaps the greatest geological force operating today, a force whose effects are apparent in less than the life span of man.

This has been no more eloquently stated than by Chamberlin and Salisbury (1904, pp. 8–9):

Of all geological agencies water is the most obvious and apparently the greatest, though its efficiency is conditioned upon the presence of the atmosphere, upon the relief of the land, and upon the radiant energy of the sun. Through the agency of rainfall, of surface streams, of underground waters, and of wave action, the hydrosphere is constantly modifying the surface of the lithosphere, while at the same time it is bearing into the various basins the wash of the land and depositing it in stratified beds. It thereby becomes the great agency for the degradation of the land and the building up of the basin bottoms. It works upon the land partly by dissolving soluble portions of the rock substance, and partly by mechanical action. The solution of the soluble part usually loosens the insoluble, and renders it an easy prey of the surface waters. These transport the loosened material to the valleys and at length to the great basins, meanwhile rolling and grinding it and thus reducing it to rounder forms and a finer state, until at length it reaches the still waters or the low gradients of the basins and comes to rest. The hydrosphere is therefore both destructive and constructive in its action.

Crustal movements of the earth have formed some of the largest, deepest, and oldest of the earth's lakes: Tanganyika in Africa and Baikal in Siberia. Other lakes lie in the craters of extinct volcanoes: Oregon's beautiful Crater Lake, its comparatively recent formation told in the legends of the Klamath Indians. But by far the greatest number of lakes have been formed by action of ice or water itself: glaciers excavating basins or depositing debris and damming streams; water dissolving underlying rocks and so leading to formations like limestone sinks; changing river channels leaving oxbow lakes. Other processes alter the morphology and substrate of lakes once they are formed. Around lake margins, the *littoral* areas, shelves begin to form, the result of mineral debris deposited by tributaries and eroded from banks. Here plants may begin to grow and add their organic materials to the sediments, so hurrying the process. And everywhere, in deep and shallow water alike, there rains to the bottom life's organic debris, forming sediments upon which benthic organisms depend and which return nutrients to the overlying waters. These processes, so important to the biological community of a lake, may through sedimentation lead ultimately to the lake's extinction.

Streams and rivers, returning water to the sea, flow in channels generally narrower than the basins of lakes. Though their waters are not deep, the channels they cut may become great canyons, for the water of rivers and streams, moving with the force of gravity, carries the mineral and organic debris of the earth away to the sea, where deltas are formed. In this way, streams and rivers deepen, widen, and even lengthen their channels. The substrates of streams and rivers are primarily of native or imported mineral materials. But such organic materials as do accumulate are important to the economy of stream communities.

The floor of the sea has its plains, mountains, and canyons, just as does the surface of the land; movements of the earth's crust and volcanic activity have been involved in their formation. *Continental shelves* appear to have been formed in much the same manner as the littoral shelves of lakes: through the deposition of mineral debris eroded from the land by rivers and wave action. In the sea as in lakes, plant and animal remains continually settle to the bottom, there to build organic sediments upon which benthic organisms depend. In shallower waters, particularly over continental shelves, nutrients from these sediments are readily returned to the cycle of life above, so making these waters more productive.

The nature of the substrate of a body of water is determined primarily by the native and imported mineral materials and the extent and kind of water movement. Secondarily, organic materials from the biological community may accumulate, even to the extent of burying deeply the primary substrate. Whatever its origin and nature, the substrate is a powerful determinant of the kinds of organisms that can live there. Beaches and stream bottoms of sand, moving and grinding with the force of water, are not very hospitable environments. But where water movement is less and silts and organic materials accumulate, more kinds of animals can find a home and food. Rocky bottoms are stable in turbulent streams or on wave-dashed coasts, and the kinds and numbers of organisms adapted to live in these places are many.

We see then that the morphology of basins and channels—their shape, depth, and extent—has much to do with the movements of their waters and so with the circulation of nutrients and the possibility of life. The nature of substrates, too, will determine what life exists.

Water Movement

The continuous turbulent flow of stream waters is the dominant characteristic distinguishing them from lakes and ponds. Though sometimes called standing waters, lakes and ponds are by no means always still, even below their wind-swept surfaces. Winds lead not only to horizonal but also to vertical movements of lake waters. Surface water, driven by wind from one shore to another, must sink along the leeward shore and be replaced with water rising from greater depths along the windward shore. The velocity of the horizontal movement decreases with increasing depth until it is zero at the *shearing plane*. Below this, the direction of flow is opposite to that at the surface; and the velocity first increases and then decreases, finally to zero, with further increases in depth. Lakes have other kinds of water movements: spring and fall circulation of lakes occurs with the breaking up of thermal stratification, to which we will return; and standing waves, known as *seiches*, occur in lakes.

In coastal marine environments, especially bays, fjords, and estuaries, tidal currents, which change direction with the ebb and flood tides, can attain very high velocities. Not only winds but also internal forces cause the great currents of the oceans. As we have explained, changes in temperature and salinity change the density of water; this leads to movement and currents. The direction of such currents is influenced by the rotation of the earth, their flow being deflected to the right in the northern hemisphere and to the left in the southern hemisphere. The Gulf Stream and other far-reaching oceanic currents determine not only submarine but coastal climates.

The velocity and turbulence of the movements of water are among its most important environmental characteristics. In waters of all kinds, there are plants and animals that depend upon water movement to deliver materials for their nutrition and respiration. Few factors are as important in determining where and what kinds of animals live in streams. Some stream animals obtain their food by straining it from moving water.

Some depend on the current for renewal of the water bathing their respiratory surfaces, water that brings oxygen and carries away metabolites.

Water movement can be a destructive force in the lives of aquatic organisms: it may tear them from their substrates, move them to unsuitable locales, or fragment them. But, in each aquatic environment, it prepares and maintains their place of life in ways for which they are adapted. Short of catastrophe, it is another force by which they live.

Light and Suspended Solids

Radiant energy from the sun, the energy of life, heats and lights the land and waters of the earth. Even the heat and light from hydroelectric power or fossil fuels had their source in the power of the sun. Photosynthesis, the primary source of organic materials for the structure and energy of life, is driven by solar radiation. In light it all begins.

Depending on the angle at which light strikes a water surface, a greater or lesser proportion is reflected and so does not penetrate. Where the angle of incidence is 60° from the perpendicular, only 6 percent of the light is reflected; but with the angle increasing to 80°, this becomes 35 percent. Light penetrating water does not pass through unaltered; and even in the purest lake water, only about 40 percent penetrates to a depth of 1 meter. That which does not either is dispersed or is absorbed and so converted to heat energy.

The spectral composition of the penetrating light changes due to differential absorption of its various wave lengths. Distilled water absorbs most of the light at the red end of the spectrum within the first meter of depth; little of the other light is so absorbed. But natural waters contain dissolved and suspended materials that affect the transmission of light; in these waters, transmission is usually greatest in the middle of the spectrum and declines rapidly not only toward the longer wave lengths (red) but also toward the shorter ones (violet) (Fig. 5–1). Natural waters differ in their ability to transmit various wave lengths; we are thus treated to a variety of beautiful colors. The color environment of the diver changes as he

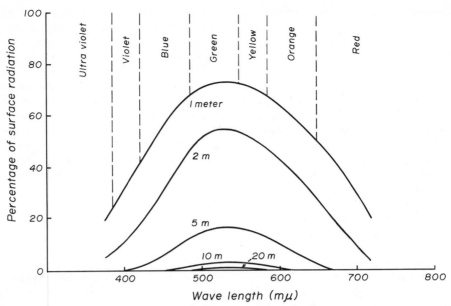

FIGURE 5–1 Average intensity and spectral composition of light at various depths in Lunzer Untersee during summer when the sun was at its mean elevation. After Ruttner (1953), based on data of Sauberer (1939).

descends into the depths. Little of the light energy reaches a depth of 20 meters, even in a moderately transparent lake like Lunzer Untersee (Fig. 5–1).

Suspended more than dissolved materials interfere with the penetration of light by scattering and absorbing its rays. Extreme turbidity, often the result of human activities, can result in the almost complete cessation of photosynthesis in aquatic environments and can in other ways affect life. Turbidity is an optical property of water and is not accurately indicative of the weight of suspended material in a given amount of water, for suspended particles differ in their properties. Nevertheless, because this optical property was measured against standards containing different weights of silica powder, turbidity was expressed in milligrams of silica per liter. Now, it is usually estimated with the aid of standards prepared from the materials causing the turbidity of a particular water, and it is expressed in turbidity units.

Even apart from scattering and absorbing light, and subsequent effects on photosynthesis, suspended matter may be injurious to aquatic life. But turbidity, being an optical property, is a poor measure of abrasive or other conditions that may be caused by suspended materials. Some materials are more harmful than others; studies of their harmful effects should relate the damage to the amount of the material involved, not to the turbidity it may cause.

Temperature

Water temperature controls the lives of most aquatic animals. They are poikilotherms: their body temperatures are near or at the temperature of their medium and change with its changes. Biochemical processes are extremely sensitive to temperature, so the rates of the processes by which these animals live change with changes in the temperature of their environment. In spite of this, poikilothermic animals have evolved ways of maintaining the effectiveness of their bodily mechanisms within the range of temperatures at which they usually must live. Still, their survival, reproduction, growth, and behavior remain very dependent on temperature. Their continued success, with solar radiation changing from night to day and season to season, is due not only to the adaptive process of evolution but also to water's remarkable temperature constancy, which originates in its heat properties.

It is a matter of no small moment that freshwaters have the unique property of reaching their maximum density at a temperature of about 4 C, density declining with further temperature reduction to 0 C, the freezing point (Fig. 5–2). On this depend the lives of aquatic organisms in lakes whose surfaces freeze in winter. As such a lake cools, the water having a temperature of about 4 C remains at the bottom, while water nearer the freezing temperature remains at the surface. Unless very shallow, a lake can freeze only at its surface. Once its surface is frozen, circulation by wind action is prevented, and further loss of heat to the atmosphere is reduced. Minimal temperatures in the depths of lakes during winter are about 4 C; at their surfaces or just under the ice, their minimal temperatures are close to 0 C.

During warmer seasons, water having temperatures higher than 4 C remains in the upper stratum, for this water too is less dense. Thus, in summer as in winter, temperature stratification occurs in most lakes. The upper stratum, known as the *epilimnion*, is rather uniformly warm; while the lower stratum, known as the *hypolimnion*, is uniformly cold. Between these two strata lies a narrow zone of rapid temperature transition: the *thermocline* or *metalimnion*. Equality of surface and bottom temperature occurs only during the spring and fall periods, after the epilimnion has warmed or cooled to the temperature of the hypolimnion. Then wind action can cause circulation throughout the entire body of water, and lake *overturn* occurs.

Whatever the climate, and whether in fresh or marine waters, maximum temperatures are rarely above 37 C; generally they do not exceed 30 C. For temperatures no higher than these in natural bodies of water, we must credit the high specific heat and the high latent heat of evaporation of water, two of its important heat properties. Sea water differs from freshwater in reaching its maximum density at a temperature lower than its freezing point. Sea water of normal salinity—about 30 parts per thousand—freezes at −1.9 C and reaches its maximum density at −3.5 C, a density it cannot reach at normal pressures. Temperatures in the great depths of the sea are, then, quite uniformly cold, from about −2 to 2 C. Still, because of the heat properties of water, the lower density of ice, and the great mass and circulation of the sea, it does not freeze to its depths. Aquatic organisms everywhere, and terrestrial ones too, owe their existence to these properties.

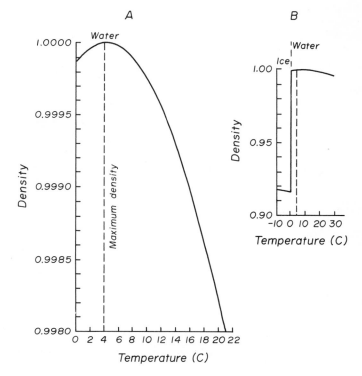

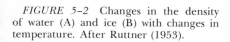

FIGURE 5–2 Changes in the density of water (A) and ice (B) with changes in temperature. After Ruttner (1953).

CHEMICAL CONDITIONS IN THE AQUATIC ENVIRONMENT

Dissolved Solids

The great solvent powers of water insure that in its natural state it contains many other substances. Rain contains the gases of the air; on contact with the earth's surface and during its trip again to the sea, water dissolves solids in great variety and amount. That this should be so is important, for the materials that dissolve in water are the stuff of life. From the beginning, the rivers of the earth have carried its salts to the sea, where they have been concentrated and remain in solution, in the life of the sea, or deposited in corals and sediments. The land is thereby the poorer and the sea the richer.

Sodium and chloride are by far the most abundant cation and anion in the sea (Table 5–1). Even though calcium carbonate is not very soluble, calcium and bicarbonate ions are the most abundant in stream waters (Table 5–2), because the more soluble sodium and chloride salts have largely been leached from the land. The calcium and bicarbonate ions in marine waters represent but a very small fraction of the calcium car-bonate which from the beginning has been dissolved from the land and carried to the sea. Most of this has been incorporated in animal structures remaining in marine sediments. Since Dittmar's (1884) careful analyses of samples collected from all the oceans during the great expedition of HMS "Challenger," it has been known that the relative amounts of the different dissolved minerals in sea waters remain quite constant (Table 5–1), even though these waters vary in salinity.

Not so the waters of lakes and streams; passing through different deposits of rock and soil, these vary greatly not only in their total content of dissolved materials but also in the proportions of different kinds. It is difficult, then, to generalize about the dissolved constituents of such waters. Hart, Doudoroff, and Greenbank (1945) estimated the median concentrations of the major ions in waters representing the total stream flow of the United States; they also estimated limits of concentrations of these ions that would include 90 percent of this flow (Table 5–2). As we have noted, calcium and bicarbonate ions are by far the most abundant in the majority of stream waters. In stream waters rich in dissolved salts, the concentration of

TABLE 5–1   *Major constituents of sea water in grams per kilogram of sea water at a chlorinity of 19 parts per thousand (g/kg) according to Dittmar's (1884) original values, according to those values recalculated on the basis of 1940 atomic weight information, and according to 1940 methods of chemical analysis. After* Sverdrup, Johnson, and Fleming (1942): *The Oceans: Their Physics, Chemistry and General Biology. Reprinted by permission of Prentice-Hall, Inc.*

Ion	Dittmar's Original Values		Recalculated Using 1940 Atomic Weights		Chemical Analysis by 1940 Methods	
	g/kg	percent	g/kg	percent	g/kg	percent
Cl^-	18.971	55.29	18.971	55.26	18.980	55.04
Br^-	0.065	0.10	0.065	0.19	0.065	0.19
$SO_4^=$	2.639	7.69	2.635	7.68	2.649	7.68
$CO_3^=$	0.071	0.21	0.071	0.21		
HCO_3^-					0.140	0.41
F^-					0.001	0.00
H_3BO_3					0.026	0.07
Mg^{++}	1.278	3.72	1.292	3.76	1.272	3.69
Ca^{++} } Sr^{++} }	0.411	1.20	0.411	1.20	0.400 0.013	1.16 0.04
K^+	0.379	1.10	0.385	1.12	0.380	1.10
Na^+	10.497	30.59	10.498	30.58	10.556	30.61
Totals	34.311		34.328		34.482	

TABLE 5–2 *Median values of concentrations of major constituents of river waters and concentration ranges that would include about 90 percent of total streamflow of the United States. After* Hart, Doudoroff, and Greenbank (1945), *based on data of* Clarke (1924).

Constituent	Median mg/l	Range mg/l
Total dissolved solids	169	72–400
Bicarbonate (HCO₃)	90	40–180
Sulfate (SO₄)	32	11–90
Chloride (Cl)	9	3–170
Calcium (Ca)	28	15–52
Magnesium (Mg)	7	3.5–14
Sodium and potassium (Na & K)	10	6–85

bicarbonate is often greater than in the sea, even though the total salinity of the sea may be 100 times more than that of these waters. High concentrations of calcium and magnesium ions, mainly responsible for the *hardness* of water, are usually associated with high concentrations of bicarbonate. Some waters—such as those containing much dissolved calcium sulfate—may, however, have high hardness without correspondingly high bicarbonate content. Waters high in bicarbonate associated with sodium are not hard. Some ground waters and artificially softened water are in this class.

Nitrogen, phosphorus, and other materials present in most waters in small amounts are exceedingly important to life. From these along with carbon dioxide and water, plants synthesize the organic materials upon which nearly all life depends. Shortage of nitrate or phosphate—frequently present only in a fraction of a milligram per liter—often limits the production of plants. Materials present in far lesser amounts than nitrogen and phosphorus are also essential to plant growth, though needed only in minute amounts. In the main, natural waters contain life's essentials; but when they do not, life is limited.

Dissolved Gases

Air having no water vapor contains on a volumetric basis about 78 percent nitrogen, 21 percent oxygen, and only about 0.03 percent carbon dioxide. Still, carbon dioxide is one of the three most abundant atmospheric gases dissolved in natural waters, the other two being, of course, oxygen and nitrogen. That much carbon dioxide is dissolved in water derives from its high solubility, its *absorption coefficient* (volume of gas absorbed/volume of liquid, when gas pressure is 1 atmosphere) at 0 C being 1.713, those for oxygen and nitrogen being only 0.049 and 0.024. Thus, at 0 C and under 1 atmosphere of air pressure, water contains on a volumetric basis about 2 percent nitrogen, 1 percent oxygen, and 0.05 percent carbon dioxide, even though these gases are represented by a much wider range of percentages of the atmosphere. Differences in solubility are the reason.

The solubility of any gas in water varies in direct proportion to its pressure in the atmosphere; it decreases with increase in water temperature or salinity. In a mixture of gases such as air, each component exerts a pressure, its *partial pressure*, which is proportional to the concentration of the gas on a volumetric basis. The total pressure of a gas mixture is equal to the sum of the partial pressures of its individual components. Air that is in contact with water contains water vapor. The partial pressure of water vapor must be subtracted from the atmospheric pressure when the partial pressures of the other atmospheric gases are computed on the basis of the composition of dry air. Vapor pressure is a function of temperature: it increases as the temperature rises, and it equals the atmospheric pressure at the boiling point of water. Thus, the amount of gas dissolved in water that is in equilibrium with air saturated with vapor decreases with rise of temperature, not only because of decrease in solubility of the gas but also because of increase in vapor pressure.

The *tension* of a dissolved gas is the partial pressure of that gas in an atmosphere with which the solution is in equilibrium. It is sometimes referred to as *partial pressure* and is commonly designated by the letter p (e.g., pO_2). At any moderate temperature, the oxygen tension of water at equilibrium with atmospheric air is about one-fifth of the atmospheric pressure—21 percent of the difference between the atmospheric pressure and the relatively small vapor pressure.

Concentrations of dissolved gases are now expressed usually in milligrams per liter of water (mg/l), often in parts per million by weight (ppm), and sometimes in milliliters or cubic centimeters per liter of water (ml/l or cc/l). When concentrations are expressed on the basis of gas volume, it is the volume the gas would have at a pressure of 760 mm Hg and a temperature of 0 C. For freshwaters, concentrations expressed in parts per million do not differ greatly from those expressed in milligrams per liter, since the weight of freshwater at ordinary temperatures is nearly 1 million milligrams per liter. Sea water and other brines, however, have densities much greater than that of pure water. Thus, a milligram of solute per liter of salt water is not nearly equivalent to a part per million by weight. Water chemists have for years carelessly continued to express concentrations of dissolved gases and other solutes in these waters in parts per million, even though they have been determining milligrams per liter and making no corrections for differences in water density. Fortunately, this practice and the use of the part per million as a unit of concentration in water chemistry are disappearing.

Physiologists often report tensions of dissolved gases rather than concentrations. Sanitary engineers frequently report oxygen values as percentages of the air-saturation value. For the aquatic biologist, consideration of gas tensions can be superfluous and even misleading. Gas exchange between the tissues of aquatic animals and their external medium usually involves no movement of molecules across gas-liquid interfaces; only the laws governing diffusion rates apply. Concentrations of dissolved gases can vary widely with water temperature or salinity while tensions remain constant. Gas concentration, in our opinion, is usually more important to the organism and more meaningful to the biologist than either tension or percentage of saturation.

The importance of carbon dioxide, abundant neither in the atmosphere nor in water, cannot be overstated. Henderson (1913, p. 133) knew this:

Two chemical individuals stand alone in importance for the great biological cycle upon the earth. The one is water, the other carbon dioxide. The one, for reasons which we have just reviewed, is the most familiar of all the varieties of matter. The other rarely is seen except by chance, and without scientific research never could have been known for what it is in value to living things. Yet these two simple substances are the common source of every one of the complicated substances which are produced by living beings, and they are the common end products of the wearing away of all the constituents of protoplasm, and of the destruction of those materials which yield energy to the body.

The properties of carbon dioxide, its great solubility and its combination with water to form carbonic acid, which ionizes to yield bicarbonate ion — to say nothing of its suitability as the primary carbon source of plants — make it remarkably fitted for its major roles in geological and physiological processes.

The concentrations of carbon dioxide in the aquatic environment are generally between a fraction of a milligram per liter and 20 mg/l. A small part of the carbon dioxide dissolved in water forms carbonic acid, with which it is in equilibrium:

$$CO_2 + H_2O \rightleftarrows H_2CO_3$$

Some of the carbonic acid dissociates to form bicarbonate ion, with which it is also in equilibrium:

$$H_2CO_3 \rightleftarrows H^+ + HCO_3^-$$

And some bicarbonate ion dissociates to form carbonate ion:

$$HCO_3^- \rightleftarrows H^+ + CO_3^=$$

At a pH below 8.3, little carbonate ion is present in water; little carbon dioxide is present at a pH above this value.

Carbon dioxide in natural waters comes mainly from the atmosphere and the respiration of plants and animals. Plant photosynthesis sometimes reduces its concentration nearly to zero. Under such circumstances, plants can obtain carbon dioxide directly from bicarbonate ions.

Hardly less important for life than water and carbon dioxide is oxygen, making the utilization of life's fuels possible. Its concentration in natural water, and hence its availability to aquatic organisms, varies from 0 to 30 mg/l or more. Temperature and salinity influence the amount of oxygen that will dissolve in water in equilibrium with air. The concentrations of dissolved oxygen present in water when it is in equilibrium with vapor-saturated air at 1 atmosphere of pressure are commonly referred to as saturation levels, but they are more properly termed *air-saturation levels* (Table 5–3). Since the amount of a gas that will dissolve in water depends on its partial pressure, the amount entering solution is a function not only of atmospheric pressure but also of the percentage of the gas present in the atmosphere.

TABLE 5–3 *Dissolved oxygen concentrations in parts per million by weight in freshwater and in sea water having different chloride concentrations, after the water has been equilibrated at different temperatures with water-saturated air at a total pressure of 760 mm Hg. Dry air is assumed to contain 20.90 percent oxygen. After* American Public Health Association et al. (1960) *from calculation by* Whipple and Whipple (1911) *based on measurements of* Fox (1909).

TEMPERATURE C	IN FRESHWATER 0 g/kg	IN SEA WATER HAVING DIFFERENT CHLORIDE CONCENTRATIONS			
		5 g/kg	10 g/kg	15 g/kg	20 g/kg
0	14.62	13.79	12.97	12.14	11.32
1	14.23	13.41	12.61	11.82	11.03
2	13.84	13.05	12.28	11.52	10.76
3	13.48	12.72	11.98	11.24	10.50
4	13.13	12.41	11.69	10.97	10.25
5	12.80	12.09	11.39	10.70	10.01
6	12.48	11.79	11.12	10.45	9.78
7	12.17	11.51	10.85	10.21	9.57
8	11.87	11.24	10.61	9.98	9.36
9	11.59	10.97	10.36	9.76	9.17
10	11.33	10.73	10.13	9.55	8.98
11	11.08	10.49	9.92	9.35	8.80
12	10.83	10.28	9.72	9.17	8.62
13	10.60	10.05	9.52	8.98	8.46
14	10.37	9.85	9.32	8.80	8.30
15	10.15	9.65	9.14	8.63	8.14
16	9.95	9.46	8.96	8.47	7.99
17	9.74	9.26	8.78	8.30	7.84
18	9.54	9.07	8.62	8.15	7.70
19	9.35	8.89	8.45	8.00	7.56
20	9.17	8.73	8.30	7.86	7.42
21	8.99	8.57	8.14	7.71	7.28
22	8.83	8.42	7.99	7.57	7.14
23	8.68	8.27	7.85	7.43	7.00
24	8.53	8.12	7.71	7.30	6.87
25	8.38	7.96	7.56	7.15	6.74
26	8.22	7.81	7.42	7.02	6.61
27	8.07	7.67	7.28	6.88	6.49
28	7.92	7.53	7.14	6.75	6.37
29	7.77	7.39	7.00	6.62	6.25
30	7.63	7.25	6.86	6.49	6.13

Concentrations of dissolved oxygen in natural waters vary widely from air-saturation levels. Plant photosynthesis and the respiration of plants and animals are most important in this. In waters having abundant growths of algae and other aquatic plants, wide diurnal fluctuations of dissolved oxygen and carbon dioxide concentrations often occur.

The oxygen concentration may rise to levels as high as two or three times the air-saturation level during daylight hours and fall to levels far below air-saturation during the night, when photosynthesis ceases and respiration continues. In the hypolimnion or stagnant deep water of a eutrophic lake during summer, dissolved oxygen may be absent because of the decomposition of organic matter. This condition persists until circulation of the lake occurs in the fall, when reaeration of this water becomes possible. During the winter stagnation period, oxygen may again be depleted in the water near the bottom. Photosynthesis can proceed near the surface under ice, but a heavy blanket of snow may interfere with the penetration of light sufficiently to permit depletion of oxygen at all depths. When large amounts of decomposable organic wastes are introduced into natural waters, severe oxygen depletion can occur, a depletion overcome only by reaeration and photosynthesis. In depriving aquatic organisms of their needed oxygen, this becomes an important kind of pollution, one which we will consider in the next chapter.

pH and the Hydrogen and Hydroxyl Ions

The *pH value* of water is a measure of the degree of its *acid* or *alkaline* reaction. It is the negative logarithm of the *hydronium ion* (H_3O^+) activity expressed in gram ionic weights, or moles, per liter. In other words, it is the logarithm of the reciprocal of that activity. The hydronium ion is commonly and conveniently referred to as the *hydrogen ion* (H^+); and its activity is still commonly, though somewhat inaccurately, referred to as the *hydrogen ion concentration*. Water ionizes slightly, the ionization yielding hydronium and *hydroxyl* (OH^-) ions. When the activities of these ions are equal, water is neither acid nor alkaline and is said to be neutral. At 25 C, neutrality is represented by a pH value of 7.0, because the product of the activities of hydrogen and hydroxyl ions is 10^{-14}, and the square root of 10^{-14} is 10^{-7}. The ionization constant of water, and therefore its exact neutrality point, varies somewhat with temperature. The neutrality point is about pH 7.5 at 0 C and about pH 6.5 at 60 C. Values of pH below the neutrality point signify an acid reaction of water, or a predominance of hydrogen ion; pH values above the neutrality point signify an alkaline reaction, or a predominance of hydroxyl ion. Any decrease of pH by one unit signifies a tenfold increase in the activity of hydrogen ion, and any increase of pH by one unit signifies a tenfold increase in the activity of hydroxyl ion. A doubling of the activity of one of the ions, accompanied by a halving of the activity of the other, is represented by a pH change of about three-tenths of a unit.

The pH of natural waters varies over a wide range. But the pH values of most stream waters in the United States range from 6.5 to 8.5; those of most ocean waters from 8.1 to 8.3. Carefully distilled water that has been exposed to air has a pH value near or slightly below 6. Its moderate acidity is due to the presence of a small amount of dissolved carbon dioxide, some of this combining with water to form carbonic acid (H_2CO_3), which is only slightly ionized. The alkaline reaction of the majority of natural waters is due mainly to the presence of dissolved salts of carbonic acid, which ionize fully and react with water hydrolytically to yield hydroxyl ion.

Limestone and other forms of calcium carbonate ($CaCO_3$) dissolve readily in water containing carbon dioxide. The carbonic acid formed reacts with slightly soluble calcium carbonate to yield highly soluble calcium bicarbonate, $Ca(HCO_3)_2$, which exists only in solution. After percolating through soils containing carbonates and bicarbonates, rainwater (which is slightly acid because of dissolved carbon dioxide) can emerge with very high concentrations of bicarbonate.

Natural waters having a low pH, an acid reaction, contain very little bicarbonate or else contain high concentrations of free carbon dioxide or organic acids. Water can be made highly acid by addition of even small amounts of a strong or fully ionized mineral acid such as sulfuric acid (H_2SO_4), if its bicarbonate content is low. It can be made highly alkaline by addition of small amounts of a strong base such as calcium hydroxide, $Ca(OH)_2$, if its bicarbonate and carbon dioxide concentrations are low. On the other hand, waters containing much bicarbonate, whether they are alkaline or somewhat acid in reaction, are well *buffered:* their pH changes relatively little upon addition of moderate amounts of strong acids or alkalies. A strong acid reacts with bicarbonates to form the weaker carbonic acid and carbon dioxide. A strong alkali reacts with any carbonic acid or carbon dioxide present to form bicarbonate ion (HCO_3^-) and with bicarbonates to form carbonate ion ($CO_3^=$). The conversion of bicarbonate ion to carbonate ion renders the water somewhat more alkaline because of hydrolysis, but the added hydroxide is largely neutralized. Upon sufficient elevation of the carbonate ion concentration, calcium carbonate may precipitate and make the water turbid. Boric acid and its salts contribute to the buffering of sea water in much the same way as do carbonic acid and its salts, but these last, because of their greater abundance, are more important buffers.

The pH of a natural water is not, then, a reliable index of its ability to neutralize strong acids or alkalies. A water containing bicarbonate ion and enough free carbon dioxide to make it moderately acid can still have a high *titratable alkalinity*, the ability to neutralize strong acids. A definitely alkaline but dilute solution of a strong base can have a much lower titratable alkalinity. The *total alkalinity* of a water is evaluated by titration with a strong acid. It is approximately equivalent to *methyl orange alkalinity*, which is determined by titration to the turning point of methyl orange indicator, about pH 4.5. The strictly correct titration endpoint for determination of total alkalinity varies somewhat with the amount of the total alkalinity to be determined, because at higher bicarbonate concentrations more carbon dioxide and carbonic acid are formed upon addition of the chemically equivalent amounts of strong acid.

The *total acidity* of a water is determined by titration with a strong base to pH 8.3, the turning point of phenolphthalein indicator. At this pH, virtually all the free carbon dioxide and carbonic acid initially present have been converted to bicarbonate ion. A water that is only slightly acid can have a relatively high titratable acidity if much free carbon dioxide or carbonic acid is present. An unbuffered water having a lower pH can have a much lower total titratable acidity if its acidity is due to mineral acid and not carbon dioxide. The words *acidity* and *alkalinity* each obviously have two different meanings in water chemistry.

Total alkalinities of natural waters, expressed as chemically equivalent concentrations of calcium carbonate, vary from almost zero to several hundred milligrams per liter.

Total acidities of unpolluted surface waters are generally low, carbon dioxide tending to escape to the atmosphere when its tension is elevated, and weak acids other than carbonic acid not being commonly present in large amounts.

Wide diurnal fluctuations of pH in natural surface waters can result from the photosynthetic activity of algae and other submerged aquatic plants. During daylight hours, these plants use free carbon dioxide and extract some carbon dioxide from bicarbonate ion. Two bicarbonate ions yield one carbonate ion, one molecule of carbon dioxide, and one molecule of water. The pH of the water then rises because of increased hydrolysis; but its total alkalinity does not increase, for one carbonate ion can react with no more hydrogen ions than can two bicarbonate ions. Indeed, precipitation and settling of calcium carbonate can result in a loss of titratable alkalinity, though usually this variation of total alkalinity is negligible, the carbonate remaining in solution or suspension until the process is reversed at night. Photosynthesis ceases at night, but respiration of plants and animals continues; increase in carbon dioxide concentration then results in decrease of pH.

With the discovery of convenient means of measuring pH there was a tendency for some to attach great significance to the many pH measurements they then could make. Alone, pH tells us little of the chemistry we need to know of waters. Nevertheless, it is a factor in the lives of plants and animals, even if not usually a dominant one. Aquatic organisms may be harmed when pollutional conditions lead to pH values much beyond the normal range for their environment. But usually it is not the change in pH itself but some associated change that is directly harmful to aquatic life. Organisms have external membranes that are quite impermeable to hydrogen and hydroxyl ions, and buffering systems control the pH within their bodies; this is important.

PHYSICAL, CHEMICAL, AND BIOLOGICAL ENVIRONMENTS

This chapter, now ending, we entitled The Physical and Chemical Environment. We have still to explain exactly what we mean by environment; we will not do so until Chapter

7. The concept of environment is elusive, and its appreciation requires ideas we have yet to present. The individual organism and its population are surrounded by a multitude of physical, chemical, and biological processes interacting with each other and with the biological system of interest. There are physical and chemical factors in an environment; but to speak of a physical and chemical environment leaves something unsaid, as though there could be an environment without food or predators or disease organisms. There is hardly one of the factors considered in this chapter which in its effects on life acts independently of surrounding biological conditions. And hardly one is unaffected by these conditions.

Physical and chemical conditions on earth are remarkably suited for life, as Henderson (1913) saw so clearly. But not only has life through evolution been molded to these conditions, these conditions have been molded by life. It is difficult to believe that anywhere in the universe there are more remarkable interactive processes. And in dealing with life and its environment, we will be in difficulty if we fail to appreciate their interactive nature. Why, then, have we chosen to devote a chapter solely to physical and chemical conditions in the aquatic environment? No similiar chapter has been devoted solely to biological conditions.

It is a matter of convenience. We have to begin somewhere, and perhaps we should begin as life did with only physical and chemical conditions. As we proceed through this book—from the individual organism through its population to its community—we will introduce increasing biological complexity, complexity of environment and complexity of biological system. In our studies it is sometimes useful to consider particular environmental factors or effects as being physical, chemical, or biological, useful so long as we remain aware of the interactive processes involved, misleading if we do not. Henceforth in this book, we rarely will distinguish between physical, chemical, and biological environments, though we often will consider such factors. Natural systems involving life do not lend themselves to rigid classifications; we should use such classifications only warily.

SELECTED REFERENCES

Frey, D. G. (Editor). 1963. Limnology in North America. The University of Wisconsin Press, Madison. [xviii] + 734pp.

Henderson, L. J. 1913. The Fitness of the Environment. The Macmillan Company, New York. xv + 317pp.

Hutchinson, G. E. 1957a. A Treatise on Limnology. Vol. 1: Geography, Physics, and Chemistry. John Wiley & Sons, Inc., New York. xiv + 1015pp.

———. 1967. A Treatise on Limnology. Vol. 2: Introduction to Lake Biology and the Limnoplankton. John Wiley & Sons, Inc., New York. xi + 1115pp.

Ruttner, F. 1953. Fundamentals of Limnology. (English translation by D. G. Frey and F. E. J. Fry.) University of Toronto Press, Toronto. xi + 242pp.

Sverdrup, H. U., M. W. Johnson, and R. H. Fleming. 1942. The Oceans: Their Physics, Chemistry, and General Biology. Prentice-Hall, Inc., New York. x + 1087pp.

Welch, P. S. 1952. Limnology. McGraw-Hill Book Company, Inc., New York. xi + 538pp.

6 Kinds of Water Pollution

Pollutants may alter the stream environments and thereby affect aquatic life in a number of ways. These environmental changes may include an increase in stream temperatures; changes in the character of the stream bottom; increase in turbidity; changes in the content of dissolved oxygen; increase in dissolved nutrients; production of undesirable growths; deposition of sludge beds; and the addition of toxic wastes. The degree or extent of the effect of these changes on aquatic life varies with the type and amount of the pollutant and the character of the receiving water.

TARZWELL AND GAUFIN, 1953, p. 295

The reader will also appreciate that most of the effects of pollution are of the same type as 'natural' phenomena, and if they are not so severe as to produce extreme conditions they serve merely to alter one sort of river environment into another. For instance a river water which is made more silty, more acid, more alkaline, less well oxygenated, warmer, harder, saltier or richer in nutrient salts, is still a natural river as long as the change is not so great as to overstep the bounds of normal variability. This applies even to some types of poisons. We return here to the great difficulty of defining pollution, and we must accept the fact that at least some man-made alterations to rivers closely resemble changes which in other rivers occur quite naturally.

H. B. N. HYNES, 1960, pp. 68–69

KINDS AND EFFECTS

We have chosen to define water pollution as any impairment of the suitability of water for any of its beneficial uses, actual or potential, by man-caused changes in the quality of the water (Chapter 2). Among the kinds of water pollution first to cause alarm were changes rendering water unfit for consumption and changes that made it unsuitable for industrial purposes. In those countries having modern facilities for treating water, the public is generally confident in the safety of water supplies; but in some parts of the world no such safety exists, and nowhere should it be taken for granted. Waters posted against recreational use are still not unusual, the possibility of the presence of pathogens making such use hazardous.

Hopefully, such posting will one day be a thing of the past. But this is the realm of public health, a realm well covered in bacteriology and other texts, one to which this book is not primarily directed. Neither are we here primarily concerned with the suitability of water for industrial or agricultural purposes. Our concern is mainly with the kinds of water pollution that endanger the plants and animals living in water, the ones man has reason to value. This is plenty for one book.

Those who accept our definition of water pollution recognize that there may be changes in water quality that, even though harmful to some species, would not be considered pollution when man has little or no interest in the affected species. But perhaps most changes in water quality that are deleterious to aquatic organisms, or favor organisms that can be a nuisance, so affect man's use of water as to be considered pollution. These changes are the subject of this book; their kinds are many. Often we speak of industrial or domestic wastes, sometimes of agricultural wastes. These expressions are useful as categories of wastes from different human activities that may be harmful to our waters. But the variety of industrial processes having wastes is great, and the industrial waste category is not very helpful in conceptualizing the problems that may be involved. Domestic wastes—sewages—are much more uniform in character; the wastes of agricultural activities are not at all uniform. Thus, it is usually more helpful to categorize wastes on the basis of the kinds of changes they bring in receiving waters: changes in oxygen concentration, plant nutrients, solids, toxic substances, temperature, and other environmental factors.

Now we are most often interested not just in the kinds of changes wastes cause in the quality of receiving waters; we are interested also in the effects these changes have on the value of the waters to man. We are here concerned with effects on the distribution and abundance of species in which man has some interest. The distribution and abundance of a population is dependent on the survival, reproduction, growth, and movement of its individuals. We must understand the effects of changes in water quality on these if we are to be most successful in controlling water pollution. Some environmental changes may be directly deleterious to the survival, reproduction, growth, or movement of a species; others may be indirectly so. A toxic substance, a change in oxygen concentration, a change in temperature, these and other conditions of life may directly affect the individuals of a species. Or, by altering the food resource of a species and in other ways changing its relations with other organisms, environmental changes may indirectly affect the success of a species. The end result is often the same: reduction in the distribution and abundance of a species of value to man.

Nuisance blooms of algae, slimy growths of bacteria on stream bottoms, the tainting of the flesh of animals man uses for food, and still other results of some changes in the aquatic environment are objectionable to man apart from changes in the distribution and abundance of valuable species. This may be so also when toxic or radioactive substances accumulate in the flesh of some species. Understanding of such changes requires the study of biology. They are not easy to understand, for biological systems are complex, and a change in one of their components leads to changes in others. This makes it difficult to develop any very useful classification of the kinds and effects of water pollution. Still, in considering some of these, we will be able to introduce certain problems of life in waters, problems to which we will return throughout this book.

ORGANIC MATERIALS AND OXYGEN DEPLETION

One very destructive kind of pollution occurs when relatively large amounts of putrescible organic materials, which require oxygen for their decomposition, are introduced into waters. The oxidation of such materials by microorganisms depends largely on dissolved oxygen already present in the receiving waters, oxygen entering from the atmosphere, and oxygen made available through plant photosynthesis. When the rate of oxidation is greater than the rate of oxygen replenishment, the concentration of dissolved oxygen in receiving waters declines. Such depletion of oxygen may be either directly or indirectly deleterious to the survival, reproduction, growth, and movement of populations of fish and other organisms of interest to man. When it is, the value of certain uses of a receiving water is diminished, and pollution may be said to occur. High concentrations of putrescible organic materials are characteristic of untreated domestic wastes and many industrial wastes, such as those from food processing and paper manufacturing. These wastes often also have toxic and other characteristics, but first let us consider only the problem of oxygen depletion.

The amount of oxygen necessary for the oxidative decomposition of a material by microorganisms is known as the *biochemical oxygen demand* (BOD) of the material. The *ultimate carbonaceous* BOD of a water or a liquid waste is the amount of oxygen necessary for microorganisms to decompose the carbonaceous materials that are subject to microbial decomposition. Oxygen is also necessary for nitrifying bacteria to oxidize *inorganic nitrogen compounds* produced in the decomposition of nitrogenous organic materials. Nitrification becomes appreciable only after most of the oxidation of carbonaceous material has been completed.

The BOD value usually reported is the amount of oxygen consumed in milligrams per liter of water or waste water over a period of five days at 20 C under laboratory conditions. This will be a variable portion of the ultimate BOD; it is dependent not only on the nature and amount of the organic materials present but also on the microorganisms and their environment in the sealed culture flask. For domestic sewage, the BOD value usually reported is about two-thirds of the ultimate carbonaceous BOD. When the water contains little or no dissolved oxygen, or when the BOD to be measured approaches or exceeds the amount of dissolved oxygen available, adequate dilution with a well-oxygenated water having virtually no BOD is necessary before incubation of samples.

The biochemical oxygen demand of a water or waste, then, is a measure of the amount of oxygen that will be required to oxidize compounds degradable by microbial action. The extent of decline of dissolved oxygen in receiving waters depends on the difference between the rates of oxidation and reaeration. All but the purest of natural waters have a measurable BOD, sometimes as high as 5 mg/l. But domestic and industrial wastes often have BOD's of several hundred milligrams per liter; when inadequately diluted in receiving waters, these wastes can lead to severe oxygen depletion. It is this that necessitates secondary waste treatment by biological processes resembling those occurring in receiving waters.

PLANT NUTRIENTS, ORGANIC AND INORGANIC

Putrescible organic substances introduced into receiving waters provide an energy and material resource for *heterotrophic organisms,* those requiring preformed organic materials. These organisms, particularly bacteria, are responsible for the oxidative degradation of organic materials, either in receiving waters or in biological waste treatment systems. But the resulting production of microorganisms may itself produce problems in receiving waters, problems secondary waste treatment helps to alleviate. Growths of bacteria and other organisms on stream bottoms, commonly referred to as *sewage fungus,* may be objectionable esthetically even when they do not interfere with other stream uses. The filamentous bacterium *Sphaerotilus natans* is often the dominant organism in these growths. Its pallid masses can be found waving in the current from stream bottoms. Clumps drifting in the current may clog water supply screens and the nets of fishermen. In changing the benthic environment, sewage fungus can render it unsuitable for the natural inhabitants of streams. Other kinds of microorganisms that live suspended in water can lead to turbidity and discoloration.

Secondary waste treatment reduces the amounts of organic materials in effluents and so reduces the energy and material resources and the production of heterotrophic organisms in receiving waters. It does not, however, prevent the enrichment of these waters with nitrates, phosphates, and other inorganic plant nutrients made available through the decomposition of organic materials. Even after secondary waste treatment, then, these plant nutrients can lead to increased production of microorganisms, particularly algae, which are *autotrophic* and use them along with carbon dioxide to form protoplasm. Thus in lakes, rivers, and estuaries, even treated effluents can lead to nuisance blooms of algae. These blooms change the aquatic environment in ways harmful to some valuable species; beautiful waters lose their transparency and become colored and unsightly; and taste and odor problems often develop. These are undesirable effects of the discharge of effluents; they represent important kinds of pollution. In rivers, estuaries, and the sea, these effects disappear soon after the discharge of effluents ceases or inorganic plant nutrients are removed from the effluents by further treatment. But in lakes, where the removal of materials by water exchange is often little, damage may be quite permanent; once introduced, nutrients can be recycled in a lake for ages.

SOLIDS, SUSPENDED AND DISSOLVED

Solid materials finding their ways into natural waters may have some undesirable effects while carried in suspension and can have other undesirable effects after finally settling to the bottom. The kinds and sources of such materials are many. Silt, eroded from land disturbed by cultivation, logging, or road building, often has pollutional effects. Mining, gravel, and other industrial operations may increase the load of finely divided, chemically inert materials carried by waters. The dredging of harbors and channels has deleterious effects on estuarine environments to which bottom materials are carried in suspension. Damage to valuable uses of affected waters makes this an important kind of pollution.

While in suspension, such solids cause waters to be turbid; reduced light penetration may restrict the photosynthetic activity of plants and the vision of animals. These finely divided materials at high concentrations are known to interfere with the feeding of animals that obtain their food organisms by filtration, and they may be abrasive to sensitive structures such as the gills of fish.

On settling to the bottom, solid materials often harm this important aquatic habitat. Oyster lands made soft with silt may no longer be able to support the weight of their product. Rocky bottoms previously providing homes for many animals may be buried. Stream gravels where salmon deposit their eggs may lose their porosity, thus reducing the movement of water carrying oxygen to the developing embryos. These are very real kinds of pollution.

Dissolved solids too may bring about undesirable changes in aquatic environments and be considered pollutants. When separately dissolved in pure water, some of the salts present in sea water are toxic to marine and freshwater animals at concentrations lower than their concentrations in the sea. Even sodium chloride is in this group. The toxicity of such solutions is due to the metal cations. When various salts are present in suitable proportions, the different cations counteract each other, and the solution is considered to be physiologically balanced. Sea water is such a balanced solution. Its harmful effects on freshwater organisms are caused by high os-

motic pressure, not by toxicity of its individual components.

Dissolved salts in natural waters are increased by many of man's activities: irrigation of land; discharge of oil field and other brines; diversion of streams and deepening of ship channels, permitting the intrusion of sea water into former freshwater areas. Water in these ways sometimes becomes no longer suitable for domestic, agricultural, or industrial uses. But here we are concerned with animals that must continue to survive, reproduce, grow, and move in changing waters, if they are to be successful and useful to man.

Freshwater organisms are adapted to living in waters of low salt concentration. Their mechanisms for maintaining water and salt balance with their environment cannot usually cope with great increases in dissolved solid concentration. So, either directly or indirectly, increases in osmotic pressure of aquatic environments may affect some stage in the life history of freshwater animals in ways decreasing their distribution and abundance and their value to man. Marine animals face similar problems of salt and water balance when unusual amounts of freshwater enter marine environments. Man's activities may cause either kind of pollution: an increase or a decrease in osmotic pressure that is deleterious to aquatic life.

TOXIC AND RADIOACTIVE SUBSTANCES

Beginning mainly with the Industrial Revolution, the introduction of toxic substances has created problems for life in many waters. Over nearly two hundred years, the discharge of toxic industrial wastes has increased greatly. Since the Second World War, the use of synthetic pesticides in agriculture, forestry, and public health work has become so general that most waters are in some degree contaminated by these extremely toxic substances. Presence of radioactive substances has also become widespread, again through the activities of man. Often the concentrations of toxic substances and sometimes the concentrations of radioactive substances are high enough to leave little doubt as to their deleterious effects on aquatic life. But more often, concentrations of these materials are not this high, and man is left wondering what if any may be their effects. Surely his investigations will increase his understanding, hopefully enough to prevent damage to his resources and himself.

Toxic substances have always been present in the earth's waters, their presence—not necessarily in harmful concentrations—being insured by natural processes. Ammonia and hydrogen sulfide from the decay of organic matter are naturally occurring toxic substances. Highly toxic organic substances produced by some plankton organisms and terrestrial plants are known to sometimes cause death of aquatic animals. Some waters normally contain fluorides, heavy metals, and other toxic minerals in concentrations harmful to life. But the activities of man have led to the larger problem and must concern us most.

We can hardly list the many kinds of industrial processes from which toxic wastes enter receiving waters. And many products of industry—including pesticides, detergents, and other chemicals—find their ways into aquatic environments, sometimes with and sometimes without man's intent. There are waters that have become all but uninhabitable for aquatic organisms because of metals entering with drainage from mines. Living in water as they do, aquatic organisms may be harmed by concentrations of toxic substances that have little effect on man, who may only drink or bathe in the water. To reduce the distribution and abundance of aquatic animals, toxic substances need not be present in concentrations that are acutely toxic; to do this, they need only be present in concentrations sufficiently high to affect directly or indirectly the longevity, reproduction, growth, or movement of these animals. Radioactive substances are no different in this regard, and, as with many synthetic pesticides, their accumulation through the food chains of animals man uses for food is a hazard yet to be fully assessed.

Biological waste treatment, originally developed for putrescible organic materials, has been found to be effective in treating many toxic substances. Many wastes requiring treatment for reduction of biochemical oxygen demand contain toxic substances that are decomposed by the treatment process. But there are many other toxic materials, including persistent pesticides, that are not

so decomposed. Advances in waste treatment technology must continue.

TEMPERATURE

Temperature is one of the great problems of life. Man, other mammals, and birds have partially solved this problem through the development of mechanisms that maintain their body temperatures relatively constant; but the body temperatures of most animals follow those of the environment. Here temperature is an ever changing thing, night to day, season to season, year to year. The temperature of water changes not so much as that of air or soil; still, its change is considerable. Man's activities have changed the temperatures of many waters beyond their normal ranges. Though aquatic organisms can adapt to temperature changes, there are limits to their adaptive capacities, limits beyond which their continued success is doubtful. Thermal pollution exists when man's activities lead to changes in temperature that decrease the distribution and abundance of aquatic animals he values. In many areas, this is becoming a serious pollution problem, one difficult to control.

Removal of forest canopies, impounding of river waters, reduction of stream flows, return of irrigation waters—so many things that man does can lead to increases in water temperatures, can lead to thermal pollution for which solutions are difficult to find. The manufacture of nuclear fuels and their use in the generation of electric power lead to much waste heat, heat taken up by cooling waters. The production of electricity through the use of fossil fuels also requires cooling water, as do many other industrial processes. As our need for electricity increases, nuclear and fossil fuel plants will become more numerous; so will thermal pollution problems, unless remedial steps are taken. Enormous cooling towers can dissipate into the atmosphere the heat of cooling waters before they are returned to their sources. This can prevent temperature increases in receiving waters. It will be the solution for many problems of thermal pollution. But few solutions, and certainly not this one, are perfect. In their operation, cooling towers disperse along with heat great amounts of water into the atmosphere, water whose vapor may contribute to air pollution and water whose loss from streams can often be ill afforded.

OTHER MATERIALS

From complex industrial societies, the kinds of materials finding their ways into fresh and marine waters seem almost limitless. In a brief chapter, we can hardly list them; were we able, the reading would be dull and largely without profit. Nevertheless, mention of a few more kinds of materials and their pollutional effects may have some value in indicating the breadth of the problem.

Fossil fuels permitted the Industrial Revolution, which has brought us so much, not all of which is good. Nuclear fuels will prove no different. But for a while we will yet depend mainly on fossil fuels, petroleum being perhaps the most important. Crude petroleum contains water-soluble toxic materials, but even those that are not toxic can have adverse effects on aquatic life. Most of these materials are resistant to bacterial decomposition; once introduced into an aquatic environment, they and their effects can persist almost indefinitely. This is particularly so when they coat the rocks, sands, and silts of the benthic environment and thus render them unsuitable for life. Oily films on water surfaces are displeasing and interfere with recreational uses. But to the plants and animals that must live in the water, these films mean much more. Their light and oxygen become restricted, and their body surfaces may become coated. The pathetic sight of incapacitated sea birds, their feathers oily messes as a result of petroleum losses from ships or drilling operations, comes to us too frequently through the news media. Serious pollution of water with oil is usually the result of carelessness, for adequate separation of oil from water is neither difficult nor costly. Nevertheless, such pollution at oil fields, refineries, and elsewhere is common.

Many materials, some of petroleum origin, have objectionable odors and can impair the flavor of aquatic animals man uses for food. Chlorinated phenolic substances are notorious in this regard. Without decreasing the distribution and abundance of valuable populations, such substances have sometimes so spoiled the flavor of food fishes as to render them useless to man.

Some materials entering waters are repellent to the animals that must live there. Fish are able to detect minute concentrations of many different substances and may avoid contaminated areas, even though their survival and reproduction would be possible there. This is sufficient to render these areas unproductive of such animals.

Finally, the color of our waters concerns us: it is an esthetic thing, whether we wish to drink or view the water. In interfering with light penetration, colored substances in water can have a serious biological impact, the reduction of photosynthesis. Materials not otherwise harmful may cause a kind of pollution if they color water.

ON CLASSIFICATION

Man is anxious to classify things. What is it? This is his first response to something unknown. Is it dangerous or is it food? For early man, it was often both, and he needed to know. This has not really changed much. Early science was mainly a system of classification; so is much of modern science. We see this in plant and animal taxonomy. Good systems of classification are useful: they bring some order to our thinking and perhaps advance it a little. But it is well to remember that classifications in themselves are not explanatory and that they may become an impediment to further thought. Man becomes peculiarly satisfied once he has categorized something.

It is helpful, perhaps even necessary, to classify kinds of pollution. But kinds of pollution do not classify well, nor do their effects. This is in part because most wastes are complex and in part because even one of their components may affect differently the various constituents of biological communities. Sewage may be considered primarily an oxygen depleting waste; but it may contain toxic substances, and toxic substances are produced in its decomposition; further, it provides food for heterotrophic organisms and mineral nutrients for autotrophic ones. Thus, the characteristics and effects of sewage are many. In this, most other wastes are not different. We must not permit our classification of wastes, classification based on either kinds or effects, to limit our possible awareness of all their actions. In the control of water pollution, we may need classifications of kinds of wastes; but more we need understanding of their effects.

SELECTED REFERENCES

Hynes, H. B. N. 1960. The Biology of Polluted Waters. Liverpool University Press, Liverpool. xiv + 202pp.

Jones, J. R. E. 1964. Fish and River Pollution. Butterworth, London. viii + 203pp.

Keup, L. E., W. M. Ingram, and K. M. Mackenthun (Editors). 1967. Biology of Water Pollution: A Collection of Selected Papers on Stream Pollution, Waste Water, and Water Treatment. Federal Water Pollution Control Administration, U.S. Department of the Interior. iv + 290pp.

Klein, L. 1962. River Pollution. 2. Causes and Effects. Butterworth, London. xiv + 456pp.

Mackenthun, K. M., and W. M. Ingram. 1967. Biological Associated Problems in Freshwater Environments. Federal Water Pollution Control Administration, U.S. Department of the Interior. x + 287pp.

Whipple, G. C. 1927. The Microscopy of Drinking Water. 4th ed. Revised by G. M. Fair and M. C. Whipple. John Wiley & Sons, Inc., New York. xix + 586pp. + 19 plates.

PART III

Morphology
and Physiology

Through evolution, each species has come to be morphologically, physiologically, and behaviorally adapted to survive and reproduce in its natural environment. No natural environment is very constant, and most individuals of successful species must be sufficiently plastic in their characteristics to adapt to environmental changes that occur throughout their lives. Physiological, behavioral, and even morphological changes in these individuals make this adaptation possible. But when environmental conditions change much beyond those experienced by a species in its evolution, most of the individuals will not have the genetic capacity to adapt, and they will fail to survive or to reproduce. Should a few individuals have the capacity to survive and reproduce, the species may evolve to be successful under the new conditions. The genetic adaptation of populations and species and the plasticity of their individuals thus permit organisms to persist in changing environments. But plants and animals are not able to adapt to many of the changes man has caused in their environments, and they do not persist in the face of these changes. Knowledge of the morphological, physiological, and behavioral capacities of organisms to adapt to changes in their environments is essential to understanding of their ability to survive and reproduce under some conditions but not under other conditions.

7   The Significance of Morphological and Physiological Change

The parts of the body are really portions of the environment of the total bodily event, but so related that their mutual aspects, each in the other, are peculiarly effective in modifying the pattern of either. This arises from the intimate character of the relation of whole to part. Thus the body is a portion of the environment for the part, and the part is a portion of the environment for the body; only they are peculiarly sensitive, each to modifications of the other. This sensitiveness is so arranged that the part adjusts itself to preserve the stability of the pattern of the body. It is a particular example of the favourable environment shielding the organism.
ALFRED NORTH WHITEHEAD, 1925, pp. 207–208

The biochemist is apt to feel that all important biological problems will be solved at the molecular or submolecular level, while the ecologist feels that the molecular biologist is preoccupied with details of machinery whose significance he does not appreciate. There is a familiar solution to this problem, widely recognized intellectually but sometimes difficult to accept emotionally. This is the idea that there are a number of levels of biological integration and that each level offers unique problems and insights, and further, that each level finds its explanations of mechanism in the levels below, and its significance in the levels above.
GEORGE BARTHOLOMEW, 1964, pp. 7–8

THE INTEREST IN MORPHOLOGY AND PHYSIOLOGY

Most populations are well adapted to existing environmental conditions in that they have evolved to live and reproduce and thus persist under those conditions. The adaptation of a population to existing environmental conditions resides in the morphological, physiological, and behavioral characteristics of its individuals, which enable them to survive and reproduce, even though the conditions are not constant. Each individual has a range of environmental conditions within which it can be successful. But not all the individuals in a population are equally well adapted to existing conditions. Those that are the best adapted will tend to leave the most progeny, and the population will continue to be well adapted to those conditions. Nevertheless, there will persist in the population some individuals whose genetically determined characteristics would suit them better for other environmental conditions. Were environmental conditions to change in some persistent way favoring these individuals, they and their progeny would tend to contribute most to successive generations.

The population could continue to persist, because, from one generation to the next, the characteristics of most of the individuals composing the population would have changed somewhat.

Individual organisms, then, within genetically determined limits, can survive and reproduce under changing environmental conditions through physiological and behavioral adaptation. This kind of adaptation contributes to the success of the population. But the population has still another means of adapting to environmental change: a shift in the genetic capabilities of its individuals through successive generations. We will be considering further these two kinds of adaptation in following sections.

The characteristics of a species permit it to persist in a particular environment and in a particular niche in that environment. Because we cannot consider adaptation apart from particular environments and niches, we must explore the meaning of these in the next two sections, before we can proceed with a more detailed discussion of adaptation.

The greatest interest in morphology and physiology lies in the adaptive value of particular morphological and physiological characteristics. The significance of morphological and physiological changes can be evaluated only in terms of whether or not they are adaptive for particular environmental conditions. In doing this, it is important to bear in mind that the individual organism, with all its characteristics, has evolved and adapts as a whole to survive and reproduce. Particular characteristics can be evaluated only as a part of this whole. This is what Bartholomew (1964, p. 8) meant when he wrote: "each level finds . . . its significance in the levels above."

Some of the most important questions in biology today are concerned with the adaptive value of morphological and physiological changes and how they arise. To the extent that such changes lead to the success or failure of populations in environments changed by man, study of these changes is also important in water pollution control.

THE MEANING OF ENVIRONMENT

The adaptive value of a biological system or a change in that system—at any level of biological organization—can only be considered in relation to an environment in which an individual, a population, or a community exists or might exist. For this reason, it is important that the concept of environment be defined in useful ways. Environment, however, is not an easy concept to define, mainly because individual organisms, populations, and communities form interacting systems with their environments. Whitehead (1925, pp. 207–208) noted another difficulty: "the body is a portion of the environment for the part, and the part is a portion of the environment for the body . . . this relation reigns throughout nature and does not start with the special case of the higher organisms."

The *environment* of the individual organism can be defined more satisfactorily than can the environment of a population or a community: *the sum of the phenomena influencing the development, metabolism, and activities of the organism.* Bates (1960, p. 554) considers this the *operational environment* and distinguishes it from the *potential environment*: the sum of phenomena that might conceivably influence the organism. For animals, he finds it useful to distinguish a third category of phenomena: the *perceptual environment.* Von Uexküll (1909) called this the *Umwelt*—the *Merkwelt*—of an organism; this concept of environment is particularly important in the study of behavior, as we will discuss in Chapter 12. Thus, for the individual organism, physical and chemical conditions, its disease organisms, parasites, symbionts, and commensals, its fellow organisms, competitors, and predators, its food resources, and still other phenomena become part of its environment.

For the individual human animal, it is more difficult to define environment. Is man's clothing part of man or part of his environment? It is a protective adaptation, just as are the hair, fur, and feathers of other warm-blooded vertebrates. And are man's religious and political ideas part of man or part of his environment? They influence him more profoundly than many important environmental phenomena. Bates (1962) writes of man's *conceptual environment.* Cultural evolution has blurred the distinction between man and his environment.

Andrewartha and Birch (1954, p. 13) regard "the population as part of the environment rather than as itself having an environment"; they prefer to define the environ-ment only for the individual organism. As we move from the individual to the popula-tion and then to the community, definition of environment is increasingly difficult. But the problem of the whole being part of the environment for the part—as the population is for the individual—exists even at the level of the individual organism with respect to its parts, as Whitehead (1925) pointed out. And Dobzhansky (1957, p. 385) argues that "a Mendelian population is an organic system which encounters and reacts to physical and biotic factors in its environment." It is useful to view the environment of a population as the sum of those phenomena to which the population as a whole and its individuals re-spond.

With the biological community, the dif-ficulty of conceptualizing separately the bio-logical system and its environment becomes most extreme. Biological communities great-ly modify and control the physical and chemical conditions and resources of their locations. Geological and climatic conditions can be distinguished as part of the com-munity environment, but even these are modified by the community. Rather than at-tempting to distinguish between the com-munity and its environment, most ecologists view the community and the conditions and resources of its location as together compos-ing an *ecosystem* (Chapter 18).

Classification of environmental factors is helpful if not necessary for physiologists and ecologists in their efforts to determine causal relations in the biological systems with which they work. It is sometimes useful to think of environmental factors as being either biotic or abiotic, but this broad separation is not so discrete as it might seem; and it does not advance far our thinking on causal relations. The ecologist can usefully separate the com-ponents of the environment into weather, food, organisms of the same kind, organisms of other kinds, and a place in which to live, as Andrewartha and Birch (1954) have done for the individual organism. Fry (1947)—in a system of value to physiologists and ecolo-gists alike—classified environmental factors not according to what they are but according to how they act on organisms. An environ-mental identity or factor would be classified as being lethal, masking, directive, con-trolling, limiting, or accessory, according to how it appeared to be acting on the or-ganism at a given time and place (Chapter 11). These and other systems of classifying environmental factors should be judged on the basis of their conceptual and analytical usefulness for particular lines of investiga-tion. No single system can prove satisfactory for all physiological and ecological studies.

The concept of environment is central to all biological thought, in spite of difficulties of conceptualization and definition. Bates (1960, p. 553) expresses this well:

The environment concept is thus a constant source of trouble, but I know of no way of getting along without it. One must go ahead and use it confidently—but also somewhat warily, keeping alert to the dangers. If we tried to avoid fuzzy and misleading words, I suspect that all verbal discourse would stop. This might make mathema-ticians and some kinds of logicians happy, but it would be hard on the rest of us.

THE IDEA OF NICHE

Each animal species has evolved as a part of an ecosystem: an ecosystem in which it occupies certain spaces during certain times; an ecosystem in which it can tolerate the ranges of physical and chemical conditions; an ecosystem in which it utilizes some species for energy and material resources and in which it is itself utilized by other organisms; an ecosystem in which it has many kinds of relations with different species and in which it can satisfy its shelter and all other needs. The *niche* of a species is the complete set in space and time of conditions and relation-ships within which a species can and must survive and reproduce, if it is to persist. A complete description of the niche of any species, which we will probably never have, would be a complete functional description of that species in terms of all its require-ments and relationships. Because no two species are morphologically, physiologically, and behaviorally identical, no two species can occupy exactly the same niche. But be-cause often two or more species in an ecosys-tem can tolerate similar ranges of physical and chemical conditions, can utilize to some extent the same energy and material resources, and can be preyed upon by the same animals, the niches of different species, particularly closely related ones, partially overlap.

The idea of niche has been important in ecological and evolutionary thought; it has been implicit if not always explicit in the writings of Darwin and many biologists since. But the idea of niche, being a complex abstraction, does not lend itself to very satisfactory definition. This has caused much of the argument concerning the role of niche in the ecology and evolution of species to be not particularly fruitful. Hoping to remove irrelevant difficulties, Hutchinson (1957b) presented a formal analysis of this idea in which the upper and lower limits of all environmental variables within which a species can persist become the dimensions of a hypervolume, the "fundamental niche" of the species. The assumptions necessary to permit such a formal definition tend to limit its conceptual value, but the clarity achieved helps to focus attention on some of the important questions.

The idea of niche is important because it helps us to conceptualize the problems faced by different species, particularly closely related ones, when they exist together. Superficially at least, such species sometimes appear to be competing for the same resources. In adapting to its environment, each species presumably must minimize its competition with others. Thus competition becomes an important evolutionary force. Much of what has been written on this problem has been concerned with whether or not two species compete for the same niche. Two species, being morphologically, physiologically, and behaviorally different, do not compete for exactly the same niche, if niche is defined to include all conditions and relations within which each species can survive and reproduce. But competition between species for space, food, or shelter does occur. Where such competition is severe, one of the species may be unable to persist. Where two species having very similar requirements persist, competition for the same resources has through evolution been minimized, this leading to specialization of each species for exploiting the environment in a distinctive way, thus reducing niche overlap.

The idea that evolution has minimized competition between co-existing closely related species is not at all recent. Darwin (1859) discussed it at length and Grinnell (1904, 1917, 1928) clearly recognized the possible outcomes of competition. Gause (1934, p. 19) stated: "as a result of competition two similar species scarcely ever occupy similar niches, but displace each other in

such a manner that each takes possession of certain peculiar kinds of food and modes of life in which it has an advantage over its competitor." This has come to be known as *Gause's principle*, undoubtedly because of the interest his experiments and mathematical analysis stimulated. But because this idea was recognized independently by others before Gause, Hardin (1960) suggested that it be called the *competitive exclusion principle*; Mayr (1963) adopted this terminology.

GENETIC ADAPTATION OF POPULATIONS

Adaptation, Evolution, and Speciation

A population is adapted to a particular set of environmental conditions to the extent that the characteristics of its individuals lead to their successful reproduction under those conditions. Each individual has a *phenotype*, the totality of its morphological, physiological, and behavioral characteristics. The phenotype of each individual is the outcome of interaction between its *genotype*—its genetic constitution—and its environment. A population is composed of individuals having many different genotypes and, in consequence, of individuals having many different phenotypes. Under a given set of environmental conditions, certain phenotypes are more successful than others in leaving progeny, and their genotypes will come to predominate in successive generations. The *fitness* of a genotype or its associated phenotype for an environment can be defined as its relative ability to contribute offspring to successive generations. Were certain environmental factors to change in ways favoring the reproduction of other phenotypes, then their genotypes would come to predominate. Thus, a population can adapt to change in environmental conditions through the differential reproductive success of genotypes existing in the population.

But new genotypes continually appear even in isolated populations through mutation and recombination. Through *mutation*, the nature of a gene changes in such a way as to lead to differences in its phenotypic expression. Although a relatively infrequent event at a particular gene locus, mutation has undoubtedly occurred many times at each locus throughout the long period of evolution of every species. Nevertheless, under environmental conditions existing when mutation occurred, the mutant gene may have been eliminated from the gene pool, only to appear again, perhaps under conditions favoring its persistence. Species reproducing asexually depend on mutation for their genetic plasticity. This plasticity provides great adaptive potential for organisms having high reproduction rates. Bacteria—whose adaptation to antibiotics is alarming—are the best example.

In sexually reproducing organisms, however, new genotypes can appear also through *gene recombination*. Genes do not assort randomly, because they are linked together on chromosomes. This linkage greatly reduces the possible combinations of genes. But during meiosis in the formation of parental reproductive cells, exchange of corresponding segments between chromatids of homologous chromosomes can occur. From this arise new gene combinations, new genotypes, and, thus, new phenotypes. Mutation and recombination, then, are the primary sources of genetic variability giving rise to phenotypic variability in sexually reproducing species. But by bringing together in many combinations the genes of parents of different genotypes, sexual reproduction even further contributes to the genetic plasticity of populations. The genetic plasticity of sexually reproducing species has probably led to the evolutionary success of sexuality throughout the biological world.

Other things being equal, the larger the population, the wider the range of genotypes likely to be represented and the greater the opportunity for new genotypes to appear through combinations of parental genotypes. Larger populations, then, have a better chance of persisting by adapting genetically to environmental changes than do small populations, except in the case of large populations having little genetic plasticity as a result of their recent development from very small populations.

It is on the various phenotypes that natural selection operates, for genotypic differences that do not express themselves in a phenotype are inaccessible to selection. The probability that an individual will leave progeny depends in part on the phenotypic expression of its genotype. "Reproductive success rather than survival is stressed in the modern definition of natural selection. A superior genotype has a greater probability of leaving offspring than has an inferior one. Natural selection, simply, is the differential perpetuation of genotypes" (Mayr, 1963, p. 183). Natural selection was the cornerstone of Darwin's (1859) theory of evolution.

Thus, those individuals phenotypically most suited to an environment will, to the extent that their phenotypes are genetically transmitted, leave the offspring most suited to reproduce in that environment. In a changing environment, a population can adapt through differential perpetuation of genotypes and their phenotypes. This kind of genetic adaptation of populations is *evolution*—the appearance and persistence of new combinations of morphological, physiological, and behavioral characteristics in successive generations of individuals, combinations that favor reproduction in the face of changing environmental conditions. Without knowledge of Mendel's (1866) discovery of the genetic basis of inheritance, Darwin (1859) saw this aspect of evolution quite clearly and explained it to the world in his inductive masterpiece *The Origin of Species.*

What Darwin did not see is how *speciation* occurs, a failure that Mayr (1963) explains (Chapter 15). Speciation, the splitting of one species into two, is now generally believed to occur only after one of the populations of a species has been spatially isolated over many generations. Through evolution during such isolation, individuals of a population may acquire characteristics that would lessen their probability of reproductive success were they to mate with individuals from other populations of their parent species; and the formation of a new species could begin. Then, were conditions again to permit individuals of the parent and the incipient species to mix, breeding between the groups would be less successful than breeding within each group. Further evolution of the two groups could lead to divergence of their genotypic and phenotypic characteristics and reinforcement of their reproductive isolation, until where there had been one species, two would exist. Attainment of our present understanding of the mechanisms of speciation is one of the great accomplishments of modern biology, an accomplishment made possible through synthesis of genetic, physiological, and ecological knowledge.

Genotype, Phenotype, and Environment

The phenotype of an organism—the totality of its morphological, physiological, and behavioral characteristics—determines its fitness for a particular environment. An organism's phenotype is an expression of its genotype interacting with its environment. Different phenotypes may develop from the same genotype under different environmental conditions. During development, the individual organism is morphologically most plastic; then, environmental conditions can most influence the ultimate phenotypic expression of genotype. The contribution of a genotype to successive generations depends on how successfully it develops into phenotypes capable of reproducing under existing environmental conditions. Because it is on phenotypes that natural selection operates, interaction of genotypes with environment during development of phenotypes is of extreme evolutionary importance.

Waddington (1957, 1959, 1960) emphasizes the importance of the interaction between genotypes and environment during phenotypic development in his conceptualization of an evolutionary system. This interaction he calls the "epigenetic system," one of four major subsystems in a scheme he (1960, p. 400) explains more eloquently than can we:

In point of fact, it would seem that we must consider the evolutionary system to involve at least four major subsystems [Fig. 7–1]. One is the 'genetic system,' the whole chromosomal-genic mechanism of hereditary transmission; the second is natural selection; a third, which might be called the 'exploitive system,' comprises the set of processes by which animals choose and often modify one particular habitat out of the range of environmental possibilities open to them; and the fourth is the 'epigenetic system'—that is, the sequence of causal processes which bring about the development of the fertilised zygote into the adult capable of reproduction. These four component systems are not isolated entities, each sufficient in its own right and merely colliding with one another when impinging on the evolving creature. It is inadequate to think of natural selection and variation as being no more essentially connected with one another than would be a heap of pebbles and the gravel-sorter onto which it is thrown.

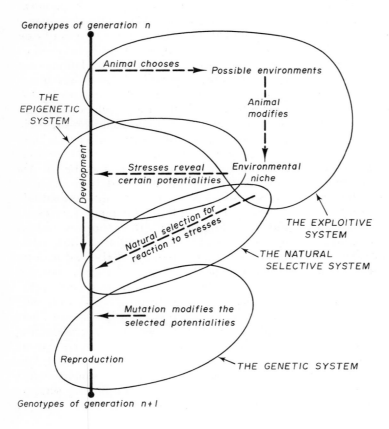

FIGURE 7–1 Waddington's conceptualization of an evolutionary system. Changes in gene frequency between successive generations depend upon the operation of four subsystems: the exploitive, the epigenetic, the natural selective, and the genetic. After Waddington (1960).

In making natural selection only one of the four subsystems in his view of evolutionary processes, Waddington did not intend to minimize its role but only to emphasize the different yet interrelated processes involved. Natural selection operates continously and over all, as his diagram suggests (Fig. 7–1).

How favorably or unfavorably a change in the characteristics of an organism affects the fitness of the organism for a particular environment depends on the effect of the change on the reproductive success of the new phenotype into which it is incorporated. Natural selection does not operate on particular characteristics alone but only as they are a part of the total phenotype. Adaptive changes are probably rarely if ever perfect solutions to environmental problems. They are usually makeshift changes in existing characteristics and become a part of the total organism whose fitness for a particular environment is expressed in its ability to pass on its genotype.

Mendelian Populations and Adaptation

"Individuals of sexually reproducing and cross-fertilizing organisms are associated in reproductive communities which share in common gene pools. Such communities are Mendelian populations" (Dobzhansky 1957, p. 385). It is in this sense that we have been using the term population when referring to sexually reproducing organisms, and it is in this sense that we define *population* in Chapter 15.

Populations have organismic characteristics that transcend those of their individual members (Chapter 15). At the population level of biological organization, the important organismic characteristics of distinctness and cohesion exist through space and time. The morphological, physiological, and behavioral characteristics of individuals that derive from genotypes persisting in the face of natural selection tend to insure this distinctness and cohesion, thus insuring the integrity of the gene pool. Behavior, particularly reproductive behavior, is the primary integrating mechanism. Individuals in the same population are genetically endowed so as to respond similarly to one another and to other factors in their environment in each succeeding generation. Natural selection acts not only to determine the characteristics of

the individuals of the population but to determine the very nature of the population itself.

Other individuals in the same population are just as much a part of the environment of a particular individual as are other environmental factors. Thus, natural selection favors the individual that is genotypically and phenotypically compatible with its fellows, just as it favors the one most suited to survive and reproduce under the other existing environmental conditions (Dobzhansky, 1957). Both kinds of genetic adaptation contribute to the integrity of the Mendelian population.

The idea of niche, which we explained in a previous section, provides a way of conceptualizing the environment to which a particular population is adapted. The niches of two closely related species may partially overlap, and there may be competition between their populations for an important resource such as food. Interspecific competition, even though slight, can become a driving force in the evolution of the two species. In each population there will be some individuals that compete less with individuals of the other population than do most of their fellows, because the characteristics of these individuals permit them to exploit the environment in a slightly different manner. Environmental conditions may be such that natural selection will favor those individuals in the two species that compete the least, and their genotypes and phenotypes will come to predominate in the two populations. Interspecific competition, then, can lead to *niche diversification* and *adaptive radiation* of two Mendelian populations.

Thus far in our discussion, we have emphasized just one of the genetic mechanisms by which populations maintain their capacities for adapting to different environments. This is by maintaining in their gene pools a wide range of genotypes, each of whose phenotypes is suited best for a somewhat different and perhaps narrow range of environmental conditions. Under different environmental conditions, natural selection will favor different phenotypes and their genotypes will come to predominate, thus insuring population persistence. But there is another genetic mechanism by which populations can maintain their capacities for adapting to different environments. Were there to appear in a population particular genotypes whose phenotypes were suited to a wide range of environmental conditions, natural selection in a variable environment would favor these phenotypes, and their genotypes would come to predominate. In extreme cases—man may be the best example—a single phenotype might have the genetic capacity to survive and reproduce over a range of environmental conditions nearly as great as the total range for all phenotypes in the population (Lewontin, 1957). In reality, both these mechanisms to a greater or lesser extent enter into the maintenance of the genetic capacity for adaptation possessed by most populations. We see extremes in the individual rigidity of bacteria and the individual plasticity of man.

Environmental Change and Genetic Adaptation

Any real environment is continuously changing, from night to day, from season to season, and from year to year. Most individuals of species having long lives must be genetically endowed to adapt to seasonal changes. But gene frequencies in populations of some short-lived species have been shown to cycle seasonally (Mayr, 1963, p. 192). Over many years, general changes in climate and other geological forces of the earth are more directional and persistent. The capacity of most populations to adapt to these changes lies more in the diversity of their gene pools than in the plasticity of their individual members, for individuals of few species live so long as to be required to adapt to such long-term changes.

Biological forces, too, change the environments of the earth. Man is one of these forces, though the changes he has wrought sometimes seem more geological in nature. Mining, forestry, and agriculture have led to changes in the temperature, silt load, and dissolved substances of natural waters. To these changes also, populations of plants and animals must adapt if they are to persist.

Mayr (1963, pp. 191–194) discusses "The Force of Selection" and points out that even slight environmental changes can favor particular genotypes by increasing their selective value by as much as 50 percent. He states: "Drastic changes in a population may occur within a few generations." We are all aware of the rapidity with which houseflies develop

resistance to DDT and pathogens become resistant to antibiotics. Ferguson and his co-workers (Ferguson et al., 1964; Ferguson and Bingham, 1966) have demonstrated that populations of several species of freshwater fish have become extremely resistant to chlorinated hydrocarbons used in cotton culture on the delta of the Mississippi.

Changes in land and water use that lead to relatively persistent changes in aquatic environments inevitably will lead to changes in the genetic capacity of populations to persist in those environments; or, just as inevitably, the populations will disappear. On the basis of laboratory studies, it is usually possible to predict reasonably well the average capacity of individuals in a population to adapt to environmental changes. It is difficult, perhaps sometimes impossible, to predict the capacity of a natural population to adapt genetically to an environmental change, for this genetic capacity may be carried by only a few individuals, even in a large population.

Practically, then, of what value in managing our aquatic resources are the ideas we have been attempting to explain? First, intelligent management of any population must be based on some understanding of the biological processes that brought it into being and may permit it to persist. Second, the genetic capacities of populations to adapt to environmental changes can and do change. We are forewarned that we will be wrong and wrong again if we atttempt to predict the fates of populations solely on the basis of laboratory studies of the capacity of a few of their individuals to adapt. And third—though we would be foolish to depend on this—the capacities that natural populations have to adapt to environmental changes caused by man will help to insure their continued utility, diversity, and beauty.

ADAPTATION OF INDIVIDUAL ORGANISMS

Homeostasis: The Tendency to Maintain a Constant Internal Environment

All living organisms must maintain some internal independence of their environment. Cellular processes can proceed only within relatively narrow physical and chemical limits; temperature, pH, ionic strength, oxygen concentration, and other conditions must be maintained within these limits. But the external environment is characterized by change, change in one or more of the conditions necessary for cellular function of almost any form of life. There have evolved, then, throughout the biological world, various mechanisms that provide the internal constancy necessary for life. Semipermeable cell walls function with internal constituents to provide the constancy necessary for unicellular organisms and some of that required by multicellular organisms. The cells of multicellular animals are usually bathed in body fluids within which the ranges of conditions essential for life also tend to be maintained. *Homeostasis* is this tendency of the living organism to maintain its internal environment within limits permitting survival and reproduction, in spite of environmental stresses that tend to displace the internal environment beyond these limits. Homeostasis is the central feature of life, which cannot exist in its absence.

Such internal constancy as animals achieve is maintained through the integrated functioning of their systems. Mediated by sensory detection and nervous and hormonal transmission, locomotor, respiratory, circulatory, digestive, and excretory systems operate to prevent environmental stresses of all kinds from displacing the internal state of the organism beyond the limits of survival and reproduction. At the subcellular level, there occur adjustments in biochemical reactions and pathways that favor the persistence of the organism and its kind.

Homeostatic responses are adaptive responses. Nearly all known adaptive responses were first noted when some change in an organism appeared to be related to some change in its environment. Even today, the mechanisms of very few of these responses are well understood. But most of them appear to have certain characteristics in common. Adolph (1964) seems to have achieved one of the better and simpler general representations of these common characteristics (Fig. 7–2). We can explain this representation no more clearly or briefly than does Adolph (1964, pp. 27–28) himself:

What intermediate steps can be recognized? Some feature A (here termed an affector) of the environment *a* plays upon a detector B within the organism. Schematically, information then passes through C to an effector D that puts forth E. But in case the response adapts, some shift in regulator R influences C or D so that output E is modified compared with the control output. This modification is the adaptate $\triangle E$. The effector modifies its response to the particular environment by virtue of the receipt in R of information, either from some sensing tissue B or from some performing tissue D, or from both. The former information (G) comes from an ordinary sensory process; its detector may or may not be the same (B) as the detector for the unmodified response E. The latter information (F) feeds back to the regulator R.

Walter B. Cannon, a great American physiologist, is generally credited with first using the term homeostasis for the adaptive responses we have been discussing. He developed in some detail the idea of homeostasis in his book *The Wisdom of the Body*, published in 1932, but both the idea and the term appear in his writings of the previous decade. The idea did not, however, originate with Cannon. Its germ certainly appears as early as the writing of Hippocrates. But

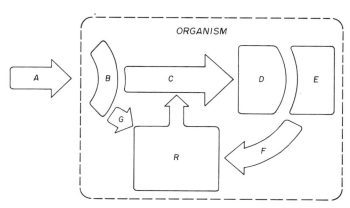

ENVIRONMENT

FIGURE 7–2 Representation of the elements involved in homeostatic or adaptive responses. A, affector; B, receptor; C, transmitter; D, effector; E, output; F, feedback transmitter; G, forward transmitter; R, regulator. After Adolph (1964).

Claude Bernard (1865; republished 1949) seems to have been the first to state explicitly the idea and to pursue its physiological mechanisms. Bernard — Frenchman, playwright, great experimentalist, and father of modern physiology — wrote: "La fixité du milieu intérieur est la condition de la vie libre."

Regulation or Conformity for Particular Environmental Variables

Prosser (1964, p. 12) has written: "No organism is equivalent in its internal composition to its environment." To the extent that it is not, homeostatic processes are regulating its internal composition. In birds and mammals, *homeotherms*, remarkable constancy of body temperature is necessary for life and is maintained in spite of wide variations of temperature in the environment. But in fish, reptiles, and most groups of animals, body temperature follows closely the temperature of the environment. These animals are known as *poikilotherms*. Still, they regulate their internal state in relation to other environmental variables. Thus an animal that is a *conformer* for one environmental variable is a *regulator* for others. Animals may be conformers or regulators with respect to environmental temperature, salinity, oxygen tension, and other factors. Even for a single environmental variable an animal may be a conformer over a part of the range of the variable and a regulator over the remainder. And even in an animal in which body fluids conform to some condition of the external environment, regulation within cells must occur to permit life.

"In general, the range of internal variation tolerated by a conformer is much greater than for a regulator. Conversely, the external or environmental variation within which a regulator can carry out normal functions is greater than for a conformer" (Prosser, 1964, p. 14). Thus we come back to Bernard's statement that the constancy of the internal environment is the condition for a free life. All animals at some metabolic cost can regulate their internal state with respect to one or more environmental variables. Those that have achieved the greatest success in this have been most successful in exploiting the resources of the earth for their own ends.

Acclimation

The response of an animal to one or more environmental factors is very much dependent on its recent history in relation to these. The recent environmental experience of an animal leads to biochemical, physiological, and behavioral adjustments that influence if not determine its response to a particular set of environmental conditions. Such compensatory adjustments to environmental variables are known as *acclimation*. Acclimation makes it possible for animals to survive and reproduce under conditions where these otherwise would be impossible. Thus, acclimation is an adaptive change in the individual organism; it involves homeostatic mechanisms.

The ways in which acclimation extends the tolerable ranges of regulating and conforming organisms differ (Prosser, 1964). Acclimation changes the limits of an environmental variable within which a regulator can control its internal state but usually does not change its optimal internal state. For the conformer, acclimation usually alters the extremes of internal state at which more or less normal function is possible, and it may shift the optimum. In determining the range of an environmental variable within which an animal can be expected to be successful, we must take into account its previous history in relation to that variable, for not only survival but also reproduction, growth, and activity have been shown to be influenced by this history.

In summary, then, an animal that has been held at one level of an environmental variable is different from an animal that has been held at another level. Grossly, this difference is apparent not only in the different ranges of the variable that the two animals can tolerate but it is apparent also in differences in their metabolic or oxygen consumption rates at each level of the variable. Such adjustments in metabolic rate aid an animal in extending its range. These adjustments may involve changes in rates or pathways of enzymatic processes as well as changes in hormonal or nervous integration. Though little understood, multiple changes undoubtedly are integrated so as to provide the highly adaptive responses of the whole organism, which we can more readily observe.

Morphological, Physiological, and Behavioral Adaptation

The conditions under which an animal is best suited to live depend on its morphological, physiological, and behavioral characteristics and their plasticity. And where it will be able to live depends not on single environmental variables acting alone but on many variables acting together. The temperature, salinity, and oxygen requirements of the American lobster (*Homarus americanus*) are interrelated (McLeese, 1956); similarly interrelated are the physical and chemical requirements of all animals. But for no animal do a few physical and chemical parameters define the environmental conditions essential for its existence. Food resources, competition, predation, and disease must also be taken into account, for these will interact with physical and chemical factors in determining the limits of existence. Each animal is adapted as a whole to survive and reproduce in a particular niche. Though it is necessary to study particular adaptations to particular environmental variables, we must endeavor to understand how these adaptations are integrated in the whole animal to provide adaptive responses to the total complex of interacting environmental variables.

The capabilities of a fully developed animal to adapt to environmental change reside mainly in its physiological and behavioral plasticity, for morphological plasticity in such an animal is not usually great. But morphological structure, from the cell to the whole organism, obviously is necessary for any physiology or behavior. And the adaptive value of any physiological or behavioral change is dependent on this structure. We will be considering physiological adaptations in Chapter 8 and behavioral adaptations in Chapter 12; these will be discussed in relation to the structures most involved. Though it may be necessary to consider individual systems of structure, physiology, and behavior, the systems of animals have not evolved individually but all together. These systems can persist only together in the whole organism, and together they determine the ability of the organism to adapt to environmental change.

Limits of Individual Adaptation

Ultimately, the limits of adaptation of the individual organism are determined by its genotype, its genetic constitution. But it is the phenotype, the total living organism, that must adapt if it is to survive and reproduce. The phenotype develops through interaction of the genotype with the environment. The environment can most affect the morphology of the individual organism during development, for then the organism is morphologically most plastic. Some of the morphological changes occurring during development may be of adaptive value in adult life; others undoubtedly are not. Even in adult life, organ systems change under stress, and these changes often are adaptive. Physiological and behavioral plasticity in adult life may also be influenced by environmental conditions existing during development.

The effects of recent environmental history on the physiology and behavior of animals are better known than are those of early environmental history. Studies of the effects of acclimation on animals are relatively simple to conduct in the laboratory. Correct interpretation of such studies is not usually so simple. Conditions in the laboratory are unlike conditions in nature, and animals in the laboratory are in some ways unlike animals in nature. By holding an animal in the laboratory at some high and constant level of temperature, it may be possible to raise considerably the maximum temperature it can tolerate. But how much increase in maximum tolerable temperature is occasioned when temperatures in nature fluctuate from low to high levels daily and seasonally and salinity and dissolved oxygen also vary? The genetically determined rates at which animals can acclimate to particular environmental variables must set limits on the extent to which animals can adapt to these variables in nature.

There is still another aspect of the problem of adaptation that should always be considered: the metabolic cost of adaptation in relation to an animal's energy and material requirements for survival, growth, and reproduction. An animal can obtain from its environment only limited amounts of the food and oxygen needed to provide energy and materials (Chapter 11). Fry (1947) viewed the optimal environment for an animal as the one in which it has the greatest metabolic scope for activity. Animals may be able to acclimate to environmental extremes. This usually involves some metabolic cost, reducing energy and materials available for life's other needs. This cost is probably not usually great. Nevertheless, in nature energy and material resources are limited. The availability of food and oxygen must set some limits on an animal's ability to adapt to other environmental variables.

There are limits to the ranges of physical conditions within which particular bio-chemical reactions inside cells can proceed. Complete failure of critical reactions would lead, of course, to death of an organism. But it is not at all clear that death at environmental extremes is always or even usually the result of intracellular conditions going beyond these limits. More often, perhaps, death ensues before these limits are reached, following more general failure of homeostatic control. "Certainly no single physiological factor has been found to be responsible for lethal limits" (Prosser, 1964, p. 20). Moreover, the natural limits of individual adaptation are those permitting not only survival but reproduction as well, and in nature physiological and ecological factors interact in determining individual success.

EVALUATION OF THE SIGNIFICANCE OF MORPHOLOGICAL AND PHYSIOLOGICAL CHANGE

Final Basis of Evaluation

The evaluation of the significance of a change in a particular biological system can finally be made only on the basis of its probable effect on the more encompassing biological system of which it is but one part. Evaluation of a particular morphological, physiological, or behavioral change must first of all be made in terms of its contribution to the homeostasis and survival of the whole living organism. But we must be more encompassing than this, for the individual organism has no biological significance to its kind except as it contributes to the persistence of its population. This contribution must come largely through reproduction for all organisms with the exception of man, a species in which the cultural contribution of an individual may be more important than the reproductive one. Thus, the significance of a morphological or physiological change lies in its influence on the survival and reproduction of the individual.

In man, except as it balances reproduction, death may not usually appear to have a biological function, though the death of a religious or political leader may in rare instances have a profound social function. But in most other species, the death of an individual often is a nutritive contribution to the persistence of other species in its encompassing biological community, even if it precludes that indivdual's reproductive contribution to its own population. Thus, adaptive changes favoring the growth and production of organisms are of biological significance going beyond their own kind even to man. Still, for a population to make this broader biological contribution, morphological and physiological changes must favor the survival and reproduction of its individuals, or the population cannot persist.

The significance of particular changes in the individual organism can sometimes be established in terms of its survival, growth, and even reproduction under laboratory conditions; this we should endeavor to do. But the ultimate importance of the individual and, hence, the ultimate importance of these changes can only be evaluated in terms of the success of the natural population and biological community from which the individual comes. And though this is a big order, it is the final basis for the evaluation of change.

Genetically Persistent and Individually Acquired Changes

Both genetically persistent and individually acquired changes in morphology, physiology, and behavior are important to the persistence of natural populations, which are our primary concern. But the way in which a particular change will be important depends on the degree to which it is genetically transmissible. In evaluating the significance of differences between individuals of a kind, we often find it helpful to determine whether the differences are primarily genetic or whether they have resulted from different environmental histories of the individuals. In doing this, it is well to bear in mind that all individual characteristics have genetic determinants, which in their expression are dependent on the environment. And we should also bear in mind that an individual can transmit to its offspring its genetically determined capacity to adapt to an environmental variable, but that it cannot ordinarily transmit an adaptive change resulting from its own environmental history.

The extent to which variations in individual characteristics are genetically determined and the extent to which they are environmentally determined can sometimes be evaluated by acclimation and breeding experiments under different conditions in the laboratory. From such experiments can come increased understanding of the biological basis of adaptation. General adaptive responses of organisms, however, usually involve more than one physiological system, and the elucidation of their genetic basis will often prove difficult. Prediction of the genetic capacity of a natural population to adapt through successive generations to an environmental change will prove even more difficult, but quantitative genetics is beginning to show the way (Falconer, 1960). Whether or not we can ever fully distinguish between the genetic and the environmental determinants of adaptive capacity, it is important that we bear in mind the possible bases of the adaptive variations we observe and that we recognize their possible significance.

Biological Performance as a Basis for Evaluating Physiological Change

The biological performance of the individual organism is ultimately expressed in survival, reproduction, development, and growth; and it can be measured in these terms. Of course, the activity and behavior of the individual are involved, but they are important only insofar as they contribute to these ultimate expressions of performance. And it is in just this way that physiological changes are important; herein lies their significance.

Changes in environmental conditions usually lead to changes in the physiology of organisms. In the laboratory, these physiological changes can often be measured and related to the causal environmental factors. And, in the laboratory, it is often possible to determine what effects physiological changes have on important aspects of an animal's biological performance. Survival is perhaps the simplest aspect of biological performance

to which physiological change can be related; it is the one usually studied. But development and growth can often be conveniently studied, and they are likely to provide more sensitive measures of the effects of physiological change than does survival. Reproduction, because of its associated behavior, is difficult to study in a meaningful way in the laboratory. Part IV of this book is devoted to study of the ecology of the individual organism, its biological performance. The significance of physiological changes must be found through studies such as those we will there describe.

But, finally, the biological performance of the individual animal as we can study it in the laboratory may or may not provide an adequate basis for evaluation of the significance of physiological change. The biological performance of an animal is really only meaningful in its natural environment, where to be successful it must contribute to the persistence of its population. We can rarely adequately model the natural environment of an animal in the laboratory. Laboratory studies of the physiology and performance of animals must be so conducted as to lead most effectively toward understanding of animals in nature. We will repeat this warning many times in Part IV of this book; in Part V, we will attempt to place the animal in the context of its population; and in Part VI, we will return it to its biological community and ecosystem.

Application of Morphological and Physiological Knowledge

Perhaps because man is conservative in his very nature, he tends to view any change with suspicion. We hope this chapter has made clear that the physiological systems of animals change continually, and that the organs and tissues of animals are not static in their structure. Adjustment to environmental change through physiological, behavioral, and even structural change is the dynamic basis of an animal's homeostasis. On this its survival and reproduction depend. This is not to say that all physiological or morphological change is good for the animal, for there are certainly limits beyond which change leads to its disintegration. But it is to say that physiological and morphological changes are essential to survival and reproduction; they are undesirable only insofar as they prevent these.

In recent years, some biologists interested in water pollution control have proposed the use of various measures of physiological and morphological change for detecting harmful changes in the aquatic environment. Rarely have these bodily changes been shown to lead to decreased survival or reproduction of animals. Thus, it has not usually been clear whether these changes are adaptive or destructive. Certainly, before physiological and morphological changes can be useful in detecting harmful environmental conditions, this must be elucidated. But, even when particular physiological responses are known to be destructive, physiological studies are usually complex, and they can be made only after the organisms have been removed from their natural environment. Studies of survival, reproduction, development, and growth are often simpler; and often they can be conducted in nature. Moreover, the information acquired pertains much more directly to the problem of determining just what is an unfavorable environment. Detection of harmful changes in the environment is not the best application of morphological and physiological knowledge.

Why, then, should biologists interested in water pollution control study the physiology and morphology of aquatic organisms? Bartholomew (1964, p. 8) gave the answer: "each level finds its explanations of mechanism in the levels below. . . ." We cannot explain the responses of the whole organism to its environment as it survives, reproduces, develops, and grows, or as it does not, unless we understand the physiological and morphological bases of these responses. But, so long as we know that an organism responds in a certain way to an environmental change, why should we seek an explanation? For this reason: we can observe the responses of organisms under only a few sets of environmental conditions; if we do not understand the physiological and

morphological bases of the responses we observe, we cannot predict with any certainty at all what the responses will be under other sets of environmental conditions; and it is toward accurate prediction that any science worthy of the name must be directed. This is why biologists interested in water pollution control must study morphology and physiology; this is application of morphological and physiological knowledge at its best.

SELECTED REFERENCES

Adolph, E. F. 1964. Perspectives of adaptation: some general properties. Pages 27–35. In D. B. Dill, E. F. Adolph, and C. G. Wilber (Editors), Adaptation to the Environment. (Handbook of Physiology, Section 4.) American Physiological Society, Washington, D.C. ix + 1056pp.

Bates, M. 1960. Ecology and evolution. Pages 547–568. In S. Tax (Editor), The Evolution of Life. (Evolution after Darwin, Vol. 1.) The University of Chicago Press, Chicago, viii + 629pp.

Cannon, W. B. 1932. The Wisdom of the Body. W. W. Norton & Company, Inc., New York. xv + 312pp.

Fry, F. E. J. 1947. Effects of the environment on animal activity. University of Toronto Studies Biological Series 55. Ontario Fisheries Research Laboratory Publication 68. 62pp.

Hutchinson, G. E. 1957b. Concluding remarks. Pages 415–427. In Population Studies: Animal Ecology and Demography. Cold Spring Harbor Symposia on Quantitative Biology, Vol. 22. xiv + 437pp.

Lewontin, R. C. 1957. The adaptations of populations to varying environments. Pages 395–408. In Population Studies: Animal Ecology and Demography. Cold Spring Harbor Symposia on Quantitative Biology, Vol. 22. xiv + 437pp.

Mayr, E. 1963. Animal Species and Evolution. The Belknap Press of Harvard University Press, Cambridge, Massachusetts. xiv + 797pp.

Prosser, C. L. 1964. Perspectives of adaptation: theoretical aspects. Pages 11–25. In D. B. Dill, E. F. Adolph, and C. G. Wilber (Editors), Adaptation to the Environment. (Handbook of Physiology, Section 4.) American Physiological Society, Washington, D.C. ix + 1056pp.

Waddington, C. H. 1960. Evolutionary adaptation. Pages 381–402. In S. Tax (Editor), The Evolution of Life. (Evolution after Darwin, Vol. 1.) The University of Chicago Press, Chicago. viii + 629pp.

8 Morphological and Physiological Adaptations

In marine creatures, also, one may observe many ingenious devices adapted to the circumstances of their lives. For the accounts commonly given of the so-called fishing-frog [angler fish] are quite true. . . . The fishing-frog has a set of filaments that project in front of its eyes; they are long and thin like hairs, and are round at the tips; they lie on either side, and are used as baits. Accordingly, when the animal stirs up a place full of sand and mud and conceals itself therein, it raises the filaments, and, when the little fish strike against them, it draws them in underneath into its mouth.

ARISTOTLE, *c.* 336 B.C., v. 2, p. 146

As we analyse a thing into its parts or into its properties, we tend to magnify these, to exaggerate their apparent independence, and to hide from ourselves (at least for a time) the essential integrity and individuality of the composite whole. . . .

. . .We may study them apart, but it is as a concession to our weakness and to the narrow outlook of our minds.

D'ARCY THOMPSON, 1942, pp. 1018–1019

FORM, FUNCTION, NICHE, AND A WAY OF LIFE

Anyone who will spend a pleasant afternoon in close observation of the life of aquatic insects in a brook or pond cannot help but begin to appreciate what we mean by form, function, niche, and a way of life. For each species, form is adapted to function and function to niche; together they constitute a way of life. If ice or freshet prevents our reader from a profitable excursion at the moment, a reading of Needham and Lloyd's (1937) delightful *The Life of Inland Waters* will be a second-best alternative, until this excursion becomes possible.

A water glass may be necessary to penetrate visually the turbulent water flowing among the stones of a brook, but with this another world with many ways of life becomes apparent. Here, we may see a mayfly nymph, aided by flattened and streamlined form, clinging to the side of a stone in rapid current. In this task, it is further aided by a suction cup formed by the ventral surface of its abdomen and the series of tracheal gill plates extending laterally along the abdomen to meet anteriorly and posteriorly. These delicate gill structures, with their ramifying tracheal tubes for exchange of respiratory gases, have been further modified in form to function also in aiding this species to maintain its position. And in its niche, our mayfly species must have food, which may be the diatoms growing on the stone. For grazing these diatom pastures, mouthparts of this nymph have a comblike form adapted to function in scraping the diatoms from the stone.

In each species of insect, crustacean, mollusc, or other animal, we have examples

81

of such adaptation of each part to one or more functions and of the total adaptation of the whole individual to a niche and a way of life. There is the larva of the blackfly, attached to the top of a stone in the fastest current by a disc of hooks at its posterior end, able to change its position safely with the aid of a silk strand along which it can move, and able to strain its food from the current with two fan-like structures of many fine rays near its mouth. And there is the larva of a species of caddis fly, living among the stones and building a silken net to strain its food from the water. But we need not go on to make our point clear: each structure is beautifully adapted for one or more functions and the organism as a whole is beautifully adapted for a way of life.

At the outset of this section, we stated that during an afternoon of observation of stream life one would begin to appreciate the meaning of form, function, niche, and a way of life. Now, we should also state that biologists who have devoted their lives to the study of adaptation only incompletely appreciate or understand its manifestations. Though they may not understand the function or functions of each structure, they are inclined to believe that it has adaptive value. D'Arcy Thompson (1942, p. 950), in his monumental book *On Growth and Form*, wrote: "That this mechanism is the best possible under all circumstances of the case, that its work is done with a maximum of efficiency and at a minimum of cost, may not always lie within our range of quantitative demonstration, but to believe it to be so is part of our common faith in the perfection of Nature's handiwork." We do not believe that adaptations are always the best possible solutions to problems of life, but we must believe that they have proved to be workable ones.

We cannot emphasize too strongly the need to view the adaptation of the whole organism to a niche and way of life; for structure upon structure, system upon system, and function upon function, it is integrated and adapted to live only as a whole. "We may study them apart, but it is as a concession to our weakness and to the narrow outlook of our minds." And, again, it is Thompson (1942, p. 1019) who expresses ideas so well:

The biologist, as well as the philosopher, learns to recognise that the whole is not merely the sum of its parts. It is this, and much more than this. For it is not a bundle of parts but an organization of parts, of parts in their mutual arrangement, fitting one with another, in what Aristotle calls 'a single and indivisible principle of unity'; and this is no merely metaphysical conception, but is in biology the fundamental truth which lies at the basis of Geoffroy's (or Goethe's) law of 'compensation,' or 'balancement of growth.'

But Thompson was far too mature a scientist to preach a holism that did not give necessary attention to parts, attention so beautifully illustrated by his own work. It is not only "as a concession to our weakness" that we must study parts; it is also because without careful study of the parts we can never achieve real understanding of the whole.

In the following sections of this chapter, we will endeavor to convey some feeling for adaptations for feeding and digestion, circulation and respiration, and osmoregulation and excretion occurring in fish, in crayfish and lobsters, and in oysters. We will also consider some of the problems temperature and toxic substances present to aquatic organisms and how they respond. Entire books have been devoted to each of the animals we will discuss and to each of these problems of adaptation. We must, of course, if we are to avoid making this chapter too lengthy and burdensome for our reader, consider these matters only rather generally, though we hope not to be too superficial.

To one interested in water pollution control, what is the value of a brief and very general discussion of such complex problems and such intricate adaptive mechanisms? Mainly this: one cannot begin to understand the possible effects of environmental changes on aquatic organisms until he has at least been introduced to the mechanisms by which these organisms maintain themselves in sensitive balance with their environment. Even in their natural environments, organisms are continually confronted with problems that threaten their survival and reproduction. Still, they have evolved in ways that favor their persistence in their natural environments. But in changed environments, their adaptive mechanisms may no longer be able to cope with the problems of life. Their ways of life may no longer be possible.

FEEDING AND DIGESTION

Feeding and Digestion in Fish

Fish, like all living organisms, require materials and energy for the structure and metabolism of their bodies. And in fish, as in other groups of animals, there have evolved many different ways of meeting these requirements. The materials and energy that fish need, of course, come from plants and animals; and most of the major groups of plants and animals present in the aquatic environment have come to be utilized by one or another species of fish. Each species of fish has acquired through evolution the structural, physiological, and behavioral characteristics that permit it to capture or otherwise obtain its food and utilize it for maintenance, growth, and reproduction. We must restrict our discussion to some of the characteristics of adult, bony fishes. Larval and juvenile fish usually differ from adults in their feeding and nutrition. Even in a restricted discussion, we can illustrate important adaptations for feeding and digestion.

Some fish strain their food from the open water or from deposits on the bottom. Others pursue and capture more motile forms of food. Some select invertebrates from the benthic environment, whereas others graze on plants attached to the substrate. For each of these sources of food, there have evolved in the fish means of detection, feeding, ingestion, processing, and utilization. These adaptations in large part determine the nature of each species, its appearance, its behavior, and its chances of persisting in a natural or in a changed environment. Most men, perhaps, can more easily appreciate the problems that animals face in the search for food than the other problems confronting them.

The search for food, with its instinctive and learned components, may be guided by directive stimuli; and the capture and ingestion of food is released by appropriate releasing stimuli. Internal motivation, sensory capabilities, and kinds of stimuli will be considered in Chapter 12. Now we need only point out that different species of fish detect their food by sight, by vibrations transmitted through the water, by touch, by taste, by smell, or by some combination of these. And, in each species, particular organs have been more or less specialized for this purpose.

Eyes and the lateral line system may need no special modifications for detection of food by sight or by vibrations it may produce. But minnows that feed on algae may have lips especially adapted for feeding by touch and suction. In catfishes, barbels around the mouth carry taste buds for locating food; taste buds are concentrated in the mouth and scattered over the surfaces of the body in many species of fish. The olfactory organs of fish are located in the nares.

The mouths of some species of fish are highly specialized, even to form beaklike and tubelike extensions. Fish that feed by grazing or sucking tend to have ventrally located mouths, whereas active predators tend to have their mouths terminally located, the size and form of the mouth being adapted to the size and nature of the food. Plankton-feeding fish are usually without teeth on the bones of the mouth. In these species, the *gill rakers*—toothlike projections on the first four gill arches—are like fine combs and serve to strain from the water the minute organisms upon which the fish feed (Fig. 8–1). The fifth gill arch in many species is modified into a jawlike structure with musculature and teeth adapted to particular kinds of food, the teeth being suited for grasping or tearing prey in some species and for grinding plants or hard-shelled animals in other species (Fig. 8–1). The bones of the jaws and mouth of predatory species often bear sharp teeth, which aid in capturing and retaining prey. Incisor-like scraping teeth are present in many species that must remove their plant or animal food from hard substrates.

We have described the mouth and pharynx; the remainder of the alimentary canal of fish can be divided into the esophagus, stomach, intestine, and rectum, further subdivision not being justifiable. Barrington (1957) has reviewed much of the literature on the structure and physiology of the alimentary canal of fish; it is apparent how much less is known of these in fish than in man and his domestic animals. Nearly all biologists simply accept that the systems of animals are generally adapted to their normal conditions of existence. But, when we come to poorly known systems in animals whose ecology also is little known, statements regarding adaptive value are tenuous. The digestive system of fish is one of these.

The usual function of the stomach of fish

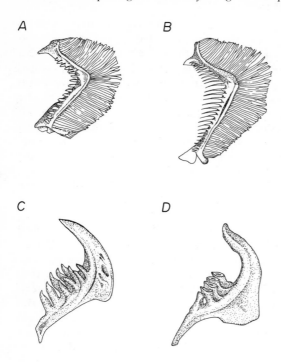

FIGURE 8–1 Morphological adaptations of the gill arches of fish that feed on different kinds of food. Gill rakers on first four pairs of arches of whitefish that feed on large prey (A) and whitefish that feed on plankton (B). Pharyngeal teeth on fifth pair of gill arches of minnows that hold and tear prey (C) and minnows that must grind their food (D). After Koelz (1927) and Grassé (1958).

is digestion of protein, this being brought about in a rather acid medium by proteases of the pepsin type. A blind sac, or caecum, present in some fish may be an adaptation permitting them to ingest and store for digestion large amounts of food when it is available. A gizzard-like modification of the stomach in some species functions in trituration of food. The absence of gastric glands in some fish suggests a loss of digestive function of the stomach. Other species of fish, particularly in the minnow family, have no stomach. In these species, the esophagus grades directly into the intestine, the bile duct entering near this juncture; pepsin and hydrochloric acid are absent. The adaptive advantages to the species in which this occurs have never been clearly explained.

The intestine functions not only in digestion but also in absorption of the digested materials; in some species absorption is aided by finger-like projections, pyloric caeca, that extend from the intestine near its origin. Intestinal digestion in most fish appears to proceed at a nearly neutral or slightly alkaline reaction, this condition depending on materials from the stomach and secretions from the pancreas, liver, and intestine. Proteases from the pancreas and from the intestine itself continue protein di-

gestion, and carbohydrases and lipases function to break down carbohydrates and fats. The secretory roles of the pancreas and intestine in the production of the latter two groups of enzymes are yet to be clarified.

There is some evidence of structural and physiological adaptation of the intestine to the food of particular species of fish. The intestine of carnivorous fish tends to be much shorter in relation to body length than does that of herbivorous fish. And in herbivorous fish, the production of carbohydrases is often much higher and the production of proteases somewhat lower than in carnivorous fish. There is no reason to suppose that individual fish do not have some capacity for structural and physiological adaptation to changes in their diet, but on this there is little evidence.

Feeding and Digestion in
Crayfish and Lobsters

Crayfish and lobsters are members of the large and diverse group known as crustacea. No one or two species can be really representative of this group, for the ways of life and the adaptations exhibited by its members are too varied, more so, for example, than among fish. With regard to feeding alone, Marshall and Orr (1960, p. 277) have written:

Almost all crustaceans have the same basic mouth parts although the diversity of form and function is enormous. The antennules are not so often used in feeding as the rest; the antennae, mandibles, maxillules, and maxillae are the main organs. In addition, the labrum and the labium may be important. The thoracic limbs are primarily locomotory in function but often aid in feeding by creating a current, by filtering, by catching and holding food particles, or by passing them forward.

We wish that we could devote the space, and take the time of our reader, to consider in some detail many of the adaptations for feeding that can be seen in the crustacea. The rhythmic movements of the limbs of the fairy shrimp, which pass food forward to its mouth, the scoop net formed by the limbs of the barnacle, which it casts to capture small organisms, the feeding currents created by the antennae of some copepods, these and many other adaptations for feeding are fascinating and instructive to study. They are instructive to study because we cannot understand the distribution and abundance of these organisms without knowledge of their modes of feeding. Although we cannot take crayfish and lobsters to be representative of the crustacea, they do provide us with examples of morphological and physiological adaptations for their ways of life, adaptations not only for feeding and digestion, but also for circulation, respiration, excretion, and osmoregulation, which we will be considering in later sections.

Not the least of the rewards of a few afternoons in the library is being reminded of how very well indeed the fine biologists of the past knew the animals they studied. Beautifully equipped laboratories of the present have certainly contributed to our knowledge of the physiology of these animals, but not so very much more. Thomas Huxley (1884), great defender of Darwin and his theory, modestly subtitled his monograph *The Crayfish* as *An Introduction to the Study of Zoology;* there is yet no better place to begin the study of this animal. And Herrick (1909) wrote his magnificent *Natural History of the American Lobster* on that close marine relative of the crayfish.

Plants as well as animals are important in the food of crayfish, whereas plants do not form a large part of the diet of lobsters. Fish, crustaceans, molluscs, and worms are the main animal foods of lobsters; to this list we must add insects for the freshwater crayfish. The food may be grasped and perhaps crushed or torn by the chela or claws, the fourth pair of thoracic appendages. The fifth and sixth pairs of thoracic appendages, which function in locomotion, are equipped with tiny chela, and these also function in holding and tearing the food and conveying it to the mouth. The mouth is an elongated opening on the ventral aspect of the head, the labrum and metastoma forming the upper and lower lips. The remaining mouth parts are formed by appendages of the last three somites of the head and the first three of the thorax. The first pair of these appendages, the mandibles, are adapted for crushing and triturating small, hard parts of food. The first and second pairs of maxillae function to masticate and transfer food, the second pair also functioning to create respiratory currents. The three pairs of maxillipeds which follow serve to break down and move foward the food, the last pair having brushes of stiff bristles for cleaning. In addition, the mouth parts have the important sensory roles of taste and touch.

From the mouth, a short esophagus extends upward to enter the ventral surface of the stomach, which is divided by a constriction into a large cardiac chamber and a smaller pyloric chamber, the latter opening through valves into the very short midgut. The stomach of the crayfish or the lobster is a most remarkable adaptive structure. From the posterior walls of the cardiac chamber, dorsal and lateral ossicles, or teeth, project and intermesh, forming with their associated musculature the gastric mill, which functions in trituration. Only very fine food particles can pass through the pyloric chamber into the midgut and thence into the hepatopancreas or into the intestine, because the lateral and ventral walls of this chamber are closely pressed into folds bearing stiff bristles, which form a filter.

The hepatopancreas of crayfish and lobsters is composed of a pair of large glandular organs, each with three lobes. These lobes of brownish color contain many caeca opening into two common ducts, which join the midgut. Through these ducts, food materials enter the hepatopancreas, which, in addition to secreting the digestive enzymes, functions as the principal site of digestion and absorption (Vonk, 1960). The hepatopancreas secretes the bile acids, proteases, carbohydrases, and lipases or esterases necessary for digestion. It may also function to store glycogen, fat, and calcium.

Feeding and Digestion in Oysters

Oysters, along with clams, mussels, scallops, and most other bivalves, belong to a group of molluscs known as Lamellibranchia. This group is characterized by greatly enlarged ctenidia or gills that have become adapted for feeding through evolution from an ancestral type in which the gills were primarily respiratory structures. In the ancestral type, cilia present on the surfaces of the ctenidia functioned primarily for cleaning, whereas in the Lamellibranchia cilia have acquired the additional functions of filtering plankton from the water and transporting the food organisms along the ctenidia to the

labial palps and mouth. The large volumes of water that must be filtered by an animal feeding in this manner are moved by the cilia.

Those who would learn something of oysters are fortunate that C. M. Yonge (1960) summarized his great knowledge of this animal in a fascinating account, *Oysters;* Orton (1937) performed a similar service for the preceding generation. And now we have Paul S. Galtsoff's (1964) wonderful monograph *The American Oyster.* Lying between the valves of its shell, the body of an oyster is next enclosed by two membranous sheets of tissue, the mantles (Fig. 8-2). Along their free margins, the mantles bear three folds. The outer of these functions in shell formation; the middle one is fringed by sensory tentacles; and the inner one is large and muscular and functions to control the inflow and outflow of water. The space enclosed by the mantles is occupied by the body of the oyster and by two chambers, the large inhalant water chamber and the smaller exhalant water chamber. The four ctenidia oc-

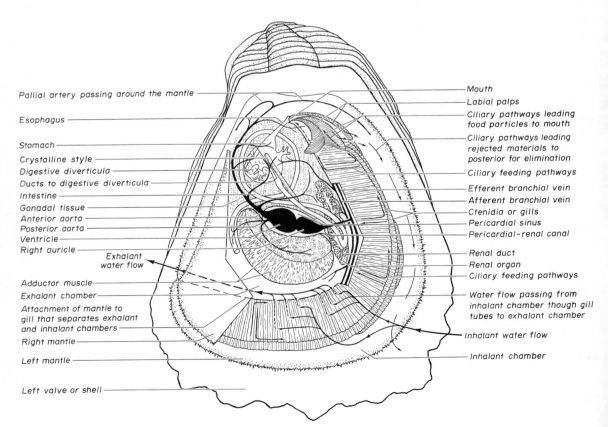

FIGURE 8-2   Anatomy of an oyster.

cupy much of the inhalant chamber, and—being attached to the mantles and to each other—they form a partition between the two chambers. Water can pass from the inhalant to the exhalant chamber only through innumerable tiny pores in the ctenidia, which thus form the filter for feeding.

The ctenidium of the oyster is another remarkable adaptive structure, functioning as it does to move large volumes of water, to filter from this water and begin sorting materials for food organisms, and to transport these to the labial palps, where they are finally sorted before entering the mouth. The ctenidia not only move the water providing oxygen, but they are also one of the mechanisms of gaseous exchange, the mantles being the other. The water currents necessary for feeding and respiration are developed mainly by only one of the kinds of cilia on the ctenidia. Other kinds function to filter the food from the water and transport it to the labial palps and mouth. Water enters the inhalant chamber through the inhalant pore of the mantles. Large particles are prevented from entering with the water by the tentacled fringes of the mantles. Current velocities are reduced in the inhalant chamber, so some heavy particles settle onto the surface of the lower mantle to be picked up by ciliary paths and carried to the mantle edge for ejection. The water then enters the ctenidia through many tiny pores and is carried by tubes within the ctenidia to the exhalant chamber, from which it leaves by the exhalant pore (Fig. 8–2).

Before the water enters the pores of the ctenidia, special large cilia filter out particulate materials and transfer them onto surfaces bearing smaller cilia. The finest particles fall into transverse grooves on the ctenidia and are moved by cilia to the bases of the ctenidia and into longitudinal grooves in which they are transported toward the labial palps and mouth (Fig. 8–2). Materials not falling into the transverse grooves are transported by small cilia to the free margins of the ctenidia and then are moved toward the palps and mouth in marginal grooves. These materials become embedded in mucus produced by cells on the ctenidia. When the masses of mucus become excessive, they fall from the marginal grooves to the mantle surface, where they are moved posteriorly to be ejected.

The ctenidia end anteriorly between the two pairs of labial palps. These structures were probably the principal organ of feeding in the ancestral bivalve type, but in the Lamellibranchia they have become adapted for sorting food from materials gathered by the ctenidia. Material transported forward by the basal and marginal grooves of the ctenidia must then traverse the ridged surfaces of the palps. Small particles and masses of mucus are passed by cilia from one ridge to the next and so to the mouth. But large particles and masses are drawn into the grooves between the ridges and conducted to the margins of the palps, then back along these margins to a place from which they fall to the mantle. There, cilia conduct these particles and masses, as well as those falling from the ctenidia and those settling from the inhalant current, along well-defined paths to the posterior edge of the mantle (Fig. 8–2). From this point, by a sudden clapping of its valves, the oyster ejects the masses of mucus as pseudofeces, so called because they contain waste materials that have not passed through the digestive tract.

Movement of materials through the digestive tract is accomplished mainly by ciliary action. Not all the materials entering the stomach of the oyster can be used as food, and here further sorting occurs. Through a complex system of canals, the larger and heavier materials, along with waste products of digestion, are passed directly to the intestine, which empties into the exhalant chamber and so to the external environment. Some digestion of food materials, mainly minute algal cells, occurs in the stomach. A gelatinous rod, the crystalline style, enters the stomach from a sheath in which it is continuously rotated by cilia. The style dissolves slowly in the stomach and contains enzymes for breaking down starch, cellulose, and fats; it also contains materials for keeping the stomach contents acid (Owen, 1966). Ducts from the stomach lead to the digestive diverticula, which form a brownish mass investing the stomach (Fig. 8–2). Minute food particles entering these ducts and their tubules are ingested by cells in which intracellular digestion of proteins and fats occurs. These substances and also glucose from the stomach are absorbed in the tubules. Other intracellular digestion occurs in blood cells that migrate into the stomach, engulf food particles, and return to the tissues and blood stream.

We have dwelt on feeding and digestion of

the oyster, as in previous sections we did on feeding and digestion of fish, crayfish, and lobsters. Our reader may well begin to wonder why, in a book on biology and water pollution control, we give detailed accounts of structural and physiological adaptations of even a few animals. Our reason is simple; perhaps we can make it clear by examining briefly the problems an oyster must face in feeding. The oyster lives in an environment in which planktonic organisms usually thrive but in which silt and debris also are often abundant. Its mechanisms of feeding and digestion are remarkably adapted for removing fine materials from the water, separating the food, and processing this to provide energy and materials necessary for maintenance, growth, and reproduction. But environmental changes caused by the activities of man can reduce the abundance of food organisms or increase the silt and debris. Then, the mechanisms of the oyster may have insufficient capacity to obtain food, because of either low food availability or excessive waste material to handle. This can lead to death or failure to grow or reproduce. Without understanding the sensitive adaptation of the oyster, or any other organism, to its way of life, we cannot predict the effects of pollutional changes on its distribution and abundance. In this chapter we cannot discuss adaptations of many animals; we can only hope to impress our reader with the importance of knowing the animals with which he may one day need to work.

CIRCULATION AND RESPIRATION

Circulation and Respiration in Fish

An animal's circulatory system with its transporting medium, the blood, moves not only respiratory gases but also nutrients, waste products, salts involved in osmoregulation, and hormones. Thus, it functions in many ways to maintain and integrate the responses and activities of the organism so as to permit it to survive, grow, and reproduce in its natural environment.

Mott (1957, pp. 82–83) has briefly and clearly described the circulatory system of fish in sufficient detail for our purposes:

The characteristic feature of the fish circulation is that the blood is pumped by the heart through the gills, where respiratory exchange takes place, and flows back to the heart through the systemic arteries and veins. There is thus a single pump and a single circuit of vessels. The *vis a tergo* propels the whole cardiac output first through the gill capillaries and then through the systemic capillaries; in addition a proportion of the cardiac output passes through the capillary beds of portal systems before returning to the heart. [Portal systems are capillary or sinus beds which occur in gills, livers, and kidneys and function to enhance transfer of gases, nutrients, metabolites, salts, and waste products to or from the blood.]

The heart consists of four chambers arranged in series. Venous blood enters the sinus venosus from the ductus Cuvieri and hepatic vein, and passes through the atrium, ventricle, and bulbus (teleosts), or conus (elasmobranchs), arteriosus. The blood delivered into the ventral aorta by the cardiac contractions thus contains little oxygen. The anatomy of the afferent branchial arteries (except in some species with accessory respiratory organs) is such that the whole of the cardiac output of poorly oxygenated blood must pass through the gills. The essential features of this arrangement are illustrated in Fig. 1 [Fig. 8–3]. Many variations in detailed anatomy occur from species to species but these are not potentially of great physiological interest except in species with accessory respiratory organs.

The blood of fish is an extremely complex mixture separable into a fluid plasma and particulate materials, mainly blood cells. The plasma contains dissolved minerals, gases, nutrients, secretions, antibiotics, and waste products. Fish have erythrocytes, or red blood cells, as well as leukocytes, or white blood cells, the latter including several cell types having different functions. The erythrocytes of fish are nucleated and contain the respiratory pigment hemoglobin. We will here mainly consider the oxygen transport capacity of the blood, which resides primarily in the hemoglobin of the erythrocytes. The integrity of the erythrocyte is important in the blood gas relations of fish, for if the hemoglobin were to be carried directly in the plasma rather than in cells, these relations would be different (Black and Irving, 1938; Black, 1951).

The blood functions to pick up oxygen and release carbon dioxide at the gills and to release oxygen and pick up carbon dioxide in the tissues. Thus, the extent to which the hemoglobin in the erythrocytes can combine with oxygen at different tensions of oxygen and carbon dioxide and at different temperatures is important in determining the ability of the animal to maintain itself. Curves representing the extent to which a respiratory pigment is saturated with oxygen

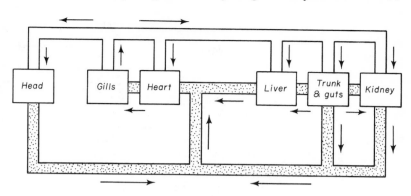

FIGURE 8-3 Diagram of the simplest arrangement of blood circulation in fish. The arrows indicate directions of blood flow, and shaded pathways indicate blood of lowered oxygen content. After Mott (1957).

at different tensions of oxygen are known as *oxygen dissociation curves*. Theoretical hyperbolic and sigmoid oxygen dissociation curves are shown in Figure 8-4. As we shall see, the form and position of such curves define characteristics of the blood that are very important to the animal. The hemoglobin in fish *loads* with oxygen at tensions existing at the gills and *unloads* at tensions existing in the tissues. If the difference between the percentages of saturation of the hemoglobin at the tensions existing at the two locations is great, as represented by the sigmoid curve (Fig. 8-4), then much of the oxygen carried

by the hemoglobin is released to the tissues. But, if the difference between the percentages of saturation at the tensions of the two locations is small, as represented by the hyperbolic curve, then little of the oxygen can be released.

For three species of fish, oxygen dissociation curves of blood samples that were equilibrated at different partial pressures of oxygen and at different temperatures are shown in Figure 8-5. Not only differences between species but also important effects of temperature are apparent. The tendency for the curves to become hyperbolic at lower

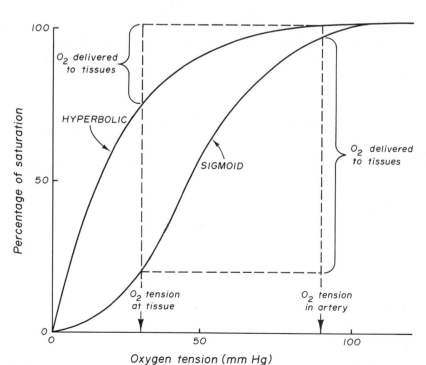

FIGURE 8-4 Conceptual diagrams of hyperbolic and sigmoid oxygen dissociation curves for two hypothetical oxygen-carrying pigments. After Florey (1966).

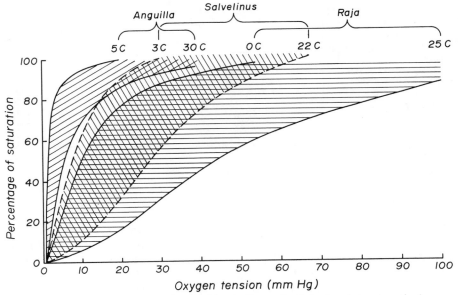

FIGURE 8-5   Oxygen dissociation curves for the blood of an eel, of the brook trout, and of a ray, at different temperatures. After Fry (1957).

temperatures could create a problem of oxygen availability at the tissues in fish maintaining high activity at these temperatures except for the effects of carbon dioxide on the form and position of dissociation curves.

As can be seen in Figure 8–6, at higher tensions of carbon dioxide higher tensions of oxygen are necessary for the hemoglobin to reach a given level of saturation. This is known as the *Bohr effect*, named after its discoverer. Since carbon dioxide tensions are low at the gills and high in the tissues, the shift in the oxygen dissociation curve caused

by carbon dioxide enhances both oxygen loading at the gills and unloading in the tissues. Krogh and Leitch (1919) first pointed out the advantages of a high Bohr effect for species maintaining high activity at low temperatures. Another effect of carbon dioxide on the dissociation curves of fish is that at higher levels of carbon dioxide the hemoglobin never becomes completely saturated with oxygen, regardless of the oxygen tension (Fig. 8–6). This was first noted by Root (1931) and has come to be known as the *Root effect*. We have been discussing

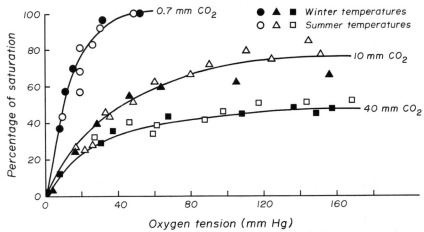

FIGURE 8–6   Influence of carbon dioxide tension on the oxygen dissociation curves of blood of brook trout (*Salvelinus fontinalis*) acclimated to winter or to summer temperatures. After Black, Kirkpatrick, and Tucker (1966).

characteristics of fish blood that have important effects on the exchange of respiratory gases. But the metabolism of the whole living animal under different environmental conditions is determined not alone by blood characteristics but also by many structural and physiological homeostatic mechanisms reacting to the environment. These have been reviewed by Fry (1957) and Hughes (1964); perhaps we can make some of them clear, even with a brief explanation.

One of the most beautifully functional adaptive structures we see in fish is the gill apparatus, operating as it does in conjunction with the branchial pump formed by the walls and floor of the oral cavity, the gill covers, and the ventral branchiostegal apparatus. Water as a medium of transport of oxygen is much heavier and contains a great deal less oxygen per unit volume than air; for solving this problem of all aquatic organisms, fish are wonderfully adapted. Though water is only intermittently taken in the mouth and expelled from beneath the gill covers, the flow of water through the gills is continuously maintained by the branchial pump, composed of a pressure pump medial to the gills and a suction pump lateral to them (Fig. 8-7). With the mouth closed with the aid of oral valves, the oral cavity is reduced in volume during the pressure pump phase. During the suction pump phase, the volume of the cavity lateral to the gills is increased with the aid of flexible extensions of the gill covers and expansion of the branchiostegal apparatus. The volume of water moved is determined by the rate and amplitude of the respiratory movements.

Pioneering studies of van Dam (1938) demonstrated how very efficient the gills of fish are in removing dissolved oxygen from respired water. Because of very small openings through which the water must pass between the gills, and because of a countercurrent system of water and blood flow, from 50 to 80 percent of the oxygen present in the water enters the blood, the percentage declining with increasing ventilation volume. Each gill arch of the four pairs of respiratory arches in bony fishes bears an anterior and a posterior column of gill filaments (Figs. 8-7, 8-8). During regular respiration, the filaments in each column are extended by muscles so that their tips touch those of the filaments on the adjacent column of the next gill arch. Dorsally and ventrally, each filament bears closely spaced transverse plates or lamellae which interdigitate with those on the filaments above and below, thus forming the small openings through which the water must pass in going from the oral cavity to the gill chambers. The lamellae are formed by respiratory epithelium of a single layer of cells and enclose blood lacunae or

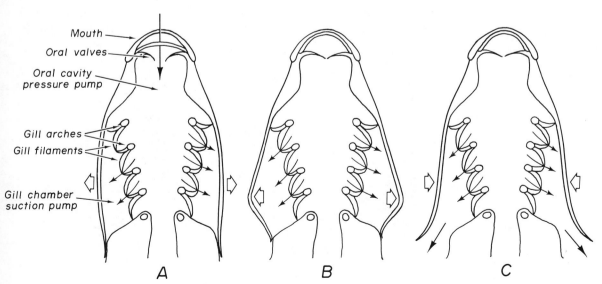

Mouth
Oral valves
Oral cavity pressure pump
Gill arches
Gill filaments
Gill chamber suction pump

A   B   C

FIGURE 8-7 Operation of the branchial pumps of fish maintaining a continuous flow of water over the gills. A. Water taken into mouth with expansion of oral cavity; flow through gills maintained by enlarging gill chamber. B. Volume of oral cavity being reduced; volume of gill chamber being enlarged. C. Water expelled from gill chamber; flow maintained by continued reduction of oral cavity volume.

spaces. Deoxygenated blood enters the lacunae from branches of the afferent branchial arteries and flows counter to the direction of water flow to leave the lacunae through branches of the efferent arteries (Fig. 8–8). The blood begins to load oxygen from water having a decreased oxygen content, but as the blood moves through the lacunae and its load increases, it receives oxygen from water having higher oxygen tensions. Thus, a steep gradient of oxygen is maintained across the respiratory epithelium at all points; hence the efficiency of oxygen uptake through the gill mechanism. In a similar manner, this mechanism favors exit of carbon dioxide from the blood.

Most species of fish, over the upper part of the range of oxygen concentrations to which they may be exposed, are able by means of homeostatic mechanisms to maintain fairly constant rates of oxygen consumption. To this extent, then, these fish are regulators. There appear to be, however, some species, such as the toad fish (*Opsanus tau*), whose oxygen consumption is dependent on oxygen concentration throughout the range of probable exposures; these species can be considered to be oxygen conformers (Fig. 8–9). We must distinguish between two general sets of conditions under which the oxygen consumption rate of a fish might be measured. If the oxygen consumption rate of a fish is measured when it is quiet and has no food, the measurement is usually called the *standard rate of oxygen consumption* (Chapter 11). If, however, the fish is forced to sustain a maximum level of activity, or if it has recently consumed much food, it is the *active rate of oxygen consumption* that is measured.

The standard metabolic rate of fish is approximately equivalent to the basal metabolic rate measured in clinical studies of man; it represents bare maintenance of bodily function. Standard metabolic rate tends to be independent of oxygen concentration down to the *incipient lethal level*, below which the fish cannot survive indefinitely (Fig. 8–9). The standard rate may, however, increase before this level is reached, mainly because increased respiratory activity is necessary at lower oxygen concentrations. The active metabolic rate is dependent on oxygen concentration up to some fairly high critical level, which Fry (1947) called the *incipient limiting level* (Fig. 8–9). Beyond this level, metabolic rate tends to be independent of oxygen concentration. Perhaps we can call these the classical ideas of relationships between metabolic rates and oxygen availability. These ideas still retain much conceptual value, but—as always with such ideas—newer information requires that we be flexible in their application. The standard metabolic rate has in some instances been shown to reach a maximum at oxygen concentrations above the incipient lethal level (Beamish, 1964a). We have considered briefly some of the circulatory and respiratory mechanisms upon which the respiration of fish depends. Now we can discuss the homeostatic feedback mechanisms that control respiration in fish.

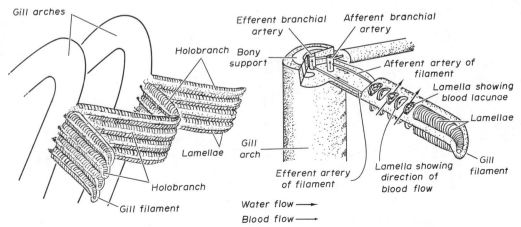

FIGURE 8–8 Diagrams showing arrangement of gill filaments on adjacent gill arches of fish and arrangement of lamellae on a gill filament. Directions of blood and water flow shown by small and large arrows. After Lagler, Bardock, and Miller (1962).

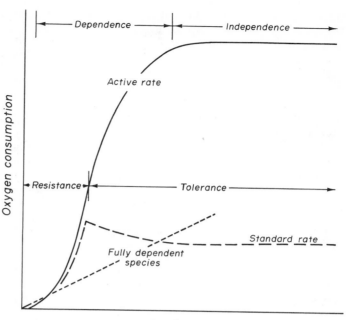

FIGURE 8–9 Generalized representation of the classic ideas concerning
relationships between the active and
standard rates of oxygen consumption
of fish and the oxygen tension of the
water. After Hughes (1964).

In his excellent paper "Fish Respiratory Homeostasis," Hughes (1964) makes clear that, even though we have a general picture of respiratory control in fish, many of the details are lacking. Figure 8–10 is his conceptual diagram of the feedback mechanisms giving fish such independence of environmental oxygen concentrations as they achieve. Homeostasis is maintained primarily by ventilation volume and secondarily by cardiac output. Oxygen lack and carbon dioxide excess act to increase ventilation volume through their effects on the respiratory center located somewhere in the upper spinal cord or lower brain (Holst, 1934). Through the cardiac center, still lower levels of oxygen and perhaps higher levels of carbon dixoide act to reduce cardiac output,

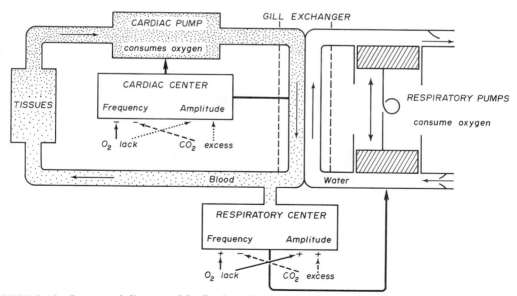

FIGURE 8–10 Conceptual diagram of feedback mechanisms for respiratory homeostasis in fish. Circulatory and respiratory pumps are involved. Nerve pathways are indicated by solid arrows for oxygen and dashed arrows for carbon dioxide. Excitatory effects are indicated by pluses and inhibitory effects by minuses. After Hughes (1964).

thus increasing passage time of blood through the gills and opportunity for oxygen uptake. This reduction of heart work also conserves oxygen.

Relationships between oxygen tension in the environment, ventilation volume, heart rate, oxygen content of the blood, and standard and active metabolism are shown in Figure 8–11. We will leave it to Hughes (1964, pp. 98–99) to explain this figure and summarize most of the knowledge of respiratory homeostasis in fish:

As yet we cannot analyse completely this complex interplay of factors. However, Fig. 12 [Fig. 8–11] summarizes data on the respiratory and cardiovascular systems from several sources, mainly experiments on the carp but some on the closely-related tench. Data of this type have not yet been obtained on any single species and hence conclusions based upon them must be tentative, yet can be of value in suggesting possible lines of future work. When such a fish is in water of high oxygen tension, the respiratory pump is able to present sufficient oxygen at the gill surface and the blood is pumped rapidly enough, to maintain a high level of saturation in the blood supplying the rest of the body. As the oxygen tension falls, however, the problem of supplying an adequate volume of oxygen becomes increasingly difficult until a point is reached when the respiratory pump is working at its maximum output and cannot pass any more oxygen over the gills. This limit is determined by the increasing proportion of the oxygen consumption that is necessary for

the respiratory muscles themselves. . . . At this limiting tension the respiration of the active fish becomes dependent and correlates with the reduction in saturation of the efferent blood. The heart rate, and presumably the cardiac output, remains constant over this whole range of tensions and only falls at a lower level. In this region of the graph, the respiratory pump provides sufficient oxygen to the gill and the same blood volume is kept fairly saturated, but later the heart rate begins to fall probably because this level of saturation cannot be maintained and this response will tend to preserve it under these conditions. The success of this regulation is evident in the afferent blood which remains at a constant level below the tension at which the heart rate begins to slow. Thus it indicates that the tissues are able to operate at a constant oxygen pressure throughout this considerable lowering of the oxygen tension in the medium, thanks to the increased ventilation volume and lowered heart rate. Clearly the amount of oxygen made available to the tissues falls as the efferent/afferent differential diminishes. Eventually, however, these homeostatic responses cannot maintain the afferent blood O_2 tension and it is at this point that even the resting level of metabolism begins to fall and it passes from the zone of tolerance to that of resistance.

Since Hughes (1964) wrote this, Holeton and Randall (1967) have published results of their fine work on the respiratory responses of rainbow trout (*Salmo gairdneri*) to decreases in environmental oxygen tensions. The general picture painted by Hughes

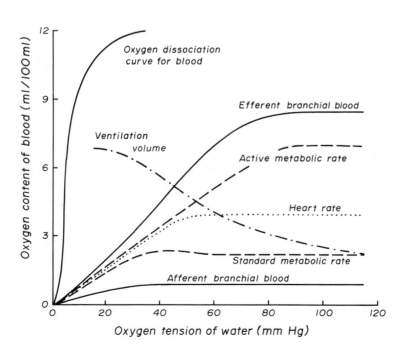

FIGURE 8–11 The effects of changing the oxygen tension of respired water on blood oxygen content, metabolic rates, ventilation volume, and heart rate of fish. Except for oxygen content of blood, scales of measurements are arbitrary. Ventilation volume and heart rate measurements on tench (*Tinca tinca*); all other measurements on carp (*Cyprinus carpio*). After Hughes (1964).

remains intact, but there is at least one nice addition. In the rainbow trout the decline in heart rate at low oxygen tensions is accompanied by a marked increase in stroke volume and some increase in cardiac output.

Circulation and Respiration in Crayfish and Lobsters

The circulatory systems of crayfish, lobsters, and other crustaceans are much more "open" than those of fish and other vertebrates. By an *open circulatory system* we mean one in which the blood passes through extensive *sinuses* and *lacunae* rather than mainly through closed, well-defined tubular vessels. Sinuses are spaces, often irregular, that are definitely bordered by membranes, whereas lacunae appear to have no such membranous borders (Maynard, 1960). Regular sinuses that clearly function only as channels for blood are sometimes called

veins, but they have no structural similarity with the veins of vertebrates.

The heart of the crayfish or the lobster has a single chamber and is located dorsally at the posterior end of the thorax. Blood enters the heart from the pericardial sinus through three pairs of lens-shaped *ostia*, or pores, having valves which close when the heart contracts (Fig. 8–12). Blood is conducted anteriorly, posteriorly, and ventrally by major arteries branching to the tissues of the organs and appendages. From these it collects in the ventral sinus opening into the gills. After passing through the gills, the blood is conducted back to the pericardial sinus and heart through well-defined sinuses.

The blood of crayfish and lobsters is perhaps more properly called *hemolymph*, for it passes into interstitial spaces and perfuses the cells, as does the lymph in vertebrates; but we will continue the more common usage. Cells present in the blood function in

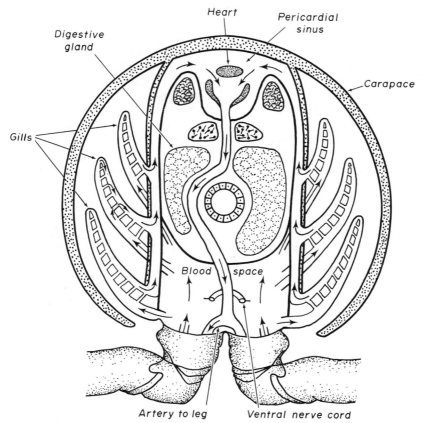

FIGURE 8–12 Cross-sectional diagram of the thorax of the lobster. Shown are pathways of blood flow from the heart to the appendages, then through the ventral sinus to the gills, from which the blood returns to the pericardial sinus and heart. After Buchsbaum (1948).

coagulation and exhibit phagocytic activity, removing harmful materials from circulation. The respiratory pigment hemocyanin present in the blood of crayfish and lobsters is carried in the fluid phase in solution rather than in special blood cells. Both temperature and carbon dioxide affect the oxygen dissociation curve of lobster blood (Wolvekamp and Waterman, 1960). Higher temperatures lead to more sigmoid curves, thus making more oxygen available at the tissues. And the blood of this animal exhibits a marked Bohr effect (Fig. 8–13).

The gills of crayfish and lobsters lie in lateral chambers enclosed by the thorax walls proper and the ventrally extending carapace (Fig. 8–12). Water enters these chambers through openings around the bases of the legs and is drawn over the gills and forward mainly by the regular respiratory movements of the second maxillae, which are adapted to form bailers. In each branchial chamber, 20 gills are present, attached to appendage bases, to articulating membranes, and to the thorax wall; but one of them is rudimentary in the lobster, as are two in the crayfish. The sinuses in the gill stems and in the many filaments are divided by septa to near their ends. Thus, blood passing through the gills from the ventral sinus enters each gill stem and each filament from one side and leaves from the other en route to the heart. Only a thin cuticle and an epidermal layer separate the blood from the water with which exchange of respiratory gases must take place.

Wolvekamp and Waterman (1960) have reviewed the rather spotty literature on metabolic rate and its control in crustacea. As incomplete as our knowledge of this subject is for fish, it is still less complete and quite unsatisfactory for crustacea, which, of course, are a more diverse group. Even the condition of the animal, whether it is in an active or a basal state, has hardly been taken into account, perhaps because in an invertebrate this presents a more difficult problem. The effects of different levels of environmental variables—even temperature and respiratory gases—on the responses of crustacea are not at all well understood.

Nevertheless, one gains the impression that crayfish and lobsters generally have somewhat less control over their metabolic rates than do fish. The lobster *Homarus gammarus* exhibits strict linear dependence of metabolic rate over the range of oxygen concentrations it is likely to encounter in its environment, the ecologically important range (Fig. 8–14). The crayfish *Cambarus bartonii* and *Procambarus clarkii*—which appear to be exceptional in having no respiratory pigment—also exhibit this dependence over a similar range (Maloeuf, 1937), though the crayfish *Astacus astacus* exhibits independence at concentrations above 25 percent of the air saturation level. There is some evidence that various species of crustacea, including crayfish and lobsters, have respiratory characteristics that are adaptive for their particular environments (Wolvekamp and Waterman, 1960), as would be expected. Such active control over metabolic rate as crayfish and lobsters may possess must reside in regulation of the flows of blood

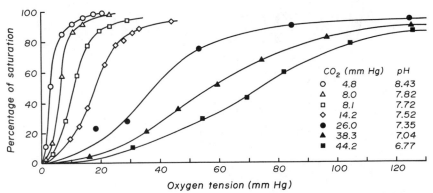

FIGURE 8–13   Oxygen dissociation curves for the blood of the lobster *Homarus gammarus* at different carbon dioxide tensions at 15 C. These curves well illustrate the Bohr effect of carbon dioxide, which enhances oxygen loading of the blood at the gills and unloading at the tissues, where carbon dioxide tensions are higher. After Wolvekamp and Waterman (1960), based on data of Spoek.

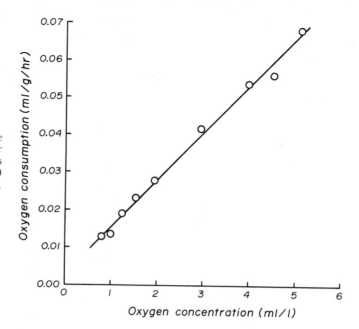

FIGURE 8-14 Linear relation between the oxygen consumption rates of the lobster *Homarus gammarus* and oxygen concentrations ranging from about 20 percent to about 100 percent of the air-saturation level at 15 C. After Thomas (1954).

and water through the gills. Increasing temperature tends to increase cardiac rate and the respiratory movements of the bailers on the second maxillae. Reduction in oxygen concentration also tends to increase the respiratory movements. The respiratory center appears to be in the subesophageal ganglion.

Circulation and Respiration in Oysters

Oysters, like crayfish and lobsters, have an open circulatory system, the blood directly bathing their organs and tissues through sinuses and lacunae rather than being retained in capillaries. Yonge (1960) provides us with a brief description of this system, Galtsoff (1964) with a very complete one. The heart, in its pericardial cavity, is composed of three chambers: two thinly walled auricles and a ventricle with thicker walls. The ventricle drives the blood through a short posterior aorta serving only the adductor muscle and the hind gut and through an anterior aorta serving the remainder of the body (Fig. 8-2). Two branches of the anterior aorta pass around the edges of the mantles; the remainder open into the sinuses. From the sinuses, blood is conducted either to the gills or to the kidney and then returned to the auricles of the heart. The sinuses present a circulatory problem that is

in part overcome by accessory hearts and pulsing radial vessels in the mantles. The blood of an oyster is a clear, almost colorless fluid. It contains two kinds of cells, one of which is phagocytic. Oyster blood has no respiratory pigment, neither in cells nor in solution in its fluid component.

We described in some detail in an earlier section the ctenidia, or gills, of oysters and the ciliary systems moving water through these animals. In oysters, the ctenidia have become highly specialized feeding mechanisms. Their relative importance in the exchange of respiratory gases has not been clearly established; the mantles too are involved. Galtsoff (1964, pp. 200–201) suggests that the mantles of the oyster are probably much less important in respiration than are the ctenidia. According to Korringa (1952, p. 275), Pederson (1947) considers the mantles to have the primary role in respiration. Ciliary activity and heart rate tend to increase with increasing temperature and thus aid in meeting the increased oxygen requirements of this poikilothermic animal at higher temperatures.

Because of the large volume of water an oyster must filter to obtain its food, relatively little of the dissolved oxygen in this water at normal concentrations need be removed to meet the small metabolic demands of such a sedentary organism. The oxygen consumption of an oyster appears to become independent of environmental concentrations when these are above some critical level (Galtsoff and Whipple, 1930; Gaarder and Eliassen, 1955).

With their valves closed and no external source of oxygen, oysters can survive for months at moderate to low temperatures after being removed from water, a capability of great importance to the oyster industry. This capability can have survival value in nature, where the oyster can, by closing its valves, isolate itself from temporarily intolerable environmental conditions. But even with its valves closed, the oyster's metabolism must proceed at a low rate if it is to survive. This is made possible by anaerobic glycolysis, the weak organic acid formed being neutralized by calcium carbonate from the valves. Under these conditions, ciliary activity ceases; it is no longer needed for feeding or respiration.

EXCRETION, OSMOREGULATION, AND IONIC REGULATION

Excretion, Osmoregulation, and Ionic Regulation in Fish

Nutrient materials, oxygen to utilize them, and water as a solvent and carrier are perhaps the first prerequisites of animal life; so now, in our discussion of morphological and physiological adaptations, we come to water. Being terrestrial animals ourselves, we have some appreciation of the problems other terrestrial animals face in maintaining suitable amounts of water within their bodies. We may not all appreciate that most aquatic animals have very similar problems. Depending on where and how species have evolved, physiological problems of water balance have been solved in different ways. Some aquatic animals are continually threatened with internal desiccation, others with internal flooding.

A *semipermeable membrane* is permeable to water but impermeable to solutes. Biological membranes only imperfectly approximate semipermeable membranes, for they to some extent pass solutes. But, as they are relatively more permeable to water than to solutes, biological membranes behave much like semipermeable ones. If the solute concentration is higher on one side of a semipermeable membrane than on the other, there will be a diffusion or net flux of water through the membrane to the side having the higher concentration of solute; this net flux of water is called *osmosis. Osmotic concentration,* the concentration of osmotically active particles, is approximately proportional to the depression of the freezing point of the solvent occasioned by the presence of these particles; it is often measured indirectly as freezint point depression (ΔF).

The *osmotic pressure* of a solution can be measured and expressed in terms of hydrostatic pressure; the movement of water across a semipermeable membrane is proportional to the difference between the osmotic pressures of solutions on the opposite sides. Florey (1966), who provides us with an excellent and much more detailed discussion of these and related matters, emphasizes the great osmotic pressures—many atmospheres—that would develop in some aquatic animals but for various adaptive mechanisms. Still, many animals do maintain by *osmoregulation* an internal osmotic concentration either higher (*hyperosmotic*) or lower (*hypo-osmotic*) than that of their external medium. Even when internal and external osmotic concentrations are equal (*isosmotic*), the concentrations of particular ions internally and externally are different because of *ionic regulation,* which is essential to life. Osmotic concentration and the concentrations of particular ions represent different problems of aquatic organisms, but osmoregulation and ionic regulation are not functionally separable. Excretion of water and ions, for example, usually contributes to the solution of both problems.

Fish are osmoregulators: they maintain their internal fluids at osmotic concentrations different from those of their external medium. The internal fluids of marine bony fish are hypo-osmotic to the surrounding saline waters; in consequence, these fish continuously lose water across exposed permeable membranes. Freshwater fish, on the other hand, maintain their body fluids hyperosmotic to their external medium and must deal with water flooding in through their membranes. The interrelated functions of excretion, osmoregulation, and ionic regulation depend primarily on the kidneys and gills in fish. Virginia Black (1957) has reviewed the more important relevant literature.

Most of the nitrogenous wastes of bony

fish are excreted through the gills, not through the kidneys as in terrestrial vertebrates. And most of this is in the form of ammonia, though some urea and perhaps other minor nitrogenous wastes are also passed through the gills. Creatine is the principal nitrogenous waste excreted through the kidneys of fish; usually in lesser amounts, ammonia, uric acid, creatinine, and other end products of nitrogen metabolism are also present in the urine. To this list, we must add trimethylamine oxide for marine fish, a material they apparently obtain from their food. Probably the most important function of the kidney of freshwater fish is osmoregulation, for it is through the production of a copious dilute urine that these fish remove from their body fluids the water flooding in primarily through their gills.

The typical vertebrate kidney is composed of many units called nephrons (Fig. 8–15). Each nephron consists of an arteriole mass (glomerulus) enclosed in a capsule (Bowman's capsule), the coelomic space of which opens into the lumen of a tubule. This tubule is differentiated into a neck segment, a proximal convoluted segment, an intermediate segment, and a distal convoluted segment; it connects with an initial urine collecting tubule. Urine formation begins by filtration of water and some solutes into the capsule, the blood in the glomerulus being

under arterial pressure. From blood vessels enmeshing the tubule of the nephron, solutes may be secreted into the tubule; the blood may pick up water and other solutes from the tubule. The kidney appears to have arisen in freshwater protovertebrates as a means of getting rid of water, and this is its primary function in freshwater fish. The return of salts from the tubules to the blood of freshwater fish is a salt conservation measure. But marine fish must conserve water and eliminate salts; some of these fish have neither glomerular filtration units nor the distal convoluted segments of the tubules from where salts re-enter the blood of freshwater fish. Thus, in aglomerular kidneys, urine is formed essentially through secretion into the tubules.

To replace the water they are losing to their hyperosmotic external medium, marine fish continually swallow water. But if they are to maintain their body fluids hypo-osmotic to the water of the sea, they must eliminate the salts they ingest with water and food. From the salts ingested, sodium, potassium, and chloride ions are absorbed from the gut; only a small fraction of the amounts absorbed can be eliminated in the small volume of urine produced. Calcium, magnesium, and sulfate ions become concentrated in the residual alkaline intestinal fluid; the cations are eliminated as insoluble oxides or hydroxides. The small amounts of

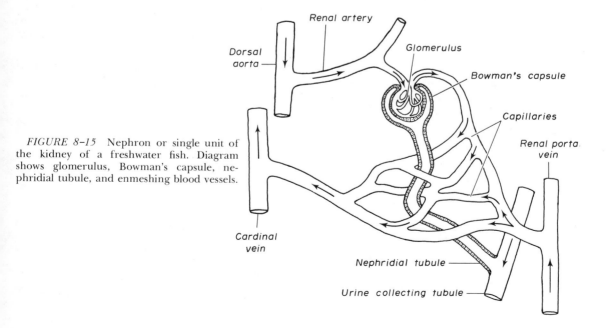

FIGURE 8–15 Nephron or single unit of the kidney of a freshwater fish. Diagram shows glomerulus, Bowman's capsule, nephridial tubule, and enmeshing blood vessels.

these absorbed are eliminated in the urine. But what about the sodium, potassium, and chloride ions?

Smith (1930) first suggested that the gills were responsible for excretion of salts; and Ancel Keys (1931) presented evidence for active secretion of salts by special cells located in the gills of the European eel (*Anguilla anguilla*). Since then, many fine studies have been conducted on the structure and function of these *chloride cells*; some of these studies have been reviewed by Black (1957). This type of cell, "a columnar, acidophilic cell found in the gill filament and localized on the side of the filament supplied with afferent blood" (Copeland, 1948, p. 216), now appears to be responsible for both excretion and absorption of ions against gradients in salt and fresh water (Copeland, 1948, 1950). August Krogh (1939), in his classic *Osmotic Regulation in Aquatic Animals*, reviews the work in which he first demonstrated active absorption of chloride, bromide, sodium, and, in one case, calcium ions by the gills of freshwater fish. The importance of this to these fish—which must replace salts lost in their copious, though dilute, urine—is obvious.

Thus, primarily through the gills and kidneys, excretion, osmoregulation, and ionic regulation are accomplished in fish, these structures being adapted to their functions in different ways for freshwater and marine life (Fig. 8–16). Though fish are osmoregulators, some can survive and reproduce only within narrow ranges of salinity and are considered *stenohaline*; others, particularly estuarine species, can succeed within wide ranges of salinity and are considered *euryhaline*. Still others spend different phases of their lives in the marine and freshwater environments. Salmon reproduce in freshwater and spend most of their adult life in the sea; they are termed *anadromous*. Fish, like the European eel and the American eel (*Anguilla rostrata*), that reproduce in the sea and spend a part of their lives in freshwater are called *catadromous*. Morphological and physiological changes permit such drastic, and successful, changes in environment. Chloride secreting cells appear in the gills of juvenile salmon at about the time they begin their journey to the sea.

Excretion, Osmoregulation, and Ionic Regulation in Crayfish and Lobsters

Ionic regulation, as it must be, is a universal phenomenon among crustaceans; osmoregulation occurs in all freshwater forms; they maintain their body fluids hyperosmotic to their environment; but most marine crustaceans are isosmotic with the sea (Robertson, 1960). Thus we are not surprised to find the freshwater crayfish to be an osmoregulator as well as an ionic regulator; nor are we surprised to find the lobster living in the sea to be isosmotic with its environment. But when the lobster moves into the brackish water of estuaries, it maintains its body fluid hyperosmotic to the external medium; it becomes an osmoregulator.

The functions of excretion, osmoregulation, and ionic regulation in crayfish and lobsters appear to depend mainly on the *antennal glands* and the gills. Thus, special excretory structures and gills function to solve nitrogenous waste, water, and ionic problems in these crustaceans, much as they do in fish. The antennal glands—variously termed renal organs, kidneys, or green glands—are paired structures situated near the bases of

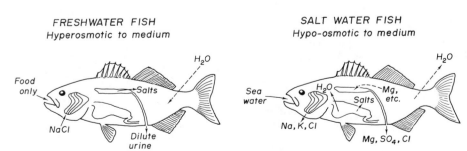

FIGURE 8–16   Schematic representation of the main paths of ion and water movement in osmoregulation of freshwater and salt water fishes. Active transport is indicated by solid arrows and passive transport by broken arrows. After Prosser and Brown (1961).

the antennae of crayfish and lobsters. Each antennal gland begins with a coelomic sac penetrated by blood vessels and sinuses. This is followed by a green tubular labyrinth connecting to a long tubule, which opens into a bladder (Fig. 8–17). Through a pore on the basal segment of the first antenna, the bladder empties to the external environment. With the exception of the bladder, these structures are well supplied with blood from vessels opening into sinuses or lacunae. Thus, each antennal gland is reminiscent of a single vertebrate nephron, which we described for fish. This similarity might cause one to suspect that the urine of these crustaceans is initially produced as an ultrafiltrate of the blood. It is not, however, clear that this is so; indeed, morphological and physiological evidence better supports the idea that the urine of crustaceans is produced by secretion (Parry, 1960).

The freshwater crayfish, maintaining itself hyperosmotic to its environment, cannot afford to lose certain ions through antennal gland secretion. There is evidence that such ions, chloride being an example, are actively returned to the blood from the nephridial canal (Fig. 8–17). The final product of the antennal glands is very hypo-osmotic to the blood and has little flow, an excellent adaptation for salt conservation, but seemingly a poor one for the crayfish to rid itself of the water flooding into its body fluids (Parry, 1960). Lobsters living in the sea, where they are isosmotic with their environment, also have little urine flow, but they have no excess water of which to rid themselves; their urine is isosmotic with their blood. Still, the ionic composition of their urine differs from that of their blood, as it does in crayfish. These animals are ionic regulators; their antennal glands function to maintain the ionic composition of their internal fluids different from that of the environment.

Lobsters maintain in their blood higher concentrations of sodium, potassium, and calcium and lower concentrations of magnesium and sulfate than occur in the sea; their chloride concentrations are nearly in equilibrium with the environment. Crayfish, being hyperosmotic to their environment, maintain higher concentrations of all of these ions in their blood than occur in their external medium. Analyses of the antennal gland product suggest conservation of these ions by means of this structure, either through secretion mainly of water into the nephridial canal or resorption of ions into the blood (Fig. 8–17). In the lobster, however, the antennal glands appear to eliminate magnesium and sulfate more readily than other ions. Ions lost through the

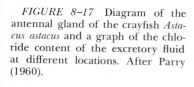

FIGURE 8–17 Diagram of the antennal gland of the crayfish *Astacus astacus* and a graph of the chloride content of the excretory fluid at different locations. After Parry (1960).

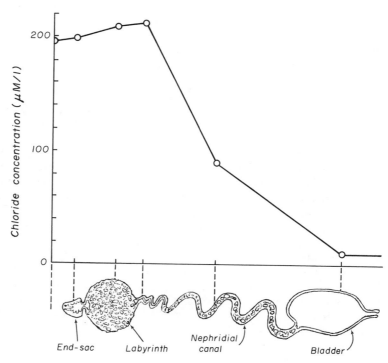

antennal gland secretions of both these animals must be replaced either from their food or by uptake across exposed membranes. There is considerable evidence that uptake occurs through the gills, either actively or passively, the manner depending on concentration gradients.

Excretion of nitrogenous wastes does not appear to be an important function of the antennal glands (Parry, 1960). The principal end product of nitrogen metabolism in crayfish, lobsters, and other crustaceans is ammonia; urea and uric acid are produced in much smaller quantities. Small amounts of all of these may occur in the urine, but ammonia and urea are excreted mainly through the gills. Uric acid may be deposited in the exoskeleton and eliminated when this is cast in the growth process.

Excretion, Osmoregulation, and Ionic Regulation in Oysters

Many species of oysters are able to survive, reproduce, and grow in estuaries where salinities range widely, not only seasonally but with the tidal cycle; they are euryhaline. The oyster is a widely recognized and studied invertebrate. This is not surprising, for it is a delicacy of great commercial importance over much of the world. Yet little is known of how it solves its problems of nitrogenous waste disposal and water and salt balance; few biologists have investigated these problems. Biologists most often study animals of particular biological interest or economic importance. These problems of oysters certainly qualify for study. But biologists also tend to select animals that are convenient for their studies; and the oyster, with its body of soft and intertwined organs enclosed in hard valves, is not an easy animal to study, particularly in relation to nitrogenous waste, osmotic, and ionic problems. Drawing as we must from knowledge of other molluscs and what little has been done on the oyster, we can perhaps reasonably conclude that the extracellular body fluids of the oyster are usually in osmotic equilibrium with the environment. There must, however, be osmotic control within its cells. And the oyster most certainly has powers of ionic regulation, for oysters are known to greatly concentrate mineral ions from their estuarine and marine environments.

The oyster, like other molluscs, has a "kidney" or renal organ that functions in the excretion of waste products and water and also in ionic regulation. This organ, in addition to having a glandular part (organ of Bojanus), appears to involve the heart and pericardial cavity as a filtration unit (Martin and Harrison, 1966). Urine may be formed initially as an ultrafiltrate of the blood passing through membranes of the heart into the pericardium (Fig. 8–2). Two tubes penetrate the pericardial sac and lead into the chamber of the renal organ, which finally empties through two ducts to the exhalant chamber of the mantle cavity. The glandular part of the renal organ, which is supplied mainly with venous blood, probably functions both in secreting substances into and in reabsorbing substances from the fluid in the renal tubules and chamber. Thus, the oyster possesses a system apparently capable of functioning not only in the elimination of water, wastes, and ions but also in the conservation of vital materials. Ammonia is probably the principal nitrogenous waste excreted by oysters.

The bay mussel (*Mytilus edulis*), like many species of oysters, is an estuarine animal adapted to widely ranging salinities. The osmotic concentration of the blood of the bay mussel follows quite closely changes in the osmotic concentration or salinity of its environment (Krogh, 1939; Prosser and Brown, 1961). With declining salinities, the blood is diluted by the influx of water, and its osmotic concentration is further reduced by excretion of chloride. With increasing salinities, osmotic equilibrium with the environment is maintained by a reversal of these processes. Insofar as this equilibrium is maintained by changes in water content, the volume and weight of the mussel change; it is an animal with imperfect *volume regulation*. Considering the small amount of relevant evidence on oysters, we can perhaps most safely assume that they too possess little ability to regulate the osmotic concentration of their blood or the volume of their bodies (Fingerman and Fairbanks, 1956). Their adaptive mechanisms probably resemble those of the bay mussel. Strips of heart muscle isolated from this animal and from the European oyster (*Ostrea edulis*) continued to function when placed in dilutions of sea water greater than those to which these animals are normally exposed (Pilgrim,

1953). There must be in these animals some kind of adjustment preventing the development of destructive osmotic pressures within their cells. Florkin (1966) has reviewed evidence that this could involve reductions in cellular concentrations of sodium, chloride, and amino acids. But to return to the homeostasis of the whole animal, mussels and oysters can isolate themselves from their environment by tight closure of their valves. By this adaptive mechanism, they can survive exposure to very low salinities for considerable periods of time.

THE PROBLEM OF TEMPERATURE

"Fish of some species or another have been observed to exist at temperatures from −2 to 40 C" (Fry, 1964, p. 715). These temperatures approximate the extremes to be found in aquatic environments. Of course, no single body of water, fresh or marine, is likely to reach both limits, and few species are likely to tolerate these temperature extremes. Nevertheless, seasonally and daily, particular species and individuals are exposed to temperatures that fluctuate widely. These species and individuals are morphologically and physiologically adapted to survive, reproduce, and grow in the face of temperature fluctuations. But in some aquatic environments, man's activities are changing the limits of temperature variation to levels for which many of the native species are not adapted.

"Temperature, perhaps more than other environmental factors, has multiple and diverse effects on living organisms" (Prosser, 1950, p. 341). Homeothermic animals—warm-blooded vertebrates—have solved some of their temperature problems by temperature regulation. Poikilothermic animals—such as fish, whose body temperatures follow closely the temperature of their medium—have solved their temperature problems in different ways. The effects of temperature on the life of a poikilotherm are profound. From enzymatic reactions through hormonal and nervous control to digestion, respiration, and osmoregulation, and to all aspects of its performance and behavior, the poikilotherm is influenced by temperature. Still, in their bodily functions, poikilotherms are often able to maintain a certain independence of environmental temperature. This is achieved by homeostatic mechanisms operating at all levels of organization of the individual: subcellular, cellular, tissue, organ system, and intact living organism. Not merely survival but reproduction and growth as well must be favored by these mechanisms, if they are to contribute to the success of the individual and its population.

High and low temperatures that are lethal to individuals of particular genetic constitutions set, of course, ultimate limits to their existence; but it is problematical just how often such ultimate limits determine the distribution and abundance of populations. Probably more often, the distribution and abundance of populations in nature are determined not by direct lethal effects of temperature but primarily by temperature interacting with other environmental factors either to favor or not to favor reproduction and growth. Nevertheless, some initial understanding of the role of temperature in the life and death of animals can be attained through consideration of lethal levels, particularly as these are influenced by the temperature acclimation or history of the animals.

In relation to survival of the individual animal, Fry, Hart, and Walker (1946) divided—at the upper and lower incipient lethal levels—the total range of temperature into upper and lower *zones of resistance* adjoining a central *zone of tolerance. Incipient lethal levels* of temperature are those levels that will eventually cause the death of a stated fraction of the test animals, usually 50 percent. The incipient lethal levels for an individual depend not only on its species and particular genetic constitution but also on its acclimation to temperature (Fig. 8–18). Within the zone of tolerance, an animal usually will not die from the effects of temperature alone. The period of time an animal can live at a temperature in either of the zones of resistance is a function of how far that temperature is beyond the relevant incipient lethal level (Fig. 8–19).

The causes of death at temperatures above the freezing level are not clearly understood. At low lethal temperatures, death may ensue from loss of nervous, circulatory, respiratory, or osmoregulatory performance (Fisher, 1958; Fry, 1964). At high lethal temperatures, enzyme inactivation, irreversible

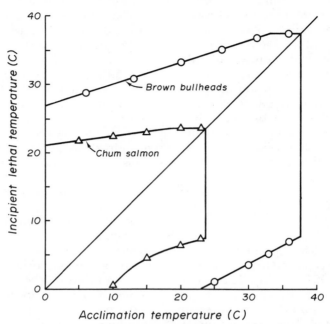

FIGURE 8–18   The zones of tolerance of brown bullheads (*Ameiurus nebulosus*) and chum salmon (*Oncorhynchus keta*) as delimited by incipient lethal temperature and influenced by acclimation temperature. After Bret (1956).

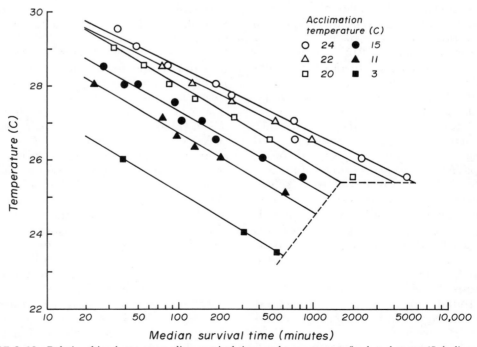

FIGURE 8–19   Relationships between median survival time and temperature for brook trout (*Salvelinus fontinalis*) acclimated to different temperatures. After Fry, Hart, and Walker (1946).

protein coagulation, or changes in lipids and cell membranes would certainly lead to death. Still, it is not usually clear whether death ensues directly from these causes or from more general loss of physiological performance, as in the case of cold death. "Whatever the mechanisms of heat death, they are certainly multiple and they are subject to alteration, as judged by resistance to heat" (Prosser, 1950, p. 346).

Through acclimation, an animal is able to adapt to changes in temperature. Animals acclimated or adapted to relatively low or relatively high temperatures may be able to survive at greater extremes of temperature. Precht (1958) termed this *resistance adaptation* and distinguished it from *capacity adaptation*. Within the range of more ordinarily occurring temperatures, animals are favored in all they do by such independence from changes in environmental temperature as they can achieve in bodily functions; acclimation making this possible is capacity adaptation. In a sense, then, resistance adaptation extends the range of mere survival; capacity adaptation favors a complete and successful life. The extent to which the mechanisms of these two kinds of adaptation differ is not at all clear, for our understanding of the processes of acclimation, in spite of the fine contributions of many biologists, is still very inadequate (Fisher, 1958; Fry, 1964; Precht, 1958). Changes in proteins and lipids may be necessary to increase resistance to heat death; but adjustments in the nervous, circulatory, and respiratory systems going beyond these changes may also be required. Enzymes and hormones are undoubtedly involved. Capacity adaptation involves the general systems of the body, but the basis of adjustment may extend from enzymes or homones through tissues and organs to the integrated activites of all systems. Acclimation to low temperatures is usually very slow, whereas acclimation to temperatures in the higher ranges may keep pace with environmental changes over a few days, a real advantage to a poikilothermic animal.

Particularly in poikilotherms, the rates of many biological processes including digestion, circulation, and respiration increase with increase in environmental temperature. This is not at all surprising; the rates of chemical reactions are dependent on temperature, and biological processes are ultimately dependent on the rates of enzymatic reactions. The magnitude of the effect of temperature increase on the rate of a process is often expressed as the Q_{10} of the process, the factor by which the rate is increased by a temperature increase of 10 C. This parameter can be estimated from the equation

$$Q_{10} = (K_1/K_2)^{\frac{10}{t_1 - t_2}}$$

where K_1 and K_2 are the rates corresponding to temperatures t_1 and t_2. Values for Q_{10} usually lie between 2 and 3, but they are very much dependent on how near the animal is to its normal temperature range and on its thermal history. The rates of respiratory processes, as measured by oxygen consumption or metabolic rate, are among those sensitive to temperature.

Some stability of metabolic rate is advantageous to an animal over its normal temperature range. "Over the range from 6° C. to 20° C. which may be taken as roughly representing the range best suited to the trout the standard metabolism of that species is approximately doubled, while over the same temperature range the metabolism of the goldfish which is adapted to considerably higher temperatures, is increased approximately five times" (Fry, 1947, p. 29).

Acclimation greatly influences the effects of temperature on rates of biological processes, and it is through this that the poikilotherm gains much of the independence of environmental temperature it can achieve. Animals adapted to low temperatures can maintain bodily functions at much lower temperatures than can warm-adapted animals, which can maintain their functions at much higher temperatures (Fig. 8–20). But either of these groups is likely to exhibit a high Q_{10}, whereas animals acclimated to the different temperatures in their normal range at which the rates of a process are measured may exhibit a much lower Q_{10} (Fig. 8–20). "In general the biological significance of a capacity adaptation . . . [resembling solid curve in Fig. 8–20] lies in the fact that this enables the animal to keep the vital functions as much as possible at a constant level independent of environmental temperature" (Precht, 1958, p. 56). By no means, however, do different biological processes and different species respond in the same manner to temperature acclimation. Precht (1958) and Prosser (1958a) have clas-

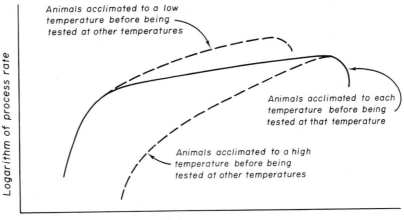

FIGURE 8-20 Hypothetical curves showing the relationships between the rates of biological processes and temperature when animals are cold adapted, when they are warm adapted, and when they are perhaps more naturally adapted to the temperatures at which the rates are measured. After Bullock (1955).

sified and discussed the possible responses. Temperature acclimation within the normal range of fish tends to stabilize their metabolic rates.

Whatever may be the acclimation changes in proteins, lipids, cell membranes, enzymes, and biological processes, it is only as these changes contribute to survival, reproduction, and growth—the performance of the whole organism—that they are adaptive. The effects of temperature acclimation on the survival of fish have been extensively studied; not so its effects on reproduction and growth. But there have been interesting studies on the effects of temperature acclimation on the swimming performance of fish, a kind of performance necessary to their survival, reproduction, and growth.

Temperature acclimation has profound influence on the swimming performance of fish at different temperatures (Fig. 8–21). But whereas the metabolic rates of fish acclimated to different test temperatures tend to be similar, no such similarity exists in swimming performance. Obviously, then, knowledge of the metabolic rate of the

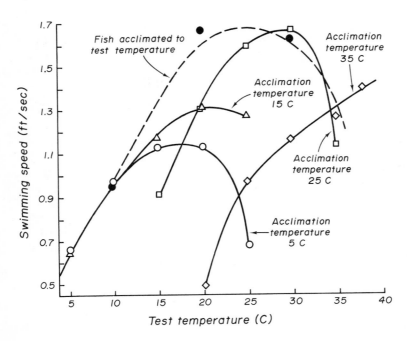

FIGURE 8-21 Sustained swimming speeds of goldfish (*Carassius auratus*) when acclimated to particular temperatures and tested at other temperatures and when acclimated to the test temperatures. After Fry and Hart (1948).

resting animal is a poor basis for predicting the influence of temperature acclimation on the swimming capacity provided the intact, living organism by the integrated functioning of all its systems. Knowledge of the metabolic rates of particular tissues or organs would be even a poorer basis. Nevertheless, if we are ever to understand the influences of temperature or any other environmental factor on the whole organism, we must study not only its integrated performance but also the activities of each of its systems. But we must be hesitant to predict from knowledge of separate systems the performance of the whole organism.

Animals are not usually passive in the face of environmental change. Through their behavior, insofar as it is adaptive, they tend to put themselves in the most favorable situation possible. Temperature change is no exception. Fish and other animals perceive differences in temperature, and they respond to these differences. Both perception and response may be influenced by temperature acclimation. We will consider the influence of temperature acclimation on the thermal preferenda of fish in Chapter 12. Even in the oyster, an animal some might think to be devoid of behavior, temperature can have profound effects through its influences on shell movements and feeding activity.

THE PROBLEM OF TOXIC SUBSTANCES

Through her book *Silent Spring*, Rachel Carson (1962) alerted a nation to the growing problem of toxic substances in its environment. True, from his earliest origins, man has lived in the presence of toxic substances, particularly plant products. The drugs man has used and will continue to use are frequently toxic. And, since the Industrial Revolution, many of the waters of the earth have borne noxious waste materials. But Miss Carson was alarmed mainly by the growing use of synthetic pesticides in agriculture, forestry, and wherever the control of animals and plants seems necessary. There is need for concern, because all life, terrestrial and aquatic, not just the species we seek to control, is faced with the problem of toxic substances.

Before man's activities became important, animals living in aquatic environments were confronted with toxic substances leached from surrounding rock and soil, arising from the decomposition of organic matter, or given off by plants and animals. Internally, every animal must contend with the toxic products of its own metabolism of proteins, fats, and carbohydrates. In each species, there have evolved mechanisms for handling the toxic substances of external and internal origin with which it has historically been faced. These same mechanisms are able to handle, to a greater or lesser extent, many of the toxic materials man has introduced into the environment. And, through recent natural selection, the ability of some species to withstand certain of these materials has increased. Thus, in a sense, animals are preadapted to tolerate at least certain levels of substances new to their species; and further genetic adaptation of their populations may begin with first exposure.

All of this is not to say that somehow all animals will survive, much less reproduce and grow, in spite of the toxic substances man introduces into their environments. They will not; for, after all, man has developed many of these substances to prevent the survival and reproduction of many species, and he has been reasonably successful in this. But for each substance, there is some level, however low, which an animal at any one time can tolerate. What this level is depends on the mechanisms the animal has available for handling the substance. This level is by no means constant; it varies with conditions in the environment and the state of the animal. Knowledge of the mechanisms with which animals handle toxic substances is necessary if we are to understand their tolerance of these substances. We must also come to understand the actions by which toxicants bring about their physiological effects, and how and to what extent physiological changes are reflected not only in survival but also in reproduction and growth.

Those of us who are aquatic biologists interested in the problem of toxic substances have yet to exploit the knowledge of human pharmacology and toxicology. We have emphasized lethal effects, even though ultimately we must be concerned about effects on reproduction and growth. Those who have studied man have had to be concerned with both lethal and sublethal effects and all the processes by which these come about. Their ways of thinking and observing have much to contribute to our studies. To those interested in the toxicology of aquatic organisms, we heartily recommend the reading of the more relevant chapters of a good pharmacology text such as Goodman and Gilman (1965).

Absorption, Distribution,
Biotransformation, and Excretion

For a toxicant to influence the survival, reproduction, or growth of an animal, it must attain an effective concentration at the site or sites of its action. Certainly for an aquatic animal, the attainment of this effective concentration will be some function of the concentration in the water and the amount of the toxicant in the animal's food. But it will also depend on the rate of absorption of the toxicant, its distribution about the body, any binding or localization in tissues, inactivation through biotransformation, and ultimate excretion (Fig. 8–22).

The absorption, distribution, biotransformation, and excretion of toxic substances all involve, directly or indirectly, passage of the substances across *plasma membranes*. This fascinating subject is adequately treated in most general physiology texts. Briefly, individual cells, their nuclei, and their mitochondria have simple plasma membranes, whereas respiratory and intestinal surfaces are covered by membranes composed of a single layer of cells. Exposed surfaces of multicellular animals are usually protected by relatively impermeable layers of cells, scales, shell, or chitin. Plasma membranes are of the order of 100 Angstroms thick and are perforated by water-filled pores, which vary in diameter from 4 to 40 Å. The membrane consists of a lipid sheet two molecules in thickness bounded on both sides by protein. Toxicants and other substances move across plasma membranes either by *passive diffusion* or *active transport*. Lipid soluble substances dissolve in the lipid membrane and their rate of transfer is proportional to the concentration gradient. Water soluble substances of proper particle size and ionization state pass through the pores along with water moved by hydrostatic or osmotic pressure, a process known as *filtration*. Active transport mechanisms requiring the expenditure of energy are able to move certain molecules across membranes against electrochemical gradients. Membrane components known as carriers are believed to form complexes with these molecules at one side of the membrane, then move to the other side, where the molecules are released.

The absorption of toxic substances by aquatic animals may be either directly from their aquatic medium or indirectly through the food they consume. The relatively extensive and permeable respiratory surfaces are perhaps usually the most important route of direct uptake. These and the intestinal surfaces have abundant internal circulation, fa-

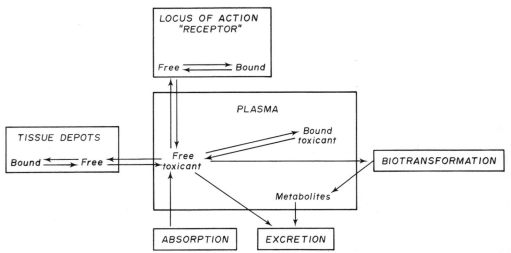

FIGURE 8–22 Processes influencing the amount of a toxicant reaching the site of its primary toxic action. After Fingl and Woodbury (1965).

cilitating absorption. Not only the concentration of a toxic substance in the external environment or in the intestine but complexes it may form with other substances can be important in determining its rate of absorption. Thus, the chemistry of the aqueous medium or of the intestinal contents—their pH and the presence of interacting substances—may determine whether or not given concentrations lead to toxic effects. The extent of fat and water solubility, size, shape, and state of ionization of a molecule are of great importance in determining its absorption.

The body may be viewed as having various fluid compartments: plasma; interstitial fluid; transcellular fluids in lumens, chambers, or bladders; and cellular fluids. To bring about their effects, toxicants must pass through one or more of these compartments. Distribution about the body may, then, be uneven owing to the differing permeability of membranes. Further, toxicants may become reversibly bound to plasma proteins, thus providing one *storage depot*. And if a toxicant is lipid soluble, considerable storage may occur in fat, from which it can be returned to body fluids and redistributed to sites of toxic action. Other storage depots can be in particular cell types, tissues, or organs, frequently the liver. Storage may occur at the locus of action, but most often it occurs at some other location.

For eliminating foreign substances from their bodies, terrestrial animals apparently need more specialized mechanisms than do most aquatic animals. Some of the materials deriving from their food, like most drugs and pesticides, are lipid soluble. Fish can excrete such substances into the open water by means of passive diffusion through the lipoidal membranes of their gills. Indeed, it will be remembered from our earlier discussion, the kidney of fish is relatively unimportant in the excretion of metabolic wastes. For the terrestrial vertebrate, the kidney must suffice; it has no external surface through which wastes can be passed in solution. But lipid soluble substances in the glomerular filtrate are reabsorbed by the kidney tubules. Thus, if the terrestrial vertebrate is to eliminate such substances, whatever their origin, they must be made more water soluble.

There occur in the *microsomes* of liver cells of terrestrial vertebrates, and in the microsomes of other cells in insects, enzymes that

are capable of altering the structure of toxic substances. Such *biotransformations* may be either synthetic or nonsynthetic. The nonsynthetic transformations involve oxidation, reduction, or hydrolysis. The synthetic transformations, sometimes called conjugations, involve the coupling of a toxicant or its metabolite with a carbohydrate, an amino acid, or a derivative of one of these. Williams (1962) has provided a pharmacological classification of the biochemical reactions involved. Some biotransformations involve first a nonsynthetic reaction and then a conjugation of the metabolite of the reaction; sometimes only one or the other step is involved. Most of these transformations result in *detoxification*, but some *activate* a substance, render it more toxic. Transformation of lipid soluble toxic substances to more water soluble metabolites enhances their excretion by reducing their binding to plasma and tissue proteins and their storage in fat as well as by reducing their ability to cross cell membranes. Reduced ability of metabolites of toxicants to cross cell membranes also decreases the amount of toxic materials reaching sites of toxic action within cells.

Biotransformation of toxic substances by aquatic organisms including fish has received little study. Brodie and Maickel (1962), in their valuable contribution entitled "Comparative Biochemistry of Drug Metabolism," have shown that important oxidative and conjugating mechanisms present in the liver microsomes of mammals, birds, reptiles, and terrestrial amphibians are absent in fish and aquatic amphibians. These aquatic animals eliminate lipid soluble toxicants through the lipoidal membranes of their gills or skin. Crayfish and lobsters—whose gills are covered by a thin chitinous exoskeleton which is relatively impermeable to such substances—have enzymes in the hepatopancreas capable of biotransformation of toxic substances.

Brodie and Maickel (1962, pp. 322–323) summarize all of this very well:

Needham, in his studies on morphogenesis, points out that the problems of water conservation and the disposal of the end products of protein metabolism are closely associated. In fish, protein metabolism ends up mainly as ammonia. This substance, though potentially toxic, causes no trouble to fish since it rapidly passes through the gills into a limitless volume of water. In terrestrial animals a limited water supply would make the ammonia a deadly side effect of eating. This problem was solved by the evolutionary development of mechanisms to convert ammonia to the less toxic urea or uric acid.

The evolution of biochemical processes to metabolize foreign compounds may be considered in a similar vein. Organic chemists have shown us that plant life contains a profusion of organic compounds, many of them lipid-soluble and not readily excretable by the kidney. Again, these substances are no problem to fish since they rapidly diffuse through the lipoidal gills. To live on land, however, animals had to develop exquisitely non-specific enzyme systems to convert these compounds to lipid-insoluble derivatives, excretable by the kidney. Without these enzymes the foreign substances would gradually accumulate to toxic levels and life would cease.

Mechanisms of Toxicant Action

Most toxicants are believed to produce their effects by combining with enzymes, cell membranes, or other specialized functional components of cells. Such interactions between cell components and toxic substances are presumed to induce other biochemical changes and the physiological effects of toxicants. Pharmacologists distinguish between *drug action* and *drug effects* in a manner that contributes to clarity of thinking. Drug or toxicant action is considered to be only the initial interaction between a substance and a cell component. The sequential events, both biochemical and physiological, that this interaction initiates are considered to be effects.

Unfortunately, the primary actions of drugs have been elucidated in only a very few cases. The more a presumed action is studied, the more likely it is to become an effect, and the real action retreats beyond our present means of discovery. In addition, the many effects of the drug must be correlated with drug action. (Fingl and Woodbury, 1965, p. 19)

Toxicants may act at any of the steps involved in maintaining normal cell functions. Their actions may enhance or prevent passage into the cell of substances required for energy production, synthetic reactions, or maintenance of the osmotic and electrical properties of the cell. Toxicants may react with enzymes or the metabolites of enzymatic reactions in ways affecting energy production or synthesis. These and other kinds of reactions, some yet to be defined, lead to the physiological effects of toxic action. Only certain cells, tissues, or organs are acted upon by some toxicants. Parathion, after biotransformation to the active paraoxon, inhibits cholinesterase activity at neuromyal junctions, thereby preserving acetylcholine and leading to neuromuscular disorders. Other toxicants may act in or on most of the cells of the body, as cyanide does in affecting oxidative phosphorylation, important in the metabolic processes of energy transfer.

The cell component directly involved in the action of a toxicant is called its *receptor,* a term usually reserved for interactions of toxicants and cell components that initiate action-effects sequences. Other interactions involving binding, storage, biotransformation, and transport are variously named. Drugs and toxicants that are able to combine with receptors and initiate action are said to possess both affinity and efficacy, or intrinsic activity. These substances are termed *agonists.* Those possessing only affinity are termed *competitive antagonists.* Agonists having a smaller effect than other agonists acting on the same receptor are considered *partial agonists.* The effects of partial and full agonists may be additive, or a partial agonist may antagonize or reduce the action of an agonist.

The actions of toxicants appear to be very much dependent on their structure, minor changes in this often having profound effects on their toxicity. It is on carboxyl, amino, sulfhydryl, phosphate, and similar reactive groups of the receptors that toxicants appear to act, when these are spatially oriented in patterns complementary to those of the toxic substances. Even though few receptors have been identified, the distinctive relationships between structure and activity and the competitive relations of chemically similar substances suggest that interactions between toxicants and receptors obeying mass law kinetics are involved in toxic action. This and other receptor theory has sometimes been extended to explain the shape of curves relating dose with effect as well as to interpret certain drug or toxicant interactions.

It deserves emphasis, however, that mechanism of drug action is defined not by an equation relating dose and effect or describing the pattern of drug interaction, but only by identifying the role of drug receptors in normal cellular function and by characterizing the action-effects sequence. (Fingl and Woodbury, 1965, p. 18)

Characterization of Toxicant Effects

The effects of toxicants can be characterized on the basis of biochemical or physiological changes, or on the basis of changes in survival, reproduction, or growth occasioned by biochemical or physiological alteration. Historically, biologists interested in water pollution have given most emphasis in their studies to the lethal effects of toxic substances, though now effects on reproduction and growth are beginning to receive the attention their importance merits. These biologists have usually concerned themselves little with action-effects sequences. Biochemists and physiologists, on the other hand, have mainly studied action-effects sequences short of ultimate changes in survival, reproduction, or growth. This specialization is understandable when we consider the complexities from initial biochemical action through physiological effects to effects on the whole living organism, and when we consider the limits time places on the knowledge and activity of an individual scientist. But any narrowness of view or any limit on the usefulness of the results of a particular investigation this specialization occasions is most unfortunate. We can best protect aquatic organisms from the effects of pollution if we understand the entire action-effects sequences of any toxicants involved.

Sequences of events leading to changes in survival, reproduction, or growth caused by toxicants begin with the initial biochemical actions of these substances. Such initial actions lead to sequences of biochemical and physiological changes or effects. These may alter digestive, circulatory, respiratory, osmoregulatory, nervous, or behavioral performance of the animals. These alterations matter not to the success of animal populations unless they affect survival, reproduction, or growth in nature. But they must be understood if we are to reliably predict the effects of toxic substances on populations.

The biochemist or physiologist may study *in vitro* the effects of toxicants on cell, tissue, or organ fractions of the whole animal. These studies are necessary. But *in vivo* studies—of the intact living animal—are also necessary. *Biological assay* of the potency of toxic substances can be made either *in vitro* or *in vivo* by measuring their effects on either the separate or the integrated systems of an animal. These effects may be expressed as a change in the activity of any system studied, or as a change in survival, reproduction, or growth if the whole animal is studied. Changes in the intensity of a given response with changes in concentration of a toxic substance can be determined by biological assay and represented graphically. The percentages of individuals responding in a given manner to different concentrations can also be determined and graphically represented (Chapter 13).

As we approach the end of a section devoted to the problem of toxic substances, we must at least mention the indirect effects toxicants can have on individual organisms by altering the biological communities within which they must live. This, of course, provides another category for characterization of effects of toxicants, a category we can consider better in Part VI of this book.

Factors Modifying Toxicant Effects

Rarely do two individuals of the same species, under the same conditions, exhibit the same response to a given concentration of toxicant; and rarely will the same individual, under different conditions, respond similarly to the same concentration. Biologists do not expect similar responses, because they are well aware that individuals differ and that many factors influence any response. In most instances, biologists do not know all the factors influencing the observations they make. But they are aware in general of the kinds of factors that may influence their results, and they endeavor to identify the ones that are operating and to determine how these interact in causing observed responses. We will consider briefly some of the factors that may be important in modifying toxicant effects, but our list should not be considered complete in any sense, for space, oversight, or lack of knowledge may leave missing factors as important as those included.

To begin, let us consider the individual organism. In sexually reproducing species, all individuals, with rare exceptions, are genetically different; this will lead to differences in their responses to toxic substances, usually in ways about which we know nothing. Differences in response may be associated with differences in sex, but most often the genetic bases of differences will be much more obscure. The entire metabolic state of the individual, as determined not only genetically but also by its age, size, present environment, and environmental history, will affect its responses to toxicants. Mechanisms for handling toxic substances may be present only in individuals in certain age groups; and body surface and metabolic rate are functions of body size. An animal's nutritional condition may make it more or less resistant to toxicants. As to environmental history, an animal at any one season or time of the year is, in a sense, a product of preceding external and internal changes, some of which may leave the animal either well or poorly equipped to cope with toxic substances. And previous exposure to the same or to a different toxicant can increase or decrease the sensitivity of an animal.

The present environment of the individual will have both external and internal effects on the toxicity of substances; water temperature and chemistry are important factors. In water having a high calcium content, fish can tolerate higher concentrations of heavy metal ions because of the antagonizing action of calcium ion. The pH of the water—so important in determining ionization of weak acids and bases—can have profound effects on the toxicity of weak acids and bases and their salts, because in a molecular state they more readily penetrate gill membranes (Chapter 13).

Perhaps never is only one toxic substance present in the environment of an animal. Different toxic substances and nontoxic materials can interact in various ways to increase or decrease the overall toxic effect. The outcome will depend on the receptor or receptors involved and on the natures and concentrations of the substances, including their receptor affinities and their intrinsic activities. When the interaction decreases the toxic effect of one of the substances, *antagonism* is said to occur. If the effect of two toxic substances is equal to the sum of their individual effects, either the term *summation* or the term *additive* is applied, according to the mechanisms and sites of action believed to be involved. When the combined effect is more than additive, it may be called *supra-additive*, though it is sometimes termed *synergism* (Chapter 13). But our reader should be warned: usage of these terms is exceedingly

ambiguous; we might better be describing our results and seeking to learn mechanisms and loci of action rather than worrying about names (Fingl and Woodbury, 1965, p. 26).

We cannot usually expect an animal to respond to a particular factor in nature as it does in the laboratory. In the laboratory we simplify nature—that we might better understand her—by controlling some factors and eliminating others. But in nature many factors interact in determining the response of an organism to any one of them. Thus, if laboratory studies are to aid us in predicting the effects of toxic substances on animals in nature, they must somehow take into account at least the more important factors modifying toxicant effects. In a way, then, the basic problem in laboratory research is the identification of those factors in nature most likely to influence the responses of organisms.

SELECTED REFERENCES

Brodie, B. B., and R. P. Maickel. 1962. Comparative biochemistry of drug metabolism. Pages 299–324. *In* B. B. Brodie and E. G. Erdös (Editors), Metabolic Factors Controlling Duration of Drug Action. (First International Pharmacological Meeting, Vol. 6.) The Macmillan Company, New York, xviii + 330pp.

Brown, Margaret E. (Editor). 1957. The Physiology of Fishes. Vol. 1, Metabolism. Academic Press Inc., New York. xiii + 447pp.

———. 1957. The Physiology of Fishes, Vol. 2, Behavior. Academic Press Inc., New York. xi + 526pp.

Fingl, E., and D. M. Woodbury. 1965. General principles. Pages 1–36. *In* L. S. Goodman and A. Gilman (Editors), The Pharmacological Basis of Therapeutics. 3rd ed. The Macmillan Company, New York. xviii + 1785pp.

Florey, E. 1966. An Introduction to General and Comparative Animal Physiology. W. B. Saunders Company, Philadelphia. xi + 713pp.

Galtsoff, P. S. 1964. The American oyster (*Crassostrea virginica* Gmelin). Fishery Bulletin Vol. 64, U.S. Fish and Wildlife Service. iii + 480pp.

Herrick, F. H. 1909. Natural history of the American lobster. Bulletin of the Bureau of Fisheries 24:149–408 + 15 plates. (Reprinted 1911. Government Printing Office Document 747.)

Huxley, T. H. 1884. The Crayfish: An Introduction to the Study of Zoology. 4th ed. Kegan Paul, Trench & Co., London. xiv + 371pp.

Krogh, A. 1939. Osmotic Regulation in Aquatic Animals. Cambridge at the University Press. [ix] + 242pp.

Prosser, C. L. 1958a. General summary: the nature of physiological adaptation. Pages 167–180. *In* C. L. Prosser (Editor), Physiological Adaptation. American Physiological Society, Washington, D.C. 185pp.

———. (Editor). 1958b. Physiological Adaptation. American Physiological Society, Washington, D.C. 185pp.

Thompson, D. W. 1942. On Growth and Form. The Macmillan Company, New York. 1116pp.

Waterman, T. H. (Editor). 1960. The Physiology of Crustacea. Vol. 1: Metabolism and Growth. Academic Press, New York. xvii + 670pp.

———. 1961. The Physiology of Crustacea. Vol. 2: Sense Organs, Integration, and Behavior. Academic Press, New York. xiv + 681pp.

Wilbur, K. M., and C. M. Yonge. 1964. Physiology of Mollusca. Vols. 1 and 2. Academic Press, New York. xiii + 473pp. and xiii + 645pp.

Yonge, C. M. 1960. Oysters. Collins, London. xiv + 209pp.

PART IV

Ecology of the Individual Organism

Understanding of the distribution and abundance of plants and animals, the objective of ecology, requires knowledge of the responses of individual organisms to factors in their environments. Of these responses, reproduction is biologically the most important, but survival, growth, and movement are generally prerequisite to reproduction. Physiological studies may explain the mechanisms of such general responses, but knowledge of an overall response is most directly helpful to the ecologist seeking understanding of the distribution and abundance of organisms. The significance of the survival and reproduction of the individual organism lies in the contribution these make to the persistence of its population in nature. Laboratory studies of the survival, reproduction, growth, and behavior of animals are abstractions from nature. These studies contribute to understanding of the distribution and abundance of organisms only insofar as their results are relevant to environmental conditions and the responses of organisms in nature.

9 The Individual Organism as a Unit of Study

The habits of animals are all connected with either breeding and the rearing of young, or with the procuring a due supply of food; and these habits are modified so as to suit cold and heat and the variations of the seasons.

ARISTOTLE, *c.* 336 B.C., v. 2, p. 122

Ecology considers not only the relations between organisms in a community made up of different species, but also the relations of each species to its environment as a whole and to each of the conditions which make up the environment. [p. 3]

. . . .

In ecology it is even more true than in work from other viewpoints, that the precision of the laboratory is necessary. Furthermore, this precision can not be attained by ignoring factors that are not especially under consideration as is common practice in various other types of laboratory research. The various factors of natural environments must be studied and duplicated so far as possible, as well as varied. [pp. 2–3]

VICTOR SHELFORD, 1929, pp. 2–3

THE INDIVIDUAL ORGANISM, ITS POPULATION, AND ITS COMMUNITY

It is the success of the population, not the individual organism, that has ultimate biological importance; and, most often, it is the population that is of practical interest to man. But the distribution and abundance of a population is determined by the survival, reproduction, growth, and movements of the individual organisms of which it is composed. To understand changes of populations in changing environments, we must know the responses of their individuals.

The whole, living, individual organism, through its homeostatic responses, operates as an adaptable, functional unit. It is perhaps the simplest unit of life that can

function to maintain itself and reproduce its kind in nature in a biologically meaningful way. The individual organism and its environment are easier to define than are populations, communities, and their environments. In nature we observe individual organisms, singly or in groups. Individual organisms can be maintained alive and studied in the laboratory under conditions representing some facets of their natural environments. For these reasons, the individual organism is a convenient and valuable unit for biological investigation. But the environments in nature to which individual organisms respond include not only physical and chemical factors but also the populations and communities of which the individuals are parts. Meaningful interpretation of research on individual organisms must always take this into account.

117

PHYSIOLOGY, ECOLOGY, AND THE INDIVIDUAL ORGANISM

It is to the responses of the whole living organism that the ecologist must look for the most direct help in his search for understanding of populations and communities. These responses represent the final outcomes of the inner workings of the individual as it tends to adapt to its environment. Biochemical and physiological studies of these inner workings are important in their own right; and they are necessary to full understanding of the responses of the whole organism. But the total integrated outcomes of internal processes are the responses of the organism that are most directly meaningful to the ecologist.

Studies of the responses of individual organisms to elements of their environments, particularly the physical and chemical ones, are often quite physiological in character. They are sometimes considered to be studies of *response physiology* or *physiological ecology.* Schröter and Kirchner (1896, 1902) coined the term *autecology* often used to cover this area of biology. When biological elements of the environment of the individual organism are taken into account, autecology grades into *synecology*, a term later proposed for studies of associations of different species.

All biologists, whatever level of organization they may study, should relate their studies to knowledge of other levels of organization. Only then can they gain the fullest understanding possible and make their work most interesting and useful. There are no sharp biological boundaries, and we should avoid attempting to make them, for "events inside the individual and outside form connected systems" (Bates, 1960, p. 549). But of information on systems below the population level of organization, that on responses of whole living individuals can be applied most directly to studies of populations and communities.

RESPONSES OF THE INDIVIDUAL ORGANISM TO ITS ENVIRONMENT

Integrated Homeostatic Responses

We presented in Chapter 7 the idea of homeostasis; in Chapter 8, particular homeo-
static responses of animals were considered. In each species, capacities for physiological and behavioral homeostatic responses have evolved, not independently but together in the total structure and nature of the individuals of the species. When an organism adapts to temperature change, adaptive changes in respiration and circulation are necessary; these and other homeostatic mechanisms, including behavioral ones, function together to favor the success of the individual and its population. Particular homeostatic responses of individual organisms to certain environmental changes have been identified and measured; we may rarely if ever learn all the internal adjustments an organism makes in the face of a given environmental change. Yet, they may all, functioning together, be necessary for its successful adaptation.

The importance of an internal or a behavioral change in an individual organism cannot be determined by merely measuring that change. It is the effect that the change has on the whole organism—on its ability to survive, reproduce, and grow—that determines its importance. Here, again, we have a principal reason for studying the individual organism as a whole.

Survival, Reproduction, Growth and Movement

Huntsman (1948) wrote that the distribution and abundance of populations can be understood only when we know how individuals in those populations respond to their environments in terms of survival, reproduction, growth, and movement. Ultimately, this is the reason ecologists study individual organisms in the laboratory and in nature. And perhaps most studies of the whole living organism are concerned with survival, reproduction, growth, or movement as overall responses.

But each of these overall responses represents the outcome of many integrated yet particular responses of the individual organism, the adaptive limits of these particular responses being determined by the structural, physiological, and behavioral characteristics of each species. Whether or not an individual survives an environmental change depends on the success or failure of its homeostatic mechanisms. Successful reproduction depends not only on development of normal gametes but on normal reproductive behavior and on suitable conditions for embryonic and larval development. For growth to occur, not only must the physical and the chemical character of the environment be suitable, but also food or energy and material resources must be sufficient. Movement or, more generally, behavior is involved in and essential to nearly all the responses of animals in their natural environments; it is necessary for survival, reproduction, and growth. Not only these overall responses but also the particular responses which lead to them are of prime concern in studies of individual organisms.

The Range of a Response to a Given Change

For a given environmental change, the response of a particular individual may usually be nearly the same, but not always, for the internal state and the needs of the organism change in time. We do not usually expect, however, the responses of two individuals from the same population to be the same or even nearly the same when they are confronted with a given change. Of course, if we test enough individuals from a population, we will find that some do exhibit the same response. When organisms have had very nearly the same environmental histories, differences in their responses can be attributed mainly to genetic differences. These genetic differences favor the persistence of their populations in the face of environmental change.

When we bring individual organisms into the laboratory to study, we hope they are representative of their population as a whole, for it is the population that is our ultimate concern. Recognizing that there is genetic variability within populations, we rarely study one individual but rather a sample of individuals. We usually determine the mean response of these individuals to a particular environmental change; we hope this mean is a good approximation of the population mean. There are, of course, statistical procedures for determining the probability of this. Because, in most natural phenomena, individual occurrences tend to group mainly around central values and to become more rare toward extreme values,

estimates of mean values are usually more precise than estimates of extreme values. The mean response of the individuals in a sample may be a reasonable approximation of how most of the individuals in a population will respond to a particular environmental change.

But genetic variability in natural populations permits some individuals to survive and reproduce in the face of environmental changes preventing the survival or reproduction of most individuals. Thus, through time, a population may be able to adapt genetically to environmental changes that originally could not be tolerated by most of its individuals. For this reason, we are interested not only in the mean response of individuals in a sample but also in their extreme responses. Information pertaining to the range of responses of individuals in a population may give us some indication of whether or not that population can persist in the face of a particular environmental change.

LABORATORY STUDIES AND STUDIES IN NATURE

The Reality and Complexity of Nature

The individual organism in nature must not only adapt to physical and chemical conditions but play its role in its population and community, which are also parts of its environment. It is here, in nature, that we really wish to understand the functioning of the individual. Only here is there ecological reality, which is in marked contrast to the abstractions we make from reality in our laboratory studies.

With this reality goes complexity of organism-environment relationships: hopeless complexity, challenging complexity, beautiful complexity, according to our viewpoint. Ecologists seek recurring patterns in these relationships. It is sometimes possible to study the functioning of individual organisms in their natural environment: we do this when we observe their behavior in nature, or when we mark them and follow their movements or growth. Such studies are particularly valuable, because the organism is in its natural setting. But they usually cannot tell us all we might wish to know, for here the organism is confronted with many varying conditions; not often can we ascertain which of these, singly or together, are leading to the responses we observe. It is for this reason that so many studies of the individual organism are conducted in the laboratory. But laboratory studies, too, have their limitations. There is no single path to ecological knowledge; hopefully, the paths will come together.

The Objectives and Possibilities of Laboratory Studies

For the ecologist, the objective of laboratory studies is to determine how environmental factors, acting singly or together, influence the performance of the individual organism. Through such studies, ecologists hope to better understand how the performance of the individual leads to the success or failure of its population. In the laboratory, the organism can be maintained alive under controlled conditions of the desired complexity. Here, the effects that controlled levels of particular environmental factors and combinations of factors have on the survival, reproduction, growth, or behavior of the individual can be determined.

The effects that physical and chemical factors have on the responses of an organism can usually be studied more readily in the laboratory than can effects of biological factors. Studies of nutrition and growth and studies of some aspects of social behavior are often better conducted in the laboratory than in nature. But most of the population and community characteristics of an organism's natural environment are difficult to model in the laboratory. Nevertheless, as we will see in Chapters 17 and 19, there has been progress in recent years in developing laboratory ecosystems in which at least some of the relations of the individual organism to its population and its community can be studied.

We must, however, continually bear in mind that our laboratory studies are, in some degree, always abstractions from the natural realities we wish to understand. We must endeavor to plan our laboratory work so as to make it most meaningful in the interpretation of these realities. This will often lead us to conduct quite complex laboratory studies. Theoretically, at least, we could design laboratory studies approaching nature in their complexity. But our laboratory studies should be no more complex than necessary to make them truly meaningful with regard to the natural systems of interest, for as laboratory studies approach nature in their complexity, they lose their peculiar advantages of simplicity and control.

Laboratory Studies and Prediction

Perhaps the severest test of man's knowledge of a complex system comes when he attempts to predict events that are determined by many interacting variables. Ecologists are not usually able to predict with useful precision and accuracy changes that occur in populations as a result of changes in interacting environmental variables. If measurements of particular populations were made over long periods of time, and if the important environmental variables were known and measured, some indication of causal relations might be found; this could be a basis for prediction, however good or poor. No matter how desirable it might be to have such measurements, we will rarely have complete historic records; and, as environments change, history is a poor basis for prediction.

In laboratory studies, we learn how particular environmental factors and combinations of factors influence some of the responses of the individual organism that determine the distribution and abundance of its population. This provides another basis for predicting the effects of environmental change, a basis that is not limited by the historic range of environmental conditions. For this reason, laboratory studies are particularly valuable as one basis for predicting the effects of man's activities on the distribution and abundance of populations.

But, as we have noted, laboratory studies are abstractions from the natural environment of the individual organism. The greater the abstractions, the poorer the bases laboratory studies provide for prediction. This we must bear in mind in planning, conducting, and using laboratory studies. If the respective strengths of laboratory studies and studies in nature are fully to materialize in providing for prediction, they must be integrated.

From the laboratory part of our investigations, it is the response of the whole living organism to factors in its environment that is most directly useful in predicting the success or failure of its population in the face of environmental change. Biochemical and physiological studies are important in their own right and can lead to understanding of the responses of the individual. But it is ultimately the response of the whole organism in its population and community that determines the success of its population. In our attempts to predict the effects of water pollution and other environmental changes on populations, knowledge of the responses of whole living organisms will be most useful.

SELECTED REFERENCES

Chapman, R. N. 1931. Animal Ecology. McGraw-Hill Book Company, Inc., New York. x + 464pp.

Huntsman, A. G. 1948. Method in ecology—biapocrisis. Ecology 29:30–42.

Shelford, V. E. 1929. Laboratory and Field Ecology. The Williams & Wilkins Company, Baltimore. xii + 608pp.

10 Development

If an animal can in any way protect its own eggs or young, a small number may be produced, and yet the average stock be fully kept up; but if many eggs or young are destroyed, many must be produced, or the species will become extinct.

CHARLES DARWIN, 1859, p. 66

It is particularly important for the physiologist in studying the fish egg to realize that the egg is only one phase of the totality of existence of his material; it is not an end in itself but merely the means of replacing the population losses from death and depredation.

SYDNEY SMITH, 1957, p. 333

THE IMPORTANCE OF DEVELOPMENT

Development is a part of the reproductive process tending to maintain populations while death inevitably and continually removes their individuals. Reproduction involves more than the development of gametes, embryos, and larvae. There is the reproductive behavior of the parents, a topic we will leave mainly for Chapter 12. Ultimately, the success of this behavior and of the entire reproductive process must be measured in terms of the persistence of the populations involved (Chapter 16). But here we will restrict ourselves mainly to the effects of environmental change on the production of gametes and the development of the individual organism.

Natural populations fluctuate, often widely, even in environments relatively unaffected by man. This is because natural variations in environmental factors lead to changing birth and death rates. But in populations that have persisted over very long periods of time, there have evolved biological ways of coping with environmental change. In time and place of reproduction, in the numbers of gametes and their characteristics, in the requirements of developing embryos and larvae, and in the extent of parental care, each species and race has come to be sensitively adapted to its environment.

Aquatic populations often cannot adapt to environmental changes caused by man, and they decline and disappear. Reproduction may be unable to balance increased death rates of later life history stages, or reproduction may be impaired even though death rates in later life remain the same. When man changes the reproductive environment of a population beyond its normal limits, then production of gametes, reproductive behavior, or development of young may be so altered as to lead to decline of a population, to reduction of its usefulness, or even to its disappearance.

In this, all animal populations are the same. But few if any animals have been studied sufficiently for us to understand fully how even the most important environmental factors act on reproduction and what may be the biological significance of such actions. More is known about the reproduction of salmon and trout than of most other species of fish, and more is known about the reproduction of oysters than of most other species of invertebrates. For this reason and because these are animals of considerable interest and importance to man, it is their development that we will mainly consider. Every species differs in its response to its environment, but these species will serve well to illustrate some of the reproductive problems of aquatic animals in changing environments.

DEVELOPMENT OF GAMETES

Reproduction in a population of animals can perhaps be taken to begin with the development of gametes in the gonads of the adult individuals. Environmental conditions permitting, this occurs annually in most species during a given season. For climatic and other reasons, the development of gametes may not be equally successful during different years, even under natural conditions. And environmental changes caused by man may interfere with hormonal control, prevent the accumulation and translocation of necessary energy stores, or in some other manner impair the development of normal gametes.

Hoar (1957a) has reviewed some of the more important contributions to the extensive literature on gamete development in fish. Most species of fish in the temperate world exhibit a seasonal periodicity in such development. In some species, this has been shown to be controlled in large part by light and temperature. But an intrinsic tendency toward reproductive periodicity also exists in many fish. Both intrinsically and externally controlled periodicity in the development of gametes are brought about through hormonal complexes. With the approach of the spawning season, there appear in many species of fish secondary sexual characteristics that are under hormonal control and are important in reproductive behavior.

Relatively large quantities of nutritive materials, mainly fats and proteins, must be stored in the eggs for use by the young during embryonic and larval development. Sometimes, as in the case of Pacific salmon, this occurs as a translocation from body stores during a period of fasting. Whatever the immediate source of these nutritive materials, they must be obtained originally by the female in the course of feeding. Environmental changes altering food avail-

ability can thus influence the development of gametes.

Reproduction may be impaired by other effects of the environment on developing gametes. Possible effects of radiation on the genetic constitution of offspring have occasioned some concern for man and other animals. Burdick et al. (1964) have shown that part of the DDT accumulated by adult lake trout (*Salvelinus namaycush*) in areas treated for insect control is transferred along with fats into the eggs. Fry hatching from these eggs suffer heavy mortality at about the time yolk absorption is complete, a mortality accompanied by a characteristic syndrome.

Temperature appears to be the most important environmental factor controlling the development of gametes in oysters. Biologists have been able to obtain for their laboratory studies the gametes and larvae of oysters at all seasons of the year by temperature control (Loosanoff and Davis, 1963). The American oyster (*Crassostrea virginica*), for example, normally spawns along the coast of New England during the warm summer months. Development of gametes can be stimulated during the winter months by bringing these oysters into the laboratory and gradually increasing the temperature of the water in which they are held. Gamete development and spawning can be induced only after the oysters have recovered from summer spawning activity. Such recovery undoubtedly involves complex physiological processes as well as accumulation of energy stores, mainly glycogen, the principal storage product. It is high glycogen content that gives oysters a creamy appearance and makes them most marketable. To obtain larvae for laboratory studies in the fall when recovery from summer spawning activity is not complete, oysters have been transplanted in May from Long Island Sound to the colder waters of Maine. Here, development of their gametes proceeds slowly but spawning does not occur. These oysters can then be brought into the laboratory and induced to spawn in the fall, when this cannot be accomplished with other oysters.

As important as temperature may be in controlling gamete development, other environmental factors are involved. Food organisms must be sufficiently available to permit the adult oysters to accumulate the energy stores necessary for this development. And Loosanoff (1952) found that the lowest salinity at which normal development of gametes proceeds in American oysters from Long Island Sound is near 7.5 parts per thousand.

The European oyster (*Ostrea edulis*)—like the Olympia oyster (*Ostrea lurida*) of the Pacific Coast—is a protandric hermaphrodite, having sex phases that alternate regularly in most individuals after the initial male phase (Cole, 1942; Coe, 1932). Under favorable conditions, some oysters may complete one male and one female phase each year. Functional hermaphrodism apparently does not exist in these oysters, for even though there may be some overlap of phases, the gametes do not ripen simultaneously. Regular sex alternation does not occur in the American oyster. This oyster may, however, spawn several times during the same season as gametes mature (Loosanoff and Davis, 1963).

Environmental conditions can, then, influence the success of animal reproduction at the very beginning of the reproductive cycle, the period of gamete development. Man's activities can so change the environment as to alter or prevent the development of gametes. Decrease in the number of gametes produced, or production of gametes from which normal young are unable to develop, can lead to the decline or disappearance of populations man values.

RELEASE OF GAMETES

Associated with the release of gametes, there may be movements of adults to spawning areas, courtship and nesting activities, and care of eggs and offspring; or, as in the case of oysters, reproductive behavior may be rather simple. The number of eggs produced tends to be inversely correlated with the complexity of reproductive behavior and the extent of parental care. For each species, reproductive behavior and the characteristics and number of eggs produced represent adaptations to the environment. Before we can consider the influence of environmental factors on the embryonic and larval development of animals, we must have some knowledge of the environments into which their gametes are released.

Many races of Pacific salmon, after feeding for one to several years along the continental shelf, enter freshwater and migrate hundreds of miles to reach their natal tributaries to spawn. Other species, sunfishes for example, may hatch, grow to maturity, and reproduce within a very restricted area. In most fish, fertilization occurs as the gametes are released into the water; but in some species, fertilization is internal and the embryos develop in specialized parts of the female reproductive system. Many species of fish deposit their eggs in nests, which may be guarded or cared for in other ways. Nests may be complex structures such as the male stickleback (*Gasterosteus aculeatus*) constructs of plant materials, or they may be simply areas of the substrate cleared of debris. The female salmon forms a depression in the streambed gravel by movements of her tail and body. Into this depression she releases eggs, which are simultaneously fertilized by the male. She then moves slightly upstream where renewed digging causes the eggs to be covered and where more eggs are released. This is repeated until spawning is complete. Cod (*Gadus* spp.), striped bass (*Roccus saxatilis*), and many other species spawn in schools, the gametes being released into open water where the embryos and larvae develop in suspension as they move with the currents.

Reproductive behavior in fish is primarily instinctive (Chapter 12). It is a response of sufficiently motivated individuals to appropriate external stimuli. Motivation is primarily determined by gonadotropic hormones. Maturation of gonads and various external stimuli such as suitable spawning habitat and prospective mates tend to increase the levels of these hormones in the fish. The final reproductive act is a response to social stimuli involving behavior of the mates, frequently their secondary sexual characteristics, and perhaps their secretions.

Salmon and other fish that produce large eggs having much nutritive material tend to have fewer eggs than do cod and other fish producing very small eggs with little nutritive material. Fish producing very large numbers of eggs could hardly provide the materials and body space necessary for these eggs to be large. And the survival advantages large eggs provide the developing young are not necessary when so many eggs are produced. Most salmon and trout produce from one thousand to a few thousand eggs, which develop into large fry in the relatively secure environment of streambed gravels. A female cod releases several million tiny eggs into the open sea. Here, they and the developing young are preyed upon by other animals, and ocean currents may carry them away from suitable nursery areas. The tiny young must soon obtain their own food from the plankton, and many may starve. Sunfishes produce more and smaller eggs than do salmon or trout, but many fewer than do cod; sunfishes give their eggs and young considerable parental care. No matter how many or few may be the eggs of fish, from the total progeny of each adult pair, only about two must reach reproductive maturity, if their populations are to remain stable.

Different species of oysters also exhibit different degrees of parental care of young; this is reflected in the numbers of eggs they produce. European and Olympia oysters retain their eggs in the mantle cavity after spawning. Here the eggs are fertilized by sperm released by the males into water that the female draws through her mantle cavity; and here the embryos develop to the larval stage that is finally released to begin the pelagic phase of the oyster's existence. Cole (1941) found the European oyster to produce from about 100,000 to 900,000 eggs each year, the number depending on the age of the female. The American oyster releases its gametes directly into the open water where fertilization and development occur. Davis and Chanley (1956) found individual females of this species to produce on the average about 50 million eggs each year.

Particular environmental factors have not been shown to provide directly the stimulus necessary to initiate spawning of oysters in nature. From Korringa's (1952) review of the relevant literature, one gains the impression that temperature controls the gonadal development leading to spawning but that the act of spawning in nature is not closely correlated with temperature. There is some evidence that lunar-tidal periodicity may be involved in the timing of spawning. Spawning does not necessarily occur as soon as the gonads have reached maturity. When it does occur, it tends to be synchronized between the sexes and in the population as a whole. Hormone-like secretions and a substance carried in the sperm provide mutual stimulation between the sexes and undoubtedly play a major role in this synchronization, the importance of which is obvious. In the laboratory, thermal shock has been used

successfully to induce ripe oysters to spawn; sometimes a suspension of gonadal materials must be added to the water (Loosanoff and Davis, 1963).

Time, place, and manner of spawning reflect adaptations of each species that help to insure its reproductive success under the natural conditions usually prevailing. Environmental changes that alter the reproductive behavior of a population can lead, as surely as can lethal conditions, to its decline or disappearance.

ENVIRONMENTAL CONDITIONS INFLUENCING SALMONID DEVELOPMENT

Knowledge of the influence of environmental conditions on the development of fish of recreational or commercial importance is reasonably extensive only for salmon and trout. One reason for this is that the embryos of salmon and trout are large and easily reared—they are excellent experimental material. Another reason is that the importance of salmon and trout to man has led to their extensive culture in hatcheries and so to much interest in conditions favoring this culture. In recent years, the most important stimulus to studies of the requirements of salmonid embryos has been the endangering of valuable populations by man-caused changes in their reproductive environments. In restricting our discussion to the influence of environmental conditions on salmonid development, we in no way mean to imply that knowledge of the embryonic requirements of other species of fish is any less important in their protection from the activities of man.

Salmon and trout usually deposit their eggs in redds or nests in the gravel of streambeds, though some species and races may spawn in lakes where there is sufficient water movement around bottom materials of rock. Here, the developing embryos and the hatching fry—which are dependent on yolk nutrients until they emerge—are protected from damaging light rays, from most predation, and usually from being carried away by currents. Freshets or the spawning of other fish may move the gravel and so disturb or dislodge the developing young and cause their damage or loss. Yet, buried in streambed gravels, they are in the most secure of stream environments. Streambed gravels do have one distinct disadvantage as an environment for these young: the developing young are dependent on water movement through the gravels to deliver oxygen and remove waste products. And water movement through streambed gravels, even under the most favorable conditions, is greatly restricted.

Actual velocities of water movement through the pores among streambed gravels would be exceedingly difficult to measure, because of turbulence; only *apparent velocities* have been estimated. Wickett (1954) has defined apparent velocity as the volume of discharge from a section per unit of time divided by the cross-sectional area of both solid particles and voids. At depths of 6 to 12 inches in streambed gravels, apparent velocities may approach zero and probably rarely if ever exceed a few hundred centimeters per hour (cm/hr). Interchange with surface water is the principal source of water flowing through streambed gravels. The extent of interchange is determined primarily by the gradient of the stream surface and the permeability, depth, and surface configuration of the gravel bed (Vaux, 1962). Silt and other fine materials deposited in the gravel bed can greatly reduce its permeability and the velocity and volume of water moving through. Reduction in volume of water reduces the oxygen resource available to the developing embryos and fry, and their respiration and the decomposition of organic materials by microorganisms may result in depletion of oxygen and accumulation of waste products.

Not only the survival of salmonid embryos experimentally buried in streambed gravels but also the size of the hatching fry have been found to be correlated with the apparent velocity and the oxygen content of water moving through these gravels (Coble, 1961; Phillips and Campbell, 1961). The very deleterious effects of silt on the emergence of salmonid fry demonstrated in early studies (Shaw and Maga, 1943) now appear to be due to actions of silt on the chorions, or shells, of the eggs as well as reductions in water movement and oxygen concentration in the redd. Koski (1966) trapped coho salmon fry (*Oncorhynchus kisutch*) emerging from 21 redds in which he had measured gravel composition, permeability, and dissolved oxygen concentration. Dissolved oxygen concentrations were high in these redds; minimum concentrations below 6 mg/l were found in but four redds, only one of these having concentrations below 4.5 mg/l. Emergence from these four redds averaged less than 4 percent of the eggs deposited, whereas for all 21 redds the mean emergence was 27 percent. Success of emergence was correlated with gravel composition; there was evidence that gravel permeability and movement also affected emergence.

The requirements of embryos and fry change with their development, and in nature it is difficult to determine at what stages environmental factors have their most serious effects. And in nature, oxygen concentrations and the accumulation of waste products are dependent on water movement, this making difficult the assessment of the relative importance of these variables. Water movement in nature has not only the function of delivering oxygen to the redd and removing waste products but also the function of insuring their sufficient transport to and from the surface or chorion of the egg. Laboratory studies have been necessary to determine the effects of water movement, oxygen concentration, waste products, and temperature on the survival, development, and growth of salmonid embryos and sac fry. For such studies, a special apparatus was constructed in our laboratory (Silver, Warren, and Doudoroff, 1963). This apparatus had six experimental chambers provided with water having desired levels of dissolved oxygen, metabolic wastes, and temperature. Each chamber was equipped with four cylinders to hold the embryos resting on porous plates through which the water was drawn upward at the desired velocities. Flow of water through the cylinders was nearly rectilinear; with 100 or less embryos resting on the plates, the velocities measured approximated true velocities. Thus, water quality and velocity could be varied independently so as to make possible the separation of their effects on the embryos.

The effects on salmonid embryos of any restriction of their rate of respiration are extremely complex and difficult to evaluate. This is so not only because the embryo is a continuously changing organism but also because the embryo is able to reduce drastically its respiration rate and still survive by reducing its growth and development rates. With very severe respiratory restriction, the embryo may succumb before hatching. Steelhead trout embryos (*Salmo gairdneri*) at 9.5 C and chinook salmon embryos (*Oncorhynchus tshawytscha*) at 11 C all died at a dissolved oxygen concentration of 1.6 mg/l (Silver, Warren, and Doudoroff, 1963). But with slightly less respiratory restriction, at 2.5 mg/l, large percentages of these embryos survived, though they hatched as very small, incompletely developed sac fry. Protected in the laboratory, such fry can survive; in nature, they may be at a serious disadvantage. Any increase in dissolved oxygen up to the air saturation level resulted in larger steelhead trout and chinook salmon fry at hatching. Any increase in water velocity up to 740 cm/hr resulted in larger steelhead fry, and any increase up to 1350 cm/hr resulted in larger chinook fry. Respiratory restriction also lengthened the period of development before hatching, an effect that could be important in nature.

Figure 10–1, which is from a study conducted by Shumway, Warren, and Doudoroff (1964), is a three-dimensional diagram showing the relationships between dissolved oxygen concentration, water velocity, and the mean dry weight of coho salmon fry, with yolk sacs removed, at hatching. Each intersection of a weight-oxygen curve with a weight-velocity curve represents the mean

weight of fry that developed at that oxygen concentration and water velocity at a temperature of 10 C. Any increase in oxygen concentration up to 11.2 mg/l resulted in some increase in size of the newly hatched fry; and the shapes of the curves suggest that further increases in dissolved oxygen would have resulted in even larger fry at all velocities tested. Increases in water velocity in the range of 3 to 800 cm/hr resulted in increases in the size of fry at hatching; the effect of water velocity was nearly as much at high as at low oxygen concentrations. A greater relative decrease in water velocity than in oxygen concentration was necessary to bring about the same reduction in growth of the embryos. The curved broken lines in Figure 10–1 indicate the various combinations of reduced oxygen concentration and water velocity that, had they been tested, would have reduced the weights of the fry by 20 percent and 33 percent of the weights of fry hatched under the most favorable conditions. Even at the highest water velocity tested, reduction in oxygen concentration to below 6.6 mg/l would have reduced the size of the fry by more than 20 percent. Water velocity would have had to be reduced below 14 cm/hr at an oxygen concentration near the air-saturation level to have had the same effect.

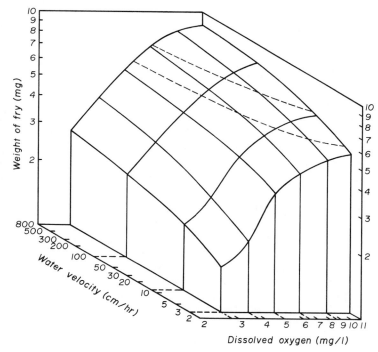

FIGURE 10–1 A three-dimensional diagram showing the influence of both oxygen concentration and water velocity on the mean dry weights of newly hatched coho salmon fry developing from embryos reared throughout development at the various combinations of oxygen concentration and water velocity represented by the intersections of the curves. After Shumway, Warren, and Doudoroff (1964).

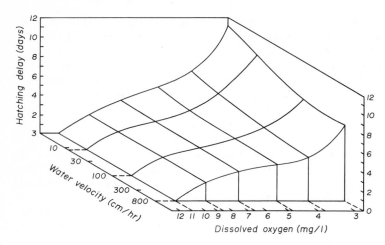

FIGURE 10–2 A three-dimensional diagram showing the influence of both oxygen concentration and water velocity on the delay in hatching of coho salmon fry developing at the various combinations of oxygen concentration and water velocity represented by the intersections of the curves. Hatching delay is expressed as the difference between median hatching time under a particular set of conditions and median hatching time under the most favorable conditions, where it was 44 days. After Shumway, Warren, and Doudoroff (1964).

Figure 10–2 shows the relationships between oxygen concentration, water velocity, and hatching delay in days in this experiment with coho salmon embryos. The hatching delay under any given set of conditions is expressed as the difference between median hatching times under those conditions and under the most favorable conditions tested. Any reduction of oxygen concentration from the air-saturation level resulted in delayed hatching of fry at all water velocities. The greatest delay, which was 11 days, occurred at 2.8 mg/l. A reduction of water velocity from 800 to 3 cm/hr did not cause any delay of hatching at 11.2 mg/l; nor did it cause any additional delay—delay not ascribable to reduced oxygen alone—at the lower oxygen concentrations of 8.6 and 6.5 mg/l. This reduction of velocity caused only a slight additional delay at oxygen concentrations of 4.9 and 3.8 mg/l; but it caused a pronounced additional delay at 2.8 mg/l.

Water movement is necessary not only to deliver oxygen to the redd and to the surfaces of the chorions enclosing the developing embryos but also to remove metabolic wastes from the redd and from the chorion surfaces. Putnam (1967), working in our laboratory with the apparatus we have described, increased the level of metabolites in water in which experimental embryos were developing by first passing this water through a large mass of developing embryos. These metabolites slightly inhibited growth of steelhead trout embryos but did not inhibit growth of sac fry. An ammonia concentration of 5 mg/l and carbon dioxide concentrations above 28 mg/l were also found to inhibit embryonic growth. After water was passed through the mass of developing embryos, it contained only 0.1 mg/l or less of ammonia and 3 to 6 mg/l of carbon dioxide. These concentrations of ammonia and carbon dioxide were insufficient to account for the growth depression observed when this water was tested. Other metabolites must have been involved. Increases in water velocity from 6 to 800 cm/hr had little influence on the effects of the metabolites on the growth of the embryos. But when dissolved oxygen concentrations were increased to 24 mg/l—much beyond the air-saturation level—the beneficial effects of high water velocities were reduced. These results, taken together, strongly suggest that the oxygen concentration rather than the metabolite concentration in the water immediately surrounding the chorion is most important in limiting embryonic growth; water movement adequate to deliver oxygen is more than adequate to remove metabolic wastes.

Because salmonid fry do not emerge from the gravel upon hatching but only after absorption of their yolk sacs, we must concern ourselves with the suitability of the streambed environment not only for the embryos but also for the sac fry. Steelhead trout embryos and sac fry have been reared at different oxygen concentrations and water velocities throughout development and up to yolk sac absorption, in our laboratory. At an oxygen concentration of 3 mg/l and a water velocity of 10 cm/hr, yolk sac absorption is completed much later and the resultant fry are much smaller than in the case of embryos and sac fry reared at an oxygen concentration of 10 mg/l and the same water velocity. At the higher water velocity of 300 cm/hr, fry reared throughout development to yolk sac absorption at 3 mg/l were nearly as large as those reared at 10 mg/l, though yolk sac absorption was completed much later. The velocity of water movement apparently is important not only to the embryo but also to the sac fry. If conditions producing small fry at hatching persist in nature, the fry are likely to be small when yolk absorption is complete, should they survive. And yolk sac absorption will be much delayed. Thus, we must concern ourselves with the possible survival and emergence of small or weak fry and with any effects that late emergence may have on their ultimate success.

Phillips et al. (1966) buried fertilized eggs of coho salmon in gravels of different size compositions contained in troughs, where the developing embryos and sac fry were held at different oxygen concentrations until emergence of the fry was complete. Emergence decreased as the percentages of fine materials in the gravels increased, apparently in part because of increased difficulty the fry experienced in making their way from the gravels. Emergence was delayed and decreased when oxygen concentrations were below 5 mg/l, and the emerging fry were smaller. The smaller fry apparently either were less able to survive through yolk sac absorption or were too weak to find their way from the gravels. These results help to explain the finding of Koski (1966) that there was little emergence from natural redds having minimum oxygen concentrations below 6 mg/l. Thus, from laboratory and field studies, it has been possible to develop a reasonably complete picture of the effects that gravel composition, water movement, oxygen concentration, and metabolites have on the success of development of salmonids.

The effects smaller size or delay in time of emergence may have on the ultimate success of fry that do emerge will be more difficult to unravel. Earlier emerging coho salmon fry were found by Mason and Chapman (1965) to have an ecological advantage over later emerging fry. The earlier emerging fry were larger at any point in time and, through social dominance over the smaller fish, were better able to maintain residence in the experimental channels in which their behavior was studied. Mason (1969) found that fry reared at low oxygen concentrations

were unable to compete successfully with larger fry reared at higher concentrations when both were introduced into experimental channels. Still unanswered is the question of whether or not this would be important if most fry in a population emerged slightly later and smaller, as presumably would occur with any general degradation of a stream. The final solution of this problem will require studies of the ultimate contribution fry emerging at different sizes and times make to the success of their populations.

We have yet to discuss the effects on developing salmonid embryos of another very important environmental factor. Temperature can greatly influence the oxygen requirements of these embryos and can itself be directly lethal. We have done only very preliminary work on this problem, work that must be repeated and extended. In an experiment conducted at our laboratory, only 25 percent of the coho salmon embryos developing at 14 C and 10.4 mg/l of oxygen hatched successfully. Though oxygen concentrations above the air-saturation level would perhaps have increased the percentage of successful hatch, temperatures this high may have a direct inimical effect. In an experiment conducted at 12.5 C, 79 percent of the embryos developing at 10.7 mg/l of oxygen hatched, 22 percent hatched at 5 mg/l, 12 percent hatched at 4 mg/l, and none hatched at 3 and 2 mg/l. At temperatures near 10 C, over 90 percent of the coho salmon embryos developing at the air-saturation level of oxygen can be expected to hatch; at concentrations from 2 to 3 mg/l, the percentage of hatch is not usually much less. The rates of development and growth of salmonid embryos increase with increasing temperature. The increase in metabolic rate that makes this possible increases their need for oxygen. The inimical effects of a given reduction in the concentration of dissolved oxygen thus become greater with an increase in temperature (Garside, 1966).

ENVIRONMENTAL CONDITIONS INFLUENCING OYSTER DEVELOPMENT

As salmon and trout are the most studied fish, so the oyster is undoubtedly the most studied shellfish. Nelson (1938, p. 1) has gone so far as to say "The oyster is scientifically the best known marine animal in the world." Interest in the culture of oysters, which are considered a delicacy the world over, has stimulated a great deal of investigation into their environmental requirements. This investigation has contributed not only much of practical value but also much of general biological interest. On the basis of his observations on oyster beds, Karl Möbius (1877) was one of the first to advance the community concept. But, unlike the culture of salmon and trout, the culture of oyster embryos and larvae is difficult and has been mastered only within the past two decades. Only during this period has there been much advance in understanding of the developmental requirements of oysters.

Loosanoff and Davis (1963) have reviewed much of the literature pertinent to the artificial culture of molluscan larvae and have described their own methods. They note that Prytherch (1924) and Wells (1927) were successful in rearing larvae of the American oyster but that other investigators were unable to reproduce their results consistently, because of poor culture methods including unsuitable food for the larvae. During the last 20 years, excellent methods for the laboratory culture of oysters and clams have been developed, in large part through the efforts of Loosanoff and Davis. Particular attention must be given to the elimination of minute amounts of toxic substances from the culture medium and even to the control of possible pathogenic organisms. But finding and providing suitable food organisms have been perhaps the biggest problems. Loosanoff and Davis (1963) have found larval growth to be best when certain naked flagellates are used for food. In the later stages of development, however, larvae can utilize some species of microorganisms having rigid cell walls. Better growth is obtained when mixtures of food organisms rather than single species are provided the larvae. Some organisms are toxic to the larvae. Although little is known of the food habits of oyster larvae in nature, there is some evidence that suitable food organisms are not always present in sufficient abundance and that toxicity problems may be associated with the presence of other organisms (Korringa, 1952). In considering the effects of environmental changes on oyster populations, we must bear in mind the possibility that these changes may influence the abundance of food organisms upon which the larvae depend.

Whether oyster larvae pass through their early stages of development in the mantle cavity of the female—as do larvae of the European and Olympia oysters—or in the open water—as do larvae of the American oyster—these stages are much like those of clams and other molluscan larvae. The oyster embryo soon develops a ciliary swimming apparatus, and then a larval shell appears. When this shell can completely enclose the larvae, its hinge side is straight; the *straight-hinge stage* has been reached. Shortly thereafter, oysters that hold their young in their mantle cavities discharge them into the open water. Here the larvae, leading a planktonic existence, feed on microorganisms and are themselves fed upon by other organisms. Here, too, they are endangered by tidal and other currents that may sweep them away from areas suitable for their further development and for their later attachment to substrate materials. Such attachment is necessary if the larvae are to change into juvenile oysters. On attachment and metamorphosis, the larvae become *spat,* the "seed" of the oyster industry. Given suitable attachment material and location, the spat can develop and grow into adult oysters.

Temperature may control the ability of oyster larvae to utilize various food organisms for growth as well as directly affect larval survival and development (Loosanoff and Davis, 1963). Larvae of the American oyster were better able to utilize *Chlorella* sp., a poor food, with increase in temperature from 20 to 33 C, perhaps because of a change in activity of digestive enzymes. At 15 C, even when good food organisms were available, no larvae developed to the straight-hinge stage, while at 17.5 C most of them reached this stage but did not grow well. In some experiments at 30 C, all the larvae reached the straight-hinge stage; only about half the larvae at 33 C reached this stage of development.

Salinities of estuarine waters vary seasonally and annually; changes in salinity may affect the survival and gonadal development of adult oysters and the success of larval development. Water resource developments can change the flow regime of rivers entering estuaries and lead to changes in the natural patterns of salinity variation. Oyster propagation may then be endangered. Davis (1958) determined the percentages of straight-hinge larvae developing at different salinities from eggs of American oysters from different waters after these oysters had developed gonads and spawned at various salinities (Table 10–1). The optimal salinity for the development of larvae to the straight-hinge stage was about 22.5 parts per thousand (ppt), whether the adults were from Long Island Sound, Peconic Bay, or Hodges Bar, so long as the adults were permitted to develop gonads at salinities of 26 to 27 ppt. This was so even though natural salinities at Hodges Bar are very low. But when Hodges Bar oysters were permitted to develop gonads at a salinity of about 8.7 ppt, the optimal salinity for larval development was between 10 and 15 ppt. This optimal salinity appeared to be determined by the salinity at which the parent oysters were held just before spawning. Figure 10–3 shows the growth of larvae at various salinities during different periods of time after they had reached the straight-hinge stage. Growth was best at 17.5 ppt, even though insufficiency of food may have limited growth at this concentration. The rate of successful metamorphosis to the spat stage was also highest at 17.5 ppt. Optimal salinities for early larval development thus appear to be higher than those for later growth and metamorphosis. But optimal conditions rarely occur in nature; the

abilities of oysters and other organisms to adapt to rather broad ranges of conditions contribute to their success.

River discharge and wind action influence the silt load of estuarine waters. Dredging operations necessary for the development and maintenance of navigational channels can at times greatly increase the silt load in these waters. This may endanger oyster propagation by affecting the survival and growth of both adult and larval oysters. In experiments conducted by H. C. Davis, 95, 73, and 31 percent of the American oyster larvae tested developed to the straight-hinge stage at silt concentrations of 125, 250, and 500 mg/l (Loosanoff, 1961). Only 3 percent of the larvae reached this stage at 1000 mg/l, and none did at higher concentrations of silt.

Pesticides in increasing variety and amount have been entering our estuaries, either by way of rivers or by direct application. Davis (1961) has studied the effects of a rather large number of insecticides, weedicides, antibiotics, bactericides, and disinfectants on the survival, development, and growth of American oyster larvae. DDT, probably the most universally distributed pesticide, was found to be one of the most toxic to oyster larvae. At a concentration of 0.025 mg/l, DDT caused a mortality of about 20 percent and drastically reduced growth of

TABLE 10–1   *Relative percentages of normal straight-hinge larvae of the American oyster developing at various salinities from eggs of oysters conditioned and spawned at a salinity of 26–27 parts per thousand (ppt) and from eggs of oysters developing gonads at 8.7 ppt and spawning after 4 days at salinities of 7.5, 10.0, and 15.0 ppt. After Davis (1958).*

SALINITIES AT WHICH LARVAE DEVELOPED (ppt)	OYSTERS CONDITIONED AND SPAWNED AT 26–27 ppt			HODGES BAR OYSTERS SPAWNED AT DIFFERENT SALINITIES		
	Long Island Sound Oysters	Peconic Bay Oysters	Hodges Bar Oysters	7.5 ppt	10.0 ppt	15.0 ppt
26–27	100	100	91	0	0	0
25.0	99	86	94	0	0	0
22.5	100	87	100	0	0	26
20.0	100	86	91	7	26	72
17.5	92	76	81	50	76	92
15.0	37	58	45	48	100	100
12.5	< 0.1	13	11	83	92	89
10.0	0	0	0	100	98	78
7.5	0	0	0	99	96	72
5.0	0	0	0	0	0	0
2.5	0	0	0	0	0	0

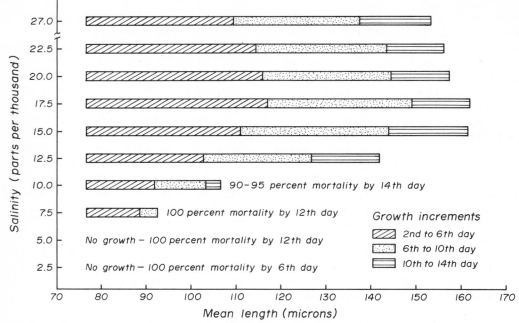

FIGURE 10–3 Growth of larvae of the American oyster at various salinities during different periods of time after the larvae had reached the straight-hinge stage. After Davis (1958).

these larvae; at 0.05 mg/l, growth was almost completely stopped and mortality was in excess of 90 percent. Sevin—an insecticide sometimes used to control burrowing crustaceans that undermine oyster beds—at a concentration of 1 mg/l caused a 40 percent reduction in the number of larvae reaching a normal straight-hinge stage.

Wastes from the pulp and paper industry enter estuaries on both coasts of North America. Puget Sound is one such area; here oyster growers have for years been concerned about the effects that wastes from sulfite process mills may have on oyster propagation. And here the native and highly valued Olympia oyster grows. Our reader will remember that this oyster broods its larvae until they reach the straight-hinge stage, and then they are released into the open water. Woelke (1960) held large groups of Olympia oysters through brood development and larval release in various concentrations of ammonia-base, sulfite waste liquor having a 10 percent solids content. He found that the mean number of larvae released per oyster was greatly reduced at 32 mg/l and that most of the larvae released at concentrations of 16 mg/l or more were abnormal (Table 10–2, Fig. 10–4). The size of the larvae released

declined as the waste concentration increased to 16 mg/l. But at higher concentrations—at which few larvae were released—the mean size of the larvae was larger, probably because of death of smaller and weaker larvae. Waste concentrations up to 32 mg/l appeared to have no deleterious ef-

TABLE 10–2 *Mean numbers of Olympia oyster larvae released per adult, percentages of abnormal larvae, and mean larval size, when adults were held at different concentrations of sulfite waste liquor prior to and during larval development. After* Woelke (1960).

Concen-tration of SWL (mg/l)	Mean Number of Larvae per Adult	Abnormal Larvae (percent)	Mean Length of Larvae (microns)
0	36,408	0.22	157.8
0	48,585	0.15	158.6
2	32,071	0.19	154.9
4	31,367	1.05	153.3
8	31,319	9.24	150.9
16	31,634	65.46	139.2
32	4826	97.22	161.2
64	910	92.48	159.2

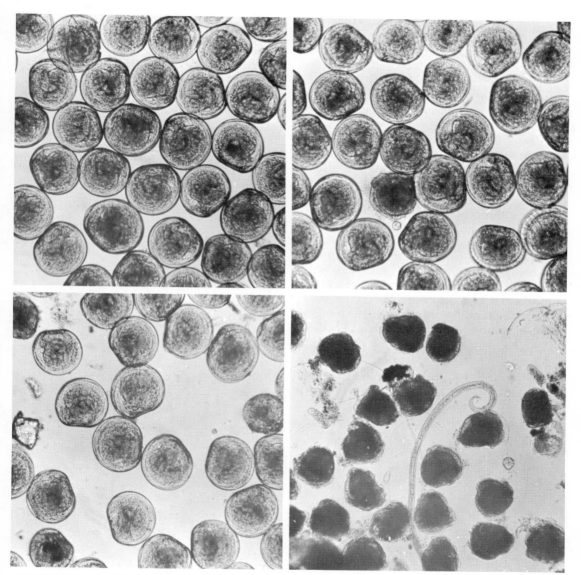

FIGURE 10–4 Straight-hinge larvae released by adult Olympia oysters held in the absence of sulfite waste liquor (upper left), held at a waste concentration of 4 mg/l (upper right), held at a waste concentration of 8 mg/l (lower left), and held at a waste concentration of 32 mg/l (lower right). Nearly all larvae released at a waste concentration of 32 mg/l were abnormal. Normal larvae measured from about 150 to 160 microns. Photographs courtesy of Washington Department of Fisheries.

fects on the later survival, growth, and metamorphosis of larvae brooded and released at concentrations up to 8 mg/l. Larvae brooded and released at 16 mg/l exhibited very low survival and no metamorphosis to spat, even when they were transferred for culturing to water containing no waste. Larvae of the Olympia oyster thus appear to be most sensitive to sulfite waste liquor during the early stages of their development.

The function of reproduction in animals is to replace population losses due to death and emigration. As much as we may know about the influence of environmental conditions on the development of oysters, salmon, and trout, we still poorly understand how developmental changes influence this replacement. And yet about so many species we know so much less. Laboratory studies are necessary to determine how environmental factors influence the development of animals. But laboratory studies alone can never provide the needed understanding, for laboratory environments are simpler than natural ones, and they are also different in other respects. Studies in nature as well as in the laboratory are necessary if we are to understand the importance of changes in the reproductive environments of species.

SELECTED REFERENCES

Coble, D. W. 1961. Influence of water exchange and dissolved oxygen in redds on survival of steelhead trout embryos. Transactions of the American Fisheries Society 90:469–474.

Garside, E. T. 1966. Effects of oxygen in relation to temperature on the development of embryos of brook trout and rainbow trout. Journal of the Fisheries Research Board of Canada 23:1121–1134.

Hoar, W. S. 1957b. The gonads and reproduction. Pages 287–321. In Margaret E. Brown (Editor), The Physiology of Fishes. Vol. 1, Metabolism. Academic Press, New York. xiii + 447pp.

Loosanoff, V. L., and H. C. Davis. 1963. Rearing of bivalve mollusks. Pages 1–136. In F. S. Russell (Editor), Advances in Marine Biology, Vol. 1. Academic Press, New York. xiii + 410pp.

Mason, J. C. 1969. Hypoxial stress prior to emergence and competition among coho salmon fry. Journal of the Fisheries Research Board of Canada 26:63–91.

_____, and D. W. Chapman. 1965. Significance of early emergence, environmental rearing capacity, and behavioral ecology of juvenile coho salmon in stream channels. Journal of the Fisheries Research Board of Canada 22:173–190.

Shumway, D. L., C. E. Warren, and P. Doudoroff. 1964. Influence of oxygen concentration and water movement on the growth of steelhead trout and coho salmon embryos. Transactions of the American Fisheries Society 93:342–356.

Smith, S. 1957. Early development and hatching. Pages 323–359. In Margaret E. Brown (Editor), The Physiology of Fishes. Vol. 1, Metabolism. Academic Press, New York. xiii + 447pp.

11 Bioenergetics and Growth

We may keep a child under observation, and weigh and measure him every day; but more roundabout ways are needed to determine the age and growth of the fish in the sea. A few fish may be caught and marked, on the chance of their being caught again; or a few more may be kept in a tank or pond and watched as they grow. Both ways are slow and difficult. The advantage of large numbers is not obtained; and it is needed all the more because the rate of growth turns out to be very variable in fishes, as it doubtless is in all cold-blooded or 'poecilothermic' animals; changing and fluctuating not only with age and season, but with food-supply, temperature and other known and unknown conditions. Trout in a chalk-stream so differ from those in the peaty water of a highland burn that the former may grow to three pounds weight while the latter only reach four ounces, at three years old or four.

D'ARCY THOMPSON, 1942, pp. 175–176

Many of our gigantic industrial organizations are extremely complex; no one can know all their details. Yet accountants, quite ignorant of these details, render intelligible and useful corporate statements. Likewise, the animal body is extremely complex; no one can know all its details. Yet, as we shall show in this chapter, the time relations of growth can be represented by intelligible, useful, and rational statements or 'laws' of growth. It may be recalled in this connection, that some of the great laws of the physical sciences, such as Newton's law of gravitation, say nothing about detailed mechanisms involved; they are only intelligible, useful, and more or less rational descriptive statements of the phenomenon. There is, of course, a wide range in rationality in many so-called 'laws' of nature. We hope that the following growth equations partake more of laws of nature and less of the accountant's purely empirical rendering of a financial statement.

SAMUEL BRODY, 1945, pp. 493–495

THE INTEREST IN GROWTH AND THE COST OF LIFE

We all have some conception of what is meant by *growth* of an individual organism. Perhaps not often do we pause to reflect on the universality or the importance of growth as a biological process. Neither, perhaps, do we often consider the conditions necessary for growth. Every successful organism must, during some period of its life, grow. And, when not growing, it must replace structural components destroyed by age or harm as long as it lives. The availability of energy and materials for these and other purposes is one of the essential conditions of the environment. But other conditions of the environment will determine an organism's ability to obtain and utilize energy and materials. Here, then, is a most general, a most important, a most difficult, and, perhaps for these reasons, a most fascinating biological problem.

The pattern of growth in time interested early biologists and mathematicians (Buffon, 1777; Quetelet, 1835); and their followers to this day have been intrigued with the possibility of mathematically formulating growth (Minot, 1908; Brody, 1927; Bertalanffy, 1938; Paloheimo and Dickie, 1965). Through the nineteenth century it was the growth of children that received emphasis, "when the exhaustion of the armies of France and the evils of factory labour in England drew attention to the stature and physique of man and to the difference between the healthy and the stunted child" (Thompson, 1942, p. 100). This, then, was interest in the effect of environment on growth in a nature changed by man. It was not until early in the twentieth century that the experimental study of growth began in earnest. We can all appreciate that food availability is one of the environmental conditions determining growth. We know of no species in which many other environmental conditions are not also determining. Growth, as a response of the integrated activities of the whole organism, is often a sensitive parameter of the suitability of the environment.

The success of any people depends upon the production of their food resources or upon their wherewithal to import food. And growth of individual organisms is the basis of the production, or total tissue elaboration, of populations. It is this growth and production of plants and animals that maintains the biological world and sustains man, through his agriculture and use of natural populations. But to grow, the living organism must first maintain the machinery of life, and this requires energy and materials. Growth is only possible after this requirement is met. It was not so long ago, perhaps in the 1920's, that students of animal husbandry, men like Samuel Brody and Max Kleiber (Kleiber, 1961a), began seriously to apply this important concept to the problem of human food production. These men realized that, to understand the growth of which an animal is capable under given environmental conditions, one must first know the fates of the energy and materials in the food it consumes. Though its conception undoubtedly occurred in the eighteenth-century experiments of Priestley, Scheele, Lavoisier, and LaPlace (Kleiber, 1961b), *bioenergetics* was nourished by animal husbandry during the latter part of its embryonic development.

Growth, being one outcome of the integrated activities of the whole organism, is dependent on the metabolic state of the animal and on energy expended in maintenance and behavior as well as on the quantity and quality of the food consumed, all functions of the environment. Bioenergetic studies give us some understanding of how environmental factors affect growth through influences on food consumption and influences on utilization of food energy and materials in the body. Brody (1945), whose monumental *Bioenergetics and Growth* is a continuing source of inspiration, laid a broad foundation for such studies. Ivlev (1945, 1961a, 1961b) paved the way to understanding of the relations between food availability and growth in aquatic communities. And Fry (1947), on the basis of his own beautiful studies on the metabolic rates of fish, suggested relationships between the environment, the metabolism, and the activity of animals that will be useful to us in developing a conceptual framework for the study of growth.

But bioenergetics, being much more than the study of growth, permits us to evaluate the cost of life under different environmental conditions. Even were bioenergetics not essential to understanding of growth, this alone would be sufficient reason for its study. For to be successful, an organism must maintain a balance between the energy and materials it can obtain from its environment and the energy and materials required to sustain the activities essential to survival and reproduction. A favorable environment permits this balance, a balance maintained in different species and under different conditions in different ways.

When man imposes changes on the environments of animals, as when he discharges his wastes into natural waters, often he influences the success—the survival and reproduction—of these animals. The environmental change may influence the quality or availability of their food, or it may influence in other ways the cost of living. If unfavorable, these influences may decrease the distribution and abundance of valuable species, and pollution may be said to be occurring. Studies of bioenergetics and growth provide a means of predicting the effects of environmental changes. Though such studies are usually most effectively pursued in the laboratory, they are of little value unless they can be related to the real problems of life in nature. This is the biologist's most difficult problem. As we shall see, there are touchstones in nature for the biologist with a bioenergetic point of view.

AGE, SIZE, AND GROWTH

The pattern of growth of the individual organism varies through time; it depends not only on environmental conditions but also on the age and size of the organism. For many years, biologists and mathematicians have sought to formulate growth mathematically as a function of age—or time—and body size. Their reasons for this search have been their own, but we can suggest some of the possible reasons why one might seek mathematical formulations of growth. When many observations on the size of animals at different times are available, the representation of these data by means of a mathematical model describing the general pattern of growth facilitates their use for comparative purposes. Such models are also of practical value in computing the total production of populations of growing individuals. Some have attributed biological significance to the

mathematical constants in their formulations, thus suggesting these formulations to be meaningful models of growth. The growth of an animal throughout its life is not likely to be so simply modeled, but the efforts of these biologists have helped to focus attention on crucial problems of bioenergetics and growth. And, with the naïveté so important to science (Lewis, 1926), biologists have perhaps most often turned to the simple beauty of mathematics in hope of finding generalizations or "biological laws" of growth.

The weight of the body of an individual animal from very early in life until its growth is completed usually tends to follow a sigmoid pattern of increase. This *growth curve* typically has its major inflection rather early in the life of the animal and then gradually approaches an asymptote. It is fitting, even in a book on aquatic organisms, that we should begin with Brody's (1927) growth curve for the white rat (Fig. 11–1), for we continue to benefit from the order this great biologist brought to mathematical thought on growth. The absolute growth per unit of time, which Brody termed the *velocity of*

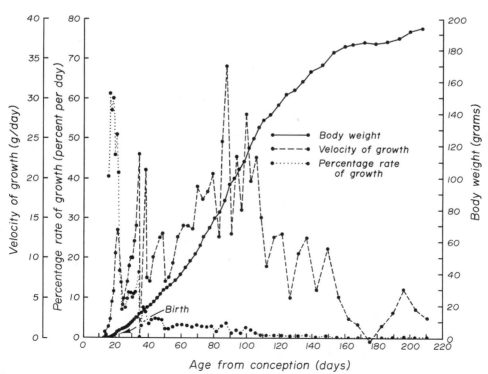

FIGURE 11–1 Sigmoid growth curve of the white rat, its velocity of growth, and its relative growth rate (percentage rate). After Brody (1927).

growth, first increases with time as body size increases, and then it decreases as relative growth rate continues to decrease. *Relative growth rate*, Brody's percentage rate of growth, is the change in body weight per unit of time relative to existing body weight. Relative growth rate tends to be high early in life and to decline with increasing age. Never in nature and rarely in the laboratory do we have daily measurements of growth of animals, and we should be warned by Brody's data that the smooth sigmoid patterns of growth our data sometimes suggest may obscure important physiological and ecological events in the lives of the animals we study. The remarkable drops in relative growth rates that Brody's rats exhibited at birth and 13 days later are hardly apparent on the sigmoid curve. Such important events in the life of an animal may be lost to our attention through smoothing of curves or mathematical models ill-equipped to deal with physiological or ecological realities.

In temperate climates particularly, the growth of fish and other poikilothermic animals varies seasonally with changes in temperature and food availability. Often, data on the size of fish are available mainly at the time of harvest. It is not unusual, then, for growth curves of fish to be based on samples separated by intervals of as much as a year; and such curves of growth may be generally sigmoid and appear quite smooth (Fig. 11–2A). Were samples to be taken at monthly, or weekly, or even daily intervals, as in the case of Brody's rats, it would often be found that over perhaps half of each year the fish grew hardly at all (Fig. 11–2A) and that the actual growth curve was far from smooth. This is not to say that over the life of a fish its pattern of growth is not generally sigmoid; it usually is for fish and most other animals. Neither is it to say that curves expressing only mean changes in body size over each year of life are of no value; they are of value. But it is to say that if we are to really understand the effects of environmental changes on the growth of animals, we must have information on their changes in body size at short and frequent intervals throughout their lives. Mathematical models describing only the general pattern of growth throughout the life of an animal contribute little to this understanding.

Most terrestrial vertebrates, having nearly attained the asymptotes of their growth

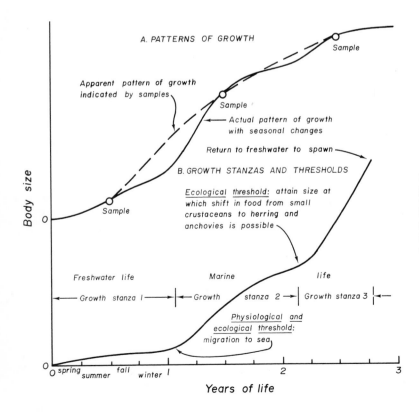

FIGURE 11–2 A. Actual pattern of growth and pattern of growth indicated by yearly sampling of some fish species. B. Growth stanzas and physiological and ecological growth thresholds for a hypothetical species of anadromous salmonid.

curves, can grow little more, even were their environmental conditions to improve greatly, for the hardening of the supportive structures of their bodies to resist gravity then permits little more than fattening. Some aquatic organisms, including fish, have no such problem; and, having nearly reached the body size asymptotic for one environment, they may again begin to grow rapidly, regardless of age or body size, if environmental conditions improve markedly. A fish that has approached through a sigmoid pattern of growth an asymptotic body size characteristic of one environment may move or be moved to a more favorable environment in which it can again begin to grow rapidly, this time toward an asymptotic body size characteristic of the more favorable environment. Fisheries biologists (Vasnetsov, 1953; Parker and Larkin, 1959) have come to consider distinctive patterns of growth in the life of a fish to be *growth stanzas*, which are separated by *physiological* or *ecological thresholds*.

Let us consider an anadromous salmonid that spends the first year of its life in freshwater and then migrates to the sea, later to return to freshwater to spawn in the third year of its life (Fig. 11–2B). For reasons not yet clearly understood, growth is likely to slow down as our salmonid approaches the end of its first year of life in freshwater; this indicates completion of its first growth stanza (Fig. 11–2B). But physiological and morphological changes in the salmonid then make it possible for it to migrate to the sea, and here food may be relatively more abundant than it was in the stream from which the fish migrated. Our fish has then crossed a physiological and ecological threshold and entered its second growth stanza. Feeding on small crustaceans, it will again grow rapidly until the cost of obtaining its food approaches the value of the food, and growth must slow down. But by now, the salmonid may have attained a size at which it can prey on small fish, giving it a more favorable energy and material balance, and it will cross an ecological threshold into its third growth stanza where rapid growth is again possible. We know all too little about the growth of aquatic animals, but the phenomenon of growth stanzas and thresholds is undoubtedly very general. When it occurs, its causal relations must be defined, if we are to understand the effects of environment on growth.

The growth rate of an animal is some function of its body size, even while being dependent upon environmental conditions. This idea is the basis of nearly all mathematical formulations of growth. The first really satisfactory formulation of growth was that of Brody (1927), a formulation that is still among the most useful. Brody recognized the generally sigmoid pattern of growth, and he believed preinflection and postinflection growth to be distinct phases in the life of an animal, phases that should be represented by different formulations. Time has shown us, as Brody thought, that growth phases must be given distinct consideration, if we are to learn much.

Brody (1927, 1945) showed, at least for the terrestrial vertebrates he studied, that growth in the preinflection part of the sigmoid growth curve is exponential and that change in body weight with respect to time can be represented by the differential equation

$$\frac{dW}{dt} = kW \qquad (1)$$

where k is the *instantaneous coefficient of growth*. When integrated, this equation yields

$$W_t = W_0 e^{kt} \qquad (2)$$

where W_0 is the weight of the animal when time $t = 0$. Written in logarithmic form, equation 2 gives us the convenient equation for a straight line:

$$\text{Log}_e W_t = \text{Log}_e W_0 + kt \qquad (3)$$

From equation 1 we can see that the *velocity of growth*, dW/dt, is proportional to the weight already attained and increases as weight increases. Rearranged, equation 1 yields

$$k = \frac{dW/dt}{W} \qquad (4)$$

and we can see that k, the instantaneous coefficient of growth, is a *relative growth rate*, a growth rate relative to body weight. If we multiply k by 100, we have the percentage rate of growth per unit time. When growth is exponential, k can be calculated from body weights at two times as follows:

$$k = \frac{\text{Log}_e W_2 - \text{Log}_e W_1}{t_2 - t_1} \qquad (5)$$

Over short time intervals, *average relative growth rate* is sometimes a convenient and appropriate statistic:

$$GR = \frac{W_2 - W_1}{0.5\,(W_1 + W_2)\,(t_2 - t_1)} \qquad (6)$$

Whereas Brody considered the velocity of growth in the preinflection phase of a sigmoid growth pattern to be proportional to the weight already attained, he considered the velocity of growth during the postinflection phase to be proportional to the growth yet necessary to reach the asymptotic weight. Thus, in this second phase of growth, Brody (1927, 1945) represented the velocity of growth by the differential equation

$$\frac{dW}{dt} = k(A - W) \qquad (7)$$

where A is the asymptotic weight. Integrating equation 7, we can obtain an equation for weight at different times during postinflection growth:

$$W_t = A - Be^{-kt} \qquad (8)$$

where B is a constant of integration having no biological significance. Instead of using B to correct for the fact that age is figured from birth whereas equation 8 is only appropriate after the inflection, we could begin counting time where an extrapolation of the postinflection curve would meet the age axis. Then, equation 8 can be written in the form

$$W_t = A(1 - e^{-k(t - t_0)}) \qquad (9)$$

a form to which we will return.

In attempting to derive a mathematical model of growth in which the constants have biological significance, von Bertalanffy (1938) started with the differential equation

$$\frac{dW}{dt} = HS - kW \qquad (10)$$

where S is the food absorptive surface of the gut, taken to be proportional to the square of body length, W is the weight, taken to be proportional to the cube of body length, and H and k are proportion constants. The idea here is that growth is dependent on the difference between food absorbed, assumed to be proportional to gut surface, and food metabolized, assumed to be proportional to weight. Ricker (1958, p. 196) has quite correctly pointed out that the surface of the gut can rarely limit food assimilation in nature, for food availability there does not permit animals the comfort of continuously full stomachs. For this reason and others (Parker and Larkin, 1959), we cannot accept equation 10 as a valid beginning for a biologically meaningful model of growth. But von Bertalanffy's equation for growth in length, derived from equation 10, has been useful in

fishery theory and management (Beverton and Holt, 1957). This equation—which has the same mathematical form as Brody's equation for postinflection growth in weight (equation 9)—yields an equation for a sigmoid pattern of growth in weight when it is transformed to express size in weight rather than in length.

Parker and Larkin (1959), attempting with a growth equation to summarize the interacting physiological and ecological factors influencing growth within a given growth stanza, began with a differential equation:

$$\frac{dW}{dt} = kW^x \qquad (11)$$

Thus, Parker and Larkin considered growth in any phase to be some function of weight already attained. They noted that many physiological processes have been shown to be proportional to some power function of body weight and that the x of W^x may be expected to have values around 0.7, according to the metabolic state of the animal in any growth stanza. They suggested that the different values of the constant k in different growth stanzas are related to differences in ecological opportunity. Parker and Larkin presented no evidence for such biological significance of these constants, but the equation for weight at different ages that they derived from equation 11 is useful for describing growth in different stanzas.

For those of us who believe that the bioenergetic point of view is perhaps most useful for increasing understanding of the physiology and ecology of growth in nature, the important papers of Paloheimo and Dickie (1965, 1966a, 1966b) are of particular interest. These authors began with the idea that the velocity of growth is equal to the difference between the rate of food consumption (R) and the total metabolic rate (T) expressed in equivalent energy terms:

$$\frac{\Delta W}{\Delta t} = R - T \qquad (12)$$

This becomes

$$\frac{\Delta W}{\Delta t} = R - \alpha W^x \qquad (13)$$

if metabolic rate is considered to be proportional to some power function of body weight, as is often done. Here, α takes on different values with different levels of metabolic rate. Careful examination of available data on the food consumption, metabolic rate, and growth of fish in the laboratory led Paloheimo and Dickie (1966a, 1966b) to suggest some interesting ideas on relations among environmental conditions, rations, body sizes, and growth efficiencies, ideas that merit careful consideration in future studies, both in the laboratory and in nature. It is of interest that the growth equation Paloheimo and Dickie derived from equation 13 and that which Parker and Larkin (1959) derived from equation 11 are equivalent. Of course, as these investigators all well know, the adequacy with which an equation describes the growth of an animal may be quite unrelated to any biological significance of its constants.

For now, let us not expect too much of available mathematical formulations of the growth of animals through time. Perhaps one day it will be possible to develop mathematical models of growth in which the values the different constants assume, when growth is defined under different environmental conditions, are related in predictable ways to particular environmental factors. But such models will never come from routine description of the growth of animals apart from detailed knowledge of the bioenergetics of these animals in relation to ecological conditions. The solution of wonderfully complex biological problems cannot depend on mathematics alone. And mathematical formulations of growth have important uses apart from the biological significance of their constants, some of which we suggested in the beginning of this section. One of these uses is to raise questions concerning growth, as Paloheimo and Dickie have done.

THE FATES OF CONSUMED FOOD

To sustain life and growth, food must fuel the body, provide catalysts and other materials essential to its operation, and furnish the building blocks of tissues. Proteins, fats, carbohydrates, vitamins, and minerals must be available not only in sufficient amount but in proper kind and balance. Amount, kind, and balance of nutrients determine the fates of food consumed and the growth possible, as do other environmental conditions and the age and size of the animal itself. To understand how environmental conditions, including the

quantity and quality of food available, will influence growth, we must first consider the possible fates of food that is consumed. The formation of tissues from the materials of digested food is growth, one of the possible fates of consumed food. But the availability of materials for this fate, growth, is very much dependent on the distribution of energy and materials among the other possible fates. This distribution will depend on the quantity and quality of the food consumed and on energy and material requirements to maintain life and necessary activities, including the search for food. To grow, as we noted in the introduction to this chapter, an organism must maintain an energy and material balance favoring growth. Environmental conditions will be primary determinants of whatever balance develops.

What, then, are the possible fates of consumed food? Very generally, they can be represented as we have done in Figure 11–3. Because some of the food consumed is oxidized to provide energy, some is lost as waste materials, and some is converted to materials of the body, it is both convenient and informative to use heat or caloric units as a common denominator. But however convenient and informative may be the use of energy terms in describing the distribution of consumed food, we must not neglect the quantitative and qualitative distribution of materials, knowledge of which is also necessary to understand metabolism and growth.

Of the food materials consumed, which we represent as their heat of combustion as determined in an oxygen-bomb calorimeter, some are *assimilated*—more properly, absorbed—through the intestinal wall after undergoing digestion in the stomach and the intestine. The remainder are passed from the intestine as *fecal wastes* (Fig. 11–3), which can also be represented in caloric units. Not all the absorbed materials are available for physiological work or growth, for some *nitrogenous materials* cannot be metabolized by the animal's body and are excreted as wastes through the kidneys or gills. The remaining *metabolizable materials* are either metabolized or appear as growth. The energy content of this category of materials is sometimes considered to be the *physiological fuel value* of the food. Physiological fuel value is not bomb-calorimeter fuel value, because nitrogenous

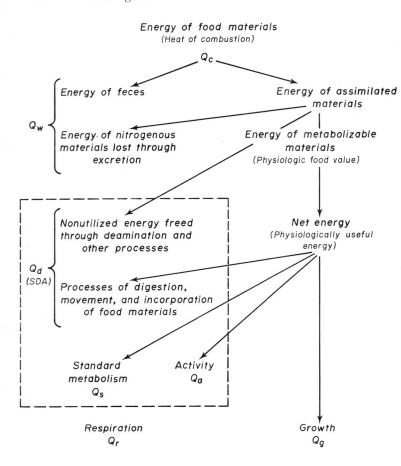

Energy of food materials
(Heat of combustion)

Q_c

Energy of feces

Q_w

Energy of assimilated materials

Energy of nitrogenous materials lost through excretion

Energy of metabolizable materials
(Physiologic food value)

Nonutilized energy freed through deamination and other processes

Q_d
(SDA)

Net energy
(Physiologically useful energy)

Processes of digestion, movement, and incorporation of food materials

Standard metabolism
Q_s

Activity
Q_a

Respiration
Q_r

Growth
Q_g

FIGURE 11–3 The fates or categories of losses and uses of the energy and materials in the food an animal consumes. After Warren and Davis (1967).

components of proteins that oxidize in a calorimeter cannot be oxidized by animals. The calorimeter combustion value of protein is about 5.65 kcal/g, whereas the bodily oxidation value is only about 4.35 kcal/g.

Winberg (1956) and some others have quite incorrectly considered the physiological fuel value category used by Brody (1945) to be equivalent to *physiologically useful energy* or materials. Not all the energy and materials represented by physiological fuel value are available for physiological work or growth. Energy released as heat in the deamination of amino acids prior to the excretion of their nitrogenous components is not generally available to animals for useful purposes. It is the physiological fuel value remaining after subtraction of this heat loss that can properly be considered physiologically useful energy and that Brody (1945) and others considered to be *net energy* (Fig. 11–3).

It has long been known that the consumption of a meal occasions an increase in the rate of oxygen consumption and heat production of an animal. This increase in metabolic rate has come to be known as the *specific dynamic action* (SDA) of the food consumed. Perhaps mainly because this seemingly simple change in metabolic rate is only the final result of many changes in intermediary metabolism caused by food intake, such understanding as we have today of specific dynamic action has had a difficult history, a history which Brody (1945) and Kleiber (1961b) have traced. It is now generally accepted that SDA mainly results from the utilization of oxygen necessary to deaminate amino acids not to be used, for whatever reasons, in the formation of protein in growth or tissue repair. Generally, an animal is unable to benefit from the energy involved or the heat released in this process. But a homeothermic animal at an environmental temperature below that of its body will exhibit a smaller increase in oxygen consumption upon ingesting a given ration than when it is isothermic with its environment, because heat released through deamination can be used to maintain body temperature, thus reducing oxidation of other materials for temperature maintenance. And an animal will exhibit a smaller increase in oxygen consumption upon ingesting a given ration when it is doing muscular work than when it is not, for deamination that would otherwise contribute to SDA can provide energy necessary for this work. Various conversions of carbohydrates and fats also involve some nonproductive release of heat, but these contribute less than do amino acids to the total SDA of food (Brody, 1945, p. 66). It is this loss of the physiological fuel value of food that prevents all of it from being physiologically useful, as we explained in the previous paragraph and illustrate in Figure 11–3.

But however little of the specific dynamic action we measure is occasioned by useful physiological processes, these processes require energy; and they are far too important to ignore, as some appear to do. The processes of digestion, movement, and storage are physiologically useful and necessary; and their costs, even though small, should properly be charged to the net energy category (Fig. 11–3). And when storage or deposition of food materials involves the expenditure of energy to organize them into storage products, cells, or tissues having higher energy levels, again physiologically useful work is done; this is the work of growth, and it is not free. Some of the confusion about specific dynamic action—or calorigenic effect, an equivalent expression—stems from the fact that it is a name historically applied to a convenient measurement of change in metabolic rate, a change some tried to attribute to physiological work and others to various causes of energy loss. Mainly, as we have tried to explain, SDA is a measure of energy loss, but we find neither simplicity nor clarity in trying to attribute it all to energy loss. We should not use *specific dynamic action*, only the name of a measurement, as an explanatory principle.

Because SDA results mainly from the deamination of amino acids, its magnitude is determined primarily by the amounts and kinds of amino acids ingested and their fates in the animal body. If the rate of ingestion of amino acids is greater than their rate of utilization in protein synthesis, excess amino

acids must be deaminated, permitting excretion of the amino fractions and biological oxidation or storage of the carbohydrate fractions. Similarly, if imbalance in the amino acids or other components in the food restricts the rate of amino acid utilization in protein synthesis, SDA will be increased. Accumulation of free amino acids in the body fluids could create a condition toxic to the animal. The rate of growth of an animal is determined not only by the quantity and quality of food available but also by its metabolic state, which is very much dependent on environmental conditions. Thus, in an unfavorable environment, SDA may be increased. Other fates of food consumed are also influenced not only by its quantity and quality but by the total environmental complex as well.

It is convenient to separate the remaining oxidation of food materials into that required for *standard metabolism* and that necessary for *activity* (Fig. 11–3). We can take the standard metabolism of an animal to be its rate of energy utilization under existing environmental conditions when it is resting and not processing food materials. As such, standard metabolism is approximately equivalent to the *basal metabolism* measured in clinical studies of man. The cost of activity we can take to be the increase in rate of energy utilization occasioned by muscular performance, energy utilization over and above that necessary when an animal is in a standard metabolic state.

Finally, then, the remaining derivatives of ingested food materials are available for utilization in the synthesis of new tissues, which we shall call *growth* (Fig. 11–3). Some prefer to distinguish between the accumulation of adipose (fat) tissue and the synthesis of other tissues, including bone or hard structures; this is a useful distinction. But however we choose to define growth, we must account for food materials deposited as fat. To understand growth, we must first know how interactions between the organism and its environment determine the quantity and quality of food consumed and the fates of this food other than growth.

A BIOENERGETIC BUDGET

The possibility of developing meaningful energy and material budgets for animals derives from the general validity of the *first and second laws of thermodynamics*. The first law of thermodynamics, the *principle of conservation of energy*, states: when the chemical energy content of a system changes, the sum of all forms of energy given off or absorbed by the system must equal the magnitude of the change. The same is said for matter in the *principle of conservation of matter*. Work requires energy, but not all the energy in a system is free or available for work, as in the case of evenly distributed heat. Any process involving the transformation of energy from one form to another, as when work is done, requires degradation of the energy level of some material and heat develops; the transformation is never 100 percent efficient and the tendency is for energy to become evenly dispersed as unavailable heat energy. Perhaps this will serve us as a statement of the second law of thermodynamics. From its food or bodily stores, an animal must obtain the chemical energy necessary for the work of the body, energy it will transform to heat through biological oxidation; and from its food an animal must obtain the materials for growth. Growth materials, waste materials, and heat produced must, over some period of time, account for the energy and materials of the food consumed. These must balance, whether all are in terms of energy or all are in terms of materials. We will use an energy budget employing caloric units.

Proof of the applicability of the first law of thermodynamics to animals is generally attributed to Rubner (1885, 1894). But, beginning with Lavoisier (1777) and his idea of animal heat, even before this law was formulated, there were many early contributions to bioenergetics. Unlike scientists in animal husbandry, ecologists were late in strengthening their field with these ideas. Ivlev (1939a, 1939b, 1939c, 1945) was perhaps the first ecologist to utilize energy budgets of individual organisms in a manner likely to yield much understanding of their population and community problems; the rest of us have yet failed to assimilate fully the perceptive approaches he suggested in his later publications (1947, 1961a, 1961b). There are difficulties in the details of Ivlev's (1945) energy budget for the individual organism, difficulties that needed to be noted (Winberg, 1956; Warren and Davis, 1967) but on which we cannot harp, with our respect for the memory of this fine Russian biologist.

Drawing heavily on the ideas of Ivlev (1945), Brody (1945), and Fry (1947), Warren and Davis (1967) proposed a formulation of the energy budget of an individual animal that will undoubtedly be improved as more is learned. We consider the formulation to be useful to the extent that its terms are independent and measurable and to the extent that it suggests ways of relating the bioenergetics of the individual organism to the problems it faces in its natural environment. On the basis of our discussion concerning the fates of food consumed, as represented in Figure 11–3, we can begin with the following equation:

$$Q_c - Q_w = Q_g + Q_r \qquad (14)$$

where

Q_c = energy value of food consumed,

Q_w = energy value of waste products in feces, in urine, and lost through gills and skin,

Q_g = total change in energy value of materials of body (growth), and

Q_r = energy metabolically degraded with or without benefit to the organism.

Now if we let

$$Q_r = Q_s + Q_d + Q_a \qquad (15)$$

where

Q_s = energy equivalent to that degraded in the course of metabolism of unfed and resting animals (standard metabolism),

Q_d = additional energy degraded in the course of digestion, assimilation, and storage of materials consumed (SDA), and

Q_a = additional energy degraded in the course of muscular activity over and above that of the resting animal,

then, substituting, we can represent the overall energy budget of an animal for any given period of time, in terms that are convenient for our purposes, as follows:

$$Q_c - Q_w = Q_g + Q_s + Q_d + Q_a \qquad (16)$$

Equation 16 can be used to describe the uses and losses of the energy and materials in the food an animal consumes over a period of time under a given set of environmental conditions. For physiologists and ecologists interested in effects of the environment on animals, equations like this are valuable when used to determine how changes in particular environmental factors or combinations of factors influence food consumption and the uses and losses of energy and materials in the food consumed. We will consider this important use of energy budgets in succeeding sections of this chapter. But first we must discuss the measurement and interpretation of the terms in equations 14, 15, and 16.

Although the growth of animals can be measured with reasonable precision both in the laboratory and in nature, only in the laboratory can we estimate with much certainty the amount of food consumed and the other fates of this food. Nevertheless, with laboratory experiments carefully designed to approximate important factors in the environment, we may be able to approximate the energy budgets of some animals in nature. This is important, because, after all, it is the animal in nature that concerns us.

In the laboratory, we can measure the amount of food an animal consumes over a given period of time. Here, too, we can measure its change in weight during the same period. Samples of the food being fed to the animal can be oxidized in an oxygen-bomb calorimeter to determine its heat of combustion; with this information, the amount of food consumed can be converted from weight to caloric units. Likewise, the change in the weight of the animal during the time interval can be converted to change in the total caloric value of its body tissues, with information on its final caloric value per unit of weight and information on the caloric value per unit of weight of a similar animal at the beginning of the time interval. We can thus represent Q_c, food consumed, and Q_g, change in body mass, in caloric units in equation 14. It may sometimes be useful, in our attempts to represent growth, to substitute two terms for Q_g, one for fat and one for protein, each of which could then be represented in caloric units. When an animal is receiving less than a maintenance ration, its body mass will decline over a period of time; and we may then wish to represent growth as a negative value, so that our equation will continue to be useful.

The determination of waste materials, both those not assimilated and those excreted as nitrogenous products after assimilation, can be a problem with aquatic animals. The fecal wastes of animals fed known amounts of food and held under

desired environmental conditions can be perhaps best determined by wet combustion methods employing powerful oxidants (Davis and Warren, 1965). Nitrogenous wastes, predominantly ammonia, may not be oxidized by these methods but can be determined separately by colorimetric tests. Once the percentages of food consumed appearing as these wastes have been determined, estimates of Q_w, in caloric units, can be made from known consumption values for use in equation 14.

By these or similar methods, we can estimate three of the terms in equation 14. The fourth term, Q_r, the total metabolic rate of the organism—its total respiration or heat production—can then be estimated by difference. Thus, generally, when we know the food consumption, growth, and waste materials of an animal in caloric terms, we can estimate its total metabolic rate.

But we will come to better understand the effects of environmental factors on the growth and cost of life of animals if we can partition their total metabolic rates among standard metabolism (Q_s), specific dynamic action (Q_d), and activity (Q_a), as in equation 15. This will require additional experiments; but if these are conducted under appropriate conditions, their results can be combined with those of experiments on the food consumption and growth of animals to yield overall energy budgets, like equation 16, for animals under environmental conditions that interest us.

We have defined the standard metabolic rate (Q_s) of an animal to be its rate of oxygen consumption or energy utilization when it is unfed and resting. For fish, at least, this can be determined by measuring their rates of oxygen consumption at different levels of activity—as when they are forced to swim at different velocities—and then by extrapolating to the zero activity level a line fitted to the lower metabolic rates at the different activity levels (Beamish, 1964b; Brett, 1964). This has been done in Figure 11–4 for juvenile coho salmon (*Oncorhynchus kisutch*) at different temperatures. The increase in standard metabolic rate with temperature is evident. If we are to make use of information on rates of oxygen consumption in equations 15 and 16, we must convert milligrams of oxygen consumed to calories of food or body material metabolized. For this purpose, an *oxy-calorific coefficient* is generally employed. Although the exact value of this

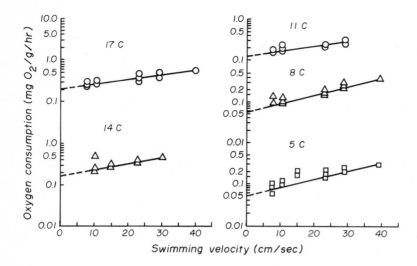

FIGURE 11–4 Estimation of standard metabolic rates of juvenile coho salmon (mean weight 8.7 g) at different temperatures by extrapolation of lines fitted to their lower metabolic rates at different swimming velocities. After Averett (1969).

coefficient is dependent on the material being metabolized—some combination of fat, carbohydrate, and protein—it will usually be 3.42 cal/mg (4.89 cal/ml) of oxygen consumed plus or minus 1.5 percent (Winberg, 1956; Kleiber, 1961a; Brody, 1945). Since equations 14, 15, and 16 usually represent energy budgets for the whole animal for some period longer than an hour or a day, it is necessary to expand metabolic rate data to the appropriate time and body dimensions.

Specific dynamic action (Q_d), the increase in metabolic rate of an animal occasioned by food consumption, can best be measured as the difference between the metabolic rate of an animal fed a known amount of food and that of the same animal in the postabsorptive condition (Fig. 11–5). To prevent variation in metabolic rate caused by random activity of the animal, it is sometimes helpful to maintain the animal at some low, fixed level of activity. Metabolic rate may be elevated for up to 24 hours after food consumption,

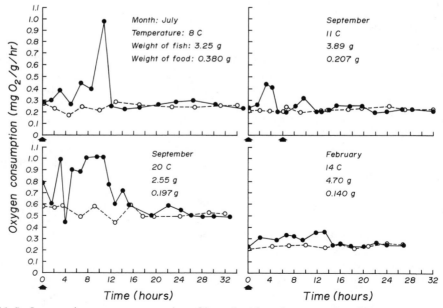

FIGURE 11–5 Increase in oxygen consumption of juvenile coho salmon occasioned by specific dynamic action of food fed at times indicated by arrows. Solid lines are for fish recently fed and broken lines are for same fish in postabsorptive condition. Areas between lines represent total specific dynamic action of food consumed. After Averett (1969).

and it is necessary to measure the total additional expenditure of energy occasioned. We must note here that higher levels of swimming activity lead to less loss of energy through specific dynamic action. Thus, measurements of the specific dynamic action of a given ration when animals are at one activity level cannot be used to estimate the specific dynamic action of a similar ration for animals at another activity level. Both quantity and quality of food, as well as temperature and other environmental factors, must be considered, if information on specific dynamic action is to be of much value.

If, as we have done, equation 16 is taken to represent all acquisitions, utilizations, and losses of energy and materials by the entire animal over some convenient time period of days or weeks, we still have the problem of representing the food consumed and its fates in rates per unit of body mass (or energy content) per unit of time. Information on total food consumption, growth, and metabolism for the entire animal is very useful, but it tells us something different than does information on food consumption, growth, and metabolism per unit of body mass. An increase in temperature may increase the rates of food consumption, growth, and metabolism and lead to a larger animal; but, then, the larger animal may eat more and grow faster because it is larger, not just because the temperature is higher. In an earlier section of this chapter, we discussed how growth was a function of body size; so also are food consumption and metabolism. If we are to distinguish the primary impetus of an environmental change on growth and metabolism from a secondary impetus, change in body mass, we must have not only an energy budget for the entire animal but also an energy budget in terms of rates per unit of body mass per unit of time.

If we assume that food consumption, growth, waste products, and the various categories of metabolism are the same function of body size, we can rewrite equation 16 in rate terms as follows:

$$A_c - A_w = A_g + A_s + A_d + A_a \qquad (17)$$

where the Q's of equation 16 and the A's of equation 17 are related by

$$Q_i = A_i W^x t \qquad (18)$$

and

$$A_i = \frac{Q_i}{W^x t} \qquad (19)$$

where

W = mean body weight or energy value,
t = time in days,
x = some mean power of W, and
i = c, w, g, s, d, or a.

Strictly, we cannot expect all the terms in equation 17 to be the same function of body size. Still, at this stage of knowledge of the bioenergetics of animals, our problem continues to be the practical one of representing our data in terms that will help us to relate the activities of animals to the problems they face in their environments. There is some evidence that food capacity, growth, and total metabolism of animals are proportional to about $W^{0.8}$ (Warren and Davis, 1967, p. 183). By using the same mean power of W — such as $x = 0.8$, or even $x = 1.0$ — we can develop simple, though not strictly correct, energy budgets of animals in rate terms. Remembering the limits such approximations must place on further theoretical development of our rate budget, we will still find it provides a useful basis for interpreting our data on the effects of environment on the bioenergetics and growth of animals. We are now ready to bring the various components of the environment into our conceptual framework for study of these environmental effects.

ENVIRONMENT, BIOENERGETICS, AND SCOPE FOR GROWTH

Our equations 16 and 17 can at one time describe the bioenergetics of an animal only under one set of environmental conditions. We must add another dimension to our conceptual framework of bioenergetics and growth if we are to understand how environmental factors, singly and together, interact with the organism to determine its food consumption, growth, and cost of life under different conditions. In this, it will be helpful if we can find a classification of environmental factors according to how they can be expected to influence the metabolism and thus the activity of animals.

Fry (1947), in one of the most useful conceptual papers in ecological literature, has provided us with such a classification and with a rationale for relating changes in an animal's metabolic capacity to changes in its environment. Fry considered the metabolic capacity that an animal has for activity under a given set of environmental conditions to be the difference between its standard and its active, or maximum, metabolic rates. This difference he considered to be the animal's *scope for activity* under the given conditions. With changing environmental conditions, standard and active metabolic rates can be expected to change, and thus the animal's scope for activity, its metabolic capacity for activity, changes. Activity, as defined by Fry, is more inclusive than the muscular activity to which we assign the metabolic cost Q_a or A_a in equations 16 and 17. According to Fry (1947), activity is what an animal does, whereas metabolism is composed of the internal energy-availing processes making activity possible. Thus defined, activity may be, as examples, food capture, migration, reproduction, excretion, or growth. Scope for activity is the metabolic capacity an animal has, under particular environmental conditions, for all its activities over and above those necessary to maintain the animal in a standard metabolic state. In a way, then, scope for activity is the total metabolic capacity an animal has available to meet the ecological realities of life in nature.

Fry (1947) classifies environmental identities as lethal factors, masking factors, accessory factors, directive factors, controlling factors, or limiting factors, according to the manner in which they may be influencing an organism at a given time and place. A particular environmental identity may act at different times as a different kind of factor, or it may sometimes act simultaneously in two or more ways. Temperature, for example, can sometimes be a lethal factor and sometimes a controlling factor, but it could simultaneously be a controlling factor and a directive factor. We will here consider in detail only controlling and limiting factors, for they are most important to our conceptual framework of bioenergetics and growth, both factors governing, in different ways, the metabolic rate. *Controlling factors* govern the metabolic rate by influencing the state of activation of the metabolites, and thus they influence both active and standard rates; temperature may act in this way. According to Fry, a material is a *limiting factor* only when its availability limits the metabolic rate. Generally a limiting factor governs only the active metabolic rate, for were it to decrease the standard rate it would become a lethal

factor. Oxygen concentration in the aquatic environment is often a limiting factor. Both temperature and oxygen, through their effects on metabolism, have much to do with bioenergetics and growth.

But in the study of bioenergetics and growth, we must consider, in addition to an organism's metabolic state, its acquisition of energy and materials and their fates. Food availability as well as temperature, oxygen, and other controlling and limiting factors will influence metabolic state and the acquisition and fates of energy and materials. Since growth can be considered an activity, Warren and Davis (1967) took a liberty with Fry's concept of scope for activity in suggesting that a *scope for growth* could be usefully defined as the difference between the energy value of all the food an animal would and could consume and the energy value of all uses and losses of food other than growth, under a particular set of environmental conditions. As defined, scope for growth is not equivalent to scope for activity. Food consumption rate (A_c) is not active metabolic rate but will be influenced by and will influence the total metabolic rate; utilizations and losses of food other than growth include standard metabolism (A_s), but also much more (A_d, A_a, A_w). Nevertheless, taking this liberty with Fry's concept permits us to take advantage of his way of relating the performance of an animal—growth, A_g, in this instance—to the action of factors in the environment. Thus, we can consider not only how temperature and oxygen concentration may influence the metabolism of an animal but also how they may interact with food availability in determining the bioenergetics and growth of the organism.

In Figure 11-6, we have attempted to illustrate the effects changes in food availability and temperature, under natural conditions, might be expected to have on the bioenergetics and scope for growth of a hypothetical poikilothermic animal. To do this, we have, at each of four levels of food availability, shown how increasing temperature might influence the terms in equation 17: food consumption (A_c), waste materials (A_w), activity (A_a), specific dynamic action (A_d), standard metabolism (A_s), and the possible growth (A_g). The total energy and materials obtained in the food at any temperature, expressed in calories per kilocalorie of body material per day, is apportioned among the

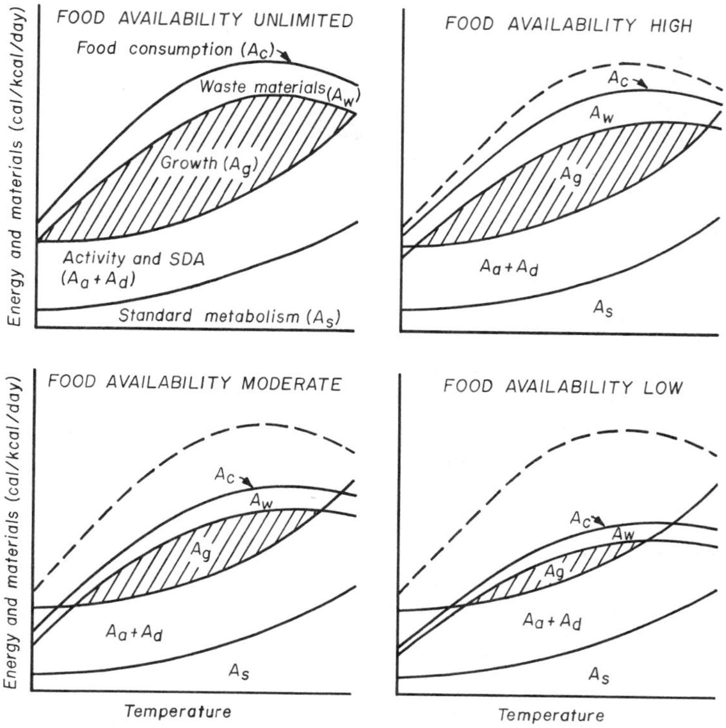

FOOD AVAILABILITY UNLIMITED

FOOD AVAILABILITY HIGH

FOOD AVAILABILITY MODERATE

FOOD AVAILABILITY LOW

FIGURE 11–6 Theoretical effects of temperature change on the food consumption, energy budget, and scope for growth of a hypothetical poikilothermic animal having food available in different amounts.

possible fates. If we assume that food consumption in nature is a function of activity, which is a function of temperature, food consumption can be expected to increase to some maximum and then decrease, with increasing temperature (Fig. 11–6). The energy value of fecal and nitrogenous waste materials must be removed from that of the food materials to obtain the energy value of assimilated materials to be metabolized or to appear as growth.

The standard metabolic rate of a poikilothermic animal generally increases with temperature and thus requires at higher temperatures more of the energy value of the food materials. Specific dynamic action tends to increase with increasing food consumption, and near the two limits of the thermal range it might increase. At any given level of food availability, activity and its cost have generally been found to be highest in the middle of an organism's temperature range. Specific dynamic action and activity in this respect, then, might to some extent tend to be compensatory. But food availability further complicates our problem of simple diagrammatic representation. With increasing food availability in nature, consumption rate generally increases; thus,

specific dynamic action increases. However, the cost of food capture may be expected to decrease with increasing food availability and in this way tend to decrease the overall cost of activity. These two components of metabolic rate may here again tend to be compensatory. Finally, as we noted under The Fates of Consumed Food, the specific dynamic action occasioned by a given ration when an animal is at a low level of activity is greater than when it is at a high level of activity. Such very tentative and complex reasoning as the foregoing not only is difficult to follow but also is of doubtful value. We only wish to suggest that these two components of metabolic rate tend to be compensatory, thus leading to a certain metabolic homeostasis. There is some evidence for this, as we shall see in a later section of this chapter. For now, we wish only to use, for diagrammatic purposes, the constancy of metabolic rate deriving from any such homeostasis. Thus, for simplicity, in Figure 11–6 we have represented the sum of the energy values for specific dynamic action and activity to be constant with changing food availability and temperature. The complex interactions of these components of metabolic rate with environmental factors probably cannot be generally and simply represented. But perhaps, so long as our reader is forewarned, we may be permitted this freedom making it possible to show total metabolic rate at a given temperature constant with changes in food availability. Then, we can explain rather simply our concept of how scope for growth may change with changing temperature and food availability.

Figure 11–6 illustrates, for our hypothetical poikilotherm, changes in scope for growth with changing temperature at each of four levels of food availability; unlimited, high, moderate, and low. If food availability were ever unlimited in nature, as it can be in the laboratory, an animal could have some scope for growth over a very wide range of temperatures. But scope for growth would be very little at the temperature extremes, and maximum scope for growth might be expected at some moderately high temperature below the upper extreme (Fig. 11–6). High but limited levels of food availability would reduce the possible rate of food consumption as well as the scope for growth at all temperatures and would decrease slightly the range of temperatures over which growth would be possible. Further decreases in food availability, first to moderate and then to low levels, would further restrict food consumption and the scope for growth at all temperatures and reduce the range of temperatures suitable for growth. At temperatures where total metabolic rate ($A_a + A_d + A_s$) becomes greater than the difference between the energy value of the food and the energy value of the waste materials ($A_c - A_w$), the animal will utilize body materials for its metabolism and lose body mass (negative growth); but over periods of time not too long, it can persist. The temperature at which maximum scope for growth occurs is likely to decrease with decreasing food availability. We have here, then, a hypothetical example of temperature acting as a controlling factor and food acting as a limiting factor in determining an animal's scope for growth, very much as Fry (1947) considered such factors to determine an animal's scope for activity.

An additional limiting factor, possibly oxygen, could limit the metabolic rate of an animal to levels below those it would attain with increasing temperature if oxygen were not limiting. In avoiding excessive "oxygen debt" and ultimate death, an animal would

presumably reduce its rate of food consumption and thus, by reducing activity and specific dynamic action, would reduce its metabolic rate. But this would further reduce its scope for growth, particularly at higher temperatures, and further decrease the range of temperatures over which growth would be possible. The brilliance of Fry's (1947) scheme is that it permits us to consider fruitfully how various kinds of environmental factors interact to determine the scope and range of activity of an animal. When coupled with a bioenergetic budget, it provides us with a useful conceptual framework for relating the bioenergetics and growth of animals to their environments.

THE EFFICIENCY OF GROWTH

The efficiency with which an animal utilizes food for growth is often of economic interest. Man stands to benefit in managing populations of domestic and wild animals so as to increase the efficiency with which they utilize their food resources in tissue production. There may sometimes be biological interest in animal growth efficiency as an indication of the suitability of environmental conditions. And, in a world where energy and materials are limited, the efficiency with which these are utilized may bear some relation to the success of a species. Yet, growth efficiencies are simply ratios, which in themselves can lead to little understanding of the success or failure of species. The efficiency of growth of an animal is determined by the quantity and quality of the food it can obtain and the cost of life, which are subject to environmental conditions. Understanding of the success or failure of a species is dependent on knowledge of how the environment influences its survival and reproduction, both of which involve bioenergetics and growth. Efficiencies are no more than summaries of outcomes, not causal explanations. For explanations we must delve deeper into the bioenergetics and growth of the organism in relation to its environment. Still, in the literature of animal husbandry and ecology, much attention is given to the question of efficiency, and we should have some rudimentary understanding of the behavior of efficiencies in the growth process.

There are two general types of efficiencies with which we will be concerned: *total* and *partial growth efficiences* (Kleiber 1961b). Each type, as we shall see, tells us something different about the efficiency of the growth process. Each may be employed with any of the food energy categories we have shown in Figure 11–3 — total, assimilated, metabolizable, or net — and it is important that we specify the type of efficiency and the food energy category when discussing efficiencies. In our discussion of total and partial efficiencies, we will be concerned with the total food energy category.

Total efficiency, E_t, is the *gross efficiency* of Brody (1945) and Margaret Brown (1946, 1957a):

$$E_t = \frac{G}{I} \qquad (20)$$

where

G = growth, and
I = food intake.

Growth and food intake may be expressed in appropriate weight or caloric terms for the whole animal (Q_i's) or per unit of body mass (A_i's). Ivlev's (1939a; 1945) coefficient of growth of the first order (K_1) is total efficiency on the basis of the total food energy category; his coefficient of the second order (K_2) is total efficiency on the basis of metabolizable food energy. Brown (1946) calculated total efficiency on the basis of assimilated food energy.

Partial growth efficiency, E_{pg}, is the *net efficiency* of Brody (1945) and Brown (1946, 1957a):

$$E_{pg} = \frac{G}{I - M} \qquad (21)$$

where

M = maintenance ration, the ration at which an animal would neither gain nor lose body mass under a particular set of environmental conditions.

This ratio expresses the efficiency with which an animal utilizes for growth the amount of food consumed that is over and above the amount the animal would require to just maintain its tissues. Brown (1946) used the assimilated food energy category in calculating net efficiency.

There is another partial efficiency, which we will only mention: *partial maintenance efficiency* is an expression of the efficiency with which an animal utilizes rations at or below the maintenance ration to maintain its tissues or to prevent them from being catabolized

(Kleiber, 1961a; Warren and Davis, 1967). Partial maintenance efficiencies have been of interest in animal husbandry. They should be of interest to ecologists, because most species grow only during a part of the year, and during the remainder of the year they must prevent too much utilization of their own tissues for metabolism.

We can now consider the behavior of growth efficiencies as food is utilized for growth. As we have indicated, the quantity and quality of food consumed, as well as other environmental factors affecting the metabolic state of an animal, will influence the efficiency with which the food is utilized for growth. Were we to take several animals of the *same genetic stock and size* and feed each a different ration of the same quality over a period of one or two weeks, we could then graph the relationship between their food consumption and growth rates. One of two general kinds of relationships can be expected, as we have shown in hypothetical examples in Figure 11–7. Sometimes, over the range of rations fed, the relationship between food consumption and growth rates can be remarkably linear; more often, perhaps, it is curvilinear (Fig. 11–7). The behavior of gross and net efficiencies with increases in ration size is determined by the form of the relationship between food consumption and growth.

If the relationship between food consumption and growth rates is linear, gross efficiency can only increase asymptotically from zero at the maintenance ration toward the net efficiency value, with increasing ration size (Fig. 11–7). And, with changes in ration size, net efficiency must remain constant. This our reader can prove for himself. If, on the other hand, the relationship between food consumption and growth rates is curvilinear, the curve bending down as ration size increases, gross efficiency will increase from zero at the maintenance ration to a maximum at some intermediate ration, and then it will decline with further ration increases (Fig. 11–7). Net efficiency must decline with increases in ration over any curvilinear part of a relationship between food consumption and growth. Our reader should perhaps satisfy himself as to the certainty of our statements, because there has been considerable confusion over the behavior of efficiencies.

For relationships between food consump-

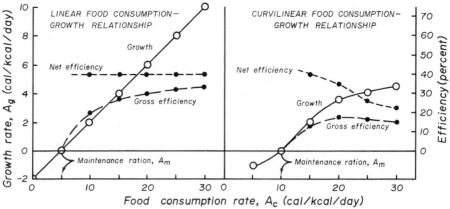

FIGURE 11-7 Relationships between food consumption, growth, gross efficiency, and net efficiency, when the relationship between food consumption rate and growth rate is linear and when it is curvilinear.

tion and growth rates to be linear, the proportion of the energy of food materials appearing as waste products, specific dynamic action, and activity must remain constant with increasing ration size above the maintenance ration. This leads to constant net efficiency. We are here assuming that temperature and other environmental factors are constant, and that standard metabolic rate therefore also remains constant. If a linear relationship were found under conditions increasing only the maintenance ration but not changing the slope of the line, net efficiency would be unchanged, but gross efficiency would be decreased. Decreasing net efficiency occurs when the relationship between food consumption and growth is curvilinear, because waste products, specific dynamic action, and activity together are requiring an increasing portion of the energy value of the food consumed as ration size increases. This accounts for the decline in gross efficiency with increase of ration from some intermediate level; but gross efficiency must increase until this ration level is reached, for gross efficiency must be zero at the maintenance ration. The quantity and quality of food available, temperature, oxygen concentration, competition, and other environmental factors determine the distribution of food energy and materials among the terms in equations 16 and 17, and thus they determine the position and form of the relationship between food consumption and growth rates. In this way, environmental factors determine the behavior of growth efficiencies.

In our examples, we have been con-

sidering animals of nearly the same body size. Were the animals to be of different body size, a large animal eating the same total ration as a small animal would be eating a smaller ration relative to its body size. Thus, the relative ration size of the larger animal would be plotted to the left of that of the smaller animal on a graph of the relationship between food consumption and growth rates (Fig. 11-7). If the animals were of approximately the same physiological age, there are reasons for believing, that their growth rates at different food consumption rates would lie along the same curve. There is evidence that this may often be so for fish. Points representing the larger fish will thus appear to the left and lower than the points representing the smaller fish eating the same total ration, on the curve expressing the relationship between food consumption and growth rates. If this relationship is linear, large fish consuming the same total ration as small fish will always have a lower gross efficiency, which will always decline with decreasing relative ration size. If the relationship is curvilinear, the gross efficiency of all sizes of fish will increase slightly and then decline toward zero as relative ration size declines from the maximum size the fish will consume. The larger fish will always have a lower efficiency than any smaller fish consuming the same total ration, so long as the fish have efficiencies on the gross efficiency curve to the left of the maximum.

Confusion has arisen over growth efficiencies when no distinction has been made between the relative and total size of rations when animals of different sizes have been studied. Paloheimo and Dickie (1966b), after examining literature on the food consumption and growth of fish, concluded that gross efficiency declines with increasing size of total ration from a maximum at very low ration levels. For fish of a given size, this cannot be so, as we have tried to explain. If fish of quite different size are being considered, and the larger ones are consuming the largest total rations, it may be so; but, then, the gross efficiencies of these larger fish are not lower because they are consuming the *largest* total rations but because they are consuming *smaller* rations relative to their body size. The importance of the papers of Paloheimo and Dickie (1965, 1966a, 1966b) justifies our recognition of this one difficulty. As these fine scientists well know, large fish do tend to grow with less efficiency than small fish. Even when the absolute availability of a given kind of food in an environment remains unchanged, the larger animal, though eating more, has difficulty in obtaining as much food as the smaller animal in relation to body weight. This in part explains the slowing down of growth of an animal with increase in size as the animal approaches in size the asymptote of any growth stanza (Fig. 11–2). If the animal crosses some ecological or physiological threshold and this permits more efficient procurement or utilization of food, its gross efficiency and growth rate may increase again.

FOOD, BIOENERGETICS, AND GROWTH

Often it is the quantity and quality of food available that determines the ability of an animal to grow under a given set of environmental conditions; food is then acting as a limiting factor. Environmental conditions other than food, through their effects on the physiology and behavior of the animal, may determine when and at what level food becomes limiting. In addition, these and still other environmental factors determine the quantity and quality of food available, a question we will consider in Chapters 17 and 19. In this chapter, we must confine ourselves to how food availability, temperature, oxygen, and toxic substances interact in determining the ability of an animal to grow. The bioenergetics of the individual organism

in relation to its environment, studied both in the laboratory and in nature, provides perhaps the most powerful approach to this fascinating problem. Our own experience has been mainly with fish, and more appears to be known of this aspect of biology for fish than for other animals living in a natural state. Accordingly, in suggesting the possibilities of this approach, we can be most helpful with examples from studies of fish. Our reader will quickly see applications of this approach to any other species in which he may be interested.

We study the growth of fish receiving rations of different size in the laboratory because we know that in nature food availability is not constant. Since the fates of the energy and materials in the food consumed and the growth possible depend very much on the size as well as the quality of the ration, it is important that we establish for each set of environmental conditions the relationships between food consumption rate, growth rate, and the rates at which energy and materials pass to other fates. The growth and other fates of energy and materials of the food consumed, expressed in terms of equation 17, are given in Figure 11–8 for a series of juvenile coho salmon fed different rations in aquaria at a temperature of 17 C (Averett, 1969). Ancillary studies were conducted to obtain the estimates for energy and materials utilized in standard metabolism and lost through fecal wastes. The energy value of the food materials used for activity or lost through nitrogenous wastes and specific dynamic action was determined by difference. The values represented by the histogram bars were taken from curves fitted to the actual data, so that, at regular increments in ration size, comparisons could be made with data for different temperatures, something we will do in the next section of this chapter. It can be seen that with increase in ration size above the maintenance level (about 35 cal/kcal/day) the proportion of food utilized for growth (gross efficiency) first increases rapidly and then declines (Fig. 11–8). Standard metabolic rate remains constant, but the proportion of energy and materials utilized or lost in activity, specific dynamic action, and waste products increases with increasing food consumption. In other ancillary experiments, the specific dynamic action occasioned by different amounts of food at different temperatures when fish were forced to swim was estimated (Averett,

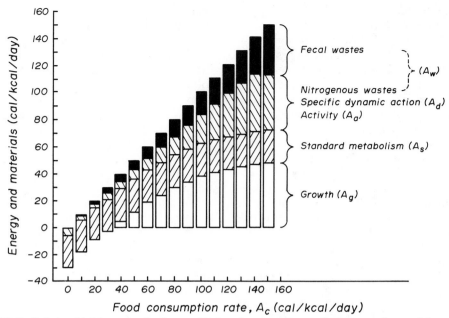

FIGURE 11-8 Relationships between food consumption, losses and uses of energy and materials, and scope for growth of juvenile coho salmon at a temperature of 17 C in an experiment conducted during April and May. After Averett (1969).

1969). But these results should not be used to determine specific dynamic action in the quieter aquarium fish, for reasons we noted earlier. Some of the salmon in this experiment had available all the food they would consume, a rare condition in nature where food is usually limited. Juvenile coho salmon in nature during April and May appeared to be able to obtain food at a rate nearer 50 than 150 cal/kcal/day. And we would be wrong to conclude from Figure 11-8 that these salmon in their natural environment could be expected to grow very well at 17 C, as we shall see.

Not always is it possible, or perhaps even necessary, to conduct the ancillary experiments that permit accounting for all the terms in equation 17. Sometimes we can feed fish different known rations and measure their growth at each level of feeding. Then, if we have information on waste materials, we can utilize equation 14, converted to rate terms, to estimate by difference energy utilization and losses through all metabolism, in terms of total metabolic or respiration rate. Lee (1969) did this with juvenile largemouth bass (*Micropterus salmoides*) fed different amounts of mosquito fish (*Gambusia affinis*) in aquaria at 20 C (Fig. 11-9). Increase in scope for growth with increasing consumption was quite striking, even though the proportions of the energy and materials utilized

and lost in metabolism and waste products also increased.

The very important question of food availability is not so easily resolved in nature as in the laboratory. We can reasonably conclude that the density and distribution of an animal's food organisms in its immediate surroundings determine the amount of food the animal can obtain in a short period of time and the energy cost of obtaining this food. The possible utility of this idea is dependent on our ability to appropriately define the food of an animal and measure its distribution and density in nature. This is sometimes possible, and remarkably interesting and useful results can be obtained. But before considering such results, let us look at results of studies on the feeding and growth of fish in simple laboratory ecosystems.

Lee (1969) maintained juvenile largemouth bass in ponds in which the populations of mosquito fish on which they preyed were kept at different, nearly constant densities by periodic replacement. The rates at which the bass consumed the mosquito fish at different densities were estimated directly from the rate of replacement and initial and terminal censuses of the prey. Growth rates of the bass were also measured directly. Waste materials were estimated on the basis of results of ancillary experiments. Lee was

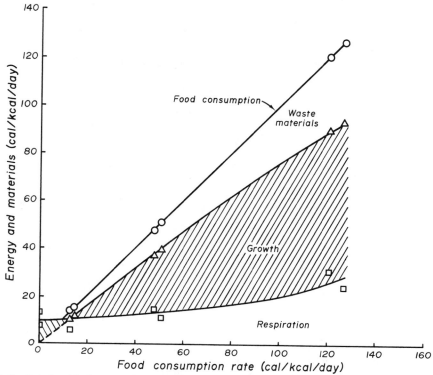

FIGURE 11–9   Relationship between food consumption and scope for growth of juvenile largemouth bass held in aquaria at a temperature of 20 C. After Lee (1969).

then able to develop energy budgets for the bass as functions of the density of the mosquito fish (Fig. 11–10). Food consumption and growth rates were simple functions of prey density and increased over the range of densities studied. At a density near 100 cal/m^2, nearly all the metabolizable energy of the food was required for respiration, and little scope for growth remained. With the temperature higher, this presumably would have occurred at a higher prey density, for more energy would then have been required for standard metabolism. The constancy of the total metabolic rate as prey density increased is of considerable interest. In the laboratory ecosystem, this appears to have resulted from decrease of the energy cost of prey capture (activity) being compensated by increase of specific dynamic action, as prey density and food consumption increased.

Direct estimates of the rates at which animals are consuming food cannot always be made even in laboratory ecosystems, and this is never possible in nature. Various indirect methods of estimating the rates of food consumption of fish have been proposed (Davis and Warren, 1968). One such method is based on the assumption that fish utilize food for growth at different consumption rates with nearly the same efficiency in aquaria as in laboratory ecosystems or in nature (Chapter 17). If we are willing to accept this assumption—there is evidence that it is sometimes valid—we can then utilize curves relating the food consumption and growth rates of fish in aquaria (Fig. 11–9) to obtain estimates of the food consumption rates of fish growing at known rates. The more nearly food quality, temperature, and other conditions in the aquarium experiments approach conditions in nature, the more reliable such estimates become.

In studies of the growth of sculpins (*Cottus perplexus*) in laboratory streams, Brocksen, Davis, and Warren (1968) used this method of estimating the rates at which the sculpins consumed midge larvae at different densities in the benthic environment, because only growth rates could be directly measured (Fig. 11–11). Information on food assimilation as well as on growth then permitted determination, by difference, of respiration rate at each food density. Energy budgets derived in this manner must, of course, remain very tentative. From Figure 11–11, it is nonetheless clear that the growth rate of

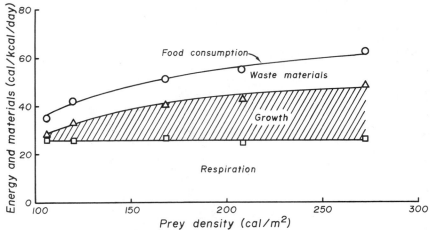

FIGURE 11-10   Relationships between prey density, food consumption, losses and uses of energy and materials, and scope for growth of largemouth bass preying on mosquito fish at different densities in experimental ponds during the summer at temperatures ranging from 18 to 23 C. After Lee (1969).

the sculpins in the laboratory streams was a simple function of the density of their prey. The prey density at which no growth was possible is of particular biological interest. It perhaps approximates the density at which sculpins would disappear from a system if that density were primarily the result of low prey production or intensive interspecific competition. Were this density primarily due to intense intraspecific competition, presum-

ably decreases in sculpin biomass would permit the prey to increase to densities permitting growth of the remaining sculpins, a matter we will return to in Chapter 17. Relationships between food density and the growth of animals may well be different during different seasons. The data given in Figure 11-11 are for an experiment conducted during the summer; lines representing data from experiments conducted

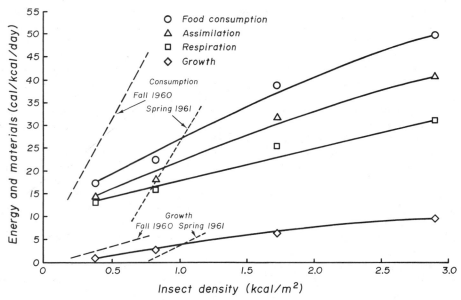

FIGURE 11-11   Relationship between insect density and growth of sculpins in laboratory stream communities during summer 1965. Food consumption, assimilation, and respiration were estimated indirectly on basis of ancillary experiments. Growth and food consumption at different food densities during fall 1960 and spring 1961 experiments are also represented. After Brocksen, Davis, and Warren (1968).

during the fall and spring show how very different such relationships can be during different seasons.

In many species, possible deleterious effects of intraspecific competition have been minimized through the evolution of social systems that lead to division of food and other limited resources among the individuals in a manner favoring the survival and reproduction of a part of the population, thus favoring its persistence. Studies of behavior, particularly when coupled with studies of bioenergetics and growth, can greatly increase our understanding of this process. There has evolved in juvenile coho salmon a territorial and hierarchical social system in which dominant fish occupying territories are able to grow and migrate successfully, whereas others disappear from the population, thus reducing intraspecific competition and insuring that the food resource remains sufficient for success of some part of the population (Chapter 12).

Carline (1968), in a preliminary way, studied the behavior, bioenergetics, and growth of juvenile coho salmon in a laboratory stream in which conditions rather closely approximated those in natural ones. A social system developed among the fish in the laboratory stream. By observing behavior, Carline was able to classify each of the individually marked fish according to its social status, from Group 1 fish, which were most dominant and occupied territories, through Group 4 fish, which were most

subordinate and held no territories. Fly larvae were floated to the fish and the number consumed by each fish was noted. Although food was provided in excess of what the fish consumed, it was unlimited only for the Group 1 fish, because they prevented the fish in the subordinate groups from obtaining all the food they would otherwise have consumed. Figure 11–12 summarizes, in terms of equation 16, the mean energy budgets of fish in each of the four social groups developing during one experiment. The overall picture is really quite clear. Group 1 fish, the dominant ones, expended the most energy for activity and specific dynamic action, obtained the most food, and grew the most. Groups 2, 3, and 4, successively, obtained less food, required proportionately more for standard metabolism, and grew less. Reductions in food availability would first prevent growth and even survival of Group 4 fish; then, presumably, Group 3 fish would become more severely affected. Mechanisms such as this prevent intraspecific competition for food from severely affecting the growth, and thus the survival and reproduction, of all individuals in a population, so helping to insure the persistence of the population.

The occurrence in nature of relationships between food density, food consumption, bioenergetics, and growth is undoubtedly very general; how often we can expect to define these relationships in useful ways is not yet clear. But where possible, their

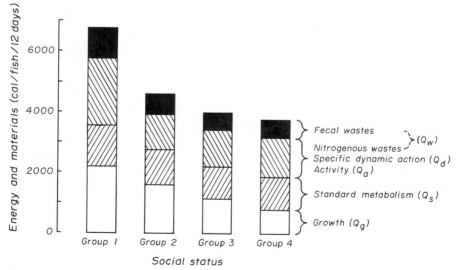

FIGURE 11–12 Mean energy budgets for juvenile coho salmon having different statuses in their social hierarchy in an experimental stream. The decreasing scope for growth with decreasing status from the most dominant individuals (Group 1) to the most subordinate individuals (Group 4) is apparent. After Carline (1968).

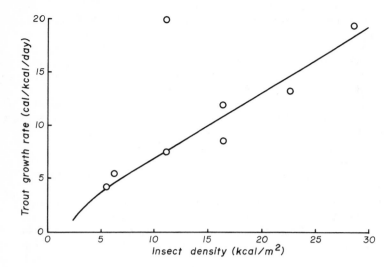

FIGURE 11–13 Growth of cutthroat trout in relation to the density of their food organisms in the benthos of the riffles of an experimental stream enriched with sucrose from October 1965 through May 1966. Trout growth rate and food organism density were estimated at monthly intervals. Unpublished data of John D. McIntyre, Oregon State University.

definition will permit analyses of important trophic processes (Brocksen, Davis, and Warren, 1970), as we will explain in Chapters 17 and 19. Now we will consider only two examples of relationships between food density and growth in nature, one for cutthroat trout (*Salmo clarki*), another for juvenile sockeye salmon (*Oncorhynchus nerka*).

Trout and other salmonids living in streams appear to feed more on insects and other animals drifting in the current than on those remaining on the substrate. Relationships between the mean density of insects drifting in the water and the growth rate of salmonids undoubtedly exist, but adequate estimates of such mean densities are difficult

to obtain, because of the variable pattern of occurrence of drifting insects. The density of insects drifting in the current must be some function of their occurrence in the benthic environment from which they come, and estimates of mean insect density in the benthos are more precise and easily obtained. Reasonably good relationships between the growth rate of salmonids and the density of their food organisms in the benthic environment are not unusual to find. Such a relationship was found for juvenile cutthroat trout in a small experimental stream enriched with sucrose (Fig. 11–13).

Juvenile sockeye salmon, living in lakes, feed upon zooplankton. These lakes often

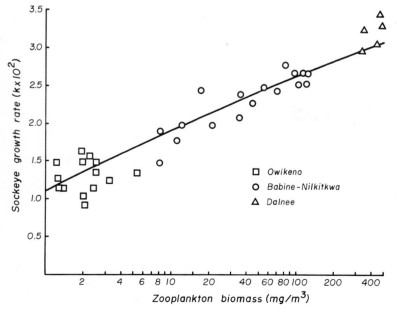

FIGURE 11–14 Relationship between the daily percentage rate of growth of juvenile sockeye salmon and the density of their zooplankton food organisms in three lake systems. After Brocksen, Davis, and Warren (1970), based on data of Johnson (1961), Ruggles (1965), and Krogius and Krokhin (1948).

differ greatly in their capacities to produce zooplankton and, thus, in the zooplankton densities maintained. The relationship between sockeye salmon growth rate and zooplankton density in three such lakes having very different zooplankton densities is shown in Figure 11–14. The proximity to one line of points representing growth rates at different food densities in three different lake systems suggests that growth rates maintained by sockeye salmon at different food densities are more characteristic of this fish than of particular lake ecosystems. Consideration of the costs of food capture and utilization in the growth process might also lead one to this conclusion.

Perhaps we can now begin to see at least one of the ways in which environmental change, whether or not it be pollution, can affect the growth and production of animals. Where such change affects the production, the density, and the availability of food, it may either increase or decrease growth. But an environmental change can also directly affect the fish so as to influence its ability either to obtain its food or to utilize food for growth and life. It is to this matter that we now turn.

PHYSICAL AND CHEMICAL CONDITIONS, BIOENERGETICS, AND GROWTH

The amount of food that an animal will consume and the fates of this food—the bioenergetics and growth of the animal—are very often determined by physical and chemical conditions of the environment. Sometimes food availability may directly determine or limit the rate of food consumption; but at other times, another condition of the environment may so affect the animal as to control or limit its rate of food consumption, whatever the level of food availability. In either instance, the consumption rate as well as environmental conditions will influence the fates of the food consumed and the growth possible. And for these reasons, we cannot consider the influence of physical and chemical conditions on the bioenergetics and growth of an animal apart from food availability and consumption rate. This is so for temperature, oxygen concentration, and toxic substances, which we will consider, and it is so for many if not for most other environmental conditions.

Temperature may control in fish the rates of food consumption and standard metabo-

lism and may influence other fates of the food consumed, growth being among these. Averett (1969) studied the influence of temperature on the bioenergetics and growth of juvenile coho salmon during the different seasons of the year. Figure 11–15 summarizes the results of his spring experiments. Most of the salmon were fed rations restricted to various levels; but for those provided unrestricted rations at the various temperatures, there was an increase in consumption rate with each increment of temperature from 5 to 17 C. Standard metabolic rate increased with temperature, but the total of other uses and losses of energy and materials tended to decrease, at any given level of food consumption. Primarily because of the increase in standard metabolic rate, then, maximum growth rate remained nearly the same with temperature increases from 11 to 17 C, even though consumption rate increased considerably. Still, one could conclude from Figure 11–15 that, if food were very abundant, juvenile coho salmon would grow quite well at temperatures as high as 17 C. But, as we endeavored to explain in the preceding section, food in nature is usually a limiting factor. We have good reason for believing that in the spring coho salmon in small streams are able to acquire food at rates nearer 50 cal/kcal/day than the higher rates at which they were fed in the laboratory. It should be apparent from Figure 11–15 that the growth of the small salmon during this time of year in nature would usually be greater at temperatures of 14 C and below than it would be at higher temperatures. During other seasons of the year, the temperature ranges most suitable for growth are different. In the summer, even if food is limited, the juvenile salmon can be expected to grow well at temperatures from 14 to 17 C, but not at higher temperatures (Fig. 11–16).

From its food an animal must obtain the energy-rich materials necessary to support all its metabolic processes, and from its environment it must draw the oxygen necessary to utilize these materials. Oxygen is not abundant in the aquatic environment, and very often the activities of aquatic animals are limited by oxygen availability. The pursuit of food and other necessary movement, the utilization of food in the growth process, and the maintenance of life at the basal level all require oxygen that must be apportioned in the animal's best interest. When oxygen becomes limiting, an animal must reduce oxygen utilization for muscular activity and

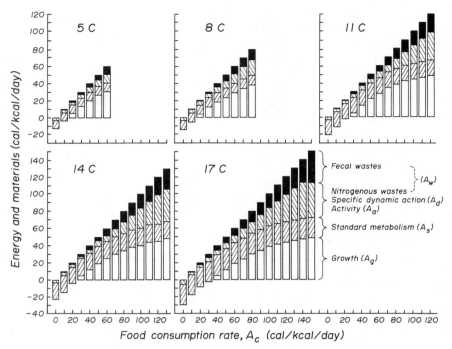

FIGURE 11–15 Relationships between food consumption, losses and uses of energy and materials, and scope for growth of juvenile coho salmon at different temperatures in experiments conducted during April and May. After Averett (1969).

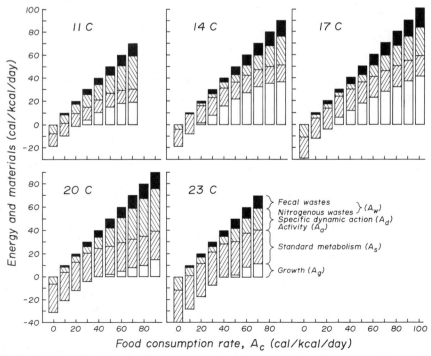

FIGURE 11–16 Relationships between food consumption, losses and uses of energy and materials, and scope for growth of juvenile coho salmon at different temperatures in experiments conducted during August and September. After Averett (1969).

food processing, or suffer an oxygen deficit, which it cannot long endure. Survival in nature requires movement, which usually can be reduced only with some danger to the animal. Reduction in growth rate too is a danger in a world having predation and competition; but over moderate periods of time, an animal may be able to reduce its food consumption and growth rates, thus conserving oxygen for movement, and better its chances for survival.

When food is not limiting, any reduction in the concentration of dissolved oxygen below the air-saturation level can be expected to reduce the rate of food consumption of fish, unless low temperatures are leading to low consumption rates. The rate of food consumption of juvenile coho salmon, forced to swim at constant low velocity, was found to increase with increasing oxygen concentration (Fig. 11–17). Growth rate increased slightly with increasing availability of oxygen, but not so much as food consumption rate, because respiration increased, primarily as a result of increased specific dynamic action. It was the ability of the fish to increase their rate of respiration with increase in oxygen availability that permitted consumption rate and, in consequence, growth rate to increase. Oxygen, here, was acting as a limiting factor.

Food in nature is often limiting, and its availability may determine the level at which oxygen ceases to be limiting. Dissolved oxygen limited the growth rates of juvenile coho salmon even when it was at or above the air-saturation level, when food availability was not restricted (Fig. 11–18). But when food was restricted, dissolved oxygen became limiting only at levels below 4 mg/l. Food availability, then, as well as energy requirements for movement and maintenance, will determine the limiting level of oxygen in nature. We will return to this, but, if we may be permitted a brief digression, there are two other interesting phenomena to be noted in Figure 11–18. First, it can be noted that very high levels of dissolved oxygen—such as occur with very high rates of photosynthesis during phytoplankton blooms in quiet waters—may depress the growth of fish. Second, results shown in Figure 11–18 suggest that growth rates may be determined more by the minimum than by the mean oxygen concentrations occurring when there are great diurnal fluctuations of dissolved oxygen, which can be expected with phytoplankton blooms.

Not only very high levels of food consumption but also very high levels of activity can lead to oxygen acting as a limiting factor, even at air-saturation levels of oxygen. Juvenile coho salmon were forced to reduce their food consumption rate to make oxygen available for the increased utilization of energy necessary when they were made to swim at increased velocities (Fig. 11–19). Reduction in the energy cost of food handling then permitted increased energy utilization for activity even though respiration rate, or total metabolic rate, remained constant. Both reduction in food consumption and increased activity were probably involved in decreasing the proportion of food energy

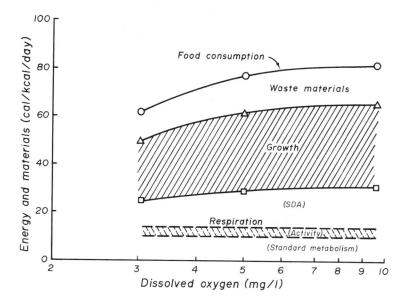

FIGURE 11–17 Relationships between dissolved oxygen concentration and food consumption, losses and uses of energy and materials, and scope for growth of juvenile coho salmon forced to swim at a velocity of 1.3 lengths per second when food was not limited. Unpublished data of Floyd E. Hutchins, Oregon State University.

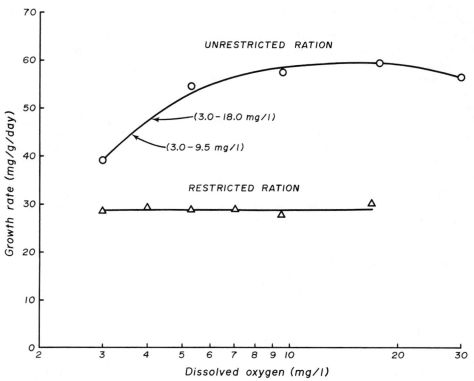

FIGURE 11-18   Relationships between dissolved oxygen concentration and growth rate of juvenile coho salmon when food was unlimited and when it was limited. Arrows indicate growth of fish when held at oxygen concentrations fluctuating diurnally between levels specified. Data of Fisher (1963).

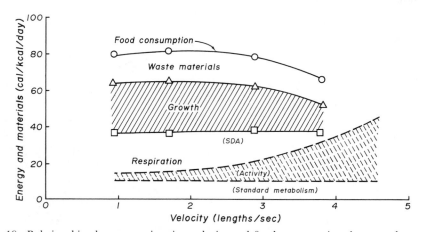

FIGURE 11-19   Relationships between swimming velocity and food consumption, losses and uses of energy and materials, and scope for growth of juvenile coho salmon when food was unlimited and dissolved oxygen concentrations were near the air-saturation level, at a temperature of 15 C. Unpublished data of Floyd E. Hutchins, Oregon State University.

lost through specific dynamic action with increased swimming velocity (Fig. 11–19), as we explained under The Fates of Consumed Food. Increases in activity beyond the level at which food consumption and growth were studied apparently would have required an increase in the total metabolic rate and would have left little or no oxygen for food consumption and processing for growth. The energy costs of activity at different swimming velocities shown in Figure 11–19 were estimated directly by measuring the rates at which unfed fish utilized their body materials. Specific dynamic action was then estimated by difference. It should be apparent to our reader, then, that environmental conditions in nature may sometimes lead to activity, food consumption, and growth being limited by oxygen availability even at rather high levels of oxygen, unless temperatures are relatively low.

Many factors in the environment interact in determining the levels of food availability or oxygen that become limiting. Not just factors that increase necessary activity but also factors that increase the standard metabolic rate are involved. In poikilothermic animals, increase in temperature leads to an increase in standard metabolism, as we illustrated earlier in this section. Other environmental stresses, such as the presence of toxic substances, increase the cost of life, and thus they may decrease the oxygen, energy, and materials available for the processes involved in growth.

Pentachlorophenol, a toxic substance frequently utilized as a preservative of wood products, is generally believed to exert its primary toxic action through uncoupling of oxidative phosphorylation. This can lead to a decrease in the efficiency with which energy is utilized to maintain life processes and activity essential for survival. Proportionately more of the materials and oxygen an animal might obtain from its environment would then be required in the support of life, and less would be available for utilization in the growth process. When groups of the tropical fish *Cichlasoma bimaculatum* were fed unrestricted rations, the cumulative amount of food consumed by those held at a concentration of 0.2 mg/l of potassium pentachlorophenate at first lagged behind that consumed by fish in the absence of this toxicant. But, before the experiment was concluded, the fish exposed to the toxicant had consumed more food than those that were not (Fig. 11–20). Growth of the poisoned fish initially lagged behind that of the others

but soon caught up, and the two groups were nearly equal in size at the conclusion of the experiment. Thus, by consuming more food, the poisoned fish were able to compensate for the decreased efficiency of energy utilization and attain the size of those that were not poisoned. But when fish of this species were fed restricted rations and, as in nature, could not obtain all the food they would consume, the story was entirely different: then, the decreased efficiency of energy utilization for necessary life processes caused the poisoned fish to fall farther and farther behind in growth (Fig. 11–20). This would appear, superficially at least, to be a relatively simple example of a toxic substance acting primarily as a stressor, increasing the cost of life and reducing the energy and materials available for growth. At concentrations little higher than 0.2 mg/l, the effects of potassium pentachlorophenate would have been lethal, the 24-hour TL_m being approximately 0.4 mg/l.

Of the hundreds, even thousands, of toxic substances known, the modes of action of really very few are understood (Chapters 8 and 13). Dieldrin, a chlorinated hydrocarbon pesticide, is extremely toxic to animals. It appears to affect many biochemical and physiological systems in elusive ways whose individual importances to the living animal may be indeterminable. Perhaps rarely can we hope to fully explain the toxicity of materials. We can, however, determine their overall effects on growth and other aspects of the performance of the living organism. When one studies the effects of very low concentrations of dieldrin on the growth of a fish (*Cottus perplexus*), one gains the picture of an animal in extremely serious difficulty (Fig. 11–21). Even when food was unrestricted, this sculpin was able to grow very little at a dieldrin concentration of 0.05 parts per billion (ppb), and at 0.5 ppb it could consume barely enough food to maintain its body tissues. The presence of dieldrin not only reduced the amount of food the sculpins would consume but also increased the amount of food necessary to provide for their maintenance. Such a simple compensatory mechanism as adjustment of the rate of food consumption, changing the energy and materials available for life and growth, appears not to be available or adequate for sculpins poisoned by dieldrin. For substances whose toxic actions are many, simple compensatory mechanisms can rarely be adequate. Dieldrin is acutely toxic to sculpins at concentrations near 2.0 ppb.

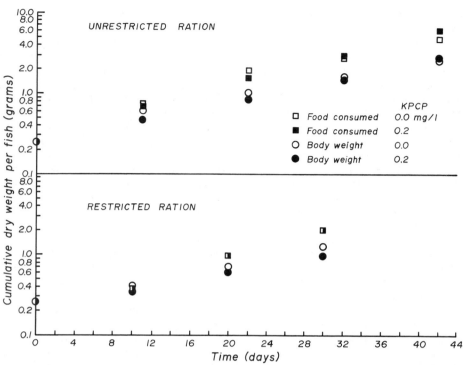

FIGURE 11–20   Cumulative food consumption and growth of cichlids through time at different concentrations of potassium pentachlorophenate, when food was unlimited and when it was limited. Data of Chapman (1965).

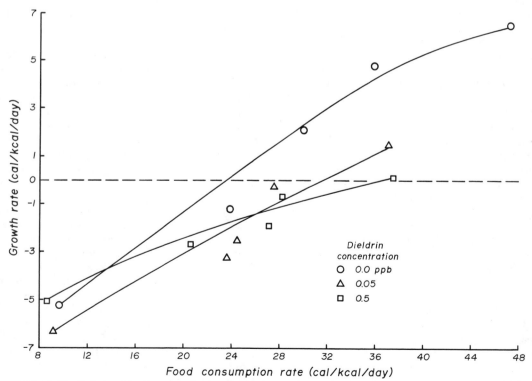

FIGURE 11–21   Relationships between food consumption and growth of sculpins held at different concentrations of dieldrin. Unpublished data of Robert Brocksen and George Chadwick, Oregon State University.

Many industrial wastes contain a variety of substances that interact in various ways in determining their overall toxicity to animals. Effluents of the kraft paper industry—in addition to having a potential for decreasing the dissolved oxygen in natural waters—contain many substances that, together, at high concentrations can be acutely toxic and at lower concentrations can exert sublethal effects on fish and other aquatic organisms. Concentrations of untreated wastes above about 0.5 mg/l BOD that are not acutely toxic usually have little effect on the rates of food consumption of juvenile chinook salmon (*Oncorhynchus tshawytscha*) in aquaria. But decreased efficiency in the utilization of energy and materials for growth tends to reduce the growth possible at given rates of food consumption (Fig. 11–22). This effect is particularly noticeable at higher rates of food consumption when the BOD concentration is about 3 mg/l, where the concentration of toxic components is not far below the lethal level.

Industrial wastes having a potential for depleting oxygen in natural waters are often biologically treated to reduce their BOD (Chapter 21). Along with BOD reduction, reduction in the toxicity of a waste may occur, as in the case of kraft process effluents. The relationship between BOD reduction and any reduction in toxicity may or may not be close, for the components responsible for these two waste characteristics may not be degraded at the same rates. Treated effluents of kraft pulp mills, when at the same concentration by volume, have much less deleterious effect on the growth of juvenile chinook salmon than do untreated effluents, as can be seen by comparing growth at 1.5 percent by volume in Figures 11–22 and 11–23. At higher percentages by volume, though the BOD levels be relatively low, treated effluents can still decrease the growth of fish (Fig. 11–23). In the future, with increased utilization of our waters for all purposes, careful attention will need to be given to all the characteristics of effluents. Their effects on the environment and its organisms are usually many. And many of these effects become apparent through studies of bioenergetics and growth.

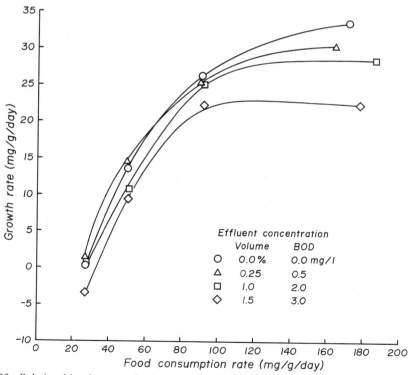

FIGURE 11–22 Relationships between food consumption and growth of juvenile chinook salmon at different concentrations of kraft process paper mill effluent that was untreated. After Tokar (1968).

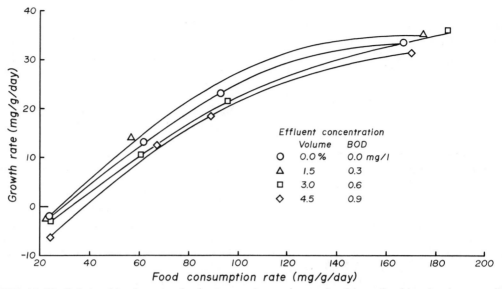

FIGURE 11-23 Relationships between food consumption and growth of juvenile chinook salmon at different concentrations of kraft process paper mill effluent treated by biological stabilization. Unpublished data of Dennis L. Borton, Oregon State University.

ENVIRONMENTAL CHANGE AND GROWTH

This entire chapter has been concerned with environmental change and growth, and it is already long. But writers keep trying to put down important ideas in just a few words, usually a few more, perhaps in fear that they may already have failed with many. The many words may be necessary to flesh the skeleton of ideas. Yet we must hope that at least the more important ideas can be simply stated.

From its environment, an animal must obtain all the energy and materials essential to its life and growth. And a change in that environment must usually mean some change in the animal's ability to live and grow. To the animal, the change may be favorable or unfavorable. To man, according to his interest, it may or may not be pollution. But for the biologist, there remains the problem of interpreting the change in terms of its probable effects on the survival, reproduction, and growth of the animal.

When the study of growth is broadened into the study of bioenergetics, the biologist has a powerful approach to understanding the effects of environmental change on the cost of life, the growth, and the success of an animal. He can then take into account physical and chemical conditions—such as temperature, oxygen, and the presence of

toxic substances—and biological conditions—such as food availability and competition—in a more illuminating way. The study of growth becomes more meaningful and more generally important; and, as an added incentive, it becomes more exciting.

Perhaps more than investigations of most biological processes we study in the laboratory, the study of growth to be very valuable must somehow take into account conditions as they actually exist in nature. The effects some physical and chemical factors have on the growth of an animal depend very much on the quantity and quality of food available. If in nature animals consume food in very different amounts than in our laboratory studies, the effects of physical and chemical factors in nature will also probably be very different. And even apart from the question of food availability in nature, we must be concerned with the behavior of the animal we study: as behavior affects the cost of life, so it affects growth; and differences between the laboratory and nature can lead us to erroneous conclusions. This is a conceptual and methodological problem for all biology, a problem that is particularly obvious in the study of growth. A satisfactory definition of the bioenergetics and ecology of growth, to which we hope this chapter is a contribution, should make clear the necessary touchstones in nature.

Studies of food availability and of the behavior of our animal in nature are essential.

And if we can gain some idea of its rate of food consumption there, so much the better, because then we can estimate its metabolic rate. This information should somehow be incorporated into the design of our laboratory studies on the effects of environmental factors on growth. Changes in some environmental factors may primarily affect food availability, and so influence growth. Other changes may have their primary effects directly on the physiology of the animal we study. Relationships between food density and growth in nature may be changed by factors directly affecting the animal; there are reasons for believing that the relationships themselves may be little changed if the effect is primarily on food availability. But most important at our present level of understanding of the effects of environmental change on the growth of animals is useful definition of the problems animals face. The bioenergetic viewpoint can help us to this definition.

We cannot complete our story of growth in this chapter: consideration of the problem of what determines food availability in nature must await Chapters 17 and 19. Then we will be able to relate growth to food availability as determined by the capacity of ecosystems to provide food and by intraspecific and interspecific competition. To predict the effects of environmental change on the growth and production of animals of interest, we must understand not only energy and material transfer within the individual organism but those within its ecosystem as well. Pollutional changes in aquatic environments are no exception.

SELECTED REFERENCES

Brocksen, R. W., G. E. Davis, and C. E. Warren. 1968. Competition, food consumption, and production of sculpins and trout in laboratory stream communities. Journal of Wildlife Management 32:51–75.

Brody, S. 1945. Bioenergetics and Growth. Reinhold Publishing Corporation, New York. xii + 1023pp.

Fry, F. E. J. 1947. Effects of the environment on animal activity. University of Toronto Studies Biological Series 55. Ontario Fisheries Research Laboratory Publication 68. 62pp.

Ivlev, V. S. 1945. Biologicheskaya produktivnost' vodoemov. Uspekhi Sovremennoi Biologii 19:98–120. (Translated by W. E. Ricker, 1966. The biological productivity of waters. Journal of the Fisheries Research Board of Canada 23:1727–1759.)

———. 1961a. Experimental Ecology of the Feeding of Fishes. (Translated by D. Scott.) Yale University Press, New Haven. viii + 302pp.

Kleiber, M. 1961b. The Fire of Life: An Introduction to Animal Energetics. John Wiley & Sons, Inc., New York. xxii + 454pp.

Warren, C. E., and G. E. Davis. 1967. Laboratory studies on the feeding, bioenergetics, and growth of fishes. Pages 175–214. *In* S. D. Gerking (Editor), The Biological Basis of Freshwater Fish Production. Blackwell Scientific Publications, Oxford. xiv + 495 pp.

12 Behavior

We have seen that activation of an instinct causes seeking behaviour which, after a series of movements of increasing specificity, ends with the accomplishment of a consummatory act. Now the adaptiveness of behaviour is to be found in the fact, stated again and again, that this mechanism enables, and even forces, the animal to do 'the right thing at the right moment.'

NIKO TINBERGEN, 1951, p. 156

It should be emphasized that the same activities are observed both in the laboratory and out of doors. The salmon's repertoire of behaviour is limited and the same behaviour patterns will be seen wherever the fish are watched. However, the interpretation of these activities is a different matter and it is only when the observations are repeated many times under a wide variety of conditions that their significance can be evaluated.

WILLIAM HOAR, 1958, p. 394

THE IMPORTANCE OF BEHAVIOR

By *behavior* we mean the integrated movements of the whole animal. An organism responds in a usually adaptive manner to stimuli that it receives from its environment through its sense organs. Through its normal behavior, an animal tends to bring itself into favorable relationships with its environment. The environment of an animal has not only physical and chemical conditions but also biological ones, which include animals of the same kind as well as organisms of other kinds. For any species, its nature—its characteristics and needs—determines which relationships are favorable. Ultimately, those relationships are favorable which favor the survival and reproduction of individual organisms and the persistence of their populations.

Any characteristic of an animal has both genetic and environmental determinants: an animal's phenotype develops through interaction of its genotype with its environment (Chapter 7). No characteristic of an animal is determined solely by its genes; no

characteristic is determined solely by the animal's environment. This is so for the morphology and physiology of an animal; it is so for the animal's behavior. Some behavioral patterns of animals appear to be passed from one generation to the next in rather fixed form; these patterns may require little or no learning by each successive generation. Other behavioral patterns must be learned by each generation of individuals. Whatever the behavioral pattern may be, both genetic and environmental determinants have been involved in its development. Behavioral patterns that appear with little or no learning are sometimes considered *instinctive* in contrast to *learned*. But probably all behavior involves some learning in its development, and learned behavior must have heritable elements. Thus, the distinction between instinctive behavior and learned behavior cannot be clear; and many students of behavior now consider this distinction to be without value. We will return to this argument in a later section. For now, we only note that early biologists showed certain behavioral patterns to be passed from generation to generation with apparently little learning, and thus they paved the way for the increased understanding and sophisticated arguments of today. And, used with some care, the distinction between instinct and learning may still be helpful in our attempt to introduce some of our readers to the complex and important phenomena we know as behavior.

Full understanding of the characteristics of animals can come only through knowledge of their development in the individual organism, their mechanisms of control, their function, and their evolution. In this, the behavioral characteristics of an animal are not different from its morphological and physiological ones. During the first half of this century, ethologists were able to explain the functions of many behavioral patterns by carefully describing and analyzing the behavior of animals, particularly birds and fish. From these studies, they drew some conclusions regarding the evolution of certain patterns of behavior, and they advanced very tentative ideas concerning physiological mechanisms of behavioral control. We can perhaps never know much about the evolution of behavior, for it cannot be reconstructed from the geological record. But one day we will know much about the physiological control of behavior. This, however, will require physiological studies, not just careful description and analysis of behavioral patterns, which can lead to understanding of the development and function of behavior.

In analyzing the functions of behavioral patterns, we hope to learn how these patterns adapt animals for life in their natural environments. The adaptive values of behavioral patterns lie in the contributions particular patterns make to the survival and especially to the reproduction of animals. We can suppose that nearly all the behavioral patterns exhibited by animals in nature have some adaptive value; otherwise, natural selection would not usually have permitted these patterns to persist (Tinbergen, 1965). Just how, then, are different patterns of behavior adaptive? We will return many times in this chapter to this question. Now we will comment only very generally.

The oriented movements of animals in response to gradients of light, temperature, current, dissolved substances, or other characteristics of their environment may act to keep them at locations in these gradients where conditions are physiologically tolerable to the animals and their survival is possible. Fraenkel and Gunn's (1940) *The Orientation of Animals* is the classic and perhaps still best review of the nature of such movements. Environmental gradients may help to guide the feeding and migratory movements of animals, favor their escape from predators, or in other ways insure the continuance of their kinds. Harden Jones (1968) has recently reviewed knowledge and theories of guidance in fish migration. Other behavioral patterns of animals are in some ways more complex than their oriented movements in response to environmental gradients (Tinbergen, 1951). Many of these other patterns are social in nature; they involve two or more individuals of the same species. Social behavior is involved in mating and in care of young. And, perhaps in most species, various higher levels of social organization exist, these resulting from antagonistic and attractive tendencies between the individuals of a species. Such social organizations may function to apportion the resources of the environment, may form protective mechanisms, or may in many other ways work to the advantage of the species in which they occur (Wynne-Edwards, 1962). The physiology, structure, and behavior of an animal together form an integrated system that operates to bring the animal into those relationships with its environment that favor its survival and reproduction and the persistence of its population.

Much of what we know about the behavior of animals, particularly fish, has resulted from laboratory studies like those of Tinbergen (1942) on the stickleback (*Gasterosteus aculeatus*) and those of Baerends and Baerends–van Roon (1950) on various species of cichlid fishes. Relatively little is known of the behavior of most important species of food and game fishes, though, as we shall see, there has in recent years been increased interest in their behavior. But laboratory studies of aquarium species have led to the development of methods of behavioral research and have provided needed understanding of the significance of general patterns of behavior that occur in many other species. The behavior of animals often remains remarkably unchanged when they are removed from their natural environment. And, without laboratory studies, the control necessary to study some behavioral patterns cannot be obtained. Nevertheless, the final interpretation of the significance of an animal's behavior must be based on careful observations made in nature.

Knowledge of the behavior of an animal is essential to understanding of how changes in its environment influence its distribution and abundance. This will become clear as we proceed through this and later chapters. Unfortunately, animal behavior has yet to receive the attention that systematics, morphology, physiology, or even ecology has received. In consequence, we know little about how changes in environment, pollutional or otherwise, influence the behavior of animals.

Complete understanding of the behavior of an animal—understanding of its development, control, function, and evolution—is a goal toward which students of animal behavior strive. It is a goal most of them realize they will never fully attain, for such is the nature of any science. And those of us who are not primarily students of animal behavior, but who seek to understand the distribution and abundance of this species or that, cannot study the behavior of our species with the intensity of the student of behavior, for our objectives may be different. But if, as we go about our other studies of a species, we never miss a chance to observe its behavior or to conduct more detailed behavioral studies when this is possible, we will find solutions for problems that could not have been solved in other ways.

Most of our discussion in this chapter will be devoted to fish. Behavior of fish is better known than that of most other aquatic organisms. And the importance of behavior to the success of an animal can be as well illustrated with fish as with any other group. We in no way wish to imply that studies of fish behavior are more important than those of the behavior of other animals. Though the behavior of an oyster may be simpler than that of a fish, it is no less necessary to its survival, reproduction, and growth.

SENSORY CAPACITIES, STIMULI, AND INTERNAL MOTIVATION

Sensory Capacities

The sensory capacities of an animal determine the environmental circumstances to which it is potentially able to respond,

though, as we shall see, an animal in nature may not respond to all the stimuli its sense organs are capable of receiving. Careful studies of the sensory capacities of many species have shown that rarely if ever do two species have exactly the same capacities. Von Uexküll (1909) stated that each animal has its own perceptual world—its own *Umwelt* or *Merkwelt*—and that this world is different from its environment as we perceive it. Thus, its *Merkwelt* differs from our own or that of other species. It is important in our studies of animal behavior that we know not only which properties of the external world can influence a species' behavior but also which cannot.

The perceptual world of an animal is determined by the *sensitivity* and the powers of *discrimination* and *localization* of the sense organs it possesses. The sensitivity of an organ has limits of both intensity and quality. Discrimination, the ability to distinguish between stimuli belonging to the same sensory modality, also has intensity and quality limits. Tinbergen (1951) suggests that "higher" and "lower" sense organs may be distinguished by their powers of spatial localization, which has both direction and distance aspects.

The *conditioned responses* of animals have made it possible to survey their sensory capacities. Bull (1957) has discussed the nature of this kind of response and reviewed some of the literature on the methods and results of studies on fish. To briefly explain how conditioned responses are used to study sensory capabilities, we can do no better than to quote from the lucid writing of Tinbergen (1951, p. 16):

The first task, therefore, when tackling the study of a new species is a careful examination of the capacities of its sense organs. The classical method used for this purpose is the conditioning method, which has been developed admirably by von Frisch and his school. Von Frisch reasoned that if an animal's sense organs are affected by a change in the environment the animal can be conditioned to show a response to it. The general procedure is the following. A simple reaction, for instance escape, or feeding, is conditioned to a definite, simple change in the environment, for instance to the experimenter blowing a whistle. When this has been accomplished, the next step is to ascertain to which sensory modality the conditioned stimulus belongs. In the present case, if a response to sound is wanted, one has to be sure that it was actually the sound the animal was reacting to and not, for instance, the experimenter's movement in bringing the whistle to his mouth. This is done by comparing the reaction to the complete situation (blowing a whistle) with that to the visual part of it, or rather to the complete situation minus the auditory part. This is done by bringing the whistle to the mouth but not blowing it. If the animal reacts exclusively to the first situation it is obvious that the response was auditory. Thus, by systematically probing the animal's potential reactivity to many different environmental influences, a survey of its sensory capacities can be made.

Studies on the sensory capacities of fish and their physiological and morphological bases are many. Here, we can only hope to indicate very briefly the extent of sensory development that has occurred in fish as a group. Different species of fish have become specialized along particular sensory lines, and the perceptual world of no species extends to the limits of the entire group. Four articles in Volume 2 of *The Physiology of Fishes*, edited by Margaret Brown (1957), provide a very helpful review: Brett (1957) discusses the structure and physiology of the eye; Lowenstein (1957), the structure and function of the inner ear and the lateral line; Hasler (1957), the olfactory and taste senses and their structural bases; and Bull (1957), the sensory capacities of fish as studied by conditioned response techniques.

Different species of fish have been shown to be able to distinguish between objects by their color, brightness, form, size, number, and position. Fish have been trained to respond to the presence of the four recognized tastes—sweet, sour, salty, and bitter—and threshold concentrations have been found in some instances to be far lower than those of man. The sense of smell appears to be exceedingly important in the lives of many fishes; it is extremely well developed in some species, these being able to detect some naturally occurring and other substances at concentrations as low as 1×10^{-11} mg/l. Various species of fish have been found to be sensitive to sound in the frequency range of from 16 to 7000 cycles per second, to be able to discriminate between pitches, and to have the ability to locate sounds from different sources. The lateral line organs of fish do not appear to participate to any significant extent in hearing but function to detect irregular water movements caused by other animals or inanimate objects. They may also detect reflected waves caused by a fish's own movements, thus permitting echo-location of

objects. Fish have not been shown to be able to detect changes in current velocities; they appear to maintain their positions in currents by means of tactile or visual orientation with fixed objects. Changes in the direction of current flow when in excess of 90° can be detected by some species. Fish have also been shown to be able to detect small changes in hydrogen ion concentration, salinity, temperature, and pressure. These studies to date have shown fish to have a rich perceptual world.

Different species of fish, then, in differing degrees, have these and other sensory capacities. The capacities of each species have evolved over the ages and provide a sensory basis for its behavioral adaptation to its natural environment. These same capacities may in some instances permit behavioral adaptation to recent changes man has caused in aquatic environments.

Stimuli

The adaptive behavior of an animal is usually a response to some external condition created by a change in its environment, this condition being the *stimulus* releasing or directing the behavior. But, as we shall see, internal factors are involved in determining the responsiveness of the animal to a stimulus.

We must distinguish between two kinds of stimuli: *releasing stimuli* and *directing stimuli*. Whereas a particular behavioral pattern may be elicited by a releasing stimulus, its orientation is under the control of directing stimuli. An example from Tinbergen (1951, p. 82) will perhaps help to make this distinction clear:

Daphnia, swimming in water with a high carbon dioxide concentration, gather near the surface. The function is obvious: in polluted water the surface layer, being in contact with the air, is relatively rich in oxygen. Analysis shows that two stimuli play a part in this reaction: chemical stimulation by the carbon dioxide and a visual stimulus. In a glass jar lighted from underneath *Daphnia* will swim downwards as soon as the CO_2 is added. The carbon dioxide merely *releases* the response, which is *directed* by the light.

The releasing and directing stimuli for a given pattern of behavior may be distinct environmental identities, as are carbon dioxide and light in the case of *Daphnia*; or, for other patterns, the same identity may act simultaneously as a releasing stimulus and a directing stimulus. Marler and Hamilton (1966, p. 229) distinguish between the facilitating and inhibiting effects of external stimuli. Their book provides an excellent review of much of present knowledge of the nature, detection, and functional roles of stimuli in the control of behavior.

Many but not all stimuli are physical or chemical characteristics of the environment. Seeing its prey may both release and direct the behavioral patterns of a predator that lead to the capture of the prey. Other stimuli serve to establish and maintain relationships between individuals of the same species. These may have either releasing or directing functions, or both, and can be called *social releasing stimuli* and *social directing stimuli.* Various structural, physiological, and behavioral characteristics have evolved in different species that can provide visual, acoustical, chemical, or tactile stimuli to which individuals of the same species respond in adaptive ways.

Knowledge of the capacities of the sense organs of an animal and of *potential stimuli* in the environment does not enable us to identify the *actual stimuli* releasing or directing particular behavioral patterns. An animal does not respond in any given instance to all the stimuli its sensory organs are capable of receiving. As an example, Tinbergen (1951, pp. 25, 27) states that:

> . . . the carnivorous water beetle *Dytiscus marginalis*, which has perfectly developed compound eyes and can be trained to respond to visual stimuli, does not react at all to visual stimuli when capturing prey, e.g. a tadpole. A moving prey in a glass tube never releases nor guides any reaction. The beetle's feeding response is released by chemical and tactile stimuli exclusively; for instance a watery meat extract promptly forces it to hunt and to capture every solid object it touches.

Responses to stimuli comprised of one or a very few characteristics of an environmental situation frequently can lead to errors in behavior. Such errors are characteristic of instinctive behavior, which is perhaps not surprising, if one considers the problem of transmitting by means of the genetic code responsiveness to entire environmental situations that are variable. Of course, an animal that does not utilize a particular sensory organ for controlling one pattern of behavior undoubtedly utilizes this organ for controlling other patterns.

Internal Motivation

The threshold stimulus necessary to release a particular pattern of behavior has often been observed to vary under otherwise constant environmental conditions. Other times, the intensity of the behavioral response varies even though released by a stimulus of constant magnitude. One can conclude that somehow the readiness of animals to perform particular behavioral patterns changes. This readiness of an animal to perform a pattern of behavior we can call *internal motivation.*

Hinde (1966) has discussed both internal and external causal factors that appear to determine the levels of internal motivation of animals. Hormones may increase the level of motivation of an animal by acting directly on the central nervous system or indirectly through peripheral organs; they may also increase motivation in nonspecific ways. Hormone level may be affected by environmental stimuli such as temperature and light, by social and other stimuli, and by internal factors. Internal sensory stimuli play a role, though not necessarily an essential one, in determining some kinds of internal motivation. Hunger in man, for example, is in part explained by rhythmic contractions of the stomach wall giving rise to hunger pangs. There is some evidence of the existence of central nervous mechanisms that influence the responsiveness of animals to environmental stimuli, and the activity of these mechanisms may be influenced by hormone level.

We should introduce the concept of *drive*, though we do so with hesitancy, for—as Thorpe (1956) has explained—the use of this term by different behavioral sciences and scientists has led to much ambiguity and argument, and we cannot here adequately develop its many connotations. Yet, the term and its associated concepts are difficult to avoid in any discussion of behavior. Hinde (1966) has given much thought to the benefits and dangers of using various concepts of drive; his book is well worth reading. Thorpe (1951, p. 37) defined drive in its widest sense as "The complex of internal and external states and stimuli leading to a given behaviour." In general agreement with Thorpe, Baerends (1957, p. 259) defined drive as "the total amount of information available that, independently of its origin, determines the kind and the intensity stage of the behavior activated." Thus, releasing stimuli, which influence the kind and intensity of the behavior activated, appear to be involved in this concept of drive. Thorpe (1956) distinguishes between *drive* and *internal drive*, and it is internal drive that is equivalent to *internal motivation*, the term we shall use.

INSTINCTIVE AND LEARNED BEHAVIOR

Instinctive Behavior

In making a distinction between instinctive and learned behavior, we are well aware that many if not most students of behavior now do not like to make this distinction (Schneirla, 1951; Hebb, 1953; Lehrman, 1953; Hinde, 1959; Tinbergen, 1963). Their reasons are good, and our reader should be forewarned. Perhaps those of our readers who know animal behavior will be unhappy because we persist in using this distinction; these readers undoubtedly do not need the brief introduction to animal behavior that is the purpose of this chapter. It might be possible to introduce our other readers to animal behavior with one brief chapter making no reference to instinct or learning, but we are not the ones to write such a chapter. The development and control of behavior are complex and poorly understood, though important advances are being made (Marler and Hamilton, 1966). There can be neither simple nor general explanations, and the arguments of today must seem inscrutable to those first exposed. The original distinction between instinct and learning helped make possible present understanding and arguments. Even though all behavior has genetic as well as environmental determinants, ethologists agree that some components of behavior are passed from generation to generation in remarkably fixed patterns that are not primarily determined by learning. Because we can devote so little space to this topic, we believe the historic approach may most help our reader to appreciate the mechanisms and significance of behavior.

In his classic synthesis of ethological knowledge, *The Study of Instinct*, Tinbergen (1951, p. 112) tentatively defined an *instinct* as:

... a hierarchically organized nervous mechanism which is susceptible to certain priming, releasing and directing impulses of internal as well as of external origin, and which responds to these impulses by coordinated movements that contribute to the maintenance of the individual and the species.

This was two decades ago, and Tinbergen (1963) and other ethologists have changed their views on instinct, as he knew they would. Nevertheless, let us begin with this definition, leading us to another definition: *instinctive behavior* is behavior that is controlled by an instinct.

Spalding (1873), on the basis of his observations on newly hatched chicks, early concluded that there are important behavioral patterns that need not be learned by each individual but that are passed from generation to generation, even as physical characteristics, by inheritance. Whitman (1899) insisted not only that some behavioral patterns are inherited by the individual but also that these patterns are a part of the origin and history of the animal's organization and reveal adjustments between the animal and its natural environment. This idea of inherited behavior was accepted by many but not all who studied animal behavior. Classical ethologists like Heinroth (1911), Lorenz (1935), and Tinbergen (1942), continued in the tradition of Whitman to develop understanding of the behavior of animals through meticulous description and analysis of particular behavioral patterns.

Early students of animal behavior noted that certain, often complex, patterns of behavior were characteristic of each species and exhibited extremely limited variability between individuals of the species, thus suggesting that individual learning might play a small role in the development of these patterns. Similar patterns were often observed in closely related species, but they were usually found to have specifically different modifications. Baerends (1957, p. 231) wrote:

These behavior patterns develop in each individual of a species in the same way and with the same final result, even if the fish are reared from the earliest stage in complete isolation, e.g. a male *Lebistes reticulatus* [guppy] kept alone from the moment it was born, shows all activities of the courtship ceremony when, after maturation, it is confronted with a female. This means that genetic factors determine the form of these activities in the same way as they determine the development and final form of organs. Learning, from observing the behavior of another member of the species, cannot be shown to play a role here.

It was behavior of this kind that at first most interested those students of behavior known as *ethologists*. They sought to describe its manifestations, determine its causes, and understand its ecological significance and evolutionary development.

Lorenz (1937) distinguished between two types of coordinated muscle activity in patterns of instinctive behavior. Form and coordination of one component of such a pattern is quite fixed, though its intensity can vary; a releasing stimulus is usually necessary for it to appear. This component came to be known as the *fixed action pattern*. In the other component of the total behavioral pattern, the coordination is less rigid and certain adaptive changes determined by the properties of the stimulus are possible. It is this component of the behavioral pattern that is responsive to directing stimuli and serves to orient the fixed action component in an appropriate way in the environment.

Tinbergen (1942), on the basis of his studies on sticklebacks, recognized the functional organization of behavior to be hierarchical. Reproduction in the stickleback involves movement to a suitable area, selection and defense of a territory, nest building, mating, and care of offspring. Before any of the behavioral patterns in each of these groups can take place, the internal motivation for reproduction must be increased by seasonal changes in light and temperature. But once a suitable territory has been located, which group of behavioral patterns will appear is determined by the external stimulus situation: another male to fight; nest materials; a female to court; or young needing care. And which of the particular behavioral patterns belonging to a group will appear is determined by the exact nature of the stimulus: whether a challenging male attacks or flees; whether a female is receptive or not. Tinbergen (1942) recognized that particular behavioral patterns, when grouped according to their reproductive function, appeared to be hierarchically organized (Fig. 12-1). Warning us of its very provisional nature and dangers, Tinbergen (1951, p. 122) went on to provide a conceptual framework for future work by suggesting a hierarchical system of nerve centers underlying the behavioral manifestations of major instincts. Nervous organization is undoubtedly hierarchical, but a functional organization of behavior may not correspond closely with the organization of its nervous basis. Present knowledge of behavior does not confirm early models of its causal mechanisms (Hinde, 1966; Marler and Hamilton, 1966; Klopfer and Hailman, 1967).

Sir Charles Sherrington (1906)—in his monumental *The Integrative Action of the Nervous System*—and Wallace Craig (1918) early distinguished two interrelated phases of behavior as a whole. The introductory phase is often quite variable and is striving or search-

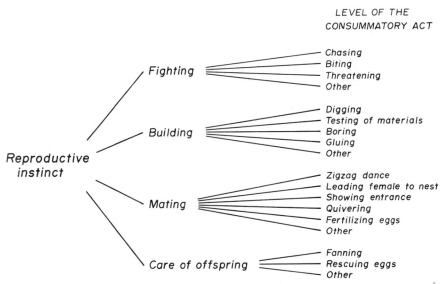

FIGURE 12–1   The hierarchical organization of the reproductive behavior of the male stickleback on the basis of function. After Tinbergen (1942).

ing in nature, the animal appearing to seek some goal. Craig called this *appetitive behavior*. A prolonged period of searching behavior is frequently followed by a relatively simple and stereotyped act that seems to reduce the internal motivation leading to the appetitive behavior. Such acts Craig called *consummatory actions*. It now appears that a complex behavioral pattern may involve a chain of these two kinds of behavior in which a given act is consummatory for the preceding link and appetitive for the following. There are other reasons to suggest that the distinction between these two kinds of behavior may not be so clear as first suggested (Thorpe, 1956; Marler and Hamilton, 1966). But the original distinction was brilliant and furthered the analysis of behavior.

Ethologists quite early observed that responses of the whole animal were often set off by elaborate stimuli involving features such as form, color, and movement. Such stimuli came to be known as *releasers* and, when they served to coordinate the activities of individuals of the same species, as *social releasing stimuli*. The specificity of releasing stimuli necessary to release particular behavioral responses led McDougall (1923) and Lorenz (1935) to postulate a series of mechanisms in the central nervous system, each sensitive to a particular releasing stimulus and bringing about a particular response of the animal. From the German *das angeborene auslösende Schema*, these have in English been called *innate releasing mechanisms*. Here again, another useful concept has been refined as knowledge has accumulated. Not only central nervous system mechanisms but also receptors and afferent nerve pathways now appear to be involved in the response of the organism to only particular stimuli (Marler, 1961).

Before proceeding to a brief discussion of learned behavior, we must try to make clear some of the reasons why many students of behavior now avoid making a distinction between instinct and learning. Hinde (1966, pp. 316–317), in his excellent synthesis of ethology and comparative psychology, has expressed very well three of the difficulties associated with this distinction:

In the first place, the innate/learnt type of dichotomy can lead to important environmental influences on development being ignored. This can occur if it is assumed that the factors influencing the development of behaviour are of two types only—genetic, and those associated with learning. . . . To take an extreme example, a given genetic strain of *Drosophila melanogaster* may be capable of full normal flight, or of erratic flight, or incapable of flying at all, depending on the temperature during development (Harnly, 1941): the difference can hardly be ascribed to learning. If all environmental influences on development are said to operate through learning (as could be inferred from some definitions of learning), then the concept of learning becomes so broad as to be valueless.

A second objection to such 'either-or' classifications is that in practice innate behaviour is defined solely in negative terms: a response is said to be innate or unlearned just so long as no learning process (or other environmental influence) has been identified in its ontogeny. Often, indeed, behaviour is labelled as innate on the basis of a 'deprivation experiment'—animals which develop in an environment lacking some factor believed to be important are compared with animals living normally. But no series of deprivation experiments can exclude *all* environmental influences. . . .

Thirdly, dichotomies between learnt and innate behaviour involve the assignment of units of behaviour into one or the other category, rather than an analysis of the factors and processes involved in their development. Since the appearance of any response depends on *both* nature and nurture, attempts to classify responses according to one source or the other are doomed to failure.*

Biologists all recognize that the characteristics of animals are determined by interactions of their genotypes with their environments. Yet, there are excellent biologists who persist in believing that the distinction between instinctive and learned behavior is still useful, in spite of associated conceptual and methodological difficulties (Lorenz, 1961; Thorpe, 1963a). The continuing contributions of these scientists are evidence that for some this distinction remains a basis of fruitful points of view.

The interactions of genotype with environment in the development of behavior are unquestionably complex, but the story of genetics has been the distinguishing of the heritable from the nonheritable characteristics of individuals. Ethologists, psychologists, and physiologists who believe the distinction

---

*From Hinde, R. A. 1966. *Animal Behaviour: A Synthesis of Ethology and Comparative Psychology*. Used with permission of McGraw-Hill Book Company.

between instinct and learning has outworn its usefulness know that some behavioral patterns require more learning than others. Hinde (1966, pp. 316–317) himself wrote: "The form of some movements is so stable that it cannot be modified by any environmental conditions within the viable limits: in particular, modifications cannot be effected by learning (Lorenz, 1961)." But these biologists believe the important question is not whether a particular behavioral pattern is instinctive or learned but how this pattern develops in the individual.

Many are now interested in the morphological and physiological basis of behavioral control, and ethologists continue to be interested in the adaptive functions and evolution of behavior. Earlier, ethologists—on the basis of their beautiful and meticulous descriptions and analyses of behavioral patterns—advanced very tentative ideas about behavioral control. More recent physiological and ethological studies have shown some of these ideas to be inadequate, as was expected by those who advanced them (Tinbergen, 1951, pp. 122, 127). Description and analysis of behavioral patterns may identify physiological problems and even suggest physiological control mechanisms (Hinde, 1966, p. 7; Marler and Hamilton, 1966, pp. 739–740), but physiological explanations require the study of physiological mechanisms. Studies of the development, control, and evolution of behavior are of interest to all biologists; studies of the adaptive functions of behavioral patterns are of particular interest to ecologists. Space and knowledge demand that in this chapter our emphasis be on the function of behavior.

Learned Behavior

We shall define *learning*, however imperfectly, as internal changes in the individual animal that result from its experiences and that are often manifested by adaptive changes in its behavior. *Learned behavior* is, then, that part of an animal's behavior that is determined by learning. Little is known of the internal changes that make up learning, so we must usually study learning by observing the changes that occur in an animals behavior as a result of its experiences.

For animals that have neither man's length of life nor his means and capacities of learning, instincts may play a greater role than learning in the control of behavior. Instincts and the behavior they control have through the ages evolved because of their adaptive value, but they alone do not permit the individual animal to adaptively modify its behavior as a result of its experiences. Learning permits the individual to better adjust to its environment by taking advantage of its own experiences, or those of others in the case of man; herein lies the adaptive value of learned behavior. Most of the primarily instinctive behavior of animals is undoubtedly refined or in other ways modified by learning.

For the ecologist, it is not so important to know what an animal can learn in the laboratory as it is to know what an animal can and does learn in its natural environment. But it is difficult to determine the extent or the importance of learned behavior in nature. Not only is it difficult to make careful observations on the daily lives of animals in their natural environments, but also it is almost impossible to determine which of the multitudinous stimuli an animal is experiencing are leading to changes in its behavior. The relative rigidity of instinctive behavior, both in nature and in the laboratory, permits better analysis of its extent, significance, and causal stimuli in the natural environment of an animal. Nevertheless, we can suppose that learning is of importance to nearly all animals.

One of the difficulties in defining learning is that the kinds of internal changes it must entail and the kinds of behavior by which it is manifested appear to be very diverse, so much so that perhaps they should not be brought together under one term. From an ecological point of view, if not always from behavioral, physiological, and psychological ones, it is nevertheless helpful to do so. Here we can only indicate some of the kinds of learning that have been identified and suggest their ecological significance. The ecologist will find particularly helpful the fine books by Thorpe (1956, 1963b) and Barnett (1963), which treat the significance of learning in the behavior of animals in which instincts have an important role.

Kinds of learning are sometimes classified into two categories: associative learning and insight learning. We shall define *insight learning* as the internal organization of previous, separate experiences in a manner that permits the solution of a new problem in a new way. We cannot suppose that much of the behavior of the lower vertebrates, including fish, will be found to be controlled by insight learning as we have defined it. But any classification of learning must be artificial and imperfect in our present state of knowledge; there is evidence that insight learning involves some elements that are similar to kinds of associative learning (Thorpe, 1956).

Associative learning can perhaps be defined as changes in the individual animal that relate particular experiences to their outcomes. Some of the kinds of learning or manifestations of learning that have been placed in this category are habituation, conditioned responses, trial and error learning, latent learning, and imprinting.

Habituation, which in many respects is the simplest form of learning, has been defined as "the relatively permanent waning of a response as a result of repeated stimulation which is not followed by any kind of reinforcement" (Thorpe, 1956, p. 54). Thus, if a response upon repeated performance is not reinforced either by the individual's avoidance of harm or by its attainment of some goal, the response gradually disappears. The adaptive value of such learned behavior is the reduction of needless activity. Habituation should be clearly distinguished from physiological fatigue.

A *conditioned response* can be defined as an act elicited by a previously indifferent stimulus as a result of this stimulus being applied repeatedly at about the time of application of the releasing stimulus for a similar act. We have already described the use of conditioned responses for study of the sensory capacities of animals. Conditioned responses of an animal in its natural environment may permit reduction in the scope of the appetitive behavior preceding particular behavioral patterns.

Trial and error learning occurs when in the course of its appetitive behavior an animal becomes more effective in attaining a goal by gradually eliminating ineffective ways it has tried. This kind of learning may involve elements of habituation and conditioning, but, since it results from appetitive or investigative behavior, the possible behavioral patterns to which it can lead are of greater variety. And thus it leads to the selection of efficient patterns of behavior by animals in complex environments.

Latent learning appears to occur without reinforcement, or reduction of internal motivation, during the course of an animal's exploration of its environment, but it is not manifested until some later time, when it has adaptive value for the animal. It is an acquiring of information on environmental relations, information that may be later put to use in making the behavior of an animal more adaptive.

Imprinting is a rapidly learned association a young individual establishes between itself and another individual, usually of its own species. A young individual may be capable of this kind of learning only over a very short, critical period of its development. Interestingly, it was the erroneous attachment young geese hatched in isolation from adults establish with their human keepers that first drew attention to this kind of learning (Heinroth, 1911; Lorenz, 1935), which may influence behavioral patterns developing later in life. The attachment the young of some species establish with a location during a brief juvenile residence, an attachment that is manifested only in later life when as adults they return to reproduce, can perhaps also be considered imprinting. And in this there would appear to be affinities with latent learning (Thorpe, 1956). Sluckin (1965) has reviewed much of the theory and research relevant to imprinting. For young animals that must rapidly and usually correctly establish appropriate relations with their environments, the adaptive value of imprinting is apparent.

The importance of instinct in the lives of animals having less learning capacity than man is undoubtedly very great. For animals in their natural environment, it is difficult to know how great may be the role of learning or how it interacts with instinct. But however great the role of learning in other activities of salmon, for example, we cannot overestimate the importance of the ability of salmon to return to their natal streams to reproduce, an ability based on learning.

BEHAVIORAL ADAPTATIONS TO THE ENVIRONMENT

Territorial Behavior

Animals usually have the reproductive potential to populate the environment beyond its capacity to support them. Such crowding could lead to disruption of reproductive or other activities, to the division of limited resources among too many individuals, or even to the depletion or destruction of food resources. Predation, disease, and starvation may limit to some extent the size of animal populations, but these of themselves do not usually provide adequate control of animal numbers: witness man. In many animals, various social mechanisms of population control have evolved. One of these is territorial behavior.

Some animals, including some species of fish, seek and defend territories. The number of their kind a given area of habitat will support is thus determined by the number of suitable territories it can provide. Animals not obtaining a territory are displaced to new areas or otherwise lost from the population. Territorial animals have probably evolved to defend areas of the sizes they need for reproduction, feeding, or other purposes. Wynne-Edwards (1962) has suggested that some animals may compete directly for space rather than for food or other resources. We will first consider territorial behavior in the reproduction of some species of fish. Then, we will discuss territorial behavior that appears to be related to the feeding requirements of the juveniles of some species.

In his classic study and review of the reproductive behavior of sunfishes, Breder (1936) noted that the males usually defend nesting territories against intrusion of either males or females by threatening displays, chases, and fights. Only the persistence and special behavior of a female that is ready to mate appears to make possible her entry into the defended nest. After mating, the male continues to defend the nest, thus protecting the young that hatch.

The observations of Fabricius and Gustafson (1954, 1955) on the char *Salvelinus alpinus* and the grayling *Thymallus thymallus* indicate that in these species also it is the male that originally establishes the reproductive territory. The male defends its territory against both males and females until a pair is formed with one female, which may then enter into the defense of the territory. Much of this defense does not involve actual bodily contact between antagonists, but is carried on through threat displays. Such displays, described by Fabricius and Gustafson, have components of form, color, and movement and function as social releasing stimuli that reduce the actual fighting and the possible damage to the participants. After spawning, one or the other of a pair may continue to defend the territory for a period of time.

It is the female of the coho salmon (*Oncorhynchus kisutch*) and the chinook salmon (*O. tshawytscha*) that selects and defends the reproductive territory (Briggs, 1953). One or more males of the same species are permitted to occupy the territory, but the intrusion of no other females is permitted; neither sex of the other salmon species is permitted within the territory. Among the males occupying the territory, threat displays, chasing, and fights lead to the development of a hierarchical social order having dominant and subdominant fish. Fighting among male brown trout (*Salmo trutta*) occupying the territory of a female also leads to dominant and subdominant males, but much more damage is inflicted among individuals of this species than among coho or chinook salmon. Jones and Ball (1954) have described in detail the threat displays, chasing, and fighting of brown trout. Female coho salmon for as long as two weeks after spawning and female chinook salmon for as long as four weeks after spawning continue to defend their reproductive territories, until they are too weakened by ensuing death.

Territorial and hierarchical behavior, then, bring a kind of social order and stability to the reproductive activities of some species of fish, thus minimizing conflict between participants and possible interference with reproductive success. Territorial behavior helps insure that all suitable spawning areas will be utilized and helps decrease the probability that later spawning fish will disturb the nests of those spawning earlier.

Kalleberg (1958), who has studied territorial behavior in juvenile Atlantic salmon (*Salmo salar*) and brown trout, considers the areas defended by these stream-dwelling young to be feeding territories. Juvenile coho salmon, rainbow trout (*Salmo gairdneri*), and brook trout (*Salvelinus fontinalis*) have also been shown to have very similar territorial behavior (Hoar, 1951; Chapman, 1962; Newman, 1956).

Keenleyside and Yamamoto (1962) have described in detail territorial defense by juvenile Atlantic salmon. The intrusion of one fish into the territory of another is a stimulus that releases in the defender either a "frontal display" (Fig. 12–2A) or a "charge," which may be followed by a "nip." These instinctive

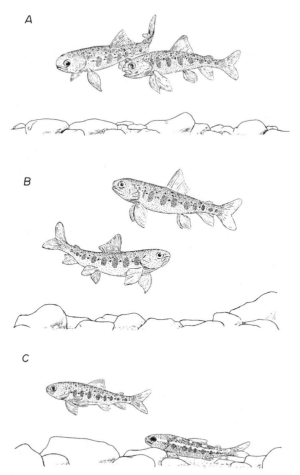

FIGURE 12-2   Juvenile Atlantic salmon in displays that function as social releasing stimuli in territorial behavior. A. Fish on right is in frontal display as it approaches a laterally displaying fish. B. Fish displaying laterally toward each other. C. Lower fish is in submissive attitude that stimulates no response from other fish in attitude typical of active territory holder. After Kennleyside and Yamamoto (1962).

acts indicate a high motivation for attack; and they may release in the intruder either a "lateral" display (Fig. 12-2B) or "fleeing," which indicate different levels of motivation for escape. Two fish in which attack and escape tendencies are similar may both exhibit lateral displays, the display intensity being greatest in the fish having the greatest motivation for attack. The intensity of displaying and nipping is greatest when the individuals are most nearly matched in size and aggressiveness. The loser of an encounter may either flee from the territory or sink to the bottom in a submissive attitude (Fig. 12-2C), which causes no further aggression on the part of the other individual.

These displays of salmonids have compo-

nents of form, color, and movement; they are highly conspicuous and function as social releasing stimuli that tend to reduce the actual fighting occurring in the defense of territories. The displays are quite similar in different species of salmonids, and their recognition between individuals of different species is apparent (Newman, 1956; Kalleberg, 1958). Thus they may function in the competition for space not only within species of salmonids but also between species of salmonids occupying the same stream.

The outcome of this seeking and defending of territories is to divide the suitable stream habitat into a territorial mosaic having a pattern determined by the outcome of conflicts between the holders of ad-

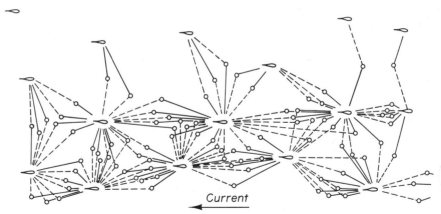

FIGURE 12–3 Aggressive contacts between juvenile Atlantic salmon holding territories in a stream aquarium. These contacts, which were observed during 212 minutes, indicate the boundaries within the territorial mosaic. Unbroken lines show directions of movements of intruding fish and broken lines show directions of movements of defending fish. After Kalleberg (1958).

jacent territories (Fig. 12–3). The size of the territories is determined at least in part by the innate behavior of the fish, their sizes, and the conformation of the streambed. Insofar as these are feeding territories, the behavior that determines their sizes may have evolved in particular streams in response to general levels of food availability over long periods of time (Chapman, 1962), though there may be adaptive changes in behavior in response to short-term changes in food availability (Mason and Chapman, 1965).

Whatever their evolutionary and ecological bases may be, territorial and hierarchical behavior provide a social structure and stability in stream populations of some species of salmonids and tend to control the size of these populations. Chapman (1962) has shown how these kinds of behavior may result in displacement from populations of juvenile coho salmon of individuals unable to maintain territories or low in hierarchical social position. Territorial and hierarchical behavior have been found in other species of freshwater and marine fish (Greenberg, 1947; Kirchshofer, 1954). Schooling and other kinds of social behavior also serve to organize populations in ways well suited to meet the needs of the species in which they occur.

Reproductive Behavior

Reproductive behavior in fish may involve migration to a spawning area, selection and defense of territories, building of nests,

courtship, spawning, and care of nests and young. We will be discussing migratory behavior in a later section and we have just discussed territorial behavior; here we will briefly consider the behavior of nesting, courtship, spawning, and care of young in sunfishes, salmon, trout, chars, and grayling. The reproductive behavior of these species has been studied in more detail and is in some ways more complex than that of species like the striped bass (*Roccus saxatilis*) and cod (*Gadus* spp.), which spawn in schools and release their sex products into the open water to drift with the currents. But environmental conditions suitable for normal reproductive behavior and the development of young are no more important for one species than for another.

Though here we will be most concerned with the habitat and social stimuli that direct and release reproductive behavior, and with this behavior itself, we must emphasize that numerous physical, chemical, and biological factors control the development of the gonads and the internal motivation that leads to reproductive behavior. Among these, light and temperature are often mentioned as being important, but this should not cause us to ignore the roles of biological factors and water quality characteristics other than temperature in the development of gonads and internal motivation (Chapter 10).

Breder (1936) has reviewed much of the literature and his own observations on the reproductive behavior of the Centrarchidae, the family of fishes that includes the black basses (*Micropterus* spp.) and sunfishes (*Lepo-*

mis spp. and other genera). Reproductive behavior in the many species of this family of North American fishes is remarkably uniform, in part, no doubt, because they all occupy rather similar habitats—relatively quiet fresh waters that are warm in summer. With increasing day length and temperature in the spring, these fish move into shoal areas; there the males establish territories and begin nest construction. According to the nature of the habitat and the species and size of the fish, nests may be constructed in water as shallow as a few inches or as deep as 10 feet. Sunfish reproductive activity appears to be greatest on sunny days; it may be interrupted at any stage even by a passing cloud.

Sandy, gravelly, or solid substrates are necessary for nesting, because the male cannot successfully form the nest depression or clean the substrate for egg deposition when the substrate is soft. The nest depression is about twice the length of the male and is prepared by movements of his tail and by occasional removal of larger particles with his mouth.

Recognition and acceptance of a female that is ready to mate is apparently based more on her characteristic movements and persistence than on secondary sexual characteristics, for the male repulses entry into the nesting territory by individuals of either sex, initially even his prospective mate. In the mating act, the female inclines sideways and quivers while the male remains upright. The eggs and the milt containing the spermatozoa are released simultaneously during separate mating acts extending over a period of perhaps two hours, after which the female leaves the nest not to return. The male, remaining over the nest, fans the eggs and drives off intruders, thus preventing the eggs from being smothered by silt or consumed by other fish. For a period after hatching, the young remain under the protection of the male in a school near the nest.

Once the internal motivation for reproduction in a fish has reached at least the necessary minimal level, some combination of releasing stimuli is required for the initiation of nest building, courtship, or mating activities. Particular characteristics of the habitat and the social behavior of potential mates provide the needed stimuli. Fabricius (1950) has considered the *heterogeneous summation* of stimuli that leads to release of reproductive behavior in pike (*Esox lucius*),

whitefish (*Coregonus*), and char. For release of spawning behavior in pike, both the presence of vegetation and a rise in water temperature are necessary. These qualitatively different stimuli sum in a compensatory way in releasing this behavior, either sparse vegetation at high temperature or dense vegetation at a low temperature providing a sufficient total stimulus. Reproductive behavior appeared to be released in the whitefish and char only when declining water temperatures reached some variable level and suitable gravel or rubble bottoms were found. These species may spawn in flowing or standing water and at a variety of depths when a suitable combination of bottom type and temperature is present.

The combinations of stimuli necessary to release spawning behavior in salmon and trout are not well known. Temperature is known to be important: species that spawn in the spring generally do so when rising temperatures have reached some variable level; species that spawn in the fall generally do so when temperatures have declined to some variable level. The nature of substrate materials is also known to be important. Mixtures of gravel and rubble, from the size of a pea to larger than a softball, are used by different species for nest preparation. Larger individuals and species generally select nest sites having coarser mixtures. Considerable differences in water movement and depth at the sites selected by populations of the same species are known to occur; temperature and substrate character may be the most important elements of the total releasing stimulus necessary for spawning.

Grayling differ strikingly in reproductive behavior from salmon, trout, and char, which have quite similar behavior even though there are generic and specific differences. The female grayling is tolerated in the territory of the male only during a mating act, and she may later mate with other males (Fabricius and Gustafson, 1955). At the time of the mating act, the movements of both fish dislodge substrate gravel while the eggs and milt are released into the gravel interstices. Thus, digging and release of sex products are combined in a continuous behavioral performance. After spawning with one or more females, the male continues to defend his territory. This results in a kind of parental care that protects the developing young from being dislodged by the activities of other fish.

In salmon, trout, and char, it is the female that digs the nest, which is more extensively prepared than in the case of grayling; nest-building and the mating act are distinct behavioral performances. Nest or redd preparation in many species of salmon, trout, and char—including coho and chinook salmon (Briggs, 1953), steelhead trout (Needham and Taft, 1934), Atlantic salmon (Belding, 1934), brown trout (Jones and Ball, 1954), and arctic char (Fabricius and Gustafson, 1954)—has been described and not found to differ strikingly. After selection of the nest site by means of visual and tactile examination of the substrate (Fabricius and Gustafson, 1954), the female brings the flat of her tail fin against the substrate by a lateral inclination and curvature of her body. Then, with a rapid straightening of the body, the tail is moved upward, hydraulic suction dislodging the substrate materials, which are displaced downstream when current is present. This movement is repeated several times in rapid succession; then the female may rest briefly before repeating the whole performance. The larger species may dig until the depth of the depression reaches a foot or more.

Courtship and spawning of salmon, trout, and char involve acts that function as social releasing stimuli. In general, the mating pair come to occupy a position side by side near the bottom of the nest depression. Then, with characteristic movements of the fish, eggs and milt are simultaneously released to settle into the gravel or rubble interstices. Courtship and actual mating tend to be distinct behavioral performances, each composed of different acts functioning as social releasing stimuli. Certain acts, like "quivering," may appear in both kinds of performance. As Tinbergen (1942) did for the stickleback and Fabricius and Gustafson (1954) did for the arctic char, Jones and Ball (1954) described the behavioral acts involved in fighting, nest building, courtship, and mating of Atlantic salmon and brown trout. On the basis of function, Jones and Ball classified these behavioral acts into subinstincts of the reproductive instinct to which the acts belong (Fig. 12–4).

Whereas char perform several spawning acts before the female covers the eggs, salmon and trout cover the eggs after each act. The female salmon or trout moves somewhat upstream and for one or two minutes executes rapid and continuous digging movements. These dislodge substrate materials that are carried by the current into the nest and cover the eggs. The female may continue digging at this location to form the next nest depression, which, in any event, will be located slightly upstream from the deepest part of the previously used depression. This process is continued until all the spawning acts are completed, the eggs having become more deeply buried with each successive digging operation. As we have already noted, female coho and chinook salmon continue to defend the nest

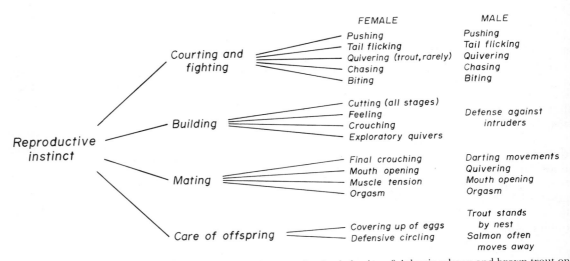

FIGURE 12–4   The hierarchical organization of the reproductive behavior of Atlantic salmon and brown trout on the basis of function. After Jones and Ball (1954).

territory for a period of from two to four weeks. For variable periods of time, trout and char—usually the female but sometimes the male—similarly defend their nests. Such behavior serves to prevent disturbance of developing young, particularly during the early period when they are most sensitive to shock.

Feeding Behavior

Obtaining of energy and materials is a fundamental problem confronting all living organisms. Animals have evolved along morphological and behavioral lines that have led to the exploitation of perhaps most of the food niches the earth affords. The many species of fish inhabiting fresh and marine waters exhibit diverse ways of obtaining their food, which may be plant or animal or some combination of both. Fish that feed on plankton often swim in schools and pass great amounts of water through their mouths and over their gill arches, on which fine, toothlike protuberances serve to filter the food from the water. Other fish sort suitable food from detrital bottom deposits by other means. There have evolved in some species chisel-like teeth that permit them to scrape plants and animals from hard substrates. Fish that feed on other fish have mouths and ways of life permitting them to exploit their food resources. Sculpins, living on the bottom of streams, possess no air bladder and have prehensile mouths adapted for removing food organisms from the substrate. Trout, living in the same streams, are able to stay up in the current, because of the buoyancy their air bladders provide; and their mouths are more suited for capturing drifting food animals. Within its morphological and behavioral capabilities, when, where, and how a species of fish feeds are perhaps determined primarily by the availability of its food organisms and by its own physiological needs. Nevertheless, its feeding will be influenced by its social relations with its own kind and by its relations with competitor and predator species. Most species of fish are opportunistic in their feeding, considerable plasticity in their feeding behavior making this possible.

Our knowledge of the feeding behavior of most fish is based largely on analyses of their stomach contents, there having been very little detailed observation on the behavior itself. General knowledge of the behavior of some species and detailed knowledge of their morphology have been correlated with information on stomach contents to piece together a tentative, incomplete picture of their feeding. Darnell (1958) did this in his excellent study of the food habits of 31 species of fish in an estuarine community. Sometimes information has been obtained on the abundances of species that potentially are food organisms; ratios of these abundances to the abundances of the same species in the stomach contents of fish have been used to represent the availability of particular organisms and the selectivity of the fish (Allen, 1951). But none of these procedures is a satisfactory substitute for direct observation of feeding behavior, which is usually difficult when we are studying fish.

Little is known of the development of feeding behavior in young fish. Kalleberg (1958) observed the feeding of fry of Atlantic salmon and brown trout soon after they emerged. At first these fry exhibited no perceptible choice of prey: they captured air bubbles and debris as well as small food organisms, if these were in motion and not too large. Feeding, which was originally directed primarily toward the substrate, became more oriented toward the open water as the age and size of the fish increased. During the first two weeks, the fry became more selective in their feeding behavior and surer to choose edible objects. Both instinctive and learned behavior must be involved in the feeding of fish. It would appear that the releasing and directing stimulus necessary for the original pattern of instinctive behavior in these fry is a moving object of small size. This behavior must then be refined by learning, until suitable food organisms are selected most often.

The feeding behavior and food resources of most species of fish change in different periods of their lives. Coho salmon, which feed primarily on insects as young fish in streams, feed upon small crustaceans after entering the sea. But, having attained sufficient size in the sea, they begin to feed mainly on herring and other small fish. Darnell (1958) noted similar changes in several of the estuarine species he studied. Such shifts in food habits permit the fish in different stages of their development to exploit the relatively most available food resources consistent with their size, morphology, and behavior, this tending to increase their efficiency in obtaining and utilizing food.

Behavioral plasticity permits different populations of the same species of fish to exploit different food resources. A species of trout that utilizes insects when in streams may when in lakes utilize crustaceans or other fish; thus it takes advantage of the different food resources available in the two habitats. Savvaitova and Reshetnikov (1961) found that three different forms of the Dolly Varden char (*Salvelinus malma*), all of which occurred in the same lakes at one time or another, utilized different food resources in these lakes. The anadromous form utilized mainly insects, the lake form was predaceous on sticklebacks, and the stream-lake form utilized mainly molluscs. Thus, specialization in feeding behavior here reduces competition for food between different populations of the same species.

The same population of fish may utilize different food resources during different seasons of the year, this probably resulting primarily from differences in the relative availability of potential food organisms. There are other kinds of periodic changes in the feeding of fish. Most species of fish are more active during the time when there is some daylight, and this coupled with diurnal changes in the behavior of their prey often leads to daily feeding patterns. In many temperate zone fishes, there are also seasonal changes in feeding activity, feeding ceasing or nearly ceasing during some seasons. Seasonal changes in light and temperature are probably primarily responsible for such changes in feeding, but there is some evidence that endogenous rhythms of endocrine activity are also operating.

Social behavior, both territorial and hierarchical, tends to replace direct competition for food with competition for space or social position (Chapman, 1966). This leads to more efficient utilization of the food resource, displacement of some individuals, and population regulation. Carline (1968) has shown that the most aggressive juvenile coho salmon obtain the most food and grow the fastest. Aggressive behavior between species also may operate in competition for food and space (Kalleberg, 1958). Differences in habitat selection by brown trout and arctic char (*Salvelinus alpinus*) have been shown by Nilsson (1955) to reduce competi-

tion for food between these species in lakes in Sweden.

Biologists interested in water pollution have properly been concerned that environmental changes could lead to the destruction of the food resources of valuable fish populations. Here, the plastic and opportunistic feeding behavior of most species of fish may be a saving grace. If environmental changes lead only to changes in the species composition of food resources—as pollutional changes so often do—and not to a decline in the total abundance of suitable food organisms, these changes need not lead to a decline in the food consumption, growth, or production of fish (Katz and Howard, 1955; Nilsson, 1955; Ellis and Gowing, 1957; Warren et al., 1964). Indeed such changes can lead to increases in these parameters.

Migratory Behavior

Regular movements of fish and other animals during different seasons to particular areas for feeding or reproduction are a very general occurrence. As in the case of sunfish moving into shoal areas to spawn, these movements may be relatively short. Or, as in the celebrated cases of the American and European eels (*Anguilla*) and the different species of Pacific salmon (*Oncorhynchus*), these movements may be very extensive. Though not really different in kind or function, the relatively longer movements are perhaps more often referred to as migrations. The evolution of such movements and migrations has permitted different species of fish to take advantage of particular habitats for feeding or reproduction, thus filling ecological niches where opportunity existed. These movements are normally essential to the completion of the life histories of the species and the success of the populations that participate in them.

Salmon, trout, and char are believed to have evolved from ancestral troutlike fish that dwelt in streams (Tchernavin, 1939). Many of these species have developed the tendency to migrate to sea, where more extensive pastures permit them to attain greater size. But all must return to freshwater to reproduce.

The internal motivation for the return migration to freshwater must be based on endocrine changes that accompany maturation of the gonads. Yet, for salmonids that have moved hundreds and even thousands of miles from their natal tributaries through many stream confluences and into the sea, the return journey presents a complex problem, which the fish solve by means that biologists do not understand.

The environmental cues or directive factors that guide these fish at least through the freshwater part of their return journey may be olfactory. As we have noted, the sense of smell in fish is extremely acute. Perhaps, during their residence in streams as juveniles, Pacific salmon learn the odors characteristic of their natal streams by imprinting. This learning might be retained by the fish as latent learning and used much later in their lives to trace the much diluted olfactory substances back through rivers and tributaries to the natal streams for which the substances are characteristic. Hasler and Wisby (1951) postulated such an explanation and showed that juvenile coho salmon can be conditioned to discriminate between the waters of two streams by olfaction. They also captured adult coho salmon that had ascended two forks of a stream, occluded the nasal sacs of half of these fish, and returned them to a point below the confluence of the forks (Wisby and Hasler, 1954). Most of the normal fish returned to the stream fork of their first choice, whereas those deprived of olfaction returned to the two forks in a random manner. Lorz and Northcote (1965) performed similar experiments with kokanee—lake resident sockeye salmon (*Oncorhynchus nerka*)—returning to a natal tributary to spawn, and they obtained results like those of Wisby and Hasler. For the freshwater part of the homeward journey, these experiments suggest a possible directive factor. But in the sea, whether salmon stay within water masses containing the freshwaters from which they have come, or whether they wander until they find these water masses, or whether they employ some means of navigation—such as the celestial navigation apparently used by certain birds—is not known. There is some evidence that juvenile sockeye salmon, migrating great distances to lake outlets, and other species of fish are capable of navigation employing sun-compass orientation coupled with some as yet to be demonstrated time sense (Brett and Groot, 1963; Hasler et al., 1958).

Fry of pink salmon (*Oncorhynchus gorbuscha*) and chum salmon (*O. keta*) move down to the estuaries very shortly after they leave the streambed gravels in which they developed. Soon after emergence sockeye salmon fry move into lakes, where they lead a pelagic existence for a year or more before descending to the sea. The fry of coho salmon live in their natal streams for about one year before migrating. Hoar (1958), on the basis of his own work and that of others, has discussed evolution of the seaward migration of these four species, which are assumed to have evolved from an ancestral stream-dwelling type.

Three steps appear to have been necessary in this evolution: the development of smolt transformation, an increase in nocturnal activity, and the appearance of schooling behavior. *Smolt transformation* involves physiological and morphological changes, including development of a tolerance for seawater, that adapt the fish for a marine existence. No smolt transformation occurs in pink and chum salmon, for these species have evolved to the extent that their fry can pass into estuaries immediately after emerging from the gravel. They exhibit marked nocturnal activity and schooling and represent the furthest step in evolution from the ancestral, stream-dwelling form. Sockeye salmon fry also early exhibit nocturnal activity and schooling behavior, but these are adaptations for pelagic existence during their period of lake residence; smolt transformation, which prepares them for entry into the sea, occurs after one or two years. Of the four species, the juvenile coho salmon remain most like the ancestral stream-dwelling type and are well adapted to stream existence. They exhibit little nocturnal activity or schooling behavior during this period of their life. Moreover, in juvenile coho salmon aggressive and territorial behavior are well developed. Smolt transformation occurs after about one year, and with it come increases in the nocturnal activity and schooling behavior of these fish, just prior to their journey to the sea. Hoar (1958) has summarized some of the behavioral tendencies that adapt pink and chum salmon for early marine existence, sockeye salmon for one or two years of lake residence prior to their seaward migration, and coho salmon for a year in streams before they descend to the sea (Table 12–1).

TABLE 12–1   *Behavioral characteristics of juveniles of four species of Pacific salmon that appear to adapt them to their particular patterns of freshwater and marine existence. Intensity of activity indicated by number of + marks; 0 indicates activity not observed. After Hoar (1958).*

| | SPECIES OF SALMON | | | |
BEHAVIORAL CHARACTERISTIC	Coho	Sockeye	Chum	Pink
River residents	+	0	0	0
Lake residents	+	++	0	0
Fry prefer sea water	0	+	++	++
Fry hide under stones	+++	+++	++	+
Nipping behavior	+++	++	+	0
Territorial behavior	++	+	0	0
Schooling behavior	+	+++	++	++++
Smolt transformation	+	+	0	0
Increased activity at night	0	+	+	+
Swim into surface film in darkness	0	0	0	+
Positive rheotaxis	+	++	++	+

The seaward migration of particular populations of juvenile salmon occurs during a relatively short period of the year. The internal motivation for the appetitive behavior leading to this migration most probably has an endocrine basis, which in pink and chum fry is present upon emergence; this develops in sockeye and coho fry at smolt transformation. The migration, which occurs mostly at night, may involve both active downstream swimming and downstream displacement of fish whose nocturnal activities have caused them to lose visual and tactile contact with the streambed.

Avoidance Behavior

Animals to be successful must position themselves in relation to the physical and chemical characteristics of their environments so as to favor their survival. The senses of different species of fish have attained different degrees of development, but most species appear to have rather rich perceptual worlds that provide the stimuli permitting the fish to locate themselves favorably in their environments. This is accomplished through behavior that leads fish to avoid locations that afford some sets of stimuli and to occupy locations that afford other stimuli. *Avoidance behavior* and be-

havior that tends to locate individuals in preferred stimulus situations—*preferenda*—are not necessarily distinct aspects of the complex of behavioral reactions that lead individuals to favorable locations. Nevertheless, biologists have tended to investigate the behavioral responses of fish to certain environmental stimuli, particularly temperature, to determine the stimulus situations the fish prefer, whereas their investigations of other stimuli have usually been directed toward determining the threshold concentrations of substances that fish will avoid. This perhaps not very important difference in viewpoint has led to minor differences in experimental procedures and treatment of data.

When fish are placed in a tank of water having a temperature gradient, they will distribute themselves so that most of the time spent by individual fish or so that most of the individuals of a group will be at a particular temperature, which is called the *temperature preferendum*. This preferendum is rather characteristic of the race of fish, but it is in part determined by the previous temperature history of the individuals. Groups of fish that have been acclimated to different temperatures prior to being placed in a gradient will have different preferenda. But if fish are left in a temperature gradient for a long period of time, they will ultimately tend to congregate at the temperature where their preferred and acclimation temperatures would have been equal. This temperature is called the *final preferendum*. Fry (1965) has reviewed these concepts. Though in nature the distribution of fish is influenced by food and many other factors in addition to temperature, fish do distribute themselves in relation to temperature in nature, and the responses there are not different in kind from those observed in the laboratory. Ferguson (1958) summarized evidence for correlation between responses of 21 species of fish to temperature in nature and their final preferenda as determined in laboratory studies. For two species of char, a hybrid char, and *Coregonus clupeaformis*, he found close correlation. But for the other species, final preferenda were higher than were temperatures at which the fish were usually found to occur in nature.

Doudoroff (1938) and Norris (1963) found that young of the opaleye (*Girella nigricans*), after having been exposed to normal, variable temperatures of sea water, selected temperatures near 26 C in experimental gradients. Doudoroff observed that this temperature was much higher than the average water temperature in the inshore area from which these fish were collected. He therefore concluded that the selected temperature was not the usual temperature in the habitat of this species and could not be regarded as representing a natural habitat preference. Norris (1963), however, came to a different conclusion on the basis of his detailed observations in nature. He found that young opaleye inhabiting the intertidal zone definitely tend to move into water having temperatures near 26 C in tide pools during low tide. He also found the appetite of the fish in the laboratory to be greatest at temperatures near 26 C. Norris came to the conclusion that reactions of opaleye in experimental gradients after these fish had been exposed to normal, variable temperatures of sea water were similar to their reactions in nature, where these reactions probably have survival value.

Doudoroff (1938) and Norris (1963) also observed that young opaleye acclimated in the laboratory to any constant temperature between 20 and 30 C selected temperatures considerably less than 26 C. As we have already indicated, fish from sea water in which the temperature varied selected tempera-

tures of about 26 C, even after they had been held in the laboratory in flowing sea water for long periods. Thus, constant-temperature acclimation of *Girella nigricans*— and perhaps other species (Ferguson, 1958)—appears to destroy normal reactions to temperature gradients. Here is a warning for all who would, as they finally should, use their laboratory findings to increase our understanding of nature.

Adult Pacific salmon have been shown to exhibit a marked alarm reaction and to reduce their rates of migration through a fish-way when rinses of mammalian skins, including those of man, bear, and seal, were introduced into the water; these responses appeared not to be learned (Brett and MacKinnon, 1954; Brett and Groot, 1963). One can understand how such avoidance behavior might evolve, because these mammals are important predators on adult salmon, and their detection from some distance downstream would have considerable survival value. Likewise, one can understand the evolution of behavior causing fish to avoid low concentrations of dissolved oxygen or high concentrations of carbon dioxide or ammonia, for these substances are present in the natural environment of fish and can occur in combinations that are injurious or lethal. But fish have been shown to avoid very dilute solutions of a great variety of other substances such as copper and lead, substances that are unlikely to have occurred in harmful concentrations in the environments of most races of fish. The evolution of behavior leading to the avoidance of such substances is perhaps more difficult to understand, unless we postulate that the avoidance of certain groups of strange conditions tends to favor survival and that this has led to the evolution of a generalized avoidance response. Of course, some conditions may be avoided simply because they are irritating or in other ways lead to discomfort.

An environmental factor avoided by fish may act both as a releasing stimulus and as a directive stimulus, or perhaps only as a releasing stimulus. Jones (1952, 1964) and Höglund (1961), who review their own extensive studies and those of others, both concluded that low oxygen concentrations have no directive influence. They suggested that a fish happening into water of low oxygen content experiences respiratory discomfort, which releases appetitive behavior in the form of nondirected movements that cease if the fish again wanders into water where it experiences no respiratory difficulty. Whitmore, Warren, and Doudoroff (1960) found juvenile coho salmon, chinook salmon, and largemouth bass (*Micropterus salmoides*) avoided to some extent water having lowered oxygen concentrations that produced no acute respiratory distress. The coho salmon exhibited some avoidance of concentrations as high as 6 mg/l (Fig. 12–5). Fish in the apparatus used in this study were able to swim about in the main part of the tank and enter at will at one of its ends channels having water of lowered oxygen content or channels having the same oxygen concentration as was present in the main part. Fish frequently changed their direction of swimming immediately upon crossing the sharp boundaries of oxygen concentration at the channel entries; Whitmore and his co-workers suggested that the fish could detect lowered oxygen concentrations before experiencing discomfort.

Since Shelford and Allee (1913) first studied the movements of fish in water having gradients of dissolved gases in laboratory apparatus, many modifications of their apparatus and other apparatus have been employed by different workers to study the avoidance behavior of fish. Nearly all workers have recognized the restriction that

conditions in the laboratory place on interpretation of their findings with respect to the behavior of fish in nature. But it has perhaps not been sufficiently emphasized that the conditions in different kinds of apparatus and the experimental procedures employed may lead to differences in behavioral responses even under laboratory conditions that superficially appear to be similar.

Höglund (1961), like several others before him, found that fish avoid waters having increased concentrations of carbon dioxide. He concluded that carbon dioxide acts as a directing stimulus as well as a releasing stimulus and that change in pH does not direct or release the reaction. Höglund discusses the possibility of there being chemoreceptors sensitive to carbon dioxide, located in the gill region of fish. In our laboratory, we have found that juvenile coho salmon respond to differences in carbon dioxide concentrations as low as 1 mg/l at temperatures around 20 C, but at temperatures around 8 C, they may not respond to differences as high as 30 mg/l. In nature, when oxygen concentrations are low, carbon dioxide concentrations tend to be high. We have found juvenile chinook salmon in the laboratory to exhibit much greater avoidance of water when the carbon dioxide concentration was 20 mg/l and the dissolved oxygen concentration was 2.5 mg/l than when both carbon dioxide and oxygen

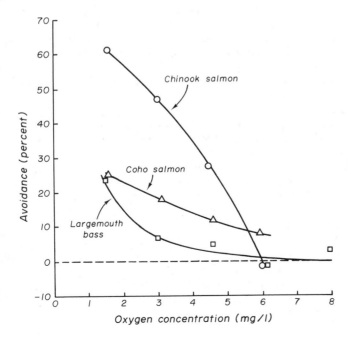

FIGURE 12–5 Avoidance by juvenile chinook salmon, coho salmon, and largemouth bass of water having various concentrations of dissolved oxygen. Data of Whitmore, Warren, and Doudoroff (1960).

concentrations were high or when both were low. This suggests an interaction that intensifies the avoidance reaction and would increase its survival value in nature.

Sprague (1964a) has shown that juvenile Atlantic salmon markedly avoid copper solutions at concentrations as low as 9 micrograms per liter—a concentration but two tenths of the incipient lethal level—and almost completely avoid concentrations of 90 micrograms per liter—about two times the incipient lethal level. Figure 12–6 shows the movements of the fish across the sharp boundary of copper concentration when different concentrations were present. These movements and other reactions of the fish suggest they could rapidly detect the copper, as Sprague perhaps correctly concluded. But, early in the experiment, the fish could have experienced discomfort after passing a certain point in the apparatus and so have been conditioned not to continue going beyond this point. Such learning might lead to results that could be interpreted as evidence of rapid detection. This could have led Whitmore, Warren, and Doudoroff (1960) to conclude, perhaps erroneously, that fish can rapidly detect differences in oxygen concentration. A careful examination of the movements of Sprague's fish might suggest to one that learning was occurring, especially since Figure 12–6 shows these movements for only the last 10 minutes of each 20-minute test period. But even the demonstration of learning need not

preclude the possibility of rapid detection. And however fish avoid situations that could be harmful, this avoidance is likely to be adaptive.

Fish have been shown to avoid many naturally occurring substances as well as substances that man introduces into streams; Jones (1964) and Höglund (1961) provide us with excellent reviews of research on this problem. Fish have a tendency to avoid many conditions that would be harmful to them, but, as studied in the laboratory, this avoidance is not always complete. And, in some instances, they may not avoid conditions that could prove rapidly fatal to them (Sprague and Drury, 1969). Unfortunately, our knowledge of the movements of fish in relation to harmful environmental conditions is largely based on laboratory findings, and again and again we are faced with a problem of interpretation.

It is for this reason that careful observations on the movements of fish in relation to harmful environmental factors in nature are of considerable interest; no opportunity to make such observations should be missed. When we consider the importance of knowledge of the responses of fish to harmful conditions, we find it rather surprising that so infrequently have biologists availed themselves of these opportunities. Sprague, Elson, and Saunders (1965) took advantage of an opportunity to compare the migratory movements of adult Atlantic salmon in the Northwest Miramichi River before and after

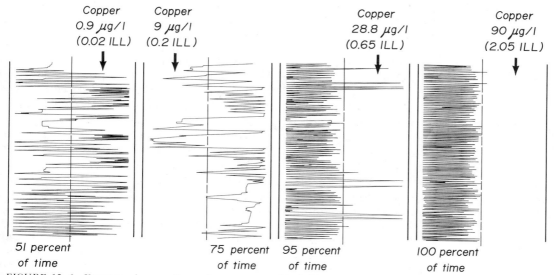

FIGURE 12–6 Kymograph recordings of movements of individual juvenile Atlantic salmon in relation to copper ions present in the water in one half of a trough designed to study avoidance behavior. After Sprague (1964a).

drainage from a mine introduced copper and zinc into this river. Of salmon ascending the river, 10 to 22 percent turned and descended during four years after drainage from the mine began; over a period of years before the mine was established, only 1 to 3 percent of the salmon turned to descend. This avoidance behavior occurred when the concentration of copper and zinc was greater than 0.35 to 0.43 of the incipient lethal level of the mixture for juvenile salmon (Chapter 13). Juvenile salmon in the laboratory avoided concentrations greater than 0.02 of the incipient lethal level.

Collins (1952) took advantage of the spring run of alewife (*Alosa pseudoharengus*) and glut herring (*A. aestivalis*) in the Herring River of Massachusetts to make carefully controlled observations of the choices these fish made between two experimental channels having different temperatures, carbon dioxide concentrations, and other water quality characteristics. When temperature differences continuously exceeded 0.5 C, 77 percent of the fish entered the warmer channel. When carbon dioxide concentrations differed by more than 0.3 mg/l, 72 percent of the fish selected the channel having the lower concentration. Often, natural conditions can be modified to the extent necessary to make controlled observations while much of natural reality is retained.

In the course of their other management and research activities, fisheries biologists have for generations observed the movements of fish and been somewhat aware of existing environmental conditions. We do not mean to belittle the knowledge deriving from this accumulated experience. A great deal is known of the timing of migrations of anadromous fish and something of river conditions when migrations occur. Biologists know that fish sometimes avoid polluted waters that would harm them and that populations of fish often persist in waters receiving harmful substances in moderate to small amounts. But such general knowledge is not sufficient for the solution of particular pollution problems. The responses of fish to different levels of harmful substances and to substances that would not be harmful, were fish to continue using waters contaminated by them, should be known. These responses can only be known through careful observation, both in nature and in the laboratory.

ENVIRONMENTAL CHANGE AND BEHAVIOR

Through evolution, the problems faced by all species in surviving, reproducing, growing, and moving have been solved in different ways. But each species now can be successful only within a limited range of environmental conditions and only through behavioral responses appropriate for that species. Each species, through its structure, physiology, and behavior, is sensitively adapted to the environment in which it has evolved. And, since even the natural environment of a species is quite variable in space and time, its physiology and behavior have come to be sufficiently plastic to permit it to adapt over some range of change in environmental conditions. But when a change in environmental conditions is much beyond the range within which a species has evolved, individuals of that species may fail to adapt and their population may not persist. If, however, there are some individuals in a population that have the genetically determined capacity to survive and reproduce under the changed conditions, the persistence of their phenotype will be favored over that of phenotypes not having such capacity, and genetic adaptation of the population may occur. But here we are concerned with the potential of the individual organism for behavioral adaptation.

In considering changes in behavior, we must distinguish between those that are adaptive and can favor the survival, reproduction, and growth of an animal in a changed environment and those that are nonadaptive or pathological and could lead to failure of the animal in an environment not physically or biologically greatly different from its natural one. The presence of a very low level of a toxic substance could quite conceivably so change the behavior of an animal as to decrease its survival or reproduction in an environment where it would have been successful had its behavior not been changed (Brockway, 1963; Ogilvie and Anderson, 1965; Warner, Peterson, and Borgman, 1966; Anderson, 1968; Anderson and Peterson, 1969).

In the previous section, Behavioral Adaptations to the Environment, we grouped certain behavioral responses into categories: territorial behavior, reproductive behavior, feeding behavior, migratory behavior, and avoidance behavior. These and other kinds of behavior bring individual animals into relationships with their environments that tend to insure the continuance of their kinds. The movements of fish in nature are essential to their survival, reproduction, and feeding. Territorial behavior tends to favor optimal use of environmental resources. And other forms of social behavior in other ways regulate the activities of individuals to the ultimate good of their kind. We may never understand fully just how all the behavioral responses of animals to their environments favor the persistence of their species. But our understanding of evolution leads us to believe that they are important in adapting each animal to its environment and that nonadaptive or pathological changes in behavior are to be viewed with concern for each species. Pollutional changes to which animals cannot adapt physiologically and behaviorally or which bring about nonadaptive behavior are likely to lead to the loss of affected populations.

We have devoted a rather long chapter to the study of behavior. We have discussed internal motivation, releasing and directing stimuli, instinctive and learned behavior, and some behavioral adaptations to the environment. In a book relating biology to water pollution control, we are concerned about probable effects of environmental change on aquatic organisms of interest to man. But little is known about the effects of environmental changes on the behavior of fish and other aquatic organisms. Even though behavior is as much a part of animals and as important to their persistence as are their morphological structures, the study of behavior has been neglected. This makes it particularly necessary that, in relating biology to water pollution control, we stress the importance of knowledge of behavior. And, hopefully, one day a book will be written that will review much research on the extent of possible behavioral adaptation to pollutional changes and the significance of nonadaptive changes in behavior.

SELECTED REFERENCES

Anderson, J. M., and Margaret R. Peterson. 1969. DDT: sublethal effects on brook trout nervous system. Science 164:440–441.

Baerends, G. P. 1957. Behavior: the ethological analysis of fish behavior. Pages 229–269. *In* Margaret E. Brown (Editor), The Physiology of Fishes. Vol. 2, Behavior. Academic Press Inc., New York. xi + 526pp.

Barnett, S. A. 1963. The Rat: A Study in Behaviour. Aldine Publishing Company, Chicago. xvi + 288pp.

Brown, Margaret E. (Editor). 1957b. The Physiology of Fishes. Vol. 2, Behavior. Academic Press Inc., New York. xi + 526pp.

Fabricius, E., and K.-J. Gustafson. 1954. Further aquarium observations on the spawning behaviour of the char, *Salmo alpinus* L. Fishery Board of Sweden, Institute of Fresh-Water Research (Drottningholm) Report 35:58–104.

Fraenkel, G. S., and D. L. Gunn. 1940. The Orientation of Animals. Oxford, at the Clarendon Press. [viii] + 352pp.

Fry, F. E. J. 1964. Animals in aquatic environments: fishes. Pages 715–728. *In* D. B. Dill, E. F. Adolph, and C. G. Wilber (Editors), Adaptation to the Environment. (Handbook of Physiology, Section 4.) American Physiological Society, Washington, D.C. ix + 1056pp.

Harden Jones, F. R. 1968. Fish Migration. St. Martin's Press, New York. viii + 325pp.

Hinde, R. A. 1966. Animal Behaviour: A Synthesis of Ethology and Comparative Psychology. McGraw-Hill Book Company, New York. x + 534pp.

Hoar, W. S. 1958. The evolution of migratory behaviour among juvenile salmon of the genus *Oncorhynchus*. Journal of the Fisheries Research Board of Canada. 15:391–428.

Höglund, L. B. 1961. The reactions of fish in concentration gradients. Fishery Board of Sweden, Institute of Fresh-Water Research (Drottningholm) Report 43. 147pp.

Jones, J. R. E. 1964. Fish and River Pollution. Butterworth, London. viii + 203pp.

Kalleberg, H. 1958. Observations in a stream tank of territoriality and competition in juvenile salmon and trout (*Salmo salar* L. and *S. trutta* L.). Fishery Board of Sweden, Institute of Fresh-Water Research (Drottningholm) Report 39:55–98.

Klopfer, P. H., and J. P. Hailman. 1967. An Introduction to Animal Behavior: Ethology's First Century. Prentice-Hall, Inc., Englewood Cliffs, New Jersey. xiv + 297pp.

Marler, P., and W. J. Hamilton III. 1966. Mechanisms of Animal Behavior. John Wiley & Sons, Inc., New York. xi + 771pp.

Norris, K. S. 1963. The functions of temperature in the ecology of the percoid fish *Girella nigricans* (Ayres). Ecological Monographs 33:23–62.

Sluckin, W. 1965. Imprinting and Early Learning. Aldine Publishing Company, Chicago. x + 147pp.

Sprague, J. B., P. F. Elson, and R. L. Saunders. 1965. Sublethal copper-zinc pollution in a salmon river — a field and laboratory study. International Journal Air and Water Pollution (Pergamon Press) 9:531–543.

Thorpe, W. H. 1951. The definition of some terms used in animal behaviour studies. Bulletin of Animal Behaviour 9:34–40.

———. 1963b. Learning and Instinct in Animals. 2nd. ed. Methuen and Co. Ltd., London. x + 558pp.

Tinbergen, N. 1951. The Study of Instinct. Oxford, at the Clarendon Press. xii + 228pp.

———. 1965. Behavior and natural selection. *In* J. A. Moore (Editor), Ideas in Modern Biology. Proceedings 16th International Congress of Zoology 6:519–542.

Warner, R. E., Karen K. Peterson, and L. Borgman. 1966. Behavioural pathology in fish: a quantitative study of sublethal pesticide toxication. Journal of Applied Ecology 3(Supplement):223–247.

13 Tolerance of Lethal Conditions

The definition of waters as suitable for aquatic life is complicated still further by the fact that various species of fishes and other aquatic animals and even individuals of different ages of the same species have different degrees of tolerance to deviations from the ideal environment, and to the cumulative effects of many stream pollutants. Consequently, the presence or even the survival for a time of fishes in waters suspected of pollution does not in itself constitute evidence that these waters are either satisfactory or safe for fishes.

M. M. ELLIS, 1937, p. 367

In the past, most studies have involved the killing effects of pollutants on selected living organisms. Useful information has been derived from such studies. The evaluation of a water toxicant in terms of kill or no-kill, however, is somewhat circumscribed in effectiveness. Many delicate responses of organisms are never observed. All too often crude data, based on 50 percent kill or on time to kill 50 percent of the test animals, are used to arrive at rather broad conclusions.

CHARLES WILBER, 1965, p. 326

LETHAL CONDITIONS, TOLERANCE, AND BIOASSAYS

A lethal condition is a characteristic of some environment. Tolerance is a characteristic of some organism. But an environment can only be lethal to an organism; and an organism can only be tolerant of an environmental condition. Organisms may tolerate for a while a lethal condition, but this they cannot do indefinitely, or the condition would not be lethal.

We wish to know what conditions would be lethal to organisms in aquatic environments, and we wish to prevent the occurrence of such conditions. Obviously we need to know much more, for it is not enough to prevent the occurrence of lethal conditions. We need to know and provide conditions that are in every way suitable for aquatic life.

Still, prevention of lethal conditions is important. A lethal condition is not usually due to any single characteristic of an environment. The lethal condition of an environment often derives from two or more environmental factors interacting so as to make that environment lethal to some organism. Some single environmental factor may indeed be the lethal agent, but the level of that factor causing death will usually depend on the levels of other environmental factors, physical, chemical, and even biological. It will also depend on the tolerance of the organism.

The tolerance of an organism for a lethal condition derives from the organism's genetic constitution and its environmental history, because these interact to make the organism what it is at any given moment. Thus, tolerance is usually no very constant attribute. It changes with the development of the organism and may change with age throughout the life of the organism. Tolerance may depend not only on an organism's recent experience with the lethal agent but also on its early experience. For these reasons, no two organisms, even very closely related ones, are likely to have exactly the same tolerance for a lethal condition.

The word tolerance carries a certain negative connotation, a connotation that the organism somehow gets along, that it merely tolerates conditions. That an organism is even able just to tolerate existing conditions may sometimes be important, so long as things get better; but better they must get if the organism is to be very successful. So far in our consideration of the individual organism, we have emphasized the importance of normal development, growth, and behavior. In earlier chapters of this book, we touched on the importance of normal survival or longevity. A chapter on survival now might seem more appropriate than a chapter on tolerance. But though an individual organism has a capacity for survival, it is the mean survival rate of individuals in a population in nature that is biologically important. The question of survival rate, as well as of birth rate, pertains more to the population than to the individual organism; it is in the next part of this book—the one on populations—that we can really consider the question of survival (Chapter 16, Population Dynamics).

In evaluating the tolerance that organisms have for lethal conditions, we observe survival or death. Working with samples of organisms under different conditions, we observe the percentages of these organisms alive or dead at intervals of time. Our observations really, then, are survival or death rates. But these are survival or death rates occurring in small samples of organisms under peculiar conditions in the laboratory. Usually tolerance experiments are of short duration, not beginning to approach the possible life span of the organisms tested. Such experiments are not intended to tell us much about survival rates in nature. Their objective is to evaluate the tolerance of organisms for lethal conditions. Tolerance measured in terms of survival rate in the laboratory may indicate whether or not survival in nature is likely to be unfavorably influenced by a potentially lethal agent, but this is all.

The idea of tolerance involves a question of time. For how long can an organism tolerate a given lethal level of an environmental factor? Within some range of levels of an environmental factor, an organism can tolerate the factor for an indefinite period of time; beyond that range, the organism can tolerate the factor only for a period of time that is a function of the level. Fry, Hart, and Walker (1946) called the range of an environmental factor—they were studying temperature—within which an organism would not usually die as a result of that factor alone the *zone of tolerance*. The zone of tolerance is separated by the *incipient lethal level* from a range of the factor they termed the *zone of resistance*. Within the zone of resistance, an organism will eventually die as a result of the environmental factor. The organism may live almost indefinitely if the factor is near the incipient lethal level; but as the factor goes beyond this level, death comes progressively sooner. It is within the zone of resistance that most studies of tolerance are conducted, for only here can lethal effects be measured. The incipient lethal level—the boundary between the zones of resistance and tolerance—must often be estimated by extrapolation.

Henceforth in this chapter, we will use the terms *tolerance and resistance synonymously*. The distinction Fry, Hart, and Walker (1946) made between the meanings of these words was useful for their purposes. But the expression *median tolerance limit* to designate lethal levels in the zone of resistance has been in use for a long time. Median tolerance limit and median resistance limit are equivalent expressions. Thus, it would not be helpful for us to distinguish between tolerance and resistance.

Now, we have been writing as though tolerance were a matter of life or death. In a chapter on tolerance of lethal conditions, this is probably permissible, and here we will continue to be concerned primarily with living and dying. Moreover, with the negative connotation we associate with the word tolerance, perhaps life and death are its most appropriate criteria. But there are certainly other indications of the tolerance organisms may have for environmental conditions. Among these are reproduction, development, growth, and behavior, responses to the environment that we consider in other chapters.

When we determine the tolerance of an

organism for an environmental factor, the result will depend on the nature of the organism, not only its species but its individual genetic constitution and its early and recent environmental history as well. Whether the factor is salinity, temperature, dissolved oxygen, or a toxic substance, the organism is our measure of tolerance levels. The condition of the measure will influence our results. This we must always bear in mind.

Whatever the environmental factor, when we use an organism to determine tolerance levels, we are in a sense making a biological assay. Nevertheless, the expression *bioassay* is usually applied to the evaluation of the potency of drugs or toxic substances. In this, the organism becomes the reagent for determining the amount or activity of a substance. Such a biological reagent is generally more variable in quality and sensitivity than most chemical reagents; we must be more concerned about the condition of our reagent. Still, biological measurements of the amounts or activities of toxic substances not only are necessary but also are often more reliable than physical and chemical measurements. Death is the response usually observed in bioassays of toxic substances conducted as a part of water pollution investigations: these are usually *acute toxicity bioassays,* with observations limited to a few days. But long-term tolerance and reproductive, developmental, growth, and behavioral responses are sometimes measured. These are more sensitive responses of great biological importance; their use in the toxicology of aquatic organisms will increase.

The purpose of most early studies of the toxicity of water pollutants was to explain observed fish mortalities or to determine concentrations of toxic substances not likely to cause such mortalities. Control of waste disposal was usually limited to prevention of sporadic gross pollution that could not escape public attention. Many years ago, Léger (1912) reasoned that concentrations of toxic substances not causing mortality in streams within one hour would usually have little or no effect on fish populations. With intermittent discharge and mixing and dilution of toxic wastes, harmful concentrations supposedly would not persist longer. As Doudoroff and Katz (1953, p. 817) pointed out, some leading European investigators not so many years ago still favored toxicity tests lasting only six to eight hours. With increases in biological knowledge and regulatory stringency, and with advances in waste treatment technology, this view has fortunately become obsolete. Persistent pollution of our waters with low concentrations of toxic substances may be even more deleterious to fish production than sporadic occurrence of high concentrations causing only occasional fish mortalities. Nevertheless, terrible mortalities caused by sporadic pollution still occur; their prevention remains an important objective of water pollution research and control. This is often a reason for performing acute toxicity bioassays.

Bioassays are sometimes used in determining the components responsible for the toxicity of complex wastes. The toxicities of the waste components—tested separately and in various combinations, including the waste itself—can be compared to determine the relative contribution of each component to the toxicity of the waste. Interactions of the components may make it difficult to attribute the toxicity of the waste simply to one or more components, but this approach is still likely to be practical and profitable. The components of a waste responsible for its acute toxicity are not necessarily the ones responsible for any effects it may have when further diluted in the receiving water. Nevertheless, until more is known, it may be useful to suppose them to be the same.

It may sometimes be reasonable to assume that physical and chemical factors in the environment that influence the acute toxicity of a substance will similarly influence its sublethal effects. Such an assumption may make the results of studies of how salinity, temperature, dissolved oxygen, or particular minerals influence the acute toxicity of materials even more useful.

Bioassays have been used to determine the concentrations of materials for which there are no adequate methods of chemical assay. The response of the organism to a standard series of known dilutions of a material is established before the unknown solution is tested. In this way, Doudoroff (1956) was able to estimate, tentatively but quite accurately, low concentrations of molecular hydrocyanic acid in certain complex cyanide solutions, before reliable chemical methods for its determination were developed (Doudoroff, Leduc, and Schneider, 1966).

Finally, we usually know very little about the concentrations of toxic substances that may in nature have deleterious effects on the survival, reproduction, growth, and behavior of animals we value. Until that distant day when we know much of what we should, knowledge of acutely toxic concentrations of substances may be all that we will have on which to base our protection of aquatic resources from harmful effects of some substances. Perhaps small though arbitrary fractions of acutely toxic concentrations can be permitted in receiving waters, until we know more. This is a question with which we will struggle in the next chapter.

We cannot hope in the present chapter, which is moderate in length, to tell our reader much about the tolerance of animals for lethal conditions. If our reader believes we are audacious in attempting to cover as much as we do in so little space, we would be inclined to agree with him. Tolerance studies have been and will continue to be a large and important part of water pollution investigations. Their results occupy the greatest part of the biological literature on water pollution. Here, we can only hope to give our reader some feeling for the rationale of tolerance studies and for the ways in which they are conducted. Because we know our reader can turn to the fine work of other men, our task is made easier. The many excellent reviews of the tolerance literature, some of which we list at the end of this chapter, will be his best introduction to the methods and results of tolerance studies. And our readers who undertake tolerance studies of their own will become engrossed in the original writings.

EVALUATION OF LETHAL CONDITIONS

Tolerance Experiments

Properly designed experiments on the tolerance of animals for lethal conditions should permit us to determine the *level of a lethal factor* that can be tolerated by a given percentage of the animals for a given period of time, or, conversely, the *length of time* a given percentage of the animals can tolerate a given level of a lethal factor. *Lethal factor level* and *survival time* become, then, the variables of interest in tolerance experiments. This is only to say that the level of the lethal agent and exposure time are both involved in any harm coming to an animal. When working with groups of animals, as we

usually do, we must note the percentage of survival or mortality, because individual animals differ in their tolerance.

Lethal levels of temperature are sometimes determined by gradually warming water until the test animals die. Similarly, lethal levels of oxygen are often determined by allowing the test animals to deplete the oxygen in water in sealed vessels until the animals die of oxygen lack. Perhaps it is the simplicity of such procedures that has encouraged their use in the determination of lethal levels, for they are recommended by little else. The two variables of interest—level of the lethal factor and exposure or survival time—change simultaneously. This makes it impossible to determine either the level of the lethal factor the animals can tolerate for a given period of time or the length of time the animals can tolerate a given level of the lethal factor. Thus, data from such experiments do not permit us to establish the most interesting and valuable relationship, the one between survival time and lethal factor level. Lethal levels determined in this way may be of comparative value when the rates of change of the lethal agent are the same and survival times are noted. Since these conditions are rarely satisfied, it is hard to justify publication of most information of this kind.

To compare the tolerance of different species for some lethal factor, or to compare the lethal effects of different factors, survival time at a given level of the lethal factor or factors is sometimes recorded. If the procedure is to be accurate and convenient, the level of the lethal factor or factors tested must cause death neither too rapidly nor too slowly. Differences in lethality or tolerance estimated only on the basis of survival time can be misleading. Two species of fish having greatly different survival times at the same level of a lethal agent may have lethal levels that are little different. For practical as well as theoretical reasons, differences not only in survival times but also in lethal levels should be taken into account in the evaluation of relative tolerance.

Tolerance experiments conducted in connection with water pollution research or control usually have one or both of two immediate objectives. One is to establish the relationship between survival time—usually median survival time—of a species and lethal factor level. This relationship can be used to estimate the length of time the species could tolerate any given level of the lethal agent, or the level that could be tolerated for any given exposure period. From the relationship it may also be possible to estimate the *incipient lethal level*, the level the species could tolerate indefinitely. The other objective of tolerance studies is to provide a basis for comparing the lethality of different agents or the tolerance of different species. The information needed to satisfy these two objectives is not really different. Neither are the procedures for obtaining this information.

Groups of animals are exposed to a series of levels of the lethal factor. One can then obtain the *median survival time* for the animals at each level. This can be obtained in one of two ways. If we are patient and have the time, we can stay in the laboratory and note the time of death of every animal at each lethal level. Then we can calculate the mean or median survival time of the animals at each lethal level.

But the easier of the two procedures—usually equally satisfactory—is to make our observations only periodically at times separated by fixed hourly or daily intervals or by a logarithmic series of intervals. We can observe and record the percentage of animals surviving at the end of each interval at every level of the lethal factor. From such data, by graphical interpolation, we can estimate the median survival time of the animals at any level of the lethal factor; and we can also estimate the median level of the factor the animals can tolerate for any given period of time. The latter is known as the *median tolerance limit* (TL_m) for the specified period of time—e.g., the 24-hour, 48-hour, or 96-hour TL_m. Some now prefer to represent median tolerance limit by TL50 rather than by TL_m. This newer usage permits more consistent representation of different tolerance limits, such as the tolerance limit for 10 percent of the animals (TL10) or for 90 percent of the animals (TL90).

Should our objective be only to compare the lethality of two or more agents under acute conditions or to compare the tolerance of two or more species for some lethal agent at rapidly lethal levels, then we would probably determine median tolerance limits for one to a few days. But if our objective is to estimate the incipient lethal level or survival times at different levels, then we would establish either the relationship between median survival time and lethal agent level or

the relationship between median tolerance limit and exposure time. The range of levels of the lethal agent we would test and the frequency and duration of our observations on percentages of survival would be determined by the objective of our experiment. Nevertheless, for either of the aforementioned objectives, the nature of our experimental design and observations would be fundamentally the same.

Median Tolerance Limits and Median Survival Times

No two animals are exactly the same in their responses to environmental conditions. They will be more alike if they are of the same species, or of the same race, or if they have the same parents. But still, genetic differences and differences in both their early and recent environmental histories make them different. These differences may complicate the work of a biologist, but they are the strength of the thread of life represented by any persisting population. Populations persist in the face of changing environmental conditions because their individuals respond differently to these conditions (Chapter 7). Were this not so, natural selection could not operate; populations could not adapt to changing conditions.

Were all individuals of a kind alike in their tolerance of a lethal agent, we could learn this tolerance by testing but one individual. But the tolerances of individuals differ, whether we measure tolerance in terms of tolerable level of a lethal agent or in terms of survival time. Thus we must deal with the percentages of animals that can tolerate a certain level or survive for a certain time.

Were we to be able to take a moderately large sample from a natural population of animals and determine individual tolerance limits of the animals in the sample, we would find the frequency distribution of these individual tolerance limits to have a maximum near some central value and to descend toward zero with increasing and with decreasing tolerance limits. Were we then to take an even larger sample from this population and similarly determine individual tolerance limits, we would probably find the extreme individual tolerance limits of the larger sample to be higher and lower than the extremes in the smaller sample. But the median tolerance limits of individuals in the two samples—the values above and below which 50 percent of the individual tolerance limits occurred—would not be greatly different. Estimates of median tolerance limits have the advantage of reasonable precision even when sample sizes are not large. It is difficult if not impossible to precisely estimate the highest level of a lethal factor that will kill no animals in a natural population or the lowest level that will kill all the animals. Thus, median or mean tolerance limits are most commonly used as estimates of the tolerance characteristics of populations.

Sometimes the distribution of individual tolerance limits will be a *normal frequency distribution*. This is a symmetrical bell-shaped curve that asymptotically approaches zero at its ends. It is not just any symmetrical bell-shaped curve but only one having certain mathematical characteristics. When the level of a lethal factor is best expressed—best in relation to its activity—as a concentration, our bell-shaped frequency distribution is likely to be asymmetrical or skewed if the level of the lethal factor is plotted on an ordinary arithmetic scale. This is because the absolute increase in concentration of a substance is not so nearly proportional to its effect on the organism as is its percentage increase. An increase in the concentration of a toxicant from 0.01 to 0.02 mg/l can be

expected to have more effect on an organism than does an increase from 1.01 to 1.02 mg/l. For this reason, concentrations of toxicants are usually plotted on a *logarithmic scale*. By thus plotting, in effect, the frequency distribution of the logarithms of individual tolerance limits, we are more likely to get a symmetrical curve approaching a normal frequency distribution curve. A frequency distribution that becomes symmetrical and strictly normal upon such logarithmic transformation is known as a *log-normal distribution*.

Now, if our data indeed do conform to a log-normal distribution, and if we plot the percentages of survival at different concentrations of a toxicant laid off on a logarithmic scale, then we will get a symmetrical sigmoid curve, a *log-normal cumulative frequency distribution curve*. As in Figure 13–1, our data will rarely if ever conform exactly to such a distribution; the reasons for this may be biologically important and should not be ignored. On the other hand, only random variability may be involved. How well our data fit the theoretical distribution can be tested by plotting mortality or survival percentages on a *normal probability* or *probit scale*. If a straight line reasonably well fits the data so plotted, then we can with some confidence use this straight line to esti-

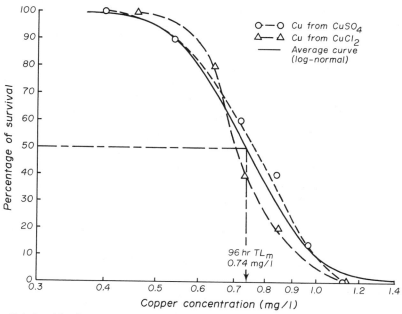

FIGURE 13–1   Relationships between percentage of survival of bluegill for 96 hours and concentration of copper ion from copper sulfate and from copper chloride, and the average theoretical log-normal relationship from which the 96-hr. TL_m is estimated. After Trama (1954).

mate the median tolerance limit for the length of time at which the percentages of survival were observed.

Trama (1954) found that the percentages of survival of bluegill (*Lepomis macrochirus*) in solutions of copper sulfate and in solutions of copper chloride deviated somewhat from the theoretical cumulative frequency distribution curve. We know of no reason why copper ions from these different compounds should affect the fish differently in the very dilute solutions prepared with a natural water (Doudoroff and Katz, 1953). Moreover, Trama showed that both sets of data fitted reasonably well a single straight line on a graph of survival probits on an arithmetic scale and concentrations on a logarithmic scale. Perhaps, then, the data can best be combined to estimate the 96-hour median tolerance limit of copper ion concentration for bluegill (Fig. 13–1).

To our reader, all this may seem rather involved, though we have tried to explain as simply and clearly as possible the reasoning on which procedures for estimating median tolerance limits are based. We have not wanted to create the impression that the estimation of median tolerance limits involves much theory and very complex procedures, for nothing could be farther from the truth. These are simple procedures and they yield simple information. Statistical theory and methods should help us to make the most of our observations without becoming awkward burdens that stretch the observations thin. Sometimes straight-line interpolation between mortalities just above and below 50 percent is quite good enough for estimating median tolerance limits, even without resorting to fancy scales. This part of a cumulative frequency distribution is usually quite linear whatever the scale. The design of experiments and the analysis of results need be no more involved than is necessary to attain our objectives, which are quite modest when we seek median tolerance limits. We only wish to know levels of environmental factors that will lead to death. Tolerance experiments can tell us little more. Now we must explain estimation of that other measure of tolerance: median survival time.

Even brief consideration will lead our reader to the conclusion that, in its effect on the organism, time of exposure to a lethal factor is more likely to be better represented on a logarithmic scale than on an arithmetic scale. Some scales are convenient in that they lead to linear relationships, as we explained for concentrations of toxicants—for which logarithmic scales are used—and for percentages of mortality—for which normal probability or probit scales are used. Percentage increase rather than absolute increase in exposure time more nearly represents the effect on an organism of increasing its exposure to a lethal factor. An increase in exposure time of one hour will probably have more effect on an animal that has already been exposed only for one hour than on an animal that has already been exposed for 100 hours. The exposure time of the first animal was increased 100 percent; that of the second animal only 1 percent. With death, survival time and hence exposure time end. *Survival time*, or we may wish to call it *resistance time*, is usually plotted on a logarithmic scale.

To estimate *median survival time* of a species at some lethal level of an environmental factor, we can expose a group of individuals to that level and observe the percentages surviving at different times. The relationship between percentage survival and survival time will usually be reasonably linear if the former is plotted on a normal probability or probit scale and the latter on a logarithmic scale. From this relationship, we can graphically estimate the length of time just 50 percent of the individuals were able to survive, the median survival time. We can repeat this procedure at a series of levels of the lethal factor to determine the median survival time at each level.

Shepard (1955) established the relationships between percentage of mortality and resistance time of brook trout (*Salvelinus fontinalis*) at ten lethal levels of dissolved oxygen (Fig. 13–2). Median survival times—or median resistance times, there is no difference—can be estimated from eight of the ten relationships as the time at each oxygen concentration when 50 percent of the individuals would have been dead. At the two highest concentrations of oxygen, 50 percent of the individuals did not die in 5000 minutes and perhaps would not have died with much longer exposure.

Here we have, then, ways of estimating median tolerance limits and median survival times, useful and different measures of the tolerance of animals for lethal conditions. With them we can explore relationships be-

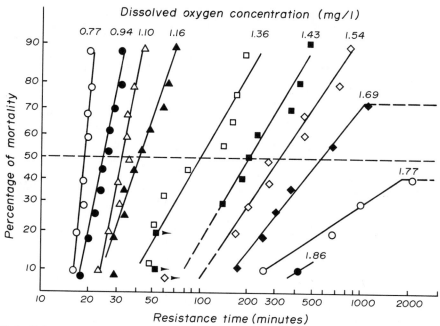

FIGURE 13–2 Relationships between percentage of mortality and resistance or survival time of juvenile brook trout at ten lethal levels of dissolved oxygen after the trout had been acclimated to a dissolved oxygen concentration of 10.5 mg/l. Median survival time at each of the eight lowest concentrations can be estimated as the time at which just 50 percent mortality existed. After Shepard (1955).

tween lethal levels and survival times as well as interactions of environmental factors leading to lethal conditions. We will take the space to consider a few uses of these measures and relationships that are of practical importance in water pollution control.

Relations Between Lethal Levels
and Survival Times

There is interest in the forms of relationships between lethal level and survival time; these forms may be characteristic of different species and lethal factors. We can, of course, from such relations estimate the length of time animals can be expected to survive at a given lethal level; or, if we wish, we can estimate the level of the lethal agent the animals can tolerate for a given period of time. Perhaps a relationship between lethal level and survival time can be used to estimate reliably the *incipient lethal level* or *lethal threshold*, the level of a lethal agent the animals could be expected to tolerate indefinitely. This is a common use of relationships of this kind, but it is a use usually involving extrapolation with all its uncertainties. Because these relationships involve both time and lethal level expressions of

tolerance, they provide a more general means of comparing the lethality of a factor under different conditions than do individual median tolerance limits. Sometimes sharp inflections in curves relating survival time to lethal level are supposed to indicate changes in the mode of action of the lethal agent or, if more than one toxicant is present, a change in the lethal agent. We should not overlook such indications, for the search for understanding is slow at best. Neither should we become very excited about indications alone; they may be artifacts of our imperfect procedures. We should begin to be excited and satisfied by our results only after we have made further observations explaining why there should be a change in mode of action or a change of lethal agent.

So many different kinds of relationships between lethal level and survival time have been described, we find it hard to believe that much of importance can be inferred from the different forms the curves assume, perhaps the forms we give them. Of course, our data become more useful when curves can be fitted to them reasonably well. But we dislike seeing biologists spend too much time doodling. We can expect relationships to have different forms under different conditions; but often curves appear to have different forms even when we can find no reason for the difference. Then, unidentified and uncontrolled variables are the cause, so we can hope to learn little from the differences we observe or imagine.

In Figure 13–3, we show a few relationships between lethal agent level and median survival time that resemble curves that have been used by biologists to fit their data. Relationships A and B have simple mathematical formulations, which are convenient if our data can be reasonably so described. Unfortunately, sets of data that in their entirety conform to simple mathematical relationships are not so very common. Then we must go to relationships, like C or D or E, that have no simple formulations.

Curve A in Figure 13–3 is of a kind that can be described by the equation:

$$(C - a)^n (T - b) = K \qquad (1)$$

where C is concentration of the toxicant, T is exposure time, a is the incipient lethal level, or threshold concentration, b is the

threshold reaction time, and the exponent n and the product K are constants. The *threshold reaction time* is the minimum length of exposure the animal can tolerate before reacting by dying or collapsing, no matter what the level of the lethal agent may be. The particular curve of this kind that we show as A in Figure 13–3 has a value of 2 for the constant n and a value of 10 for the constant K. The incipient lethal level, a, is 0.1 mg/l; and the threshold reaction time, b, is 4 minutes.

All or a portion of a relationship between lethal level and survival time may be rectilinear when logarithmic scales are used for both the axis of abscissas and the axis of ordinates. Such a relationship is described by the equation:

$$C^n T = K \qquad (2)$$

where, again, C is concentration, T is exposure time, and n and K are constants. For relationship B in Figure 13–3, the exponent n has a value of 4 and the product K has a

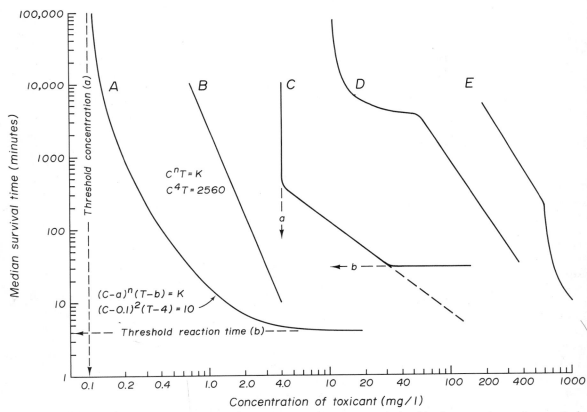

FIGURE 13–3 Various kinds of curves that have been used to define relationships between the median survival times of animals and the concentrations of toxicants.

value of 2560. Were we to have reason for assuming that such a linear relationship would persist with indefinite extension of exposure time and reduction of concentration, we then might have some confidence in estimating with the aid of equation 2 the lowest concentration that would be lethal within the normal life span of the organism. Such extrapolation, which must usually be far beyond available data, is ill-advised. Moreover, the absence of a lethal threshold—some level of a lethal agent an animal can tolerate indefinitely—has never to our knowledge been conclusively demonstrated in any study of fish toxicology.

Perhaps it would be helpful to our reader for us to examine briefly just how well these equations and some of the other kinds of relationships shown in Figure 13–3 appear to fit the data various investigators have obtained on the tolerance of fish for toxicants. First, let us look at cyanide. Some investigators have found the tolerance of fish for cyanide to be described reasonably well by equation 1 (Wuhrmann, 1952; Burdick, 1957; Burdick, Dean, and Harris, 1958). But Herbert and Merkins (1952) believed that equation 2 adequately described their data on the toxicity of potassium cyanide to rain-

bow trout (*Salmo gairdneri*). Doudoroff, Leduc, and Schneider (1966), who studied the relationship between the survival time of bluegill and the concentration of molecular hydrocyanic acid in simple and complex cyanide solutions, found equation 2 apparently applicable over a range of exposure times up to 400 minutes but not beyond this time. Thus, neither equation 1 nor equation 2 was adequate for describing their results.

Mathematical description of only a portion of a curve usually has little real value. And subordination of data to obviously inappropriate mathematical relationships is silly. Nevertheless, to the kinds of relationships we have shown in Figure 13–3, toxicity data often conform well enough to make it worthwhile for our reader to be generally familiar with them. A few examples will help. Lloyd (1960) used equations 1 and 2 in order to fit two curves and a straight line to his data on the toxicity of zinc ion to rainbow trout at three levels of water hardness (Fig. 13–4). The data appear to be described reasonably well by these equations. It is interesting that a linear relationship was found at the lowest level of water hardness, whereas the relationships were curvilinear at the intermediate and high levels. Here, then,

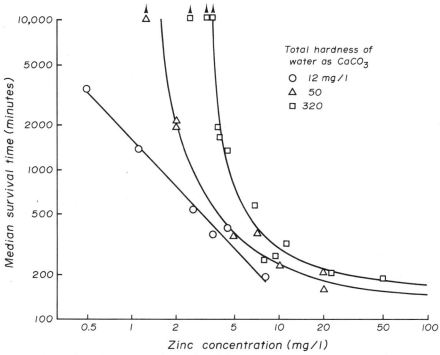

FIGURE 13–4 Relationships between median survival time of rainbow trout and concentration of zinc ion in waters having different total hardness. After Lloyd (1960).

we have an example of the toxicity of an ion being sometimes best described by a rectilinear relationship and sometimes by a curvilinear relationship, when both axes are logarithmic. Interaction with another environmental factor, as occurs so often, is here changing the relationship between survival time and toxicant concentration. The increasing tolerance of the fish for zinc with increasing water hardness was undoubtedly due to the presence of greater amounts of the divalent hardness cations, mainly calcium. Calcium ions are known to antagonize or counteract the toxic action of heavy metal ions.

Sprague and Ramsay (1965) have presented data showing the relationship between median survival time of juvenile Atlantic salmon (*Salmo salar*) and the concentration of copper to be linear at concentrations above the incipient lethal level, which is clearly denoted by the abscissa at the sharp inflection of the line (Fig. 13–5). This is like relationship C in Figure 13–3, except that concentrations of copper high enough to determine the threshold reaction time were not tested. The graph Jones (1964, p. 132) drew on the basis of data Tarzwell and Henderson (1957) obtained on the toxicity of

dieldrin appears to be an example of relationship C in its entirety. Before leaving this relationship, we should call our reader's attention to the difference in tolerance of copper ion exhibited by Atlantic salmon tested at different times in waters of slightly different hardness (Fig. 13–5). A small increase in hardness apparently caused changes in the incipient lethal level and in the slope of the line fitting the data but not in the kind of relationship found. But some factor other than water hardness may have been involved.

Toxicity relationships resembling D and E in Figure 13–3 are sometimes found (Jones, 1939; Anderson, 1944). Inflections of curves depicting such complex relationships are commonly supposed to indicate changes in mode of action of the toxicants or their interaction with other environmental factors. Lloyd and Jordan (1964) found the relationship between survival time of rainbow trout and hydrogen ion concentration to be quite linear when the concentration of carbon dioxide was 1.5 mg/l (Fig. 13–6). But at a carbon dioxide concentration of 50 mg/l, they found a complex relationship. The two relationships were not essentially different at pH values up to about 4; with further increase in

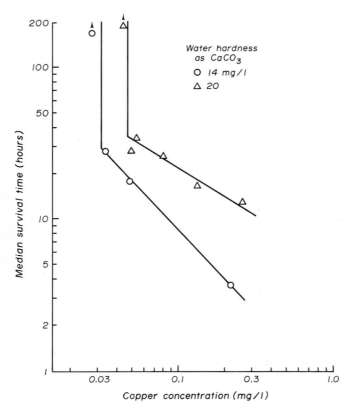

FIGURE 13–5 Relationships between median survival time of juvenile Atlantic salmon and concentration of copper ion in waters having different total hardness. After Sprague and Ramsay (1965) and Sprague (1964b).

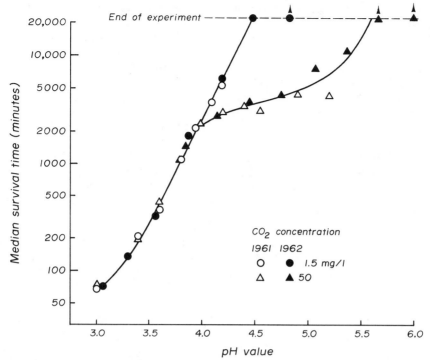

FIGURE 13–6 The effect of carbon dioxide concentration on the relationship between median survival time of rainbow trout and pH value of the water. After Lloyd and Jordan (1964).

pH, however, survival time was greatly reduced in the presence of the higher concentration of carbon dioxide (Fig. 13–6). At pH values higher than 4, the fish at the 50 mg/l concentration of carbon dioxide were apparently less able to maintain the buffering capacity of their blood than were the fish at the 1.5 mg/l concentration. Acidemia appears to have been the cause of death. At pH values below 4, the concentration of carbon dioxide in the water apparently had little effect on the hydrogen ion tolerance of the fish. Here, then, we have an example of a complex toxicity relationship, one for which we have a reasonable explanation. The explanation involves interactions, the subject of our next section.

INTERACTIONS INFLUENCING TOLERANCE

Experimental Animals

Tolerance is a characteristic of the organism; the organism is our only measure of tolerance. Nevertheless, the tolerance of an organism for a particular lethal agent is determined not only by its own characteristics but also by the set of environmental conditions with which it is confronted along with the lethal agent. Before considering interactions between lethal and other factors in the environment that influence the responses we observe, let us review briefly the nature of the organism that responds. This is all-important, whether we are using this organism as a means of increasing our knowledge of the tolerance of its kind for conditions in nature or only as a biological reagent for measuring some toxicant.

Ultimately, all the characteristics of an animal are genetically determined, and rarely do any two individuals have exactly the same genetic endowment. This is a wonderful thing, but it complicates our laboratory studies. The responses we observe depend on the genetic constitution of the animals we study; in planning and conducting our work, we must always take this into account. But the expression of an animal's genetic endowment depends on environmental conditions throughout its development and life (Chapter 7). We cannot usually know this endowment or even what environmental conditions have been throughout the life of an organism. We know only that they will determine its tolerance.

We usually have some control over the recent environmental history and the age and size of our experimental animals. These also are important determinants of the responses we observe. The nutrition of our animals and their previous exposure to the lethal agent as well as other environmental factors must be taken into account before we perform tolerance experiments.

An animal's recent experience with a lethal factor is often particularly important in determining its tolerance. Previous exposure to a toxicant may in some cases render an animal more sensitive to later exposure. But we usually find the tolerance of an animal for a lethal factor increased by recent exposure to levels of the factor that did not kill the animal but that caused some stress. Under stressing conditions, there usually occur within an animal adaptive responses rendering the animal more tolerant not only of these conditions but also of more critical ones. These adaptive responses are *acclimation processes.* An animal made more tolerant of environmental conditions by these processes is generally said to be *acclimated* to these conditions. In Chapters 7 and 8, we discussed adaptive responses and acclimation in considerable detail; we have no intention

of here burdening our reader much more with this subject. But since tolerance of lethal conditions is so affected by acclimation, and also since such tolerance is perhaps the most used measure of acclimation, we cannot avoid further consideration.

Fish are able to acclimate to temperature changes; investigations that have established the possible rates and extent of temperature acclimation are among the most extensive and finest acclimation studies. The ability of fish to acclimate to changes in dissolved oxygen is also generally known. Fish and most animals can undoubtedly acclimate to a wide variety of other factors and conditions existing in their environments. Toxic substances are among these, but information on the substances to which fish can acclimate and on the extent of any acclimation is rather spotty. Still, there is some: fish have been shown to acclimate to phenol (Bucksteeg, Thiele, and Stöltzel, 1955), synthetic detergents (Degens et al., 1950; Lemke and Mount, 1963), hydrogen ions (Jordan and Lloyd, 1964), ammonia (Lloyd and Orr, 1969), cyanide (Neil, 1957), zinc (Lloyd, 1960; Edwards and Brown, 1967), and other toxic substances.

The extent of acclimation is a function not only of the level of the factor to which the animal is exposed but also of the length of exposure. Rates of acclimation to different

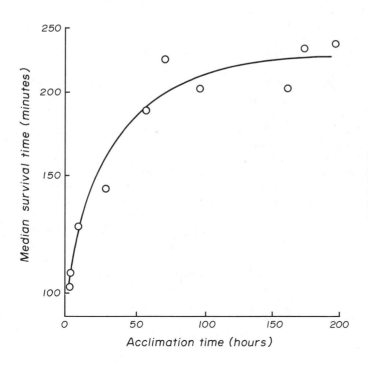

FIGURE 13-7 Increases in the median survival time of juvenile brook trout exposed to a lethal level of dissolved oxygen (1.06 mg/l) after the trout had been acclimated for different periods of time at a dissolved oxygen concentration of 3.8 mg/l. After Shepard (1955).

lethal factors vary greatly. The gain in heat tolerance of fish usually is quite rapid and may be nearly complete within one day; loss of heat tolerance at low temperatures is usually much slower. Doudoroff (1942) found opaleye (*Girella nigricans*) to gain cold tolerance at low temperatures and to lose this tolerance at high temperatures at slow but nearly equal rates. Acclimation to low oxygen levels is probably slowed by low temperatures. Even at moderate temperatures, Shepard (1955) found full acclimation of brook trout to low oxygen concentrations to require several days (Fig. 13-7). The incipient lethal level is usually made more extreme by acclimation. When determined after maximum acclimation, it is known as the *ultimate incipient lethal level.*

In conducting laboratory studies, we should always bear in mind how really little we know about our experimental animals. They are often obviously stressed by handling and confinement, even to the extent of dying in the absence of known lethal agents. How lesser and unnoted stresses resulting from routine laboratory procedures may influence our results is difficult to determine. But interacting with other variables, such stresses may cause the results of apparently beautiful experiments to be misleading.

Interactions of Lethal and Nonlethal Factors

We like Fry's (1947) method of classifying environmental factors not according to what they are but according to how they are acting under a given set of circumstances (Chapters 7 and 11). Thus, within some range, temperature acts primarily to control the metabolic rate of fish and is here best considered a *controlling factor.* But should temperature press the metabolic rate of a fish above the rate at which it can obtain oxygen from its environment, the fish will die, not because the temperature was at a lethal level but because the oxygen supply was insufficient. Still, temperature was involved and might best be considered an *accessory factor* leading to death; oxygen deficiency was the *lethal factor.* There are levels at which temperature acts as a lethal factor, leading either to heat death or to cold death. Most environmental factors can under dif-

ferent circumstances act in different ways, their action depending on their level and other environmental conditions. Factors at levels well within tolerable ranges may still act so as to decrease the tolerance of an organism for a lethal agent. Sometimes they act so as to increase this tolerance.

We have already given examples of chemical constituents of natural waters physiologically counteracting—antagonizing—the toxic actions of zinc and copper ions (Figs. 13-4, 13-5). The divalent ions calcium and magnesium are particularly effective in anatagonizing the toxic actions of heavy metals on fish. Toxicants can also be rendered less effective by chemical reactions taking place in the medium external to the aquatic organism. Some of these reactions lead to the formation of relatively harmless compounds.

Many toxic pollutants are weak acids or bases that penetrate readily the external membranes of fish and other aquatic animals only when in the undissociated or molecular form. Since the ionization of weak acid or base molecules to form anions and cations—which do not readily penetrate membranes and hence are not very toxic—is very dependent on the pH of the water, the toxicity of weak acids and bases to aquatic organisms is also very dependent on pH. Nevertheless, within the pH range in which the molecular form greatly predominates, changes in pH have little effect on the toxicity of the solution. Thus, the toxicity of hydrocyanic acid or of sodium cyanide changes little with variations of pH between 6 and 8, because here the cyanide already exists mainly in its most toxic form, molecular hydrocyanic acid. Ammonia, in contrast, occurs mainly in its harmless ionic form within this pH range; and the concentration of its toxic molecular form will increase nearly tenfold with one unit increase of pH. Here, then, we have an environmental factor, pH, sometimes not affecting and other times affecting the toxicity of substances, even though that factor is well within the range of tolerance of most aquatic organisms.

Complex metallocyanides provide us with an extreme example of changes in the toxicity of solutions with changes in pH (Fig. 13-8). Doudoroff (1956) found that the toxicity of solutions of the nickelocyanide complex—formed by combining sodium cyanide and nickel sulfate—increased more than ten-

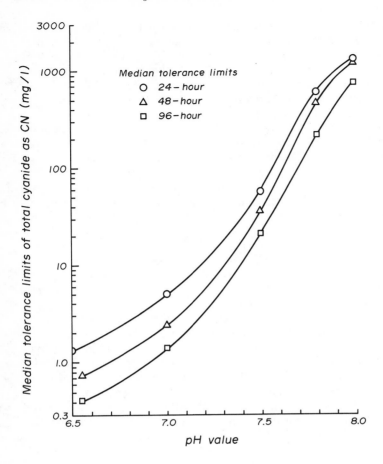

FIGURE 13–8 Influence of pH on the toxicity of solutions of the nickelocyanide complex to fathead minnows (*Pimephales promelas*). This complex was formed by combining sodium cyanide and nickel sulfate. After Doudoroff (1956).

fold with a doubling of the hydrogen ion concentration, a pH decrease of only three tenths of a unit. This was due to increased dissociation of the almost nontoxic nickelocyanide complex and formation of highly toxic molecular hydrocyanic acid.

Many environmental factors may be involved in determining the tolerance of an organism for a toxicant. The tolerance of rainbow trout for ammonia appears to be dependent on pH, bicarbonate alkalinity, temperature, and the concentrations of carbon dioxide and oxygen (Lloyd, 1961a). The curves shown in Figure 13–9 illustrate how much pH and bicarbonate alkalinity can influence the toxicity of a solution of ammonia.

Temperature variations well within the ordinarily harmless range can greatly influence the tolerance of most aquatic organisms for lethal agents, as we suggested in beginning this section. Increase in temperature accelerates metabolic processes and rates of gill irrigation. In consequence, aquatic animals are usually killed by toxicants more rapidly at moderately elevated temperatures than at lower temperatures. Median tolerance limits determined for short exposure periods therefore usually decrease as temperature increases. Incipient lethal levels, however, may increase, decrease, or remain unchanged as the temperature is raised. Increasing temperature can accelerate not only penetration and harmful action of a toxicant but also adaptive responses, including its elimination from the body. Thus, constant pollution of water with a toxicant is not necessarily more dangerous at high temperatures than at low temperatures; the reverse can be true. In the laboratory, however, if for comparative purposes we wish to measure promptly and adequately the toxicity of a substance, we must usually perform our experiments at moderately high and uniform temperatures.

As do increased temperatures, reduced dissolved oxygen concentrations also require accelerated irrigation of respiratory surfaces.

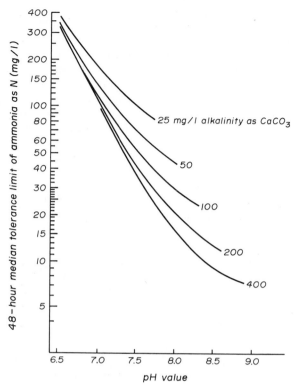

FIGURE 13–9 Influence of pH and bicarbonate alkalinity on the toxicity of ammonia to rainbow trout at 20 C. After Lloyd (1961a).

This brings to the respiratory surfaces a larger quantity of any toxicant present in the water. At lower oxygen concentrations, a toxicant may be absorbed more rapidly or may sooner damage respiratory structures. In an important contribution to the toxicology of aquatic animals, Lloyd (1961b) demonstrated that the toxicity of several substances to fish increased almost equally when oxygen was reduced by the same amount. His conclusion that these increases in toxicity were primarily due to increased respiratory irrigation bringing more of the toxicants to gill surfaces is probably widely but not always applicable.

Some toxic materials can enter fish more readily by the gastrointestinal route than through the gill membranes. Ingestion of sea water by marine bony fishes—essential to their osmoregulation and survival in their natural medium—can speed such absorption of toxicants.

The natural medium of an aquatic organism is not necessarily the medium most favorable for its resistance to lethal agents. The tolerance of freshwater forms for toxi-

cants may be favored by unusual salinity or pH of the water. Extreme low temperatures may be more rapidly fatal to freshwater and marine fish in their natural media than in dilute sea water nearly isotonic with their blood. Doudoroff (1945) found that the marine fish *Fundulus parvipinnis* when acclimated at 20 C died within four days at a temperature of 3 C in sea water; the fish lived for six days at 3 C in sea water diluted to 45 percent of its normal salinity. This lethal low temperature evidently rendered normal osmoregulation impossible, and osmotic dehydration of the tissues followed. Dehydration apparently was the eventual cause of death at normal salinity; but we can hardly consider normal salinity to be the primary lethal factor. It was the low temperature, for the fish eventually died at 3 C even in the diluted sea water, where they suffered no dehydration. Failure of osmoregulation, probably a consequence of damage to the nervous system, was only one of the low temperature effects that could lead to death; it was only the first to prove fatal when the fish were living in their normal medium of undiluted sea water. Brett (1952) found juvenile sockeye salmon (*Oncorhynchus nerka*) to be more tolerant of lethal low temperatures in dilute sea water than in fresh water, their normal medium. So much for interactions of lethal and nonlethal factors; we must now conclude our discussion of interactions influencing tolerance with a brief section on interacting lethal factors.

Interactions of Lethal Factors

When the disintegration known as death occurs, it is difficult to learn the most immediate cause; indeed, we rarely do. For that matter, the demarcation between life and death is not at all clear; this gives us difficulty in determining survival time accurately and precisely. Death comes, the animal's systems run down and stop. The immediate cause of death, which system stopped first, just when the animal is to be considered dead, these questions may seem rather insignificant in contrast to the very significant fact of death. Just when an animal is to be considered dead may be a technical problem in our laboratory; it is also raising social questions in modern medicine. But, more important, we must continue to seek understanding of the actions and effects of lethal factors that lead to death.

We have chosen to distinguish, however imperfectly, between interactions of lethal and nonlethal factors leading to death and interactions of lethal factors leading to death. For interactions hurrying death, this may to our reader seem an overly fine distinction, particularly when we do not know the lethal actions or the final causes of death. The distinction may be more easily appreciated for those interactions of lethal and nonlethal factors that slow death. We will make the distinction on the basis of whether or not any harmful effect at a given level of a second or third factor can be demonstrated in the absence of a known lethal factor. If not, we will consider the interaction to be between lethal and nonlethal factors. But if a second or third factor is at a level that would be harmful even were the lethal factor to be absent, then we will consider the processes leading to death to be interactions of lethal factors.

This is the general statement of our distinction. We make general statements to

be helpful, but sometimes they cause difficulty. A possible difficulty with our statement, a possible exception, may occur when several toxic substances are individually at low and apparently harmless levels but together they lead to the death of an organism—their effects are somehow additive. This may, however, be only apparently an exception to our general statement. Such addition of toxic actions or effects may occur only when individually the toxicants have harmful actions or effects, however small and whether or not we can detect them. There is much room for growth of toxicology as a science; little is known of the actions and effects of most toxic substances.

In Chapter 8 in the section entitled The Problem of Toxic Substances, we sketched some of the adaptive responses of animals, the idea of action-effect sequences, and the problem of toxicant interactions. We would be derelict in our responsibility to our reader were we to describe toxicant interactions no more fully than we did there. Perhaps a diagram and a brief explanation will be of some help. Figure 13–10 is an attempt to illustrate possible kinds of interactions between two toxic substances. As abscissas we show different combinations of solutions of toxicants A and B, solutions prepared to have equal toxicity when tested separately. The ordinates are relative toxicities—reciprocals of median tolerance limits—of combinations of the two solutions.

When two equally toxic solutions of different toxicants are combined in various proportions, a *strictly additive interaction* is revealed if all the mixtures prove to be just as toxic as either of the solutions would be were it tested separately (Fig. 13–10). *No interaction*—strictly independent action—can be considered to occur if, when equal amounts of the two solutions are mixed and tested, the toxicity of the mixture is equal to exactly one half of the toxicity of either solution tested separately (Fig. 13–10). In this case, the toxicant in each solution is diluted by the other solution to only one half of its initial concentration; there being no interaction of the two toxicants, either could be responsible for the toxicity of the mixture. When the mixture is a combination of other proportions of the two solutions, its toxicity will be determined by the substance whose solution is present in the greatest proportion in the mixture. Since still there is no inter-

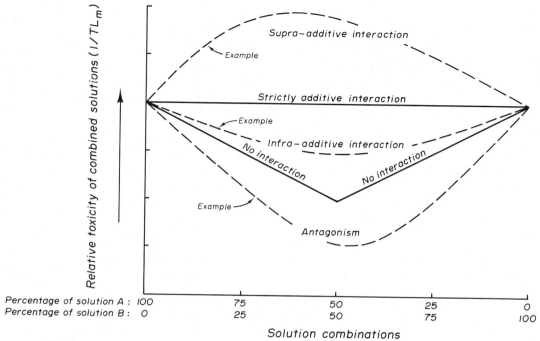

FIGURE 13–10 Possible kinds of interactions between two hypothetical toxicants, A and B. Solutions of these toxicants are assumed to have been prepared so as to have equal toxicity when tested separately. For each kind of interaction, there are shown the changes in relative toxicity (reciprocal of median tolerance limit) of the combined solutions that will occur when the two solutions are combined in different proportions.

action of the two toxicants, mixture in any proportion results only in dilution of the solution of each toxicant.

But if the two substances, A and B, have an additive interaction that is less than strictly additive, then we can consider the two substances to have an *infra-additive interaction* (Fig. 13–10). And if the two substances have an additive interaction that is more than strictly additive, we can consider the two substances to have a *supra-additive interaction.*

Antagonism is indicated only if the toxicity of the mixture is less than would be expected were there to be no interaction at all between the two toxicants. Thus, in our view, antagonism is an interaction of two substances that leads to a reduction of toxicity greater than would be expected were one toxic solution merely to be diluting another (Fig. 13–10). Unfortunately, some systems of classifying interactions fail to make this distinction (Jones, 1964), a distinction that helps to clarify our thinking. The term antagonism is sometimes applied to chemical and physical reactions that occur in the external medium of an organism and reduce the toxicity of the medium. Again to clarify our thinking, it would be better to reserve the term antago-

nism for physiological phenomena. We should remind our reader that nontoxic substances can antagonize toxicants, and they can interact with toxicants so as to increase toxicity. In presenting this classification of interaction of toxic substances, we must not allow these two important facts to be forgotten.

Now we have not used the terms *synergism* and *potentiation*, we believe for the best of reasons. Their ambiguous use has contributed more to confusion than to clarity of our thinking or to our understanding of the actions and effects of interacting toxicants. Gaddum (1953, pp. 479–480) considered synergism to be any nonantagonistic interaction of toxicants; any supra-additive interaction he called potentiation. Fingl and Woodbury (1965, p. 26) noted that supra-additive interactions are sometimes called synergism but recommended that this term be reserved for interactions increasing the effect of one substance by another which alone would have no effect. They also recommended that the term potentiation be abandoned. These men were clear and they recognized distinctions between kinds of interactions; in these respects, many others have failed. When men, for their own very good reasons, use the same term for many different ideas, that term loses most of its symbolic value. It becomes a hindrance, not a help, to our thinking and understanding. When this occurs, the best course would seem to be abandonment of the term; this course we choose to adopt. After all, it is understanding of actions and effects of toxicants that we seek, not systems of terminology. Thought processes are too sensitive and too important to be needlessly impeded. Words, particularly ambiguous words, are often impediments to clear and penetrating thought.

But, at this point, our reader may fairly object to our use of the term *antagonism,* which also has come to be ambiguous. Some authors have included under the designation antagonism the interaction that we consider infra-additive. If we are to drop the use of the term synergism, should we not also drop the use of the term antagonism? We would drop the term antagonism too, were we to have a convenient and equivalent descriptive expression; but others will want to continue using both terms. Synergism and antagonism seem only to be terms for opposite kinds of interactions of toxic substances, the one increasing and the other decreasing toxic actions and effects. If used in just these senses—in our thinking and in our verbal and written discourse—then these terms should lead to no difficulty. But a problem arises when we try to decide and agree on when substances are not interacting and when they are interacting, either to increase or to decrease toxic actions and effects. If it can be accepted that no interaction of substances occurs only when the relationship is similar to the one we have so labeled in Figure 13–10, then toxicities below this line reveal antagonistic interactions and toxicities above this line reveal synergistic interactions. And then infra-additive interaction, strictly additive interaction, and supra-additive interaction become *infra-additive synergism, strictly additive synergism,* and *supra-additive synergism.* Such a system of terminology may satisfy some but not all, and the problem of ambiguity will remain.

Recent studies of British investigators have suggested what may be a very practical approach to defining the toxicity of combinations of toxicants occurring in polluted waters. Although, as is ever so, this approach has its roots in many earlier ideas and investigations, it has grown to bear fruit mainly as

a result of intensive studies by British investigators beginning with Lloyd (1961c). This approach, which has been explained in detail by Brown (1968), is based on considerable data showing that even toxicants having different modes of action very often have sublethal effects that combine in a seemingly strictly additive way to kill fish. We would not expect this to be so; the approach has little theoretical basis; yet it is proving useful.

This system of estimating the toxicity of a mixture of toxicants is based on the *lethal threshold concentrations*—the incipient lethal levels—of the toxicants. The toxicity scale is a biological one in which the lethal threshold concentration of any toxicant in a mixture or of the mixture itself is taken to be 1.0 unit. The units of toxicity contributed to the mixture by each toxicant, which Sprague and Ramsay (1965) have called *toxic units,* can be calculated by dividing the actual concentration of each toxicant by its lethal threshold concentration. Fish often have been found to be killed by mixtures of toxicants when the sum of the concentrations of the individual toxicants expressed in toxic units has been greater than 1.0 toxic unit. Except when individual toxicants are present in concentrations much above their lethal thresholds, the survival time of fish in mixtures of toxicants appears to be inversely related to the total concentration of the toxicants in the mixture expressed in toxic units (Herbert, 1965).

Basically the method described here assumes that all poisons can contribute in a similar manner to the overall toxicity of a mixture, although it is appreciated that *a priori* it is illogical to expect poisons of different toxicological properties and different concentration-response curves to summate in this manner. Nevertheless, the method has been found to work empirically. . . . If, for the present, the poisons can be regarded as agents producing stress. . ., each of which produces a degree of shock with resulting non-specific effects, it might be considered reasonable that summation of the overall stress is possible. (Brown, 1968, p. 731)

Here is a good place to leave interactions influencing tolerance. We must not let theory obscure experience. We need theory to guide our thinking; but we must be careful with biological theory, for we really know so little about biological systems. Theories of toxic action and effect undoubtedly have some basis in facts that must be cherished and developed. We would not want observations that we do not understand and that may only superficially conform to some system of measurement to in any way undermine the development of sound theory. Nevertheless, neither do we want to discard systems of measurement that promise to be useful, even if they do not conform to present theory. The solution of water pollution problems requires tools whose making cannot await full understanding of the actions and effects of toxic substances.

TOLERANCE STUDIES IN PERSPECTIVE

Before attempting to place tolerance studies in some perspective in the realm of biology, we should perhaps again try to make clear the idea we here associate with tolerance. We have taken the theme of this chapter to be the tolerance animals exhibit for lethal conditions in their environments. Animals may survive under lethal conditions for a while but not indefinitely; otherwise the conditions are not lethal. Tolerance, then, is measured in terms of the level of the lethal factor and the length of time the animal survives. We do not suppose an animal is getting along well under conditions it merely tolerates for a while. We take tolerance to mean only that the animal is somehow surviving.

Now we know that survival is not the most sensitive measure of how well an organism is getting along under a given set of environmental conditions. Reproduction, development, growth, and behavior are much more sensitive measures. There is no reason whatsoever why any of these should not be used to measure the tolerance of an organism for environmental conditions. It is just that we prefer not to refer to studies of these other biological responses as tolerance studies. They are much more than this. They are studies of all the interactions between organisms and their environments that lead to the success or failure of their populations. These studies are very important; they need more attention; and they receive most of the attention in this book. But, classically, tolerance studies have not encompassed so much, and we prefer to restrict the idea of tolerance to matters of life and death over periods of time that are not very long.

This is enough semantics. Of what use are tolerance studies? We can suggest four general categories within which most uses would fall. Tolerance studies can be used to determine incipient lethal levels delimiting the ranges of environmental factors within which animals might be able to survive and their populations could possibly persist. Tolerance studies are often used to determine levels of environmental factors that soon lead to death, levels we certainly wish to prevent in nature. With death as a convenient end point, tolerance studies may be used to learn how several environmental factors interact to determine the effect a particular factor will have on an organism. Finally, we have the classical bioassay use of tolerance studies when the organism is used only as a reagent in the estimation of the amount or activity of a toxicant. In one way or another, each of these uses of tolerance studies has played a role in physiology, pharmacology, or ecology. In one way or another, each is relevant to the problem of water pollution control.

It is not usually easy to experimentally determine for a species the incipient lethal level of an environmental factor, the level that will just begin to kill the animals, no matter how long their exposure. Time or investigational resources may be the problem, but there are others. Suffice it to say that the interest in ways of estimating incipient lethal levels from short-term tolerance tests is not without basis. Extrapolation usually involves assumptions we would rather not make, but it can often be justified as a first approximation. There is importance in knowing the levels of toxic or other environmental conditions that animals can tolerate indefinitely. Since the experimental determination of incipient lethal levels is no more difficult than determination of levels not affecting reproduction, development, growth, or behavior, we wonder if perhaps more attention should be given to experimental determination of incipient lethal levels and less to obtaining tenuous estimates based on extrapolation. Then, again, if we are to go to the trouble of long-term experiments, perhaps our attention would better be directed toward those other responses that are more sensitive than mere survival. Still, there are many environmental factors that interact to determine the effective levels of any particular factor. Something should be known of all these interactions; we sometimes become discouraged when we think of the improbability of learning of them through long-term studies alone. Short-term tolerance studies have their place.

Most tolerance studies have been concerned with the levels of environmental factors that can lead to the death of the test animals in no more than a few days. This is not surprising, because biologists, like others, are anxious to get results. We could argue that the amount of information on rapidly lethal conditions is out of all proportion to its intrinsic interest or practical significance. But this may not be so, even when we recognize that much better conditions must in nature exist most of the time, if animals are to remain. We do not want conditions that would soon lead to the death of animals to persist in considerable areas of water for long enough to cause mortality; we should know these conditions. If, during periods of time when maximum waste concentrations occur, these conditions are not approached too closely, conditions may most of the time be quite satisfactory for species of value. This argument is contrary to our whole philosophy of biology and water pollution control, as our reader may sense. It is not an argument we would wish to press very far. Still, in a terribly practical world, this argument cannot be ignored. But there are other interesting and practical reasons for all this information on lethal conditions.

Experiments with rapidly lethal conditions cannot last very long. We can conduct many of them and can study many combinations of these conditions in the time that would be required for a single experiment on long-term effects of environmental factors. An earlier section of this chapter was devoted to interactions influencing tolerance. Short-term tolerance studies will continue to be the most exploited approach for determining how changes in the animal and in factors in its environment alter the effectiveness of a particular factor. Most important interactions influencing tolerance occur within the organism; but some, particularly chemical reactions, occur in the water medium and aggravate or reduce the effect of the lethal agent. In either case, survival or death is a convenient measure of the effect of the interactions on tolerance, though we may not understand the protective or harmful mechanisms, or even the disintegrative processes leading to death. Without such understanding, it is difficult to know whether or not interactions influencing tolerance of lethal conditions would similarly influence biological responses other than death. In water pollution control, we often need to make decisions involving such matters. Unless or until we know differently, we will probably continue to assume that interactions influencing tolerance of lethal conditions will at levels not lethal similarly favor or harm reproduction, development, and growth.

When we use an organism only as a reagent to determine the amount or activity of a toxicant, we are using a bioassay in the classical way. This is an important use of tolerance tests. But, conceptually and practically, this use of tolerance tests is very different from the ones we have been considering. Until now, we have been concerned primarily with determining, on the basis of laboratory studies, conditions in nature that might in some degree be suitable for animals, conditions that at least would not be lethal. Now let us consider only the problem of measuring the amount or activity of a toxicant. Chemical tests are usually adequate for measuring the amount of a material; only the organism can be a suitable reagent for measuring the biological activity of a toxicant under different conditions. But chemical reagents have an advantage over biological ones. They are not so variable in performance. Thus, there has for a long time been interest in the standardization of bioassay methods.

In controlling water pollution, it is often necessary to know the toxicity of substances or effluents. Even when the toxic substances are known, chemical methods of analysis may not be available or suitable. And even when chemical measurement is feasible, it is not a direct measurement of toxicity. Thus, interest continues to develop in the use of organisms as measuring tools. Doudoroff et al. (1951) proposed standard bioassay procedures, which were later incorporated with little change in *Standard Methods for the Examination of Water and Sewage* (American Public Health Association et al., 1960). When the objective of bioassays is to obtain entirely reproducible analyses of the amounts or activities of toxicants, bioassay procedures should be standardized even much more than was proposed. A single species—even a single race—of test animals should be agreed on, as should the age, size, sex, and environmental history of the animals. Only a standard water at a standard temperature should be used. We could go on, but perhaps our point is made. If the organism is to be used only as a reagent, we must go as far as necessary to get reproducible results. There is need for this approach when we wish to compare the toxicities of different substances and effluents.

But this is not usually the objective of the so-called bioassays used in water pollution investigations. And the procedures proposed by Doudoroff et al. (1951) were not primarily intended for this purpose. These procedures were more to determine the tolerance of endangered species for effluents entering receiving waters than to compare the toxicities of different effluents. To reach this other objective, tolerance studies should not be highly standardized. Good procedures should be used, as are those that were proposed. But the test animals, waters, and temperatures should be similar to the ones where the pollution problem exists. Otherwise, the toxicities measured might not be relevant to the problem. The objectives of our tolerance studies—whether or not these tolerance studies are bioassays in the strict sense—should be foremost in our minds when we consider the desirability of standardization of procedures.

With so many kinds of changes coming in the aquatic environment, and with time, money, and talent forever limited, short-term tolerance tests will continue to play a major role in water pollution investigation and control. They are of value, they are needed, and they are a beginning for almost any investigation. It is nearly always of interest to relate the levels of environmental factors that have no effect on reproduction, development, growth, or behavior to the levels that would soon lead to the death of animals. Not usually knowing the concentrations of toxic wastes that would in no way harm animals in nature, we may often need to assume that some small fraction of the concentrations animals can tolerate for a while would not interfere with their biological success. We may need to use this fraction as a *bioassay application factor* in regulating waste discharges. It is hard to get excited about this kind of biology, but water pollution biology necessarily has this aspect. We all want to protect as best we can our resources; we cannot wait until as biologists we are satisfied with the state of knowledge. The story of application factors is a tenuous one, which we will leave for the next chapter.

SELECTED REFERENCES

American Public Health Association et al. 1960. Standard Methods for the Examination of Water and Wastewater Including Bottom Sediments and Sludges. 11th ed. American Public Health Association, Inc., New York. xxi + 626pp.

Brett, J. R. 1956. Some principles in the thermal requirements of fishes. Quarterly Review of Biology 31:75–87.

Brown, V. M. 1968. The calculation of the acute toxicity of mixtures of poisons to rainbow trout. Water Research (Pergamon Press) 2:723–733.

Doudoroff, P., B. G. Anderson, G. E. Burdick, P. S. Galtsoff, W. B. Hart, Ruth Patrick, E. R. Strong, E. W. Surber, and W. M. Van Horn. 1951. Bio-assay methods for the evaluation of acute toxicity of industrial wastes to fish. Sewage and Industrial Wastes 23:1381–1397.

———, and M. Katz. 1950. Critical review of literature on the toxicity of industrial wastes and their components to fish. 1. Alkalies, acids, and inorganic gases. Sewage and Industrial Wastes 22:1432–1458.

———, and ———. 1953. Critical review of literature on the toxicity of industrial wastes and their components to fish. 2. The metals, as salts. Sewage and Industrial Wastes 25:802–839.

———, and D. L. Shumway. 1970. Dissolved Oxygen Requirements of Freshwater Fishes. European Inland Fisheries Advisory Commission, Food and Agricultural Organization of the United Nations. In press.

Fingl, E., and D. M. Woodbury. 1965. General principles. Pages 1–36. *In* L. S. Goodman and A. Gilman (Editors), The Pharmacological basis of Therapeutics. 3rd ed. The Macmillan Company, New York. xviii + 1785pp.

Fry, F. E. J. 1947. Effects of the environment on animal activity. University of Toronto Studies Biological Series 55. Ontario Fisheries Research Laboratory Publication 68. 62pp.

———. 1964. Animals in aquatic environments: fishes. Pages 715–728. *In* D. B. Dill, E. F. Adolph, and C. G. Wilber (Editors), Adaptation to the Environment. (Handbook of Physiology, Section 4.) American Physiological Society, Washington, D.C. ix + 1056pp.

———, J. S. Hart, and K. F. Walker. 1946. Lethal temperature relations for a sample of young speckled trout, *Salvelinus fontinalis*. University of Toronto Studies Biological Series 54. Ontario Fisheries Research Laboratory Publication 66:9–35.

Gaddum, J. H. 1953. Pharmacology. 4th ed. Oxford University Press, London. xviii + 562pp.

Jones, J. R. E. 1964. Fish and River Pollution. Butterworth, London. viii + 203pp.

Sprague, J. B. 1969. Review paper: Measurement of pollutant toxicity to fish—1. Bioassay methods for acute toxicity. Water Research (Pergamon Press) 3: 793–821.

———. 1970a. Review paper: Measurement of pollutant toxicity to fish—2. Utilizing and applying bioassay results. Water Research (Pergamon Press) 4: In press.

———. 1970b. Review paper: Measurement of pollutant toxicity to fish—3. Sublethal effects and "safe" concentrations. Water Research (Pergamon Press) 4: In press.

14 *Application of Toxicity Bioassay Results*

Permissible concentrations of toxicants in waters receiving industrial wastes are those which can be tolerated indefinitely by all individuals (and not only 50 percent) of all significant species of aquatic organisms, including organisms which serve in the food chains as well as fish and others of direct economic and recreational importance, in all stages of their life history. Therefore, the concentration of toxic industrial wastes should never be more than a small fraction of that concentration which under experimental conditions is demonstrably fatal within a limited period of time to 50 percent of the test animals used in the bioassay.

AQUATIC LIFE ADVISORY COMMITTEE, ORSANCO, 1955, p. 330

. . . but no simple formula of this type ever can be developed which will take fully into account all possible peculiarities of the toxicity relationships of the great variety of toxic substances and of the physiological responses of different organisms. . . . The formulae are intended to make possible the derivation of a permissible concentration from the smallest amount of experimental data which can be regarded as at all adequate to permit making any such estimate. The application of these formulae is not to be regarded, however, as an entirely satisfactory substitute for more detailed investigation which will yield more reliable results.

HART, DOUDOROFF, AND GREENBANK, 1945, p. 132

THE PROBLEM

Acute toxicity bioassays—the results of which are frequently expressed as median tolerance limits—provide measures of the relative toxicities of different materials and the relative sensitivities of different species under conditions that rapidly lead to the death of the test animals. But they provide not much more: acute toxicity bioassays can tell us little about the maximum concentrations of toxic materials that will not affect either directly or indirectly in some deleterious manner the success of species in nature. Knowledge of such harmless concentrations can come from laboratory studies of the effects of toxic substances on the survival,

reproduction, growth, and behavior of species. Nevertheless, it is in nature that we wish the substances to be harmless; it is to nature that we must finally turn in our studies. Toxic materials may directly affect a species in which man is interested, or they may affect it indirectly through changes in its biological community, changes in its food resource or other relations. It would be nice if there were easy ways to learn the concentrations of toxic materials that are harmless to life, but there are not, no matter how much man may need this knowledge.

Probably no biological approach to the solution of pollution problems involving aquatic resources has received as much attention as the acute toxicity bioassay, except perhaps

214

the use of biological indices. This is not at all surprising, for the satisfactory conduct of such bioassays requires relatively little in the way of time, experience, or facilities, and the information yielded often is directly relevant to water quality problems involving toxic substances. But, even with all of this, biologists have long recognized that acute toxicity information gives us a poor basis for determining the concentrations of toxic substances that would have little or no effect on the distribution and abundance of species in nature. Certainly concentrations of substances determined in the laboratory to lead to the death of half the animals in one to a few days—the median tolerance limits (TL_m's)—will not permit in nature the survival of most of their kind over long periods of time. And concentrations so low as to not decrease the life span of a species may yet depress its reproduction or growth, either directly or through community changes. Biologists have emphasized that only some fraction of a TL_m would be harmless for a species in nature. This has called attention to the limits of utility of acute toxicity information and has identified the need to determine for different pollutants and species what these fractions might be. Such fractions have come to be known as *application factors:* the decimal fractions by which one would multiply the TL_m's of species for different pollutants in order to estimate the concentrations of these pollutants that would be harmless to these species in nature.

We can conceptualize an extensive biological investigation that could lead to the development of an application factor which, when multiplied by the median tolerance limit, would yield a reasonable estimate of the maximum concentration of a substance that would not decrease the distribution and abundance of a species in a particular environment. But such an investigation would be no ordinary undertaking. Man's problems being great and his time being short, perhaps application factors developed through lesser investigations can be justified socially, even though their scientific justification may be subject to serious question. But some methods of arriving at application factors have practically no scientific basis; their use could lead to a false sense of security and ultimate damage to aquatic resources.

Wastes discharged into surface waters usually contain many substances, some toxic and some not; their nontoxic effects on aquatic communities can hardly be predicted on the basis of their toxicity. But when toxicity is important among the effects of a waste on a species of interest, its acute toxicity is relevant to the problem of water pollution control. Appropriate use of an application factor so as to avoid acutely toxic concentrations in the receiving water can prevent short-term lethal effects. Yet, for a species to continue to be successful in nature, toxic substances must not decrease its life span, its reproduction, or its growth: there must be neither acute nor chronic effects on any of its life history stages. Use of an application factor and acute toxicity information could provide the needed protection. Unfortunately, however, there may be no constant relationship between the lowest concentration of a waste that is acutely toxic and the lowest concentration that has some other adverse effect; the effects may be caused by different components of a waste. The use of an application factor along with acute toxicity information should also insure protection of the food resources of the species of interest. But the substances responsible for the acute toxicity of a waste to one species may or may not be the ones responsible for harm to other species, including its food organisms. To provide a margin of safety for the species of interest and its community in the face of changing effluent and receiving water quality, the most unfavorable conditions likely to exist must be taken into account in selecting application factors. Thus, though superficially the application factor idea is simple, the known and unknown relationships it encompasses are complex indeed. How useful the idea may be in solving particular waste disposal problems will depend on the complexity of these relationships.

Daily, water pollution biologists are faced with the need for answers to these problems, immediate even if imperfect answers. If a waste is to be discharged, what treatment and dilution are to be required when not all of its possible effects are known? Perhaps some waste concentration that is but a fraction of the concentration acutely toxic to the species of most concern can be permitted as an expedient until more is known. In the absence of knowledge, that fraction becomes anyone's guess. Some biologists, hoping to make the selection of suitable application factors something more than guesswork, have undertaken research to determine the levels of toxic substances that are safe for aquatic life and to determine the relationships between these levels and acutely toxic ones. In examining some of this research, perhaps we can further acquaint our reader with the problem of toxic substances and with problems biologists face in learning the effects of these substances and making recommendations for their safe disposal.

THE DEVELOPMENT OF APPLICATION FACTORS

Development of application factors involves, then, the acquisition of two kinds of information. First, we must know something about the concentrations of toxic substances that will not lead to decreases in the distribution or abundance of the species we value. Let us call these *safe concentrations*. For concentrations of toxic substances to be safe, they must not affect adversely the survival, reproduction, growth, or movements of the species we seek to protect. To determine the maximum concentration of a substance that would be safe for even one species would be a difficult task requiring much research, both in the laboratory and in nature; there is no other way.

Now, if we can determine such a safe concentration of a toxicant, we are in a position to explore the possible usefulness of an application factor. This will require our finding whether or not, under varying conditions, the concentration that is safe is a nearly constant fraction of an acutely toxic concentration, say the 96-hour TL_m. Should this fraction be fairly constant, it may prove useful in predicting, on the basis of acute toxicity bioassays, the safe concentrations of wastes containing the toxic substance. Experience may teach us that such a fraction is useful as an application factor also for similar toxic substances and similar species. But we are here supposing that the research has been done and that we have learned the truly safe concentrations under different conditions. Not often will this be the case. By a variety of means, biologists have sought indications of what the safe concentrations of toxic substances may be and how these concentrations might be related to acutely toxic ones. In broad outline, then, this is the development of application factors. Let us now examine these efforts more closely.

Perhaps the earliest formal method for estimating safe concentrations of toxic wastes was proposed by Hart, Doudoroff, and Greenbank (1945). This proposal was made at a time when biologists were suggesting more extensive use of acute toxicity bioassays as being better than chemical analyses for predicting possible toxic effects of many wastes. To biologists, however, it was apparent that most toxic substances can have harmful effects at concentrations far below those known to be acutely toxic. But, at that time, even as now, there was little knowledge of chronic effects of toxicants on aquatic organisms. How, then, were organisms in waters receiving toxic wastes to be protected from these other and unknown effects?

Hart, Doudoroff, and Greenbank (1945) suggested a formula incorporating a factor and an exponent to be used in estimating, from 24- and 48-hour median tolerance limits, concentrations of wastes that might be presumed safe until more was known. The factor was intended to make allowance for possible variations in the tolerance of organisms and for instability of environmental conditions. The exponent was to extrapolate from acutely toxic concentrations to concentrations that would have no long-term deleterious effects. But at the time this formula was suggested, there was practically no knowledge of the long-term effects of toxic substances on valuable aquatic organisms and their communities, no knowledge of what safe concentrations might be. The values that Hart, Doudoroff, and Greenbank tentatively proposed for the factor and the exponent in their formula were selected arbitrarily. Equally arbitrary was the application factor 0.1 — with which to multiply the 48-hour TL_m in estimating safe concentrations — suggested by the Aquatic Life Advisory Committee of the Ohio River Valley Water Sanitation Commission (1955). But, in the absence of knowledge, what was to be suggested, what was to be done? Toxic wastes were being discharged into receiving waters. Better to regulate the concentrations of toxic substances in receiving waters to some seemingly reasonable fraction of acutely toxic concentrations than to regulate not at all, because the discharge of such substances could not be prevented at that time and cannot be entirely prevented now.

Hart, Doudoroff, and Greenbank (1945) knew that arbitrary application factors used with acute toxicity data cannot yield reliable estimates of concentrations of toxic substances safe for aquatic life. They knew that only research determining safe concentrations could ever lead to the development of reliable application factors. Some of this research could be done in the laboratory, but, ultimately, research in nature would be needed.

One might reason that the soundest method of determining safe concentrations of a particular toxicant or kind of effluent would be to study its effects on biological communities of receiving waters over periods of years when known dilutions have been maintained. If concentrations appearing to be safe for the species of interest turned out to be some relatively constant fraction of the median tolerance limit of the species, this fraction would seem to be a reliable application factor. Henderson (1957) noted the possible utility of such an approach to the development of application factors. This approach is being followed with some success in Great Britain (Edwards and Brown, 1967; Herbert, Jordan, and Lloyd, 1965). It does, however, present many practical difficulties. Seasonal and annual changes in effluent characteristics and volume and in the quality and discharge of the receiving stream are likely to make difficult the determination of conditions which did or did not lead to undesired changes in an aquatic community. And really good studies of populations and communities in nature are difficult. Moreover, long periods of time would be required, time during which considerable damage to a community could occur. There would still remain the need for a reasonable trial application factor. Rather than as a method for developing application factors, such studies would perhaps better be considered as means for testing application factors derived by other procedures.

Successful reproduction and growth of animals are of vital concern if populations of valuable species are to be maintained in waters receiving toxic wastes. To serve their intended function, application factors must insure such biological performance. For species of fish that will grow and reproduce under laboratory conditions, Mount and Stephan (1967) have suggested that the maximum concentration of a toxic substance depressing neither growth nor reproduction during long periods of exposure under these conditions be tentatively accepted as a safe concentration in receiving waters not differing materially from the water used in the experiments. Observations on the effects of

a substance on the growth, reproduction, and spawning behavior of adult fish and on the viability of eggs and the growth of the progeny over at least one generation would be taken into account. Mount and Stephan recognized that changes in community composition, including food availability, could occur at concentrations of a substance judged to be safe on the basis of such experiments, certainly if the tested species were not the most sensitive ones in the community. Nevertheless, this approach affords the opportunity to observe chronic effects of toxic substances on the biologically important phenomena of growth and reproduction. Experience may teach us that, for some species, concentrations that are safe in the laboratory are safe in nature as well. Perhaps concentrations not affecting species that can be studied in the laboratory can be used for developing tentative application factors for valuable species that cannot be so studied, but only until we know better the requirements of the species that concern us.

Effluents from the kraft process of manufacturing pulp and paper, like many other wastes, are potentially toxic as well as oxygen depleting, and they may, through enrichment of receiving waters, lead to the production of nuisance organisms. Some years ago, we undertook studies of the feasibility of developing application factors of value in the safe disposal of kraft process effluents (Warren and Doudoroff, 1958). For this industry, and probably for many others, the answer to this problem is elusive. At concentrations no longer severely oxygen depleting, these effluents may yet be toxic and lead to the production of *Sphaerotilus* and other benthic sliming organisms. We suspected then, and we are convinced now, that the acute toxicity of kraft process wastes cannot be a reliable index of all their potentially harmful effects; we should not expect it to be. During the ensuing years, biological treatment of these wastes has become generally required. This has yielded effluents that are no longer acutely toxic but that are still harmful where insufficiently diluted. Wastes must be acutely toxic if application factors are to be used for predicting safe concentrations. Progress, then, has made our initial goal obsolete, as is so generally true in the course of human affairs. But knowledge of the concentrations of wastes that are safe for aquatic life will not become obsolete so long as wastes are introduced into natural waters. A brief examination of the evolution of our studies on kraft process effluents may be worth our reader's time.

As initially conceived, our studies of kraft mill effluents were to determine the fraction of the 24-hour TL_m for a test fish—the guppy, *Lebistes reticulatus*—that would yield concentrations not influencing the survival of selected species of fish and invertebrates in laboratory streams (Warren and Doudoroff, 1958). Different fractions of this median tolerance limit were maintained in five streams during experiments lasting one month; one stream received no kraft mill effluent. The salmonids in the streams were fed, and measurements of their growth were obtained; only the survival of crayfish, stonefly naiads, caddisfly larvae, and snails was determined (DeWitt, 1963). These animals were separated from the fish in the streams and depended on algae and midge larvae for food. Since the effluents studied in this phase of our work had not been treated and were acutely toxic to the guppies, it was possible to determine an application factor that would yield concentrations not demonstrably influencing the survival or growth of the salmonids or the survival of the invertebrates. But we became concerned about other effects of the effluents, even at concentrations not influencing survival of the animals or growth of the fish. There were effects on the plant community and on the behavior of the invertebrates, effects that might influence stream communities in ways reducing fish production.

To study these other effects under laboratory conditions, we needed laboratory stream communities capable of supporting the food chains of fish and the growth and production of the fish themselves. Several years of research were required for us to gain the understanding necessary to maintain laboratory stream communities capable of supporting fish production under different experimental conditions. Then we were able to return to the problem of kraft mill effluents. But with increased understanding of the biological basis of fish production (Chapter 17), we could no longer justify our earlier simple rationale. We found it necessary to deal more discretely with direct effects of the effluents on growth of the fish and indirect effects through changes in the food chains. And we became much more concerned with the problem of

extrapolating our findings to nature. Thus, we began to study the direct effects of the effluents on the growth of salmonids receiving different rations in aquaria (Chapter 11). Concentrations of effluent that were found to have little or no direct effect on the growth of the fish were then introduced into laboratory streams having communities able to support fish production. Sometimes these low concentrations decreased salmonid production in the streams; sometimes salmonid production was increased, the difference in part being due to seasonal changes in the composition of the communities (Chapter 17). The unnatural simplicity of the laboratory communities led to these differences. Studies of fish production in large experimental stream channels resembling natural streams were necessary if we were to determine the highest concentrations of kraft mill effluents that would not in nature be deleterious to fish production. Such studies on untreated and on biologically stabilized kraft mill effluents have now been undertaken.

Now, what are we to say about application factors for kraft mill effluents? For untreated effluents, those acutely toxic to fish, perhaps application factors derived through our studies could reasonably be used for estimating safe concentrations in receiving waters. But acute toxicity may bear no closer relationship to threshold effects in nature than do other characteristics of these effluents: it may be only an indication of effluent strength. Still, such application factors may be useful in the disposal of untreated kraft mill effluents that have not been intensively studied. For biologically stabilized effluents that have no acute toxicity, the story is entirely different. These effluents may even at low concentrations have deleterious effects on stream communities and fish production. Determination of safe concentrations of such effluents can probably be made through studies like the ones we have described, but effluent characteristics other than acute toxicity will need to be used in this determination.

There will in the future undoubtedly be more studies of the sort we have outlined. But time and scientific resources will not always permit such extensive experimental investigations. And there will remain the need to estimate, however crudely, safe concentrations of wastes about which we know little. It was, indeed, primarily for this purpose that application factors were originally suggested. This will remain their primary function, because when we know the safe concentration of a waste, we no longer need an application factor with which to estimate this concentration from acute toxicity data. If we refuse to acknowledge the validity or utility of purely arbitrary application factors, what can be suggested that is short of a full-scale ecological investigation and retains at least some scientific basis? Painfully aware of the need for interim decisions and of the technological and biological complexity involved, we are both compelled and hesitant to comment further.

No single fractional factor, when used with acute toxicity information on different substances or complex wastes, can be expected to yield concentrations harmless to different aquatic communities in receiving waters, unless that factor be so small as to require waste disposal regulations unjustifiably stringent in some cases. This possibility can be dismissed. But some substances and mixtures of substances are known to act on aquatic organisms in similar ways. An application factor developed for one waste might be used for the safe disposal of a similar waste, until more is known. If not for every waste, then, an application factor perhaps can be developed for every class of wastes.

How else can we proceed? We cannot test all substances and combinations of substances; neither can we test all species of organisms likely to be affected. Perhaps we should concentrate on the species of primary economic or recreational interest: usually fish in freshwater, and fish, molluscs, and crustaceans in the sea. The effects substances representing each class of wastes have on sensitive species—effects on their survival, reproduction, growth, and behavior—could be studied. Maximum concentrations of these representative substances that have no effect on the test species could be related to median tolerance limits so as to obtain application factors for the various waste classes. Determined for a few species under laboratory conditions, such application factors would not necessarily afford adequate protection for all species in the communities in nature. Until further research or experience teaches us better, we would need to assume that application factors suitable to protect some species from some wastes in the laboratory would be of value in protecting other species from similar wastes in nature. For

biologists, this is hard to swallow. But the information really needed will be a long time in coming; the plants and animals biologists love cannot wait.

Even these minimal studies would be a big undertaking, but the value to society of the resources better protected would more than justify the necessary expenditures. And, as always, experience would be the severest test of the application of the knowledge gained, but only if we are prudent and observe carefully the fates of the species we would protect. Truly adequate application factors will prove elusive. We must place little confidence in the ones we develop and be ready to change them as research and experience dictate. And, as our knowledge of safe levels of toxic substances increases and our waste treatment technology improves, application factors, having served a useful purpose, may one day become obsolete.

ACUTELY TOXIC AND OTHER CONCENTRATIONS

We will here consider *acute toxicity* to be "rapid damage to the organism by the fastest acting mechanism of poisoning, fatal unless the organism escapes the toxic environment at an early stage" (Sprague, 1969, p. 809). Mortality resulting from acute toxicity usually occurs within 96 hours; it is this kind of toxic effect that is measured by acute toxicity bioassays. *Chronic toxicity* may influence the ability of the organism to reproduce, grow, or behave normally, but probably is not often a direct cause of death in nature. It can make an animal more susceptible to predation or other hazards and so lead indirectly to death, long before the animal would die from the poison alone. Toxicants neither acutely nor chronically toxic to a particular species could still decrease its distribution and abundance in nature by decreasing its food resource or by changing in other ways its relations with other organisms.

In recent years, more general concern over the fates of animal populations in polluted waters has led to much more research on the effects toxic substances have on reproduction, growth, and behavior. Very often, studies of the acute toxicity of these substances are simultaneously conducted. And, very naturally, there has developed the practice of expressing the maximum concentration of a toxicant not affecting growth or some other response as some fraction of an acutely toxic concentration. Thus, fathead minnows (*Pimephales promelas*) were found to reproduce successfully at a copper concentration of 0.03 of the 96-hour TL_m (Mount, 1968). In a sense, then, this is expressing concentrations in biological units rather than in chemical units. This measurement system provides a common scale for comparing the highest concentrations of different substances that have no effect on the biological performance of the same or different species. For reasons not really understood, the toxicities of mixtures of substances are often equal to the sum of the toxicities of their components, so long as these toxicities are expressed as fractions of the median tolerance limits for sufficiently long periods of exposure (Lloyd, 1961c; Brown, 1968). Such toxic components appear, then, to be simply additive in effect. Even if this is only superficially and approximately true—as we expect must usually be so in instances of this sort—it can be very helpful in dealing with pollution problems involving several toxic substances. Sprague and Ramsey (1965) have suggested that the units of measurement be called *toxic units* when concentrations of toxicants are expressed as fractions of the lethal threshold concentrations, which often are equal to or not much less than the 96-hour TL_m's.

The *safe concentration* based on some particular response or on all observable responses of a species may be expressed as some fraction of a TL_m. If we are willing to assume that such a fraction of the TL_m would provide adequate protection for this species or other species in nature, we might use that fraction as an application factor to approximate the concentration in nature that would really be safe, a concentration that only a great deal of research could determine. An application factor, then, is simply the decimal fraction of a TL_m or a *toxic unit* that we are willing to consider a reasonable estimate of a safe concentration in nature.

Acute toxicity bioassays are usually relatively easy to perform; they require little in the way of time, materials, or experience. Not so simple are laboratory studies of reproduction, growth, or behavior; and meaningful studies of the effects of toxic substances on the food resources and production of animals are most difficult of all, whether conducted in the laboratory or in

nature (Chapter 17). This is why it would be very useful to know for many kinds of toxic materials quantitative relationships between their acutely toxic levels and their minimum concentrations having other effects. When it is only possible to determine the acute toxicity of a material, some fraction of a determined acutely toxic concentration might be tentatively judged harmless to a species in its natural environment. But, first, some knowledge of the fractions of acutely toxic concentrations of similar materials that would not affect the survival, reproduction, growth, or behavior of various species in the laboratory becomes necessary. Second, possible effects on the food resources of the species of concern should be taken into account. Ultimately, we must know how well things are going for our species in its natural environment. Let us now examine, from a literature that is not rich, a few examples of the presence and apparent absence of effects of concentrations of toxic substances that are but fractions of acutely toxic ones.

The influence of toxic substances on the growth of individuals of a species can usually be studied most conveniently in the laboratory. Some species reproduce under laboratory conditions, thus permitting study of the effects of toxic substances on development of gametes, fertilization, embryonic development, hatching, and survival of young. But for many species, such laboratory studies are not possible. Behavioral studies are difficult, whether in the laboratory or in nature, and their meaning is not always clear. We have taken from the literature a few examples of the concentrations of certain toxic materials—expressed as fractions of median tolerance limits (TL_m's)—having definite effect and having little or no observed effect on the growth and reproduction of various species of fish (Table 14-1). If the growth of some species of fish is to be unimpaired under laboratory conditions, the concentrations of typical organic pesticides, copper, and even kraft mill effluents can be but a few hundredths of the 96-hour TL_m's. For pentachlorophenol and cyanide, the fraction can be larger. Reproduction, in some cases, appears to be unimpaired at concentrations safe for growth. Were we willing to assume that a concentration of a toxicant permitting growth and reproduction of a certain species in the laboratory would in no way influence its distribution and abundance in nature, we then might choose to employ as an applica-

TABLE 14–1 *Concentrations of different toxic materials that have been observed to cause little or no inhibition or to cause definite inhibition of the reproduction and growth of various species of fish in the laboratory. Concentrations are expressed as decimal fractions of median tolerance limits.*

Toxic Material	Species of Fish	Response	Concentration Causing Little or No Inhibition	Concentration Causing Definite Inhibition	Reference
Pentachlorophenol	Steelhead (alevins)	Growth	0.1(96-hr TL_m)		Chapman (1969)
Pentachlorophenol	Guppy	Growth	0.12(96-hr TL_m)	0.25(96-hr. TL_m)	G. Chadwick (unpub.)
Pentachlorophenol	Cichlid	Growth	0.25(36-hr TL_m)	0.5(36-hr TL_m)	Chapman (1965)
Pentachlorophenol	Cichlid	Reproduction	0.12(48-hr TL_m)	0.87(48-hr TL_m)	Brockway (1963)
Cyanide	Cichlid	Growth	0.5(48-hr TL_m)	0.7(48-hr TL_m)	Leduc (1966)
Cyanide	Cichlid	Reproduction	0.15(48-hr TL_m)	0.7(48-hr TL_m)	Brockway (1963)
Dieldrin	Sculpin	Growth		0.02(96-hr TL_m)	G. Chadwick (unpub.)
Dieldrin	Chinook salmon	Growth	0.05(96-hr TL_m)	0.17(96-hr TL_m)	G. Chadwick (unpub.)
Dieldrin	Guppy	Reproduction		0.2(96-hr TL_m)	T. Roelofs (unpub.)
Ester of 2,4-D	Fathead minnow	Reproduction	0.05(96-hr TL_m)	0.25(96-hr TL_m)	Mount and Stephan (1967)
Malathion	Fathead minnow	Reproduction	0.02(96-hr TL_m)		Mount and Stephan (1967)
Copper	Fathead minnow	Reproduction	0.03(96-hr TL_m)	0.07(96-hr TL_m)	Mount (1968)
Untreated kraft effluent	Chinook salmon	Growth	0.05(96-hr TL_m)	0.1(96-hr TL_m)	Tokar (1968)

tion factor the fraction of the TL_m yielding the concentration found harmless in the laboratory study.

Such an assumption may make possible first approximations of suitable application factors, but it is a big assumption, one we will wish to examine further. Chlorinated hydrocarbons like dieldrin and DDT are terribly persistent. Fish may accumulate them not only from the water but also through their food. Even when the concentrations of these pesticides in the water are very low, fish can accumulate in their tissues sufficient amounts to have harmful effects on survival, reproduction, and growth. Being fat soluble, chlorinated hydrocarbons have been found to accumulate in the eggs of lake trout (*Salvelinus namaycush*) and to lead to heavy mortality of hatching fry (Burdick et al., 1964).

We should also be warned by other effects of toxic wastes, even direct effects on the organism, let alone effects on its community that may influence it indirectly. The normal behavior of an animal is essential to its existence in nature. Laboratory studies have shown that the behavior of animals can be altered by exceedingly low concentrations of toxicants (Warner, 1967). Application factors taking effects on behavior into account could

be developed. This should be done, when we are sure of the importance of the kinds of behavior we study in the laboratory. But usually we are not at all sure. Still, we cannot safely ignore behavioral studies. Cyanide impairs the swimming performance of salmonids at 0.1 of the 48-hour TL_m (Neil, 1957). The swimming of cichlids appears to be affected at lower cyanide concentrations than does their growth (Leduc, 1966). Unfortunately, we know little about the swimming performance required of fish in nature. In developing application factors through studies of the individual organism, we could take into account many other sublethal responses: changes in histology, blood characteristics, enzyme activities, or respiratory patterns. Understanding as little as we do about the importance of such changes to the whole organism, we should perhaps first direct our attention to the responses of survival, reproduction, and growth.

This brings us finally to the problem of protecting the biological community upon which any species depends for its existence, particularly for its food. We can determine the toxicity of substances to important food organisms of the fish we wish to protect. This has been done for stonefly naiads (Jensen and Gaufin, 1966) as well as for other invertebrates and the algae upon which some invertebrates feed. The toxicities of wastes to particular species of food organisms could be related to their toxicities to fish in order to develop application factors. But there are so many kinds of food organisms, one is inclined to seek an all-encompassing community approach. In this, we can perhaps begin in the laboratory, though we must surely end in nature.

We have studied the effects of toxic substances on the food chains, growth, and production of sculpins and salmon in laboratory stream ecosystems. Dieldrin reduced the abundance of midge larvae and the growth and production of sculpins (*Cottus perplexus*) at 0.25 of the 96-hour TL_m for the sculpins, but not at 0.025. Untreated kraft mill effluents reduced the production of juvenile chinook salmon (*Oncorhynchus tshawytscha*) at 0.14 of their 96-hour TL_m, but not at 0.08. Yet, when biologically treated so as to have no acute toxicity, these effluents sometimes reduced salmon production at concentrations as low as 1.5 percent by volume. It would be nice if our laboratory stream studies could yield reliable estimates of safe concentrations, but they cannot, because laboratory communities are too simple. These studies are helpful, but they are only one step along the pathway back to a nature so difficult to study on her own terms.

It is to nature that we must turn finally for the answers we need, answers she seems to yield with the greatest reluctance. In nature we have interacting systems of the greatest possible complexity, systems that tend through their interactions to resist changes, even in the presence of very toxic materials. But change they do, as do the systems within the individual organism. Our problem is to identify which changes in communities are important to the distribution and abundance of the species in which we have interest. Further, we must identify those changes in the physical and chemical environment that lead to undesired changes in communities. We are concerned here with toxic substances man introduces into our waters. Anyone who has followed for a while through time and space the changes in the concentration of some material in a lake, a river, an estuary, or the sea is painfully aware of how great these changes can be and how difficult it is to know just what changes in concentration lead to changes in the community. Dilu-

tion capacities of receiving waters change, as do the amounts of toxic materials man introduces. What levels of these materials, then, are responsible for effects man does not desire? This question needs answering if we are to know the safe concentrations of toxic substances in nature.

In England, Edwards and Brown (1967) found fish populations in a river at locations where the total concentration of all toxic substances was less than 0.4 of the acutely toxic level, concentrations at these locations being mainly between 0.1 and 0.3 of this level. From studies of this sort, Herbert, Jordan, and Lloyd (1965, p. 579) concluded that "it should usually be possible to maintain a fishery of some kind in waters where the toxicity is kept below, say, 0.2 of the expected threshold concentration, provided that the dissolved-oxygen concentration is also . . . satisfactory. . . ." This is the method of experience based on some systematic observation, a method that is hard to beat when man is sufficiently concerned to maintain an awareness of the more important changes taking place in populations and communities. But usually when man is concerned, he seeks to maintain more than "a fishery of some kind." He is concerned about fisheries of particular kinds; he wants them to be highly productive and wishes to avoid conditions that would lessen this. The mere existence of some species of fish in waters receiving toxic substances is not what he seeks. From the fine beginnings biologists have made, studies in nature must progress to knowledge of the effects of toxic substances on the survival, reproduction, and growth of valuable species, effects on their distribution and abundance, if such studies are to lead to the kinds of application factors upon which man can rely.

THE USE OF APPLICATION FACTORS

Application factors of the sort we have been discussing are intended only to provide some measure of protection for valued populations that must exist in waters receiving toxic wastes. This protection is intended only to prevent possible toxic effects of materials in wastes being discharged. Wastes often contain materials whose harmful actions are not toxic: excessive enrichment, oxygen depletion, or physical alteration of the aquatic environment. The use of application factors, by requiring greater dilution of wastes in receiving waters, may reduce harmful effects of wastes that are not due to their toxicity. But application factors may not adequately provide for this other protection. For this reason, other characteristics than the toxicity of effluents and other effects than their toxic ones must be understood and monitored. Our stool needs more than one leg.

We have through this chapter sketched the problems with which application factors are intended to deal; we have traced the ways in which these factors are developed; we have considered relations between acutely toxic concentrations of substances and those that have other effects; and, throughout, we have expressed hopes and we have expressed doubts, there being quite valid bases for hopes and doubts. We have done these things because application factors are being used. There is a need, they will continue to be used, and we should have some understanding of their weaknesses as well as their purposes and strengths.

In Holland, Germany, and Switzerland, application factors of 0.05 to 0.1 of concentrations lethal to 50 percent of the test animals in 20 days are generally considered acceptable in receiving waters (Warner, 1967). For use in the United States, the National Technical Advisory Committee (1968), under the auspices of the Federal Water Pollution Control Administration, has recommended application factors intended to protect aquatic organisms from some toxic substances being discharged into receiving waters. According to the substance and its characteristics, the recommended application factors range from about 0.1 to 0.002 of acutely toxic concentrations. Available information was inadequate to recommend application factors for many substances, and it was not really adequate for the recommendations that were made; still, these recommendations were necessary. Where, as is so often the case, many toxic substances are present, it was recommended that

$$\frac{C_a}{L_a} + \frac{C_b}{L_b} \ldots + \frac{C_n}{L_n} \leq 1,$$

C_a, C_b, and C_n being the measured concentrations of toxic substances in a receiving water, and L_a, L_b, and L_n being the concentrations permissible for each substance indi-

vidually. The utility of this formula is dependent on the toxic effects of the materials being simply additive. This condition may be sufficiently satisfied often enough to make this formula useful in dealing with the complex problem of mixtures of toxicants in the aquatic environment.

Biological communities are complex; the toxic wastes man introduces into the aquatic environment are complex; the actions toxicants have within the individuals of each species and the changes they cause among populations of different species are complex. We will never fully understand the resulting interactions that lead to changes in the distribution and abundance of species man values. Application factors—based on whatever understanding man may have—are a tool for dealing with potentially harmful circumstances. One may well wonder whether or not they are a reliable tool. But they are a tool, a tool in a meager collection, a tool that should be used when appropriate and not carelessly discarded. Still, they should not be used with the complete confidence that all will be well.

Man must wait neither for full understanding nor for his judgment day before taking those actions that may delay that day. To any scientist, and certainly to a biologist, the assumptions implicit in the development and use of application factors must appear very big indeed. The investigations necessary to examine these assumptions will extend far into the future. But all, scientists and nonscientists alike, are becoming painfully aware of the need to act now. The scientist would only warn that there may be danger in poorly understood approaches, that the search for understanding must go on, even as we act on the basis of what we know today.

SELECTED REFERENCES

Aquatic Life Advisory Committee, Ohio River Valley Water Sanitation Commission. 1955. Aquatic life water quality criteria: first progress report. Sewage and Industrial Wastes 27:321–331.

Brown, V. M. 1968. The calculation of the acute toxicity of mixtures of poisons to rainbow trout. Water Research (Pergamon Press) 2:723–733.

Edwards, R. W., and V. M. Brown. 1967. Pollution and fisheries: a progress report. Water Pollution Control (Journal of the Institute of Water Pollution Control) 66:63–78.

Hart, W. B., P. Doudoroff, and J. Greenbank. 1945.

The Evaluation of the Toxicity of Industrial Wastes, Chemicals and Other Substances to Fresh-water Fishes. Waste Control Laboratory, The Atlantic Refining Co., Philadelphia. 317 + [14]pp.

Henderson, C. 1957. Application factors to be applied to bioassays for the safe disposal of toxic wastes. Pages 31–37. *In* C. M. Tarzwell (Editor), Biological Problems in Water Pollution. Transactions of the 1956 Seminar. R. A. Taft Engineering Center, U.S. Department of Health, Education, and Welfare. 272pp.

———, and C. M. Tarzwell. 1957. Bio-assays for control of industrial effluents. Sewage and Industrial Wastes 29:1002–1017.

Herbert, D. W. M., Dorothy H. M. Jordan, and R. Lloyd. 1965. A study of some fishless rivers in the industrial midlands. Journal of the Institute of Sewage Purification 1965(6):569–579.

Mount, D. I., and C. E. Stephan. 1967. A method for establishing acceptable toxicant limits for fish — malathion and the butoxyethanol ester of 2,4-D.

Transactions of the American Fisheries Society 96:185–193.

National Technical Advisory Committee. 1968. Water Quality Criteria. Report to the Secretary of the Interior. Federal Water Pollution Control Administration, Washington, D.C. x + 234pp.

Sprague, J. B. 1970b. Review paper: Measurement of pollutant toxicity to fish — 3. Sublethal effects and "safe" concentrations. Water Research (Pergamon Press) 4: In press.

———, and B. Ann Ramsay. 1965. Lethal levels of mixed copper-zinc solutions for juvenile salmon. Journal of the Fisheries Research Board of Canada 22:425–432.

Warren, C. E., and P. Doudoroff. 1958. The development of methods for using bioassays in the control of pulp mill waste disposal. Tappi 41(8):211A–216A.

Woelke, C. E. 1968. Application of shellfish bioassay results to the Puget Sound pulp mill pollution problem. Northwest Science 42:125–133.

PART V

Population Ecology

From a strictly biological point of view, it is the population of organisms that is important, the individual organism being important only insofar as it contributes to the persistence of its population. The importance of the individual is a not too generally applied concept of recent man. The population represents the continuing, evolving race. From a practical point of view, it is the population that is important to man, either because of its value or because it is a nuisance. But the distribution and abundance of a population are determined by the integrated responses of individual organisms to their environment. We may measure changes in the size or changes in the production of a population facing environmental change. But to explain these changes, we must know how factors in the environment are influencing the survival, reproduction, growth, and movements of the individual organisms in the population. Moreover, populations are interacting parts of biological communities. They cannot be understood except as parts of communities. The most general significance of a population lies in the role it plays within its community.

15 The Population as a Unit of Study

There have been philosophical concepts in the past (and several are powerful determiners of present attitudes) that either emphasize the individual to the exclusion of the group entity or emphasize the group system to the exclusion of the individual entity. In my opinion, both extremes are scientifically untenable, whether applied to biological or to human systems. Dichotomies are often treated as mutually exclusive, but in this instance there is much evidence of complex transactions between the individuals associated in more inclusive group systems and between the whole inclusive population and its component individuals.

ALFRED EMERSON, 1960, p. 308

Typological thinking is . . . [a] . . . major misconception that had to be eliminated before a sound theory of evolution could be proposed. . . . According to this concept the vast observed variability of the world has no more reality than the shadows of an object on a cave wall, as Plato put it in his allegory.

. . . .

The replacement of typological thinking by population thinking is perhaps the greatest conceptual revolution that has taken place in biology.

ERNST MAYR, 1963, p. 5

THE INTEREST IN POPULATIONS

From a practical point of view, it is the population of plants or animals that is of importance to man, either because it is useful or beautiful or because it is a nuisance. A single tree or fish is not a resource to utilize or enjoy; a few mosquitoes are rarely a problem.

From a biological point of view, it is the success of the population and not of the individual that is important, except as the individual contributes to the success of the population. The importance of the individual is an ethical concept of recent man, which, unfortunately, is not too generally applied; and among men certain individuals have had a lasting impact. It is the population that

through successive generations tends to persist, sometimes as a distinguishable race, adapting by means of changes in its gene pool to changing environmental conditions. Populations separated by geographic barriers may develop inheritable mechanisms of reproductive isolation and emerge as new species. The genetic adaptation of populations and the adaptation of the individual organism were considered briefly in Chapter 7, for we could not delay the introduction of these important concepts until this point in our book.

Biologists have still other reasons for being interested in populations. The student of evolution views the species as being composed of numerous local populations. To the ecologist, the population represents

227

the total integrated activities and the total existence of the individual organisms of which it is composed; and communities are formed by interacting populations. Thus, there is a great deal of interest in populations; and they represent a level of organization that is convenient to study, a level of organization midway between the individual organism and the biological community. The greater difficulty in defining the limits of populations as compared to the limits of individual organisms in no way makes the population concept less important or useful. We must, however, define the term *population* in the manner conceptually most valuable.

THE MEANING OF POPULATION

Some have considered the population to be only an abstract concept, as has Thompson (1956, p. 392):

> ...the population, like the species, in itself, is merely a concept. It exists, of course, in the real world, as a collectivity, but it is unified only in the mind, and it is therefore only in the mind that it exists as an entity. To describe it as a self-regulating system, like an animal body, is merely playing with words.

Darwin (1859), too, denied the existence of nonarbitrary species; and although he explained the modification of species through time, he failed to explain the splitting of one species into two.

> By eliminating the species as a concrete natural unit, Darwin also neatly eliminated the need for a solution to the problem of how species multiply.
> . . . .
> Whoever, like Darwin, denies that species are nonarbitrarily defined units of nature not only evades the issue, but fails to find and solve some of the most interesting problems of biology.
> . . . .
> A species in time and space is composed of numerous such local populations, each one intercommunicating and integrating with the others. (Mayr, 1963, pp. 14, 29, 136)

The population is the unit of a species that adapts genetically to local conditions and that, when isolated, has the capacity to evolve into another species. To deny the existence of populations as natural units of organization is to deny the basis of much of modern biology. Most biologists, like Mayr (1963), Emerson (1960), and Nicholson (1957), accept not only that populations exist

as real natural units that must be recognized in order to explain some biological phenomena but also that the population concept is of great value in biological thought.

The word population comes from the Latin *populus*, the people; its most common usage undoubtedly is in reference to human populations, particularly now with the concern over "the population problem." According to their interests, biologists have used the term to cover various associations of organisms, sometimes associations that are different in kind. This does not make any particular usage incorrect. But at least some of the argument over whether or not populations are concrete natural units of biological organization stems from differences in definition. The point is not whether a definition is correct or incorrect but whether the definition is the one conceptually most useful in the area of knowledge in which the defined term is being used, in biology in our case.

For the biologist, we believe the most useful definition of *population* is a group of interacting and interbreeding individuals that has relatively little reproductive contact with other groups of the same species. To the taxonomist and the geneticist, this is the *local population* of potentially interbreeding individuals sharing a single gene pool (Mayr, 1963, p. 136). This is the population that adapts genetically to changing environmental conditions, the population in which individuals of successive generations come to acquire distinctive morphological, physiological, and behavioral characteristics adapting them to their environment. This is the population that through both genetic adaptation and the physiological and behavioral adaptation of its individuals responds to its environment as a concrete, integrated, natural unit of biological organization. We believe this definition to be the most generally useful for biologists of all specialties.

THE INTEGRATION OF POPULATIONS

The concept of integration involves the idea of distinctness as well as the idea of cohesion. Most individual organisms are distinct from others because of their structural boundaries; and they are cohesive as individuals because of their structural nature and their systems of nervous and hormonal communication and control. The distinctness and cohesion of populations are of a different order than those of individual organisms. Populations generally do not have structural boundaries giving them distinctness. And the cohesion of populations comes not from structural tissues and communication through nerve networks and circulatory systems. We must not look for the same order of integration at different levels of organization. Populations have their own communication systems serving to give them both distinctness and cohesion and so giving them integration.

Because populations persist through untold generations and long periods of time, we must explain their integration in time as well as their integration in space. Both are ultimately traceable to the genetic code, a system of communication between generations. By transmitting to individuals in successive generations the genetic basis of similar responses to environmental stimuli, the genetic code holds the population together through time; this is also one of the ways in which it maintains the integrity of the gene pool.

The geological conformation of the habitat is sufficient explanation for the distinctness in space of many populations; the distinctness of a plankton population in a pond may be explained in this way. Its cohesion, however, depends on the individuals in the population responding in like manner to each other and. to other environmental stimuli. But for most populations, the geological limits of the habitat are insufficient explanation either for their distinctness or for their cohesion. For highly motile organisms, both must be explained on the basis of similar responses of the individuals to environmental stimuli and conditions.

Oyster larvae, developing in tidal currents, have the genetically determined physiological capacity to survive within certain limits of salinity and temperature. If carried beyond these limits, they will be eliminated from the population. And only on some substrates can they attach and metamorphose into the adult form (Chapter 10). Responses of these kinds insure the distinctness of some populations.

Less passive responses of individuals in populations of other species lead to population distinctness. Similar genetically determined responses of individuals to gradients of stimuli in their environment bring these individuals to the same locale, thus tending to hold their population together. Learning, too, in some species plays a role in this: Pacific salmon return to their natal stream along gradients of odors they learned to recognize through imprinting that occurred during their juvenile life in the stream (Chapter 12).

Most of the cohesion of populations must be explained on the basis of the similar responses their individuals exhibit to social stimuli—to each other. Many of the responses of individuals to social stimuli are instinctive, but social relations learned by imprinting or conditioning also are often important integrators of populations (Chapter 12). The social releasing and directing stimuli involved in reproduction are particularly important. These highly specific stimuli—detected by the chemical, visual, auditory, and tactile senses—and the responses to which they lead tend to insure reproductive isolation and the integrity of the gene pool. Family behavior, territorial behavior, and grouping responses such as the schooling of fish further contribute to the cohesion of populations. Thus genetic similarity leads the individuals in a population to respond similarly to their environment, integration of the population resulting; and this, in turn, tends to insure the integrity of the gene pool.

All these considerations have led most biologists to recognize the population as a concrete, natural unit of biological organization having its own adaptive characteristics, as Nicholson (1957, p. 154) has written:

I contend that a population is something more than a concept. It has a similar objective reality to a family, or a tribe, or a nation. Because individuals within such groups react upon one another, and with the outside world, in ways they would not react were they not parts of an integrated group, each of these groups has characteristics which are more than the sum of those of the constituent individuals. In many of its relations with the outside world each such group acts as a unit; and the behavior of individuals is often strongly influenced by special interactions with other members of the same group.

THE STUDY OF POPULATIONS

The study of populations in nature is often difficult; but it is to nature that we finally must turn to find the most important questions and the real answers, wherever and however else we conduct our population studies. Questions of evolution and speciation have been apparent to biologists observing populations both before and after Darwin. Experimental studies of genetics and mathematical models of gene systems have greatly increased our understanding of some of the mechanisms of evolution; but natural selection, the determiner of evolutionary direction, is an ecological process most meaningfully studied with populations in their natural environments. Students of evolution have in the past few decades been doing just this, in most ingenious ways (Ford, 1964).

Most population studies conducted in nature are perhaps not so difficult as those directed primarily to elucidating the processes of genetic adaptation; whatever the objectives of these studies, the results often are relevant to questions of evolution. Although most biologists would like to contribute to our understanding of evolution, there are other important biological questions. And answers to some of these man needs. Many population studies are intended to increase understanding of the distribution and abundance of species man wishes to manage in his own best interest. To do this, he must know about the biological happenings leading to observed abundances of animals. He has been able to estimate rates of birth, death, immigration, and emigration of some natural populations (Chapter 16). Sometimes it has been possible to determine causal relations between particular factors in the environment and changes in populations. Perhaps more often, changes in birth and death rates have been vaguely related to population densities and groups of factors supposed to vary with these, for the complexity of natural systems usually precludes clear definition of causal relations.

In seeking better understanding of the relationships between environmental factors, the densities of populations, and their rates of increase, many biologists have studied experimental populations of small animals in the laboratory (Chapter 16). Along with these studies and studies in nature, mathematical models of populations have been developed. The theoretical ideas expressed in these models sometimes have come from laboratory and field studies and sometimes have stimulated such studies. But whether theory has preceded or followed observation

and whether the studies have been in nature, in the laboratory, or with mathematical models, the ultimate objective has usually been the same: to understand the causes of such changes and such stability as are observed in natural populations.

Man's interest in natural populations frequently derives from their utility: the tissue they produce that he is able to harvest (Chapter 17). In recent decades, ways have been developed to estimate the tissue elaboration or production of natural populations. But studies in laboratory systems have been necessary to clarify relations between energy and material resources and the production of populations. Still, the ultimate test of our knowledge of production relations is our ability to explain and predict the production of populations in nature under different environmental conditions, often different because of man's activities, as in the case of water pollution.

UNDERSTANDING POPULATION CHANGE

The individual organism can survive, reproduce, and grow in a changing environment so long as the changes are not beyond the limits of its genetically determined capacities for physiological and behavioral adaptation. The capacity of a population to adapt to environmental change over periods of time less than the length of a generation is determined by the existing capabilities of its individuals for physiological and behavioral adaptation—these, interacting with the environment, determine birth and death rates. But, over several generations, populations can adapt genetically, this being manifested by changes in the morphological, physiological, and behavioral capacities of individuals in populations to adapt to environmental changes. Biologists seek to understand not only the short-term but also the long-term responses of populations that determine population distribution and abundance. The biologist may seek only understanding, but society needs this understanding for very practical reasons. For man to manage his environment well, he must be able to predict the effects his activities will have on the plants and animals that influence him, directly or indirectly.

Prediction of the effects that environmental changes will have on the distributions and abundances of populations must be based on understanding of the capacities of these populations to adapt to both short-term and long-term changes. Such predictions cannot be based solely on historical records of changes in population numbers, even when the changes are known to have been caused by particular environmental changes. Populations are continually confronted with new environmental changes, and their responses can change. The biological basis of population adaptation resides in the genetically determined capabilities of individuals in the populations, capabilities that change through generations. This biological basis of adaptation must be understood.

We return again to Bartholomew (1964, p. 8): "each level finds its explanations of mechanism in the levels below, and its significance in the levels above." With particular reference to the population level, Huntsman (1948) argued that to explain changes in numbers of animals, we must understand how individual organisms respond to their environment in terms of survival, reproduction, growth, and movement. We cannot, then, predict how populations are going to respond to environmental change unless we know how the individuals in the populations are going to respond. This requires study at the level of the individual organism (Part IV). But in nature, individuals live only as components of populations, and populations exist only as parts of biological communities (Part VI). Our studies of populations must extend into both the higher and the lower levels of biological organization, if they are

to yield information of real predictive value.

Physiological and behavioral adaptation of the individual organism and genetic adaptation of its population may seem to be fundamental biological problems of remote interest to those whose concern is simply water pollution control and abatement. Yet, to control water quality in the environment of valuable aquatic populations so as to retain their usefulness, we must be able to predict the responses of these populations to changes in water quality. The understanding necessary for such prediction can be based only on the most fundamental biological concepts and research. It would be erroneous to believe that biological investigations directed toward the solution of water pollution problems can be more superficial than other biological investigations and still be truly useful.

SELECTED REFERENCES

Emerson, A. E. 1960. The evolution of adaptation in population systems. Pages 307–348. *In* S. Tax (Editor), The Evolution of Life (Evolution after Darwin, Vol. 1). The University of Chicago Press, Chicago. viii + 629pp.

Huntsman, A. G. 1948. Method in ecology—biapocrisis. Ecology 29:30–42.

Mayr, E. 1963. Animal Species and Evolution. The Belknap Press of Harvard University Press, Cambridge, Massachusetts. xiv + 797pp.

Nicholson, A. J. 1957. The self-adjustment of populations to change. Pages 153–173. *In* Population Studies: Animal Ecology and Demography. Cold Spring Harbor Symposia on Quantitative Biology, Vol. 22. xiv + 437pp.

16 Population Dynamics

Through the animal and vegetable kingdoms, nature has scattered the seeds of life abroad with the most profuse and liberal hand. She has been comparatively sparing in the room, and the nourishment necessary to rear them. The germs of existence contained in this spot of earth, with ample food, and ample room to expand in, would fill millions of worlds in the course of a few thousand years. Necessity, that imperious all pervading law of nature, restrains them within the prescribed bounds.

THOMAS MALTHUS, 1798, pp. 14–15

The phenomena observed in the field and laboratory determine the postulates that go into theoretical studies and, when the postulates can be realistically defined, the analytical methods of the theorist can often move rapidly ahead suggesting new experiments and new observations to be made in the field. Theoretical models of populations and actual populations will behave alike when the correct postulates are put into the model.

LaMONT COLE, 1957, p. 6

THE DISTRIBUTION AND ABUNDANCE OF ANIMALS

The distribution and abundance of animals are continually changing as the birth rates, death rates, and movements of their populations change in response to environmental variables. In an area where environmental conditions are particularly favorable for a species, a population of that species will on the average usually be high. But even within this area, from place to place and time to time, its abundance will vary, for in nature environmental conditions are nowhere uniform in space and time. Andrewartha and Birch (1954, p. 5) have written that "distribution and abundance are but the obverse and reverse aspects of the same problem." Environmental factors changing birth rates, death rates, and movements—and thus population abundance—can at some time in some part of a population's range reduce its abundance to zero and change its distribution.

Population dynamics is an expression much used by ecologists. Elton (1933, p. 48), who appears to have first coined this expression, defined it to cover the area of knowledge "concerned with rates of increase, fluctuations in numbers, and the relation of problems of numbers to the environmental factors which influence the population." In population dynamics, we deal with rates of birth, death, emigration, and immigration, and with environmental factors that influence these and so lead to changes in numbers in space and in time. In nature, populations are influenced in various direct and indirect ways by their own density as well as by factors that appear to act independently of population density. It is difficult to distinguish the environmental factor or factors actually influencing a population. Thus, biologists have sometimes found it convenient in their analyses of population changes to consider groups of factors to be *density dependent* and to analyze changes in population numbers in relation to population density. Much of the concern of biologists working with mathematical and laboratory population models has been with the effects of population density on birth rates and death rates. But the changes in the individual organism and in its environment that are caused by changes in density are manifold, and the identification of the specific factor or factors actually causing changes in birth rates, death rates, and movements is properly in the realm of population dynamics. Here, however, we also enter the realms of the ecology of individual organisms and the ecology of communities. In biology there are no clear boundaries, and there should be none in the interests of biologists.

Often we can measure birth rates, death rates, movements, and the distribution and abundance of animals. But in seeking understanding of these, as Huntsman (1948) has instructed us, we must learn how the individual organism responds to factors in its environment by surviving or not surviving, reproducing or not reproducing, growing or not growing, and moving or not moving. Except for surviving, Huntsman did not mean that these responses of the individual organism were all or nothing; it is particularly important that we realize that, in populations in their normal ranges, survival, reproduction, growth, and movement may vary between low and high values in response to environmental change. Forbes (1887, republished 1925, p. 537), too, gave us invaluable instruction when he wrote: "If one wishes to become acquainted with the black bass, for example, he will learn but little if he limits himself to that species." A population in nature cannot succeed, nor can it be understood, apart from its community: its food species, competitors, predators, and disease organisms.

The distribution and abundance of untold numbers of species have been changed by man's activities, including those that have led to changes in aquatic environments. When the production of valuable species and other uses of waters have suffered, all will agree that pollution has occurred. As an aid in preventing further pollution, we would like to be able to predict what changes in the distribution and abundance of species of interest will occur with particular alterations of the environment. When certain kinds of environmental changes have repeatedly led to changes in the distribution and abundance of particular species and this has been well documented, we may have a basis for prediction. Knowledge of the population dynamics leading to these changes would be a better basis for prediction. This, however, represents an ideal toward which we must work, for information on changes in birth rates, death rates, and movements, and on their causes is difficult to obtain, as we shall see.

BIRTH RATE

We shall use *birth rate* to mean the rate of production of young by a specified group and number of organisms in a specified period of time. Use of the broader term *natality* to cover the many ways young are produced might be more appropriate, but we prefer to use the more common expression. This should cause no difficulty: we most certainly will know the mode of reproduction of an organism we are studying, and we must always specify the life history stage at which we are measuring reproduction, whether it is young born alive, eggs laid, or young hatching.

The *crude birth rate* of a population, or the number of young produced per numerical unit of the population per unit time, is determined by the mean numbers of young the females produce during different age intervals throughout their reproductive life and by the age structure of the population. Knowing these, we can calculate the crude birth rate. It is most convenient, for some purposes, to deal only with numbers of female parents and offspring, total numbers being easily calculated when sex ratios are known.

A table giving the *age-specific birth rates* throughout the reproductive life of a group of organisms is known as an *age-schedule of births*. The age-specific birth rate is the mean number of offspring produced per female of x age per unit of time; it is usually denoted by m_x. The fecundity of females in different periods of their reproductive life, as represented by an age-schedule of births, is determined in part by innate characteristics of their species or race and in part by environmental conditions, as will become apparent in this and following sections.

Frank, Boll, and Kelly (1957) determined the effects of population density on the birth rates, death rates, and rates of increase of *Daphnia pulex* in laboratory cultures. A rather involved technique permitted them to determine these rates while maintaining population numbers constant at each density studied. Groups, or *cohorts*, of *Daphnia* one day or less in age were placed in glass beakers containing 25 ml of synthetic pond water. Temperature was maintained at 20 C, and the alga *Chlamydomonas moewusi* was provided for food. In 28 experiments, each having six replicates, tens of thousands of *Daphnia*, parents and their offspring, were censused and provided new food and water every 48 hours.

Frank and his co-workers calculated age-specific birth rates (m_x values)—the average number of births per day per female of x days of age—by dividing the total births between days x−1 and x+1 by twice the number of females in a culture on day x−1. As can be seen from Figure 16−1, the age-specific birth rates of *Daphnia pulex* for given ages decrease markedly with increasing population density. These rates reach maxima at all densities when the *Daphnia* are about 15 days old, but reproduction begins later at the higher densities. Age at first reproduction has a large effect on the rate of increase of a population, because, with earlier reproduction, the number of generations reaching

reproductive age in a given period of time is increased.

Population density as we are using it is, of course, a convenient parameter that encompasses the known and unknown environmental factors actually influencing the age-specific birth rates, when these factors are density dependent. Some animals appear to have an innate need for space, but in this case perhaps metabolite concentration or more probably food abundance was the most important factor directly influencing *Daphnia* birth rates. That growth of the individual *Daphnia* was permanently stunted at densities of eight individuals per milliliter and higher indeed suggests that food was limiting. This could decrease the energy and materials available for reproduction, decrease birth rates, and, if growth and maturation are correlated, delay reproductive maturity. Delay in the commencement of reproduction occurred at densities of eight individuals per milliliter and higher, but not at lower densities (Fig. 16−1).

The work of Frank, Boll, and Kelly (1957) provides us with a clear example of how density-dependent environmental factors can influence the age-specific birth rates and the age at first reproduction of an animal.

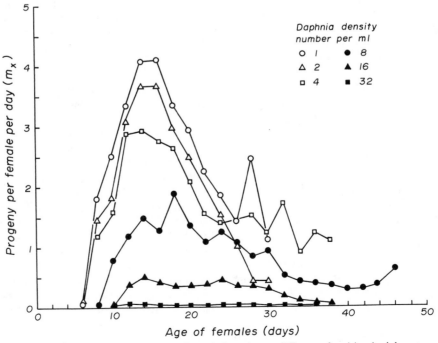

FIGURE 16−1 Age-specific birth rates (m_x) of *Daphnia pulex* at different densities in laboratory cultures at a temperature of 20 C. After Frank, Boll, and Kelly (1957).

We will return to their work when we consider death rates and how these and birth rates together determine rates of population increase or decrease. In concluding this section, we wish to emphasize that, in all organisms, age at first reproduction and age-specific birth rates have innate or genetic determinants that are very much dependent on environmental conditions in their expression. For few animals, and perhaps never in nature, can we hope to obtain such nice information as we have discussed. Still, we must come to appreciate the kinds of ways environment can influence the birth rates of the animals we do study in nature, and, as we will be discussing, what this has to do with their distribution and abundance.

SURVIVORSHIP CURVES FOR ANIMALS

The *crude death rate* of a population is the number of organisms dying per numerical unit of a population per unit time, regardless of the distribution of deaths among ages, sexes, or other classes in the population. But, as in the case of births, deaths are not usually uniformly distributed throughout the various classes of populations, and *age-specific death rates* or other specific death rates are more useful in determining the environmental factors influencing populations. A *life table* is a concise summary of vital statistics concerning mortality and survival. From an initial cohort whose individuals start life together, such a table provides for every interval of age the number of deaths (d_x), the survivors remaining (l_x), the rate of mortality (q_x), and the expectation of further life (e_x). In his very useful review of the ecological literature pertaining to the survival of animals in nature, Deevey (1947) explains how any of these life table functions can be calculated from data on the number of deaths occurring in every interval of age (d_x).

Consideration of *survivorship values* (l_x) will be most useful for our discussion of population dynamics. If we start with a cohort of say 1000 individuals, survivorship values can be determined by noting the numbers of individuals surviving to successive ages x. When these values are expressed as decimal fractions of the initial number at birth, l_x is equal to 1.0 when x=0, and the l_x values at successive ages x become the probabilities at

birth that an individual will survive to each of these ages. The probabilities of survival, along with age-specific birth rates, are most useful for calculating potential rates of population increase; for this purpose, only survivorship values for females are used.

A *survivorship curve* is formed when we plot survivorship values against age. If we plot survivors per thousand on a logarithmic scale against age as percentage of deviation from mean length of life, as has been done in Figure 16-2, then a straight line indicates constant age-specific mortality rates, and the curves for animals having greatly different mean lengths of life can be conveniently compared. Curves like the one for starved fruit flies, which all lived out their physiological life span, probably never occur in nature. The curve shown for oysters is perhaps typical for many species of marine invertebrates and fish that have pelagic eggs and larvae suffering extremely high mortality rates. Many species of fish appear to have fairly constant age-specific mortality rates after juvenile life.

The shapes of the survivorship curves for the same species under different environmental conditions can be expected to be different. The effects of numerical density on the survivorship curves of the cohorts of *Daphnia pulex* studied by Frank, Boll, and Kelly (1957) are shown in Figure 16-3. Survival at given ages increased with increasing density to eight individuals per milliliter, but it decreased with further density increases. A possible explanation may be that—as will be remembered from the previous section—the animals at the lowest densities grew faster, reached reproductive maturity earlier, and may have completed their physiological life span in fewer days. Food shortage or higher concentrations of metabolites may have adversely affected survival at the highest densities of *Daphnia*.

Many environmental factors are excluded from laboratory studies like those of Frank and his co-workers. Predation, or any other factor that might occur in nature, could change the shapes of the survivorship curves of *Daphnia* at different densities. Moreover, in the laboratory we usually attempt to control factors such as temperature, factors that in nature vary in time and can cause perturbations in survivorship curves. Rarely in nature can we hope to measure survivorship over all life history stages of a species and obtain such nice information as we can from

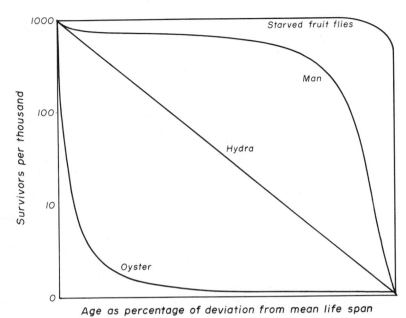

FIGURE 16–2   Survivorship curves having different general forms. After Deevey (1950).

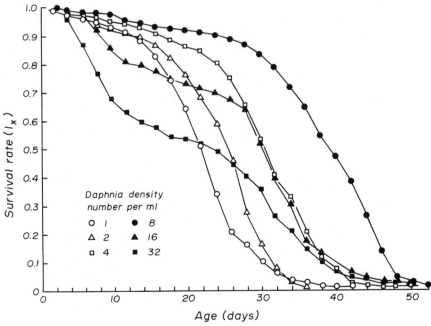

FIGURE 16–3 Survivorship curves (l_x) for *Daphnia pulex* at different densities in laboratory cultures at a temperature of 20 C. After Frank, Boll, and Kelly (1957).

laboratory studies. Nevertheless, whatever species concerns us and however difficult may be the obtaining of survivorship information, this information is very valuable in helping us to understand those responses of organisms to the environment that determine their distribution and abundance. There are various direct and indirect means of determining survivorship in nature during at least some parts of the lives of many species, and doing so can help us to understand some effects of pollution. Under some conditions, we can make reasonable inferences about survival from knowledge of the age structure of populations; we will touch on this in the next section.

AGE COMPOSITION OF POPULATIONS

The crude birth and death rates of a population will in part be determined by its *age composition*, or *age structure*, by which we mean the proportions of total population number present in various age classes. This is because, as we have seen in the last two sections, individuals in different age classes usually have different probabilities of giving birth as well as of dying. A rapidly increasing population will tend to have a greater proportion of young individuals and a rapidly declining population a greater proportion of old individuals than does a population whose numbers are tending to remain constant. Thus, birth and death rates determine the age composition of populations, and differences in age composition can obviously influence the crude birth and death rates of populations.

Lotka (1925) demonstrated mathematically that in an exponentially increasing population having constant age schedules of birth and death a *stable age distribution* would ultimately develop, regardless of the initial age composition or the initial crude birth and death rates. Once such a population had attained a stable age distribution, neither the age distribution nor the crude birth and death rates would change further so long as environmental conditions did not change. But a change in environmental conditions would lead first to a change in age schedules of births and deaths and ultimately to a new stable age distribution characteristic of the new conditions, if these remained constant. The proportions of the population present

in the various age classes of such an ideal population—one having a theoretical stable age distribution—can be calculated by methods described by Lotka (1925), Birch (1948), or Andrewartha and Birch (1954), when the age-schedules of births and deaths are known.

Constant age-schedules of births and deaths can occur only under constant environmental conditions. But since a population cannot be expected to increase exponentially for long without the density of the animals inducing changes in the environment, the ideal sequence of events Lotka demonstrated mathematically could occur only in an unlimited environment. Natural environments cannot usually be considered unlimited; and even if we sometimes might reasonably assume an environment to be virtually unlimited, environmental factors other than those that are strictly density dependent do not remain constant for long. Nevertheless, Lotka's idea is of considerable value. It suggests to us that the age structure of a population will tend to assume a form characteristic of environmental conditions, to the extent that those conditions remain constant during a few generation lengths. Lotka's idea has at least one other practical value. Because crude birth and death rates even under constant environmental conditions become constant only after the age composition of the population has become constant, the potential rates of increase of different species, or of the same species under different conditions, can be compared only on the basis of an assumed stable age distribution. We will consider this matter further in the next section.

We can conceive of a different kind of constant age distribution from the one Lotka had in mind, one that develops when the population is not increasing exponentially. A population in which the crude birth rate equaled the crude death rate would neither increase nor decrease in number. If environmental conditions were to remain constant for a few generation lengths, birth and death rates would remain constant, and such a population would assume an age structure characteristic of those conditions. Then the distribution of individuals in the different age classes at a point in time would be quite similar to the survivorship curve for the progeny produced during any one time period. We cannot always follow a single cohort to obtain survivorship information, but those

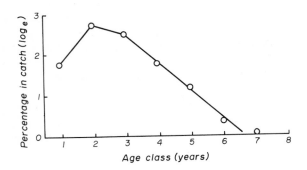

FIGURE 16–4 A catch curve showing the percentage representation of numbers of pilchards in successive age classes in the catch from California waters during the 1941–1942 fishing season. After Silliman (1943).

interested in the survivorship of fish and other animals have sometimes been able to obtain this information from the age composition of populations.

Fisheries workers have sometimes been able to estimate survival and mortality rates for fish populations from the age composition of the catch. When the logarithms of the percentage representation of successive age classes in the catch are plotted against age, a *catch curve* is formed, like the one for the California pilchard (*Sardinops caerulea*) shown in Figure 16–4. Only the straight part of the descending right limb of this curve is of value in estimating vital statistics, for its ascending left limb is due to inability of the gear to retain the same proportion of the smaller, younger fish as of the older fish. Were the seines fine enough to retain young fish below three years of age in proportion to their actual numbers in the population, then the catch curve would descend continuously with increasing age of the fish and be more representative of actual survival in the population. Constant annual recruitment and mortality are important conditions underlying such use of catch curves, but interpretation of catch curves is hazardous, and our reader is referred to Ricker (1958) for a discussion of the full rationale involved in their use.

Because in nature environmental conditions are rarely constant and because birth and death rates respond sensitively to environmental change, age-frequency distributions of populations often do not form smooth curves like the one we have shown for the pilchard. But, also because birth and death rates respond to environmental changes, we can sometimes gain insight into the impact of known environmental changes, pollutional or otherwise, by examining the age composition of populations. Interest in the age composition of populations is likely to be rewarded with increased understanding.

INNATE CAPACITY FOR INCREASE IN NUMBERS

An animal having a very high birth rate will not necessarily be very abundant in nature, for its numbers under different environmental conditions will depend on the densities it can attain before births tend to be balanced by deaths. Birth and death rates have innate or genetic determinants, but the expression of these determinants under any given set of environmental conditions is very much dependent on those conditions. Thus, for a given species, the genetically and physiologically possible age at first reproduction, births at different ages, and longevity are variables that are dependent on environmental conditions. In nature we can rarely determine what is genetically and physiologically possible for a species in a particular climate, because such environmental factors as food shortage, predation, and disease take their tolls.

Andrewartha and Birch (1954) explain how cohorts of some kinds of animals can be held in the laboratory under different combinations of such environmental variables as temperature, moisture, and quality of food, where the amount of food and the density of the animals are optimal, and where competitors, predators, and disease organisms are absent. Under these conditions, it would be possible to determine the genetically and physiologically possible age at first reproduction, births at different ages, and longevity at each combination of temperature, moisture, and food quality. The results of such a study can be most explicitly and usefully represented for each combination of conditions by age-schedules of births and deaths.

Lotka (1925), Birch (1948), and Andrewartha and Birch (1954) explain how from such age-schedules of births and deaths, the maximum possible rate of increase can be calculated for a population that has theoretically attained a stable age distribution. It is this maximum possible rate of increase that Lotka (1945) called the *intrinsic rate of natural increase* and Andrewartha and Birch (1954) called the *innate capacity for increase*. As carefully defined by Andrewartha and Birch, it is the exponential rate of increase of a population theoretically having a stable age distribution when there is no food shortage, predation, or disease. At different levels of environmental factors such as temperature, moisture, and food quality, the innate capacity for increase of a species will be negative infinity if no individuals of reproductive age are added to the population, will be zero if births just balance deaths, and will be positive if the species has a capacity for increase under the particular environmental conditions. For a species to maintain itself in nature, its innate capacity for increase must be greater than zero, because births must balance not only the deaths from genetically determined physiological causes but also those from starvation, predation, and disease.

Frank, Boll, and Kelly (1957) determined the effect of numerical density on the intrinsic rate of natural increase of *Daphnia pulex* at a temperature of 20 C. From age schedules of births and survivorship values, which we have already summarized in Figures 16–1 and 16–3, they estimated intrinsic rates of increase at different densities and found these to decrease linearly with increasing density. It is of particular interest that the rate of increase of the *Daphnia* should be highest at the lowest density, where survival tended to be quite low (Fig. 16–3). Thus, as is generally so, younger age at first reproduction and higher age-specific birth rates (Fig. 16–1) had a greater effect on the rate of increase than did survival. But, if we are to follow Andrewartha and Birch (1954) in the careful restrictions they placed

on the environmental conditions under which intrinsic rate of increase (which they called innate capacity for increase) can be measured, we would consider only the value obtained at the lowest or optimal density as being a satisfactory measure of the innate capacity for increase. Presumably, food shortage or other effects of density were preventing the genetically determined physiological capacity for increase at 20 C from being realized when densities were higher. We will here consider only the rate of increase at the lowest or optimal density as being the innate capacity for increase of *Daphnia pulex* at a temperature of 20 C.

The innate capacity for increase (r_m) is the exponential rate of increase of a population that has theoretically attained a stable age distribution under carefully specified environmental conditions. Hence, when N represents population number,

$$\frac{dN}{dt} = r_m N \text{ and}$$

$$r_m = \frac{dN/dt}{N}.$$

If b and d are the crude birth and death rates for this population, then

$$r_m = b - d.$$

The precise value of r_m for *Daphnia pulex* at 20 C can be calculated from the data of Frank and his co-workers given in Table 16–1 by solving the equation:

$$\int_0^\infty e^{-r_m x} l_x m_x d_x = 1$$

where zero to infinity is the life span of the reproductive stages. Tedious calculation is involved in the accurate solution of this equation; Birch (1948) describes arithmetic procedures that yield satisfactory approximations.

For most of us, the relationships between age-schedules of births (m_x), survivorship values (l_x), the net rate of increase per generation (R_0), the mean length of a generation (T), and the innate capacity for increase (r_m) can be more clearly seen in the following equations:

$$R_0 = e^{r_m T} \text{ and}$$

$$r_m = \frac{\text{Log}_e R_0}{T}$$

TABLE 16–1 Computation of innate capacity for increase (r_m) of Daphnia pulex *at a density of one individual per milliliter in laboratory cultures at 20 C with survivorship values (l_x) and age-specific birth rates (m_x) by using net reproduction rate (R_0) and mean generation length (T). Data of* Frank, Boll, and Kelly (1957).

Pivotal Age in Days (x)	l_x	m_x	$l_x m_x$	$l_x m_x x$
0	1.000	0	0	0
2	0.989	0	0	0
4	0.971	0.051	0.049	0.196
6	0.958	1.790	1.715	10.290
8	0.943	2.490	2.348	18.784
10	0.924	3.330	3.077	30.770
12	0.904	4.060	3.670	44.040
14	0.871	4.080	3.554	49.756
16	0.814	3.310	2.694	43.104
18	0.733	2.920	2.140	38.520
20	0.630	2.220	1.399	27.980
22	0.505	1.660	0.838	18.436
24	0.335	1.390	0.466	11.184
26	0.204	2.430	0.496	12.896
28	0.156	1.090	0.170	4.760
30	0.102	1.340	0.137	4.110
32	0.058			
34	0.035	$R_0 = \Sigma l_x m_x = 22.753$		
36	0.025	$T = \dfrac{\Sigma l_x m_x x}{\Sigma l_x m_x} = 13.836$		
38	0.017			
40	0.013	$r_m = \dfrac{\text{Log}_e R_0}{T} = 0.226$		
42	0.010			

where

$R_0 = \Sigma l_x m_x$ and, approximately,

$$T = \frac{\Sigma l_x m_x x}{\Sigma l_x m_x}.$$

Because this method of estimating T is only approximate, estimates of r_m derived from the latter procedure are not satisfactory for animals with high rates of increase. Using this procedure we obtain a value of 0.226 for the r_m of the *Daphnia* population (Table 16–1). We would have obtained a higher value of r_m had we used an approximation of the integral relationship.

Smith (1954), after examining the literature available, concluded that small animals tend to have higher values of r_m and lower values of T than do larger animals. He concluded that larger animals live in a more benign world than small animals, which are more subject to physical factors in the environment. With increasing size animals generally avoid more possible causes of death. Abundances of different species are probably not closely correlated with their innate capacities for increase in numbers, because high capacities for increase must often have evolved in harsh environments as a means of permitting the existence of some species even in low numbers. An animal cannot be expected to persist long in areas where physical conditions do not permit its innate capacity for increase to be greater than zero.

Pollutional and other environmental changes certainly can affect the distribution and abundance of a species by altering its relations with other species, its food organisms, competitors, predators, and disease organisms. But environmental changes can also affect the distribution and abundance of a species by determining its innate capacity for increase, through effects on age at first reproduction and age-specific birth and death rates. We will probably not be able to determine the effects of environment on the innate capacity for increase of most species in which we are interested. Nevertheless, even when we cannot do this, it is helpful to be able to distinguish at least conceptually between the many ways in which environmental factors and the intrinsic nature of animals interact to determine the distribution and abundance of these animals.

CHANGES IN ABUNDANCE

Everyone who enjoys nature is aware that the numbers of his favorite species are continually changing. Biologists are unable to explain most of these changes. Some of the changes are due to immigration and emigration, but the numbers of immigrants and emigrants change too. Seasonal and annual changes in abundance often are the direct and indirect results of changes in weather and climate. In addition to natural changes in environment, many of man's activities cause environmental changes that also lead to changes in the abundance of plants and animals. But even in environments that appear unchanging, interactions of some species with their environments can lead to their numbers changing continually, sometimes quite regularly. To understand the effects of man's activities on the distribution and abundance of animals, we must first understand the influence of other factors; biologists have not found this easy.

Not often would we expect animals in nature to increase in numbers at a rate near their innate capacity for increase. Here, the effects of their own numbers, predation, and disease usually tend to depress the rate of increase. In nature, the innate capacity for increase of a species must be most nearly approximated when the species first enters a new environment, or perhaps after its numbers have been severely reduced and the environment has again become favorable. Under such conditions, the population may for a period of time increase in a nearly exponential manner. But this sort of increase cannot go on indefinitely. Climatic conditions may greatly reduce population numbers, and then increase may begin again. The numbers of some species of insects tend to be controlled in this manner. In other species, including most vertebrates, changes induced in the environment and in the animals themselves by their increasing numbers tend to reduce birth rates and increase death rates until some usually unstable balance between these rates is reached, and the populations no longer increase. The numbers in such populations tend to follow sigmoid patterns of increase through time; but they may not be stable near the asymptotes, either because of instability of the physical environment or because of the natures of the populations themselves or their interactions with other populations.

The tendency of many laboratory and natural populations of organisms to follow a sigmoid pattern of growth has interested biologists for a long time. Much thought has been given to the possibility of representing the rates of increase of populations as functions of their numbers, changes in numerical density thus somehow representing changes in all the environmental factors that influence rates of increase. Although it has become clear that the factors operating on different populations are not the same and that

no single equation producing sigmoid curves adequately describes the growth of all populations following a sigmoid pattern of growth, the biological ideas involved in such equations make it worth our while to consider at least the most simple equation.

Verhulst (1838), stimulated by the ideas of Malthus and encouraged by Quetelet, his teacher and colleague, derived an equation for the growth of populations in limited environments, an equation for a sigmoid curve that he called a *logistic curve*. He assumed that there is a constant upper limit (K) to the numbers (N) in a population and that the rate of increase of the population at any particular time is linearly proportional to the difference between this upper limit and the population number at that time. The rate of change of such a population can thus be expressed:

$$\frac{dN}{dt} = rN\left(\frac{K-N}{K}\right)$$

The number in a population at any time t then becomes

$$N_t = \frac{K}{1 + e^{(a-rt)}}$$

where a is the constant of integration defining the position of the curve relative to its origin. The coefficient of increase r might approximate the innate capacity for increase (r_m) if a stable age distribution existed, because were N to be very small, the rate of increase would be near rN.

Application of the logistic equation to data on laboratory and natural populations has met with quite limited success. Many populations have been shown to have no constant upper numerical limit (K), their numbers fluctuating after reaching some level. And the relationship between density and rate of increase cannot usually be expected to be linear. Moreover, it is implicit in this equation that the animals whose numbers are specified by N each have an equal probability of giving birth or dying, and, as we have discussed in earlier sections, this is manifestly not generally true of populations having individuals of different ages. It has been suggested that were a population to maintain a stable age distribution, this difficulty would be overcome (Leslie, 1948; Andrewartha and Birch, 1954), but stable age distributions must rarely if ever persist in populations. Finally, it is implicit in the logistic equation that the animals respond instantaneously to changes in their density and environment—that there is no lag between density changes and the responses in births and deaths—an unlikely condition.

Age structures and time lags occurring in populations suggest interesting and important problems, because these characteristics of populations can cause population numbers to fluctuate regularly—to oscillate. The characteristics and responses of populations that cause some populations to fluctuate regularly and others to fluctuate irregularly must be understood before their distributions and abundances can be explained.

The introduction of the logistic equation— a model whose simplicity both gives it appeal and limits its utility—into a brief discussion of changes in the abundance of organisms may seem inappropriate. It has, however, given us a means of directing attention to several of the population-environment interactions determining abundance. And since Pearl and Reed (1920) independently derived this equation, which had lain buried for nearly a century, it has stimulated much interest in the development of mathematical models of populations and, indeed, in the study of population dynamics. Even much more complex models may never adequately define population-environment interactions. Numerical density is a poor substitute for the multiplicity of environmental factors actually operating, and appropriate interrelating of many constants representing these factors in mathematic models presents serious problems of formulation and testing. But models should not be used as substitutes for careful observation and analysis of particular responses of organisms to factors in their environment. As Frank (1960, p. 371), who reviewed some of the recent literature on population models, has written: "they function partly to clarify what might otherwise be vague and seemingly intuitive ideas."

THE NATURAL REGULATION OF ANIMAL NUMBERS

The natural regulation of animal numbers cannot, in a continually changing environment, lead to very stable numbers. And some of the characteristics of populations, as we noted in the previous section, can lead to oscillations in their numbers, even in a constant environment. Indeed, there have evolved certain patterns of interaction between some species and their environments that tend to prevent these species either from increasing without limit or from disappearing, patterns that in themselves lead to oscillations.

By *natural regulation of animal numbers*, biologists mean those interactions between a population and its environment that tend to prevent its numbers either from increasing without limit or from decreasing to zero. Such interactions supposedly tend to keep population numbers fluctuating about some mean level, which may change with general changes in the environment. Were population numbers controlled solely by variable *physical factors* such as weather, most biologists reason that periods of favorable years could lead to unlimited increase of populations and so to the destruction of their food or other resources; they further reason that periods of unfavorable years could lead directly to the extinction of populations. Over long time spans, such periods inevitably would come, and, for one reason or the other, populations would become extinct, more often than most biologists believe they do. Such considerations led Howard and Fiske (1911) and Nicholson (1933) to suggest that *biological factors* increasing percentages of mortality as population densities increase must be primarily responsible for the regulation of animal numbers. These Smith (1935) named *density-dependent mortality factors*; catastrophic factors he considered to be *density-independent*. Smith recognized that some factors (*inverse density-dependent factors*) might decrease in their effectiveness with increasing density; he further recognized that competition for favorable living space could interact with weather in a nonuniform environment so as to make weather act as a density-dependent factor.

These ideas have influenced the thinking of nearly all biologists and, with different degrees of emphasis, have been accepted by most (Solomon, 1949). Nicholson (1954) and Lack (1954, 1966) have been important proponents of this view of natural regulation. But some, perhaps most notably Andrewartha and Birch (1954), have argued that it is not necessary to invoke density-dependent factors to explain the abundance of animals.

A very convenient way of considering how density-dependent mortality factors may

influence the numbers in successive generations of animals was developed by Ricker
(1954) in his classic paper "Stock and
Recruitment." Ricker presented a series of
hypothetical reproduction curves relating
the number of eggs produced by a parent
stock in a given year to the number of eggs
produced by the progeny of that year (Fig.
16–5). These curves are meant to represent
the net effect of the sum total of all density-
dependent mortality factors acting upon
populations. In nature, density-independent
factors are also operating, and their effect
would be to displace from a curve points
representing eggs produced by successive
generations. Any density-dependent relationship would thus tend to disappear with
increasing incidence of density-independent
mortality. Since environmental conditions
are rarely constant, density-dependent factors vary in the magnitude of their effects
from generation to generation, and this too
tends to obscure density-dependent relationships.

Each curve in Figure 16–5 is drawn in
relation to a 45 degree line that would describe a stock in which filial generations
tended to equal parental generations, density-dependent mortality being unimportant.
Such a population would have no mechanism for regulating its numbers, and
variations in density-independent mortality
could eventually reduce it to zero or lead to
its unlimited increase. Any reproduction
curve lying wholly above the 45 degree line
would describe a population increasing
without limit. Thus, to describe populations

regulated by density-dependent factors,
reproduction curves must generally begin
above the 45 degree line and then cross the
line to end below it.

Ricker (1954) examined the changes in
abundance that would occur in populations
having single-aged spawning stocks and in
those having multiple-aged spawning stocks,
when only density-dependent factors are
operating, and when both density-dependent and density-independent factors are
operating. For this purpose, he assumed
density-dependent mortality factors affected
only the immature, not the mature, individuals, a reasonable approximation for
many species of fish. A population of single-
aged spawners influenced only by density-
dependent mortality factors according to
Figure 16–5A, after an initial deflection of
abundance from the equilibrium point at the
45 degree line, would through successive
generations gradually return to this point by
population adjustments following either
path a through f or path A through D. Similarly, we can examine the effects displacements from the equilibrium points have on
the abundances of successive generations of
populations having reproduction curves
resembling the others shown in Figure 16–5.
It can be seen that displacements from the
equilibrium points on curves B through D
result in returns to stable equilibria in a few
generations, whereas displacements on
curves E through H result in an oscillating
return to the equilibrium point (E), stable
oscillations (F), series of oscillations (G), or
violent oscillations (H).

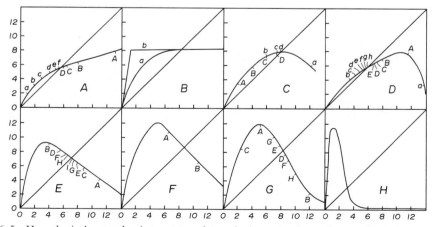

FIGURE 16–5 Hypothetical reproduction curves whose abscissas represent the number of eggs produced by a
parent stock in a given year and whose ordinates represent the number of eggs later produced by the progeny of that
year. After Ricker (1954).

When the spawning individuals belong to two or more age groups, deflections from the equilibrium points of curves A through D tend to be followed by a direct return to these points, deflections on curves E and F by damped oscillations returning to stable equilibria, and deflections on curves G and H by permanent oscillations (Fig. 16–5). The amplitude of the oscillations depends on the number of generations in the spawning stock and on the age at first reproduction. Their period depends on the mean length of time between generations.

Density-independent mortality factors operating along with those that are density-dependent tend to obscure regular oscillations, but if the reproduction curves are steep, wide fluctuations may result. Exploitation of populations through capture of mature individuals tends to reduce the amplitude and complexity of oscillations; if sufficiently intensive, exploitation can eliminate oscillations completely. Light or moderate exploitation of populations having reproduction curves with maxima above the 45 degree line will increase the abundance of the populations in successive generations; intensive exploitation will decrease it. We will return to the effects of exploitation in Chapter 17.

In addition to presenting the hypothetical reproduction curves we have been discussing, Ricker (1954) examined published and unpublished data to determine the nature of the reproduction curves of different species and to relate fluctuations in abundance of these species to their reproduction curves. He found evidence of actual reproduction curves that resemble the different hypothetical ones. Let us consider only two of these. The numbers of smolts, or downstream migrants, produced by different numbers of female coho salmon (*Oncorhynchus kisutch*) on various years are shown in Figure 16–6. The resulting curve resembles curve B in Figure 16–5, but the two curves are not strictly equivalent. The hypothetical curve relates the same stage—the eggs—of parental and filial generations, whereas the reproduction curve for the coho salmon relates adults to juveniles produced. Density-dependent factors operating between the smolt and adult stages of this salmon could lead to marked differences between the two curves. Nevertheless, Figure 16–6 clearly demonstrates density-dependent regulation of reproduction. As we discussed in Chapter 12, juvenile coho salmon during their stream residence are territorial, and this functions to regulate the number of young a stream can support at any given stream flow. If adjustments were made in Figure 16–6 for annual differences in stream flow, the curve would be even more flat and the deviations would be considerably reduced.

Figure 16–7 shows the relationships between the abundance of adult haddock (*Melanogrammus aeglefinus*) and the abundance of their progeny three years later on Georges Bank during moderate (1912–1929) and intensive (1930–1943) periods of exploitation. In effect, this graph shows the success of haddock reproduction at different densities of adults. The inverse stock-recruitment re-

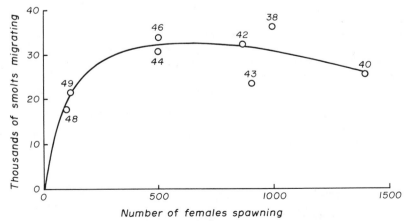

FIGURE 16–6 Smolts produced by different numbers of female coho salmon spawning in Minter Creek, Washington, during years indicated. After Ricker (1954), based on data of Smoker (1954).

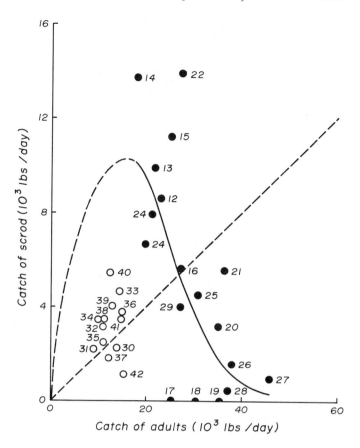

FIGURE 16-7 Relationships between the abundance of adult haddock on Georges Bank during the spawning season and the abundance of their progeny (scrod, mostly 3 years old) three years later. Data are based on adult haddock and scrod in the commercial catch on different years. The reproduction curve is drawn only for the years of moderate exploitation, 1912 through 1929. After Ricker (1954), based on data of Herrington (1948).

lationship, which resembles curve G in Figure 16-5, is believed to be due to competition for food between the young and adult haddock. Oscillations in abundance that occurred from 1912 through 1929 can plausibly be ascribed to the steep reproduction curve. These oscillations disappeared, as would be predicted, during the years of intensive exploitation, 1930 through 1943.

It is not enough to ascribe the natural regulation of animal numbers to density-dependent mortality or to analyze changes in rates of increase in terms of changes in density. It is helpful, even necessary, for us first to conceptualize how many factors operating together may influence animal abundance. But, ultimately, we must learn how individual factors act and interact in determining birth and death rates. During some periods of time, competition, predation, or disease may depend upon the density of a population in such a way as to bring some regulation to its numerical fluctuations. To predict the effects of environmental changes on the distribution and abundance of an

animal, we must know not only which factors are operating but also how they operate.

ENVIRONMENTAL CHANGE AND THE DISTRIBUTION AND ABUNDANCE OF ANIMALS

With particular reference to insects, Milne (1957, p. 253) has summarized the problem of understanding changes in most wild populations:

Unfortunately, birth, death, immigration and emigration certificates can not be issued for every member of an insect population. We can only observe and sample the population at intervals. In between, changes in numbers occur from ecological causes of which some are perceived, others surmised and the remainder unknown.

In some instances when vital statistics for populations have been estimated with fair precision, important characteristics of the environment have, through neglect, not been defined. Not only the responses of organisms but the environmental factors to which they are responding must be measured. But even when this is done, the problem of understanding remains great, because in maintaining themselves in their environments, organisms exploit intricate and sensitive physiological, behavioral, and ecological relationships, relationships that may be masked in the final outcomes of birth and death.

Populations that are of interest to man, either because of their utility or their nuisance potentials, become so because of their distribution and abundance. Through population dynamics we seek to explain these in an ever-changing world, a world where many of the changes are due to man's activities and may be pollutional. But in population dynamics, we focus our attention on the population, on its vital statistics and on its numerical changes in space and time. The causes of changes in populations are most certainly within the scope of population dynamics; yet their elucidation may require studies of physiology, behavior, or community ecology beyond its usual purview. Population dynamics brings together information on the responses of birth and death in ways of great value in describing, examining, and even predicting changes in abundance. It is immaterial where population dynamics ends and other approaches to the same biological problems begin, so long as we continue to seek the causes of responses and understanding of the significance of these responses to the population in its environment, changed or unchanged.

Implicit if not always explicit in the preceding sections of this chapter has been the idea that changes in the environment can be expected to change the distribution and abundance of an animal, "the obverse and reverse aspects of the same problem." Through their changing birth and death rates, populations are continually changing in response to natural changes in their environments. In water pollution control, we hope to eliminate the deleterious effects of man-caused changes in the aquatic environment. These changes may be so gross as to obscure the effects of natural changes, or the effects of both the activities of man and natural changes may be apparent in the dynamics of a population. Sometimes natural changes may act to increase or to mitigate the effects of those caused by man. The problem of understanding the effects man's activities may have on a population is increased by the superimposition of these effects upon those caused by natural changes in the environment. Even when both are understood, they remain difficult to separate.

We all know that pollutional changes often lead to changes in the distribution and abundance of valuable species. Population dynamics provides us with a very quantitative way of viewing these changes. But study of the dynamics of wild populations is difficult at best; we have hardly touched on the sophisticated methods employed, methods that to summarize for fish alone has required a book (Ricker, 1958). And beyond population dynamics, there continues the need for study of the physiological, behavioral, and ecological causal relations. Perhaps not for many species can we hope to fully understand distribution and abundance. Still, in seeking this understanding, even limited studies of population dynamics can be valuable. Some appreciation of the dynamics of populations can help us to conceptualize the problems we face in identifying the deleterious effects man's activities have on natural populations.

SELECTED REFERENCES

Andrewartha, H. G., and L. C. Birch. 1954. The Distribution and Abundance of Animals. The University of Chicago Press, Chicago. xv + 782pp.

Birch, L. C. 1948. The intrinsic rate of natural increase of an insect population. Journal of Animal Ecology 17:15–26.

Deevey, E. S., Jr. 1947. Life tables for natural populations of animals. Quarterly Review of Biology 22:283–314.

Frank, P. W., Catherine D. Boll, and R. W. Kelly. 1957. Vital statistics of laboratory cultures of Daphnia pulex DeGeer as related to density. Physiological Zoology 30:287–305.

Lotka, A. J. 1925. Elements of Physical Biology. Williams & Wilkins Company, Baltimore. xxx + 460pp.

Nicholson, A. J. 1954. An outline of the dynamics of animal populations. Australian Journal of Zoology 2:9–65.

Ricker, W. E. 1954. Stock and recruitment. Journal of the Fisheries Research Board of Canada 11:559–623.

———. 1958. Handbook of Computations for Biological Statistics of Fish Populations. Fisheries Research Board of Canada Bulletin 119. 300pp.

Solomon, M. E. 1949. The natural control of animal populations. Journal of Animal Ecology 18:1–35.

17 Production

THE INTEREST IN PRODUCTION

In the case of a biological resource, man is usually interested not only in its *biomass,* or the amount of its material existing per unit area, but also in its ability to replace by growth any material removed, because this ability determines how intensive man's utilization of the resource can be without danger of its depletion. From a very practical point of view, the effects of environmental change on the survival, reproduction, growth, and behavior of animals—which we have considered in earlier chapters—become of direct concern to man mainly when they influence this ability of populations of animals to re-place themselves in terms of the materials in the bodies of their individuals. This is often measured as the *production* or *production rate* of a population. Biologists usually define production as the total tissue elaboration of a population per unit of area per unit of time regardless of the fate of that tissue, whether or not it survives to the end of a period, with one exception. Tissue that is elaborated by a population but is subsequently metabolically utilized for its own maintenance, during the time period over which production is being measured, cannot correctly be included as a part of the production of the population. This biological definition of production is different from one a cattleman might find most useful. He would not be likely to include the growth of a steer that died before shipment to market as a part of the production of his herd. His estimate of production would probably be the weight increase he realized in his herd by the time of marketing. For the fisheries biologist at least, the term *yield* approximates the meaning of our cattleman's idea of production.

The yield of a fish population is only some variable part of its production. Though potential yield is ultimately dependent on the magnitude of production, actual yield depends also on the intensity of fishing, which, in turn, depends on social and economic factors. Thus, yield is not always a good measure of the capacity of a population to support fishing. And, from a strictly biological point of view, that part of the production of a population that dies and decomposes—and so returns nutrients to the ecosystem—or that part consumed by some predator other than man, is as important as the yield upon which man may depend. For reasons such as these, the biologist, though he may also be interested in yield, is interested in total tissue elaboration, the production of populations.

Though fisheries biologists had many years earlier become interested in the fishing intensities that lead to maximum equilibrium yield (Baronov, 1918), and thus in the production of fish populations, Lindeman's (1942) magnificent paper, "The Trophic-Dynamic Aspect of Ecology," fired the imagination of all ecologically minded biologists and encouraged a great deal of direct interest in production. Clarke, Edmondson, and Ricker (1946) proposed a general mathematical formulation of production based on the fates of energy and materials obtained by populations of plants and animals. Their distinction between *gross production* and *net production*—the latter being tissue elaboration, the former in addition including materials metabolized—is useful only for plants and leads to conceptual difficulties when applied to animals. Plant production has now been studied in many different kinds of waters over the earth, but we will reserve any discussion of this for Chapter 19 and here devote our attention to animal production, which presents greater analytical problems. In a most imaginative study of the production of juvenile sockeye salmon (*Oncorhynchus nerka*) in a lake, Ricker and Foerster (1948) demonstrated the use of a mathematical model (Ricker, 1946) for the computation of animal production. Allen (1951), in his study of brown trout (*Salmo trutta*) production in the Horokiwi Stream of New Zealand, provided a very useful graphical method, which was later used by Neess and Dugdale (1959) for computing the production of midge larvae. In the years that have followed, interest in measuring the production of fish and other animals has increased greatly, increased to the point where many such studies are contributing significantly to our understanding of the biology of aquatic organisms.

The first basis of production of any population lies in the physical and chemical characteristics of its ecosystem: the energy and nutrients this ecosystem provides for plants. The biological basis of animal production begins with plant production, whether the animals are herbivores utilizing plants directly, or whether the animals are carnivores benefiting only indirectly from plant materials. But determining and influencing animal production are most of life's interactions. For the individuals of an animal population, there are problems of survival, reproduction, and growth, which will be influenced by the individuals' food organisms, competitors, predators, and disease organisms, as well as by physical and chemical factors. It is never enough to seek measurements of production apart from the understanding of how production is determined and influenced by its physical, chemical, and biological bases. The production of a population is a general expression of the characteristics of its species and those of its environment. As such, production is a most convenient and useful parameter. Yet, understanding of production cannot come from this parameter alone but only from examination of a population's interactions within its ecosystem.

Most ecologists studying production processes have followed the holistic *trophic-level* approach first suggested by Lindeman (1942). As Ivlev (1945) so early predicted, this has not been fruitful. There are many reasons why this approach has not been fruitful (Chapter 19), but fundamentally it is because of the sterility of a holism that does not give due attention to the parts of a system. Before we can begin to understand energy and material transfer in entire ecosystems, we must first understand their transfer in at least some parts. Ivlev (1945) suggested that we begin with a *product of interest*, any species with which we are concerned for one reason or another. The production of this species would be explained on the basis of its utilization of food organisms for maintenance and growth. Understanding of the production of the species of interest would be further increased by likewise explaining the production of its food organisms. And so parts of an ecosystem could come to be understood, and with this understanding could come real knowledge of the ecosystem, but synthesis of this knowledge would still require some retention of a holistic view.

This book is devoted to the environmental changes we call pollution and to the biology that is relevant, though to some of our readers the biology we consider relevant must seem to overpower our discussion of pollution. We see no need today for us to beat the drum; while others do this, we will continue to concentrate on explaining the biology that is relevant to water pollution control. It is the production of an aquatic resource that makes it useful to man; any decrease in its production is likely to make it less useful. Environmental change, be it called pollution or not, can be expected to change the production of species of interest to man. Some of these changes may directly affect the survival, reproduction, and growth of a species and thus affect its production; the effects of other changes may be indirect. To know only that a pollutional change decreased production of a species in one situation may be of some value. But to know how it changed production is of much more value, because we may then be able to predict more reliably similar changes in production in other situations.

Considerable effort is necessary to study the production of even one animal species. Practical and biological considerations often warrant the expenditure of this effort. But studies of production should not be undertaken lightly, for to be valuable they are very demanding. Mainly for practical reasons, it has been the study of fish production that has received the most attention. In consequence, the important interactions determining production are better known for fish than for other animal species; knowledge of fish production provides the best basis for our discussion here. But fundamentally, the processes operating to determine fish production operate in all animal species. Those of our readers who pursue an interest in production of other species will have much to contribute and will have much enjoyment.

ESTIMATION OF PRODUCTION AND FOOD CONSUMPTION

Fundamentally, then, production is the total growth or tissue elaboration occurring in a population, regardless of the fate of that tissue, with the exception we have already noted in the case of a population utilizing previously elaborated tissue for its own maintenance. The estimation of production requires, over fairly short intervals of time, information on the relative growth rates of the individuals in each age group (their weight increments per unit of biomass per unit of time) and information on their total biomass; or it requires information on their mean weights and on their numbers, at short time intervals. The production of each age group over longer intervals of time is estimated by summing estimates of its production for short periods. And the production of a population is estimated by summing the production values for each of its age groups. Since relative growth rate is expressed in increments of body mass per unit of body mass per unit of time (mg/g/day or cal/kcal/day), and since biomass is expressed in units of mass per unit area (g/m^2 or $kcal/m^2$), production, the product of these, is in increments of mass per unit area per unit time ($mg/m^2/day$ or $cal/m^2/day$). As we have noted, *production* and *production rate* are used synonymously; but production is always a rate, though the time interval for which it is reported may be a day, a month, a year, or some other period over which measurements were made. Measurements are usually originally made in units of wet weight, but they may for convenience be converted to dry weight, caloric, or nitrogen values. To do this, reliable conversion coefficients must be developed.

The information needed to estimate production is essentially very simple; the obtaining of this information is not so simple. Estimates of the growth rate or mean size of individuals in a population are usually easier to obtain than are estimates of their total biomass or numbers. It is sometimes necessary to assume that particular patterns of growth exist in order to be able to utilize various exponential or other models of growth that may be convenient for purposes of interpolation or computation. Biomass or numbers can often be estimated directly through sampling, but sometimes they must be computed with the aid of mathematical models involving growth and survival rates, usually as exponential functions. Anyone becoming involved in the measurement of production should be familiar with these models, so as to be able to proceed as best suits his needs. Chapman (1968), in an excellent review of methodology, explained these models and discussed their possibilities and limitations. Here, we wish only to introduce our reader to the essentials without too much burdensome detail; this will permit us to devote more space to the biological basis of production and to the effects of environmental change.

Production can be estimated by means of either computational or graphical procedures. The information needed for both procedures can be derived from the same basic data: growth rates or mean weights at short time intervals and biomasses or numbers of individuals during the same intervals. Formal models of growth or survival are not essential to use of either procedure, but when a model is to be used, it should be selected on the basis of its appropriateness and convenience.

As to computational procedures, our definition of production (P) suggests that it must be computed as a sum of the products of growth rates (G) and mean biomasses ($\bar{B}$) evaluated at successive short intervals of time. For any such interval within which growth rate is quite constant:

$$P = G\bar{B}$$

The production models of Ricker (1946) and Allen (1950) are essentially just this, but they become more involved because mean biomass is estimated from growth and survival, as exponential functions, rather than directly from biomass samples. When growth is measured over short intervals, say of two weeks, it is not a matter of great moment whether it is taken to be exponential or linear. That the interval should be no longer than periods of rather constant growth rate is an important consideration, because changes in growth rates within a period—unless accompanied by parallel changes in biomass, an unlikely circumstance—will lead to erroneous estimates of production. We must point out that appropriate models of growth and survival, when incorporated into production models, can increase the length of time intervals between sampling that is consistent with reliable estimates of production. This can be a matter of considerable importance, since investigational resources are usually limited. Production must be computed for each age group separately, because the growth rates of different age groups in a population are usually different and the age groups constitute different proportions of the population biomass.

Table 17-1 is an example of the computation of production for one age group of a hypothetical population; it also contains data necessary for the graphical estimation of production, which we will consider next. The annual production (16.4 kg) is estimated by summing algebraically the monthly values. Were other age groups to be present in the population, their monthly and annual production values should be estimated in the same way, and population production values would be obtained by summing the age group values. During the months of December and January, individuals in the population are supposed on the average to have lost weight as a result of utilizing their own tissues to support their metabolism. These *negative growth rates* lead to *negative production* values for December and January (Table 17–1), which should be summed algebraically with the other monthly values to arrive at the annual production value. To be a logical extension of the computation of growth, the computation of production can be made only in this manner. An individual organism reaches a certain size by positive, zero, and negative growth increments. This size is the algebraic sum of all previous growth increments, both positive and negative. Production is only the total growth of all the individuals in a population; tissue elaborated but subsequently metabolized by the individuals during the period of time over which production is being measured can be no more a part of population production than it is of individual growth. Such tissue in its elaboration required food, but it is not available to other species, including man, and cannot lead to an increase in population biomass, by the end of this period of time. Evaluation of the food required for the metabolism and maintenance of a population can best come from estimates of its food consumption, a question to which we will soon turn.

Allen (1951) developed a graphical method for estimating production (P), a method in which the numbers of individuals (N) in a population at successive instants of

TABLE 17–1   *Computation of production of a single age group of fish for 12 months on the basis of hypothetical data. After* Chapman (1968).

Date	Mean Weight ($\bar{w}$, g)	Growth Rate (G)	Stock Number (N)	Stock Biomass (B, kg)	Mean Biomass ($\bar{B}$, kg)	Production (P, kg)
May 1	1.5		8000	12		
		0.29			10.5	3.0
June 1	2.0		4500	9		
		0.22			8.8	1.9
July 1	2.5		3500	8.7		
		0.34			9.6	3.3
Aug 1	3.5		3000	10.5		
		0.26			10.8	2.8
Sept 1	4.5		2500	11.2		
		0.37			11.6	4.3
Oct 1	6.5		2000	13.0		
		0.06			13.0	0.7
Nov 1	6.9		1900	13.1		
		0			12.4	0
Dec 1	6.9		1700	11.7		
		−0.01			10.9	−0.1
Jan 1	6.8		1500	10.2		
		−0.03			9.7	−0.3
Feb 1	6.6		1400	9.2		
		0			8.9	0
Mar 1	6.6		1300	8.6		
		0.04			8.0	0.3
Apr 1	6.9		1100	7.5		
		0.07			7.2	0.5
May 1	7.4		1000	7.0		
Annual production (ΣP)						16.4

time are plotted against the mean weights ($\bar{w}$) of the individuals at those instants. The rationale of this method is illustrated in Figure 17-1 and has been explained both briefly and clearly by Chapman (1968, pp. 183-184):

The production in a small unit of time (Δt) would be approximately equal to $N_t \Delta \bar{w}$, where $\Delta \bar{w}$ is the growth in mean weight of the population in the time interval: i.e., production is the shaded rectangle in Fig. 8–1 [17-1]. If $\Delta \bar{w}$ is made very small, approaching zero, $N_t \Delta \bar{w}$ in the very small unit of time would approach P in that time. Summing $N_t \Delta \bar{w}$ for all increments $\Delta \bar{w}$, which is equivalent to measuring the total area between the curve and the horizontal axis, gives production from t_1 to t_2.

A hypothetical example illustrating the use of this method is given in Figure 17-2. The dates on the curve are the times at which the numbers of individuals and their mean sizes were estimated. Production for the year is the area under the curve from June 1 to June 1. The production for any month, as shown for August, is the area under the curve between the inclusive dates of that month. If the abscissal scale begins at zero, biomass at any time can be represented as it has been for August 1. Figure 17-3 makes use of data we have already presented in Table 17-1 and illustrates the behavior of such a graph when growth is zero and when it is negative. All areas must be determined down to the abscissal scale. When negative production occurs, as from December 1 to February 1, this must be subtracted from the sum of the areas for the other months (Chapman, 1968, p. 190). The areas under such curves, representing production over time periods of interest, can be determined with the aid of a planimeter, or squares on the graph paper under the curve can be counted. Production values estimated by these graphical procedures do not differ significantly from those estimated by appropriate computational procedures. If the mathematical models used in computational procedures exactly describe the lines connecting the points on the graphs, production values estimated by the two procedures will, of course, be identical.

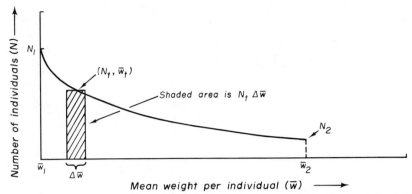

FIGURE 17-1 Graphical estimation of the production of a single age group of animals by the method of Allen (1951) explained in the text. After Chapman (1968).

The release of sex products to the environment is one of the fates of production, sometimes a major fate and always a fate of great importance to the persistence of a population. The elaboration of these products requires food, appears as increments of body mass, and may very well be considered a part of production. On the other hand, weight losses occasioned by the release of sex products to the environment do not constitute negative production and should not be subtracted in determining production values. Rather, they should be considered as the passage of some proportion of the production of one generation to the next generation.

Production is not always a good measure of the capacity of an environment to support a given species. When the biomass of a population is high, the growth rates of indi-

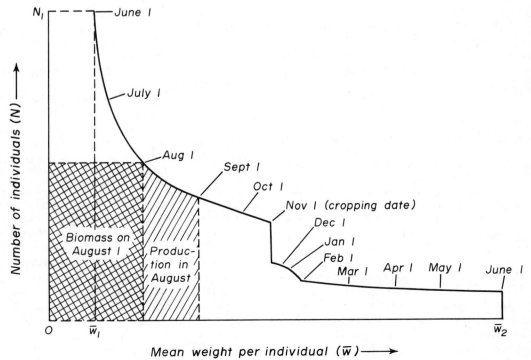

FIGURE 17-2 Graphical estimation of the production of one age group of pond fish on the basis of seasonal changes in their number and mean weight. After Chapman (1968).

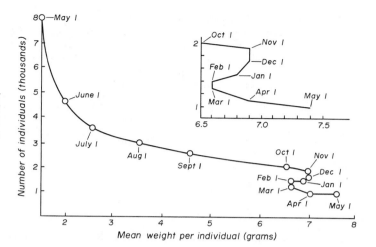

FIGURE 17-3 Graphical estimation of the production of a single age group of fish for 12 months on basis of hypothetical data given in Table 17-1. After Chapman (1968).

viduals in the population may be severely reduced; in consequence, production will be reduced. Still, the population may be consuming large quantities of food, which could lead to considerable production if the biomass of the population were smaller and not so much food were needed for the metabolism of the population. Under these conditions, information on the amount of food consumed by a population could be a better indication of the capacity of an ecosystem to support the production of the consumer than would be information on production of the consumer.

Several methods of estimating the food consumption of animals in nature have been developed, all have very real shortcomings, and none as yet can provide more than crude approximations (Davis and Warren, 1968). Perhaps the simplest and most reliable method is based on the assumption that the metabolic rate and growth of animals receiving known rations in the laboratory are the same as they would be for animals consuming similar amounts and kinds of food in nature at the same temperature. Knowing the growth rate of an animal in nature, we could estimate its rate of food consumption from laboratory data, if we choose to make this assumption.

Figure 17-4 shows the relationship between the rates of food consumption and

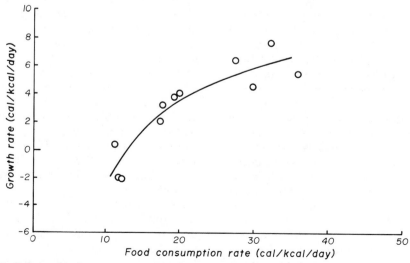

FIGURE 17-4 Relationship between rates of food consumption and growth of yearling cutthroat trout in aquaria. After Warren et al. (1964).

TABLE 17–2   *Computation of amounts of food consumed during two-week intervals, between December 28, 1960, and April 5, 1961, by yearling cutthroat trout in a section of Berry Creek, Oregon, on basis of their mean biomass and their consumption rates estimated from Figure 17–4. After* Davis and Warren (1968).

DATE	(a) NUMBER OF DAYS	(b) MEAN BIOMASS (kcal)	GROWTH RATE (cal/kcal/day)	(c) ESTIMATED CONSUMPTION RATE (cal/kcal/day)	(a) · (b) · (c) COMPUTED TOTAL CONSUMPTIONS (kcal)
12/28–1/11	14	105	0.6	13.0	19.1
1/11–1/25	14	107	−0.5	12.5	18.7
1/25–2/8	14	95	7.5	40.0	53.2
2/8–2/22	14	85	5.3	27.5	32.7
2/22–3/8	14	122	3.0	18.5	31.6
3/8–3/20	12	146	5.2	27.0	47.3
3/20–4/5	16	152	4.5	23.8	57.9
Total					260.5

growth of cutthroat trout (*Salmo clarki*) held in aquaria under temperature and light conditions resembling those in their natural environment. We discussed other curves like this in Chapter 11. Knowing the growth rates of this species of trout in a small stream, we could estimate their food consumption rates from Figure 17-4. Then, knowing the mean biomass of the trout in the stream, we could estimate the total food consumed by the population over any given period of time, as we show in Table 17-2. If the curve relating growth to food consumption has some points at which growth was zero or negative, we can even estimate the food consumed in the stream when growth and production are zero or negative (Table 17-2); often, this amount of food is considerable.

In any laboratory experiments designed to estimate food consumption in nature, it is important that the condition of the animals and the conditions of the experiment be as nearly like those in nature as practicable. Rather elaborate experiments may be necessary before the investigator is satisfied that the metabolic rates of his laboratory animals approximate those of the animals in nature. But such care has its rewards, not only in improving estimates of food consumption in nature but also in providing some indication of the metabolic rates of animals in nature. In our discussion of bioenergetics and growth (Chapter 11), we endeavored to explain how knowledge of the metabolic rates of animals in nature can help us to understand the effects of environmental changes on these animals.

BIOMASS, FOOD CONSUMPTION, GROWTH, AND PRODUCTION

Before we can consider the biological basis of production and the effects of environmental change, we must explain some very important relationships between the biomass of a population and its food consumption, growth, and production. In any ecosystem in which the limiting resource of a species is food, an increase in the number and biomass of that species will lead to a reduction in the amount of food each individual can obtain, and this in turn will lead to a reduction in individual growth rate (Fig. 17-5). We have defined the production of any age class of a population in any period of time as the product of its growth rate and its mean biomass. Thus, with growth rate declining as biomass increases, production will increase from a low level at some low biomass to a maximum at some intermediate biomass, and then, with further increase in biomass, production will decline toward zero as growth approaches zero (Fig. 17-5). This logically sound idea is supported by general biological experience and some experimental evidence (Ivlev, 1947; Davis and Warren, 1965; Brocksen, Davis, and Warren, 1968).

We have used laboratory stream communities consisting of algae, herbivorous insects, and fish to investigate the productive relations of animals. Six laboratory streams and their communities, when exposed to the same light levels and receiving the same levels of plant nutrients, can be considered to represent the same ecosystem. Then, if

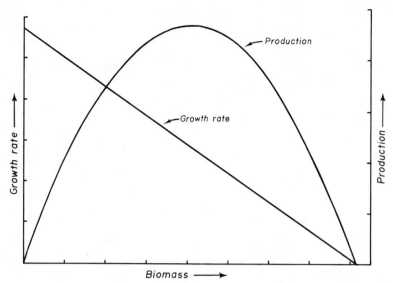

FIGURE 17–5 Theoretical relationships between growth rate and biomass and between production and biomass, production being the product of growth rate and biomass. After Brocksen, Davis, and Warren (1970).

different biomasses of fish are stocked in each of the six streams, we can determine the relationships between biomass, food consumption, growth, and production of fish in a system having a single basic capacity to produce the food organisms of the fish. With increasing biomass of sculpins (*Cottus perplexus*) stocked in the streams, the decline in growth rate and the ascending and then descending curve of production were apparent (Fig. 17–6). At the highest biomass in this experiment, the fish were unable to obtain enough food even to maintain their tissues, and a negative growth rate leading to a negative production value resulted. In nature, such high biomasses could not persist, for loss of weight and mortality would reduce biomasses to levels permitting the existence of sufficient food for growth and survival.

Now, as we have stated, with increases in population biomass, the amount of food each individual can obtain—its food consumption rate—declines. It is this decline that leads to the reduction in growth rate with increasing biomass. But, with increases in biomass from low levels, the total food consumption of all the individuals—the product of food consumption rate and biomass—first increases; then, with further increase in bio-

mass, total food consumption declines if overcropping of the food resource leads to a decline in its production. This phenomenon can be observed in the case of the sculpin (Fig. 17–6), which can reduce the biomass and production of its food organisms living on the stream substrate. Trout, when feeding solely on drifting organisms that have left the substrate, cannot similarly reduce the production of their food organisms at its source. With increasing biomass, the decline in growth rate and the production curve of cutthroat trout resemble those for sculpins (Fig. 17–7), but the curve of total food consumption for trout under the conditions specified may remain high after reaching a maximum. In the case of the sculpin, then, we have two reasons for the decline in production at high biomass levels: a larger proportion of the available food went to maintain the metabolism of the population and less was available for growth; and the total availability of food for the sculpins was reduced. In the example we have given for trout, the reduction in production was due primarily to increased maintenance costs, there being little reduction in total food availability. It is to such causes of changes in production that we turn next, to the biological basis of production.

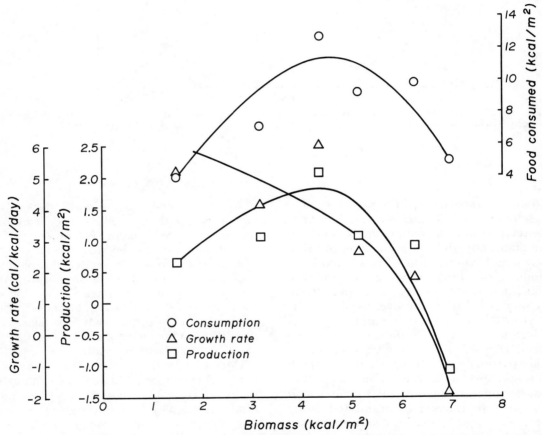

FIGURE 17-6 Relationships between the biomass of yearling sculpins and their food consumption, growth, and production in a laboratory stream experiment lasting 80 days during the spring, 1961. After Davis and Warren (1965).

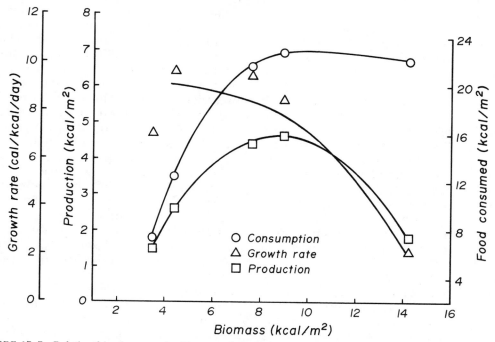

FIGURE 17-7   Relationships between the biomass of yearling cutthroat trout and their food consumption, growth, and production in a laboratory stream experiment lasting 70 days during the winter, 1964. After Brocksen, Davis, and Warren (1968).

DENSITY DEPENDENCE IN TROPHIC PROCESSES

There exists for any animal some relationship between its production and its biomass, a *density-dependent relationship*. For an animal in a particular ecosystem, we may only know the level of production at one level of biomass, but we must always remember that were the biomass to be different, the level of production would most likely be different. And, were we to be able to measure the production at a series of sufficiently different levels of biomass, the production values would form a curve of production in relation to biomass like the ones we have shown for sculpins and trout in laboratory streams (Figs. 17-6, 17-7). Were the capacity of a particular ecosystem to produce the food of an animal to change from season to season or from year to year, and were we to be able to define a production curve for this animal during each season or year, these curves would be different. In two ecosystems having different capacities to produce its food, a particular species would have different production curves. The production

curve for a species in a particular ecosystem and differences in its production curves between seasons and ecosystems are the result of density-dependent relationships between that species and its food organisms. If we can define these density-dependent relationships, we will be able to understand why in a particular ecosystem different biomasses lead to different production values for a species; moreover, we will be able to understand how differences in ecosystems lead to different production curves. This will help us to understand how environmental changes caused by man through pollution or resource management lead to differences in the production of animals of interest to man.

First, let us explain the density-dependent relations that led to the curve of production for sculpins in laboratory streams that we have shown in Figure 17-6. This curve was defined in an experiment conducted during the spring of 1961; different curves were defined during different seasons and years (Davis and Warren, 1965). The biological bases of this curve and of differences between it and the other curves can be explained (Brocksen, Davis, and Warren,

1968). For now, we will consider only the basis of the single curve obtained in the spring of 1961. In Chapter 11 we explained how differences in the density of food organisms lead to differences in the growth rate of a consumer. The relationship between food organism density and the growth rate of sculpins in the six laboratory streams during the spring of 1961 is shown in Figure 17-8. The differences in food organism density were caused by differences in the biomasses of sculpins stocked in the six streams, as we will explain. But whatever the cause of differences in food density, the important thing is that the growth rate of the sculpins was a simple function of the density of their food organisms; a simple density-dependent relationship existed.

The six laboratory streams were operated during the spring of 1961 so as to have the same basic capacity to produce food organisms for the sculpins. The only difference in treatment of the six streams was that they were stocked with different biomasses of sculpins. These different biomasses led to the different densities of food organisms that we noted above. Here again, then, we have a simple density-dependent relationship, this time between the biomass of the sculpins and the density of their food organisms (Fig. 17-9). These two density-dependent relationships—the one between food density and sculpin growth rate, and the one between sculpin biomass and the density of their food organisms—led to the relationship that existed between sculpin biomass and production during the spring 1961 experiment (Fig. 17-6). As we will explain later, these relationships between an animal and its food organisms can be incorporated into equations for the production of the animal.

We cannot often obtain well-defined production curves for animals in natural systems, because the biomass of animals cannot be manipulated so conveniently in nature as in laboratory streams. We are fortunate, however, to have Johnson's (1961) data on sockeye salmon and their food organisms in the Babine-Nilkitkwa lake system of British Columbia. This tremendous lake system is composed of rather distinct basins of water having apparently the same basic capacity to produce the zooplankton food organisms of juvenile sockeye salmon. For one reason or another, different biomasses of the juvenile salmon occur in the different basins in any given year. Thus, we have essentially a gigantic natural experiment with which to define the relationship between production and biomass of the small salmon in this system. We find a curve of production that is well defined to a point near the maximum (Fig. 17-10), which would occur at biomasses slightly higher than those found in the system. Still higher biomasses would, of course, lead to declining production values. We might well wonder if the biomass of juvenile sockeye salmon is somehow regulated so as to maintain them on the ascending limb of their production curve.

Can we now explain the production curve of the sockeye salmon in the Babine-Nilkitkwa system on the basis of density-dependent relations between this fish and its food organisms, as we did for the sculpins in the laboratory streams? Apparently we can. Calculations based on Johnson's (1961) data show a very nice relationship to exist between the growth rate of the salmon and the density of their zooplankton food organisms (Fig. 17-11). And other calculations show the density of the zooplankton to be dependent on the biomass of the salmon (Fig. 17-12). In a system having a single basic capacity to produce food organisms, then, increases in the biomass of the consumer lead to decreases in the density of the food; and decreases in the density of the food lead to decreases in the growth rate of the consumer. Thus, increasing consumer biomass leads to decreasing consumer growth rate and the curve of production characteristic of that consumer in a particular ecosystem. Nature does not always bear out so well our ideas.

We have said that the production curves of a particular species in different ecosystems will be different if these systems differ in basic capacity to produce the food of that species, so long as food is the limiting resource. Differences in the production curves for a species in different ecosystems can also be explained on the basis of density-dependent relations between the consumer and its food organisms (Brocksen, Davis, and Warren, 1970). We noted in Chapter 11 that data of Ruggles (1965) from Owikeno Lake in British Columbia, data of Johnson (1961) from the Babine-Nilkitkwa system, and data of Krogius (1961) and Krogius and Krokhin (1948) from Lake Dalnee in Kamchatka all

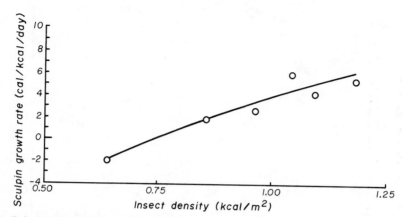

FIGURE 17–8   Relationship between the growth rate of yearling sculpins and the density of insects upon which they fed in a laboratory stream experiment conducted during the spring, 1961. After Brocksen, Davis, and Warren (1968).

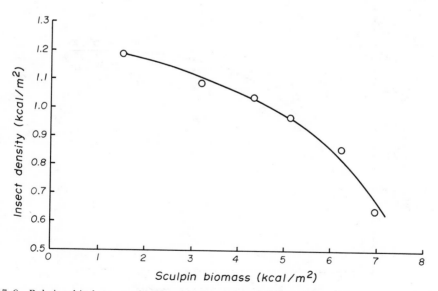

FIGURE 17–9   Relationship between the biomass of sculpins and the density of the insects upon which they fed in a laboratory stream experiment conducted during the spring, 1961. After Brocksen, Davis, and Warren (1968).

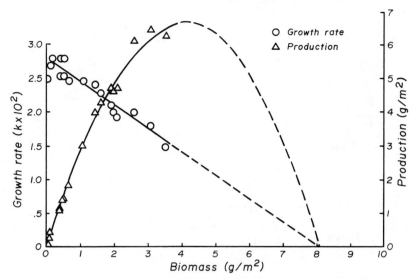

FIGURE 17–10   Relationships of mean daily growth rate and production to biomass of juvenile sockeye salmon in the Babine-Nilkitkwa lake system, British Columbia, during four-month periods from June through September, 1956 and 1957. After Brocksen, Davis, and Warren (1970), based on data of Johnson (1961).

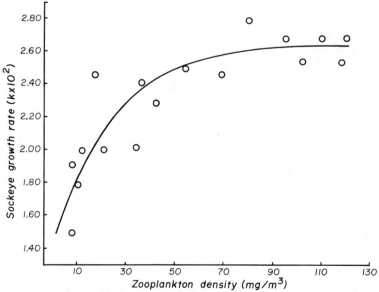

FIGURE 17–11   Relationship between mean daily growth rate of juvenile sockeye salmon and mean density or biomass of the zooplankton upon which they fed in the Babine-Nilkitkwa lake system, British Columbia, during four-month periods from June through September, 1956 and 1957. After Brocksen, Davis, and Warren (1970), based on data of Johnson (1961).

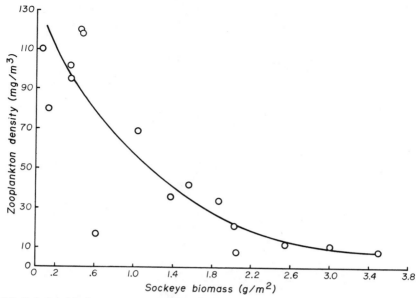

FIGURE 17-12 Relationship between mean biomass of juvenile sockeye salmon and the mean density or biomass of the zooplankton upon which they fed in the Babine-Nilkitkwa lake system, British Columbia, during four-month periods from June through September, 1956 and 1957. After Brocksen, Davis, and Warren (1970), based on data of Johnson (1961).

show that the growth rates of juvenile sockeye salmon lie along the same line, when growth is plotted against zooplankton density (Fig. 11-14). This is so even though these three lake systems have greatly different capacities to produce zooplankton. There exist very low zooplankton densities in Owikeno, intermediate densities in the Babine-Nilkitkwa system, and very high ones in Dalnee. The low growth rates of the sockeye salmon in Owikeno, the moderate growth rates in the Babine-Nilkitkwa system, and the extremely high growth rates in Dalnee can, then, be explained on the basis of differences in zooplankton density.

Now, we can logically reason that those systems having the highest capacity to produce the food organisms of a given species should maintain the highest densities of those food organisms and the highest growth rates of the consumer species, and that higher consumer growth rates should lead to higher consumer biomasses. This would suggest that, in comparing systems having different capacities to produce the food of a species, we should find those systems maintaining the highest food densities to also maintain the highest consumer biomasses. In the case of the juvenile sockeye

salmon, nature again bears out our argument (Fig. 17-13). Lake Dalnee, having the highest capacity to produce zooplankton, maintains the highest zooplankton densities and sockeye biomasses, the Babine-Nilkitkwa system and Owikeno Lake being successively lower in both respects. Thus, between systems of differing productive capacities, food density and consumer biomass are positively correlated (Fig. 17-13). This relationship is different from the relationships between food density and consumer biomass within a system (Figs. 17-12, 17-13) or between systems having equal capacity to produce the food of a consumer (Fig. 17-9), for these relationships exhibit negative correlation, as we have explained. In systems having very high capacities to produce food organisms, no such negative correlation may exist, for then consumption by a particular species may no longer be the major fate of the food organisms. Such appears to be the case for Lake Dalnee (Fig. 17-13).

Those systems having the highest capacities to produce food organisms should, then, maintain the highest food densities and the highest consumer growth rates at the highest consumer biomasses, so long as food is the resource limiting growth and

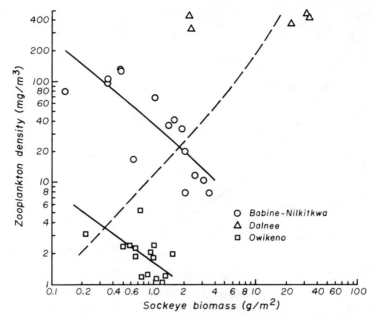

FIGURE 17–13 Relationships between biomass of juvenile sockeye salmon and the density or biomass of the zooplankton upon which they fed in the Babine-Nilkitkwa lake system, Owikeno Lake, British Columbia, and Lake Dalnee, Kamchatka. After Brocksen, Davis, and Warren (1970), based on data of Johnson (1961), Ruggles (1965), and Krogius and Krokhin (1948).

production of the consumer. And potential consumer production must be highest in those systems having the highest capacity for food production. Figure 17–14 well illustrates how great may be the differences in the curves of production of a consumer in systems having different capacities to produce its food. Lake Dalnee can maintain juvenile sockeye salmon at much higher growth rates across a very much greater range of biomasses than can either Owikeno Lake or the Babine-Nilkitkwa system; consequently, its curve of production for sockeye salmon is much higher and wider than those of the other two lake systems.

Definition of the density-dependent relations leading to the curve of production of a consumer species in a particular ecosystem helps us to see the kinds of changes in the consumer or in its environment that will change the function that the production of the consumer is of its biomass. These changes will be still easier to see if we can be even more explicit as to how the relationships between consumer growth rate and food density and between food density and consumer biomass are involved in the relationship between consumer production and consumer biomass. We are now able to formulate production equations for a population having a single age class in a particular ecosystem (Brocksen, Davis, and Warren, 1968, 1970). With more knowledge, we should be able to formulate similar equations for populations having several age groups and even able to write equations of value in predicting the effects changes in the capacity of an ecosystem to produce food will have on the production curve of a consumer (Chapter 19).

The production of a consumer having a single age group can be represented by the equation:

$$P_{cons} = G_{cons} \times B_{cons} \qquad (1)$$

where

P_{cons} = production of consumer of interest,
G_{cons} = growth rate of consumer of interest, and
B_{cons} = biomass of consumer of interest.

In Chapter 11, we presented examples in which the growth rates of consumers were simple functions of the densities of their food organisms (Figs. 11–10, 11–11, 11–13,

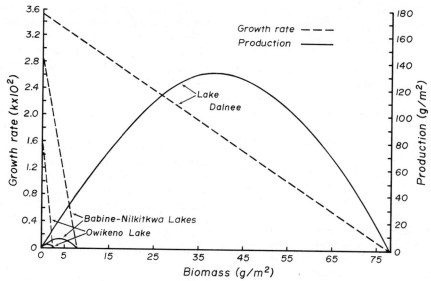

FIGURE 17-14 Comparison of relationships of the biomass of juvenile sockeye salmon to their growth rate and production in Owikeno Lake, the Babine-Nilkitkwa system, and Lake Dalnee. After Brocksen, Davis, and Warren (1970), based on data of Ruggles (1965), Johnson (1961), and Krogius and Krokhin (1948).

11-14); in this chapter, we have presented similar examples (Figs. 17-8, 17-11). There is evidence that this may be a fairly general phenomenon in nature (Brocksen, Davis, and Warren, 1968, 1970). Where it occurs and can be defined, the relationship between the growth rate of the consumer and the density of its food can be represented by the equation:

$$G_{cons} = f_1 (B_{food}) \qquad (2)$$

where

B_{food} = biomass (density) of food organisms.

The function f_1 can be expected to change if behavioral changes of the consumer alter the efficiency with which it crops its food at one or more food organism densities or if physiological or bioenergetic changes in the consumer alter the efficiency with which it utilizes for growth the food consumed at one or more food densities. For example, in different seasons of the year or in different ecosystems, the growth rates of a particular consumer species may be different functions of its food organism biomass. This function should not change with changes in food organism production or with changes in either intraspecific or interspecific competition, if the effect of the competition is *only* to change the density of the food.

By substituting f_1 (B_{food}) for G_{cons} in equation 1, we can express the production of the consumer as a function of food density and consumer biomass as follows:

$$P_{cons} = f_1 (B_{food}) \times B_{cons} \qquad (3)$$

This equation should define the production of the consumer unless behavioral or physiological changes in the consumer alter the relationship between consumer growth rate and food organism density, as we explained for equation 2.

When the consumer of interest has no important competitors for food, the biomass of the food organisms may be some fairly simple function of the biomass of the consumer, as we have shown in Figures 17-9 and 17-12 (Brocksen, Davis, and Warren, 1968, 1970). It may then be possible to represent the biomass of the food organisms as follows:

$$B_{food} = f_2 (B_{cons}) \qquad (4a)$$

Such a relationship will most often be apparent when consumption by the consumer is a principal fate of food organism production. The function f_2 can be expected to change if there are changes at one or more consumer biomasses in the cropping efficiency of the consumer, in the production of the food, or in the proportions of food passing into different trophic pathways. For a given consumer species, such changes might occur with seasonal or ecosystem changes.

When the consumer of interest has important competitors for food, the biomass of the food can perhaps be represented by some function of the biomasses of the consumer and its competitors. In our laboratory stream studies (Brocksen, Davis, and Warren, 1968), we have found the biomass of the prey organisms to be a simple function of the combined biomasses of sculpins and stonefly naiads when these two predators are present; when trout are also present, the density of their prey organisms drifting in the current appears to be a function of trout biomass and the combined biomasses of sculpins and stoneflies. Juvenile sockeye salmon are believed to compete with sticklebacks (*Gasterosteus aculeatus*) for their zooplankton food organisms, and zooplankton density may be found to be a function of the combined biomasses of these two species of fish when both are present in important numbers. Perhaps food biomass when competitors are present can be represented:

$$B_{food} = f_{12} (B_{cons}, B_{comp}) \qquad (4b)$$

where

B_{comp} = biomass of competitors of consumer of interest.

Function f_{12}, like function f_2 in equation 4a, may change if at one or more consumer or competitor biomasses there are changes in the cropping efficiency of either the consumer of interest or its competitors, changes in the production of food, or changes in the proportions of food production passing into different trophic pathways.

Now we can represent the production of a consumer having no important competitors as functions of its biomass by substituting $f_2(B_{cons})$ from equation 4a for B_{food} in equation 3, this giving:

$$P_{cons} = f_1 \left[f_2 (B_{cons}) \right] \times B_{cons} \qquad (5a)$$

This equation should define the relationship between the production of a consumer and its biomass, even in different seasons of the year or in different ecosystems, so long as f_1 and f_2 do not change, as we explained for equations 2 and 4a.

When a consumer has important competitors, we may be able to represent its production as a function of its biomass and the biomasses of its competitors by substituting $f_{12}(B_{cons}, B_{comp})$ from equation 4b for B_{food} in equation 3, this yielding:

$$P_{cons} = f_1 \left[f_{12} (B_{cons}, B_{comp}) \right] \times B_{cons} \qquad (5b)$$

Equation 5b should define the relationship between the production of a consumer and its biomass and the biomasses of its competitors in different seasons or in different ecosystems unless f_1 or f_{12} change, as we explained for equations 2 and 4b.

By bringing functions together, we can write equation 5a as:

$$P_{cons} = f_3 (B_{cons}) \times B_{cons} \qquad (6a)$$

or as

$$P_{cons} = f_4 (B_{cons}) \qquad (7a)$$

Similarly, we can write equation 5b as:

$$P_{cons} = f_{13} (B_{cons}, B_{comp}) \times B_{cons} \qquad (6b)$$

or as

$$P_{cons} = f_{14} (B_{cons}, B_{comp}) \qquad (7b)$$

We come back, then, to the idea we first introduced in the previous section of this chapter: the production of any animal is some function of its biomass. We have illustrated this idea with curves of production for sculpins (Fig. 17-6), trout (Fig. 17-7), and salmon (Figs. 17-10, 17-14). But, by the very definition of production, this idea must be true for all animals. In the examples we have given, however, we have been able to define the curve of production of each species not only in terms of its biomass but also in terms of its food resource. It may not always be possible to do this by the simple expedients of relating the growth rate of a consumer to its food density and the food density to the consumer's biomass. But where it is possible, we have gained an advantage not often possessed by biologists: we will be able to predict what changes in the food resource of a consumer and what changes in the consumer itself will lead to changes in the function that the consumer's production is of its biomass. In dealing with environmental changes of all kinds, including pollution, this can be invaluable.

Before turning to the question of environmental change, we would find it difficult to pass over another idea that intrigues biologists, an idea our reader may already have caught. For the growth rate of a consumer to be a simple function of the density of its food organisms, and for the density of food organisms to be a simple function of the consumer's biomass, there must exist *dynamic equilibria* between the consumer and its food organisms. If such equilibria did not exist, points on the graphs of these relationships would not lie so close to their respective lines. The feedback mechanisms leading to these dynamic equilibria are intriguing to contemplate, for they cause growth rate and production to be rather simple functions of consumer biomass. One of these feedback mechanisms is more readily apparent than others. If decreases in food density resulting from low production or high consumption of food cause the growth rate of a consumer to approach zero (Fig. 17-8), then consumer biomass will tend to decrease, because of weight loss or mortality. But the decrease in consumer biomass will permit the biomass or density of the food organisms to increase (Fig. 17-9), and this will tend to increase consumer growth rate. Thus, new equilibria between consumer biomass and food organism density and between food organism density and consumer growth rate become established. In mechanisms such as this lie the biological bases of the density-dependent functions we have been discussing.

ENVIRONMENTAL CHANGE AND PRODUCTION

Substantial changes in the environment of a species must almost inevitably lead either to an increase or to a decrease in the production of that species. Increases in production of valuable species are often desirable, but environmental changes caused by man that decrease the production of valuable species may be considered pollution. Production, being the product of growth rate and mean biomass, may be changed by any environmental factor influencing either growth rate

or biomass. And, because growth rate and biomass are interdependent—as we have explained—a change in one usually leads to a change in the other; these changes may tend to be compensatory. Growth rate can be altered by any factor affecting either the physiological or behavioral ability of an animal to obtain and utilize food for growth, at any given level of food availability (Chapter 11). And because growth is dependent on food availability, a factor need not affect directly the physiology or behavior of an animal to influence its growth rate. Any change in an animal's environment that brings a change in the availability of its food will alter its growth. Apart from changes caused by growth, the biomass of an animal can be changed by environmental factors affecting its reproduction, survival, or movements. Improvement or degradation of the environment of a species will lead to biomass changes, as will exploitation by man. These effects of environment on the growth and biomass of an animal, and thus on its production, are not independent and are perhaps not often caused by a single action of the environment. Recognizing this, we will still find it easier to understand the effects of environmental change on production if we try to identify the ways the environment can influence growth, biomass, and, thus, production.

Eutrophication, the increase in concentration of nitrogen, phosphorus, and other plant nutrients, is perhaps the most universal change occurring in the waters of the earth. It began as a natural process, long before the activities of man became important, and it continues today in waters yet unaffected by man. Now the activities of man have become the most important cause, hurrying the process of eutrophication, changing the character of natural waters and the environments of their inhabitants. By increasing plant production, enrichment of waters with plant nutrients can sometimes lead to increases in the production of animals of value to man. This fertilization is often practiced in the pond culture of fish. Agricultural fertilizers leaching into natural waters and even the discharge of domestic and other wastes can sometimes have the same effect. But, over a period of time, increase in plant production and the accumulation of organic material lead to other changes that are a part of the process of eutrophication of waters (Chapter 20).

These changes may be unfavorable to certain valuable species, though they may favor species not originally present or abundant. We must, then, be concerned about eutrophication and about the effects of this environmental change on plant and animal production.

Some years ago, we began studies of the effects on the production of cutthroat trout occasioned by enriching with sucrose and organic nitrogen and phosphorus a small woodland stream, Berry Creek, in the Willamette Valley of Oregon (Warren et al., 1964). Sugars present in many wastes often lead to excessive production of the bacterium *Sphaerotilus natans* on streambeds. These unsightly growths change the environment of other stream organisms so as to favor some and suppress others. We hoped to explain any effects enrichment and *Sphaerotilus* had on the production of the trout. The low concentrations of nutrients maintained in Berry Creek were insufficient to cause water quality changes directly injurious to the juvenile and adult trout studied, and changes in the production of the fish could be attributed only to changes in their food resources.

A quarter-mile section of Berry Creek was bypassed by a diversion canal, which permitted regulation of flow in the original streambed at about 0.5 cfs throughout most of the year (Warren et al., 1964). This allowed us to maintain sucrose concentrations near 4 mg/l in some parts of the experimental section and to stock and recover trout in order to measure their growth and biomass for production computations. Within the controlled section of Berry Creek, four sections—each consisting of a riffle and a pool below—were separated by screens preventing the movement of fish and largely preventing the drift of insects from one section to another. Sucrose was introduced continuously into the two lower sections. Even though the concentrations maintained were low, the total energy value of the sucrose introduced was of the same order of magnitude as the solar radiation in the visible spectrum reaching the surface of the stream. The other important source of energy and material for the stream biota—leaves entering from the forest canopy above—represented only about 1 percent of the energy value of either the sucrose or the solar radiation.

The addition of sucrose to Sections 3 and 4 of the experimental stream led to a luxuriant growth of *Sphaerotilus natans* over the streambed in these sections. Those species of herbivorous insects, particularly midge larvae (Chironomidae), able to utilize the *Sphaerotilus* as food benefited greatly from this new energy and material resource. Thus, biomasses of midge larvae and biomasses of carnivorous insects (Plecoptera and Megaloptera) utilizing midge larvae as food became much greater in the enriched than in the unenriched stream sections (Table 17-3). The food resource of the trout in the enriched sections was, then, considerably increased. In Chapter 11 as well as in this chapter we have discussed how the growth rate of fish is dependent on food availability. In Berry Creek, the growth rate of the cutthroat trout appears to have been a simple function of the density of their food organisms in the benthos (Fig. 11-13).

The greater availability of food organisms, particularly midge larvae, in the enriched sections of Berry Creek permitted the trout to consume more food and led to higher levels of trout production in the enriched sections than in the unenriched sections during the years 1963-64 and 1964-65 (Table 17-4), as it did during other years (Warren et al., 1964). Beginning in March 1965, sucrose was introduced into the previously unenriched Section 2 and this led to levels of production in this section that were higher from March through August of 1965 than during the same period of the previous year.

Food consumption by trout in the enriched sections of Berry Creek was approximately twice that in the unenriched sections (Table 17-4). Thus, the addition of sucrose about doubled the basic capacity of Berry Creek to produce trout. Production of trout in the unenriched sections was only a small fraction of that in the enriched sections, because nearly all the food consumed in the unenriched sections was necessary merely to maintain metabolically the trout in those sections; little food was left for sustaining growth and production (Chapter 11). Trout biomasses in the unenriched sections, then, were well above those optimal for production; reductions in biomasses in these sections would have led to increases in trout production, in ways we attempted to make clear in earlier parts of this chapter. Were trout biomasses in the stream sections allowed to adjust naturally to the food

TABLE 17–3   *Mean biomasses of insects and other invertebrates in the benthos of riffles of Berry Creek, Oregon, during each season, 1963–64. Some riffles were enriched with sucrose and others were not. Each seasonal value is a mean of monthly samples and is expressed in kilocalories per square meter. Unpublished data of* John D. McIntyre *and* Gerald E. Davis, *Oregon State University.*

Group	Summer	Fall	Winter	Spring
Unenriched Riffles				
Ephemeroptera	0.5	1.3	3.4	2.4
Plecoptera	0.8	0.7	0.7	0.9
Trichoptera	0.3	1.7	1.4	1.2
Coleoptera	0.2	0.6	0.6	0.3
Diptera:				
(a) Chironomidae	0.3	1.5	0.4	0.4
(b) Other Diptera	0.2	0.7	0.5	0.4
Megaloptera	0.1			
Acarina		0.2	0.7	0.6
Oligochaeta	1.6		0.3	0.2
Totals	4.0	6.7	8.0	6.4
Enriched Riffles				
Ephemeroptera	1.0	1.5	1.8	2.4
Plecoptera	2.8	2.9	1.4	2.3
Trichoptera	0.4	1.7	0.4	0.3
Coleoptera	0.2	0.6	0.1	0.3
Diptera:				
(a) Chironomidae	2.1	9.0	5.4	2.5
(b) Other Diptera	1.4	0.7	0.2	0.6
Megaloptera	1.4	1.3	0.4	0.3
Oligochaeta	2.7	2.4	3.4	1.2
Totals	12.0	20.1	13.1	9.9

resource, rather than being controlled, trout production in the enriched sections would have been only about twice that in the unenriched ones. Enrichment with sucrose, then, about doubled the basic capacity of the Berry Creek ecosystem to produce trout; it increased the height and width of the trout production curve.

We have considered the effects of enrichment and the growth of *Sphaerotilus* on only the availability of food and the production of trout. In this regard, the *Sphaerotilus* was beneficial. But had maintenance of trout populations in the experimental section of Berry Creek been dependent on natural reproduction, detrimental effects of the *Sphaerotilus* would probably have been observed. Trout embryos developing in streambed gravels are very sensitive to any restriction of water flow through these gravels caused by silt or other materials, including *Sphaerotilus* (Chapter 10). In evaluating the effects an environmental change may have on a species, we must consider reproduction and survival as well as growth and production.

Lake Dalnee is apparently much richer in plant nutrients than are the oligotrophic Owikeno and Nilkitkwa-Babine lake systems. Lake Dalnee would appear to be approaching the mesotrophic condition, which is midway between the oligotrophic and eutrophic conditions. It is undoubtedly this greater richness in plant nutrients that permits Lake Dalnee to maintain zooplankton densities and sockeye salmon growth rates and biomasses much higher than those found in the other lakes, as we explained in the previous section of this chapter. The juvenile sockeye salmon can be considered an oligotrophic animal, but even an oligotrophic animal can benefit from an increase in the capacity of its ecosystem to provide it with food, as the data from these three lake systems suggest. Thus, very early in the process of eutrophication by which an oligotrophic lake passes through mesotrophy to the eutrophic condition, the production of an oligotrophic animal may be increased, until the process leads to changes in water quality or food organisms that are deleterious to the animal. When an oligotrophic lake in Alaska was fertilized annually with inorganic nitrate and phosphate, annual in-

TABLE 17–4 Mean biomasses, growth rates, production, and food consumption of cutthroat trout in enriched and unenriched sections of Berry Creek, Oregon, during experiments conducted in 1963–64 and 1964–65. Unpublished data of John D. McIntyre *and* Gerald E. Davis, *Oregon State University.*

Month	1963–64				1964–65			
	Mean Biomass (kcal/m²)	Growth Rate (cal/kcal/day)	Pro-duction (kcal/m²)	Consump-tion (kcal/m²)	Mean Biomass (kcal/m²)	Growth Rate (cal/kcal/day)	Pro-duction (kcal/m²)	Consump-tion (kcal/m²)
	Section 1 (unenriched)							
September	2.35	2.84	0.22	1.89	2.57	2.95	0.22	1.83
October	2.67	1.10	0.08	1.50	2.92	1.77	0.15	1.86
November	2.85	−0.65	−0.05	1.35	2.83	0.00	0.00	1.78
December	3.16	0.57	0.05	1.73	2.17	0.00	0.00	1.16
January					1.56	−0.85	−0.04	0.80
February	3.11	0.55	0.05	1.76	2.68	0.53	0.03	1.10
March	2.90	1.49	0.13	1.86	3.12	1.40	0.14	2.12
April	3.13	0.00	0.00	1.56	2.84	−0.63	−0.05	1.38
May	3.03	1.65	0.16	2.09	2.80	3.88	0.25	1.74
June	2.78	0.72	0.05	1.38	3.11	−0.58	−0.06	1.80
July	1.68	0.54	0.03	1.08	2.62	−0.99	−0.07	1.18
August	2.40	−2.53	−0.17	0.96	3.32	0.68	0.07	2.03
Totals			0.55	17.16			0.64	18.78
Means	2.73				2.71			
	Section 2 (enriched beginning March, 1965)							
September	2.72	−1.89	−0.17	1.36	1.61	0.64	0.03	0.92
October	2.46	−8.43	−0.56	0.40	3.08	−0.56	−0.05	1.56
November	3.07	−1.81	−0.15	1.28	2.83	0.10	0.01	1.79
December	3.38	0.85	0.08	1.89	1.93	1.25	0.07	1.18
January					2.23	0.90	0.06	1.35
February	2.98	0.93	0.08	1.75	2.87	−0.33	−0.02	1.08
March	2.94	2.61	0.23	2.09	3.75	1.08	0.13	2.46
April	3.86	2.40	0.25	2.44	4.78	7.17	0.96	4.95
May	3.68	2.04	0.24	2.65	3.87	9.66	0.86	4.14
June	2.70	1.48	0.10	1.44	3.52	1.81	0.21	2.58
July	2.45	−1.73	−0.14	1.25	3.44	1.94	0.18	2.08
August	2.20	−2.76	−0.17	0.85	2.73	2.13	0.18	1.93
Totals			−0.21	17.40			2.62	26.02
Means	2.95				3.05			

TABLE 17–4   Continued.

MONTH	1963–64				1964–65			
	MEAN BIOMASS (kcal/m²)	GROWTH RATE (cal/kcal/day)	PRO-DUCTION (kcal/m²)	CONSUMP-TION (kcal/m²)	MEAN BIOMASS (kcal/m²)	GROWTH RATE (cal/kcal/day)	PRO-DUCTION (kcal/m²)	CONSUMP-TION (kcal/m²)
				Section 3 (enriched)				
September	3.60	4.97	0.59	3.56	2.48	4.31	0.31	2.01
October	3.08	0.60	0.05	1.61	3.30	2.30	0.22	2.20
November	3.96	7.29	0.78	3.99	3.78	0.54	0.07	2.51
December	3.16	−2.94	−0.26	1.19	2.65	3.90	0.30	2.09
January					2.75	2.55	0.21	1.96
February	3.52	2.74	0.28	2.48	3.43	8.47	0.61	2.97
March	3.83	4.35	0.50	3.23	3.87	5.01	0.62	3.75
April	3.66	9.11	0.90	4.32	4.68	10.68	1.40	6.62
May	4.36	6.88	0.96	5.01	4.16	10.87	1.04	4.90
June	3.70	4.65	0.43	2.71	3.38	2.33	0.26	2.59
July	3.44	0.88	0.10	2.28	3.68	−1.61	−0.16	1.56
August	2.42	4.28	0.29	1.90	3.28	0.59	0.06	1.98
Totals			4.62	32.28			4.94	35.14
Means	3.52				3.45			
				Section 4 (enriched)				
September	3.15	3.27	0.34	2.64	1.88	6.24	0.34	1.85
October	2.89	1.03	0.08	1.59	2.38	1.16	0.08	1.44
November	2.32	−1.44	−0.09	1.00	2.86	1.65	0.16	2.12
December	3.44	−0.93	−0.09	1.61	2.88	5.99	0.50	2.76
January					2.53	11.33	0.86	4.04
February	3.48	1.59	0.16	2.17	3.80	2.88	0.23	1.96
March	3.07	4.13	0.38	2.55	3.94	4.68	0.59	3.69
April	4.83	7.21	0.94	4.89	4.70	4.26	0.56	3.68
May	4.77	6.35	0.97	5.22	5.82	9.26	1.24	5.96
June	3.47	5.76	0.50	2.82	3.01	3.32	0.33	2.53
July	2.81	2.91	0.27	2.30	3.14	0.24	0.02	1.59
August	2.33	6.44	0.42	2.25	3.52	−0.92	−0.10	1.83
Totals			3.88	29.04			4.81	33.45
Means	3.32				3.37			

creases in the gross production of phytoplankton resulted (Nelson, 1958). And the growth rates of the four age groups of juvenile sockeye salmon present in the lake were closely and positively correlated with the algal production (Fig. 17–15). Juvenile sockeye salmon feed on zooplankton, not phytoplankton, so we must suppose that the increases in phytoplankton led to increases in zooplankton that increased the growth and, undoubtedly, the biomass and production of the salmon. But the process of eutrophication, if continued too long, will not continue to benefit an oligotrophic species. Further increases in the density of zooplankton in Lake Dalnee are not likely to benefit the sockeye salmon, and other environmental changes deleterious to this species could well begin to appear.

This is the story of Lake Zürich in Switzerland and Lake Erie in the United States, a story we will tell in Chapter 20. In these lakes, the process of eutrophication through the introduction of domestic and other wastes went too far for the oligotrophic salmonids and whitefishes. These have been replaced by mesotrophic perch and other species; carried further, the process of eutrophication will lead to dominance of eu-

trophic minnows. For those species able to reproduce and survive well in more eutrophic environments, there are available greater resources of energy and materials. These resources permit higher growth rates and biomasses and, thus, higher levels of production (Fig. 17–16). Should the value of mesotrophic or eutrophic species be as great as that of oligotrophic ones, man may benefit. Unfortunately, this is not always the case.

Not often are the wastes introduced into natural waters so simple in their composition and effect as was the sucrose introduced into Berry Creek. Here, we were dealing with a material that had little or no direct effect on the fish studied. The increases in trout production resulted from an increased capacity of Berry Creek to produce trout food, the sucrose leading to *Sphaerotilus* production, which was utilized by midge larvae for food. Many domestic and industrial wastes contain materials that can increase the productivity of waters. But these wastes in addition often contain materials exerting some toxic action directly on fish or on organisms in their food chain. Effluents from the kraft process of making wood pulp and paper are such wastes. As we explained in Chapter 11, both

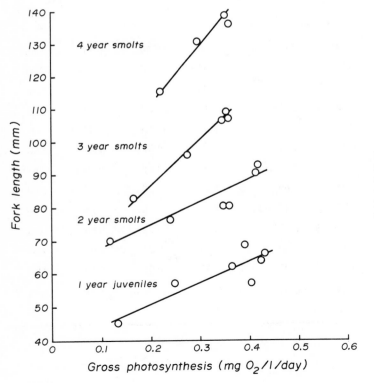

FIGURE 17–15 Relationships between the mean size of individuals in four age groups of juvenile sockeye salmon and the rate of photosynthesis in an oligotrophic lake, which was fertilized with inorganic nitrogen and phosphorus on successive years from 1950 through 1956. After Nelson (1958).

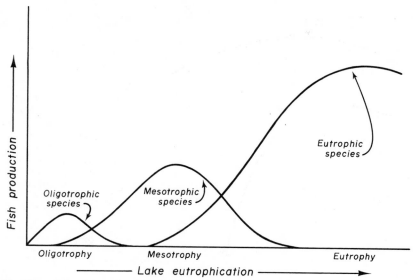

FIGURE 17-16 Changes in the production of oligotrophic, mesotrophic, and eutrophic species of fish that would theoretically take place as a lake passed from the condition of oligotrophy to the condition of eutrophy through the process of eutrophication.

treated and untreated kraft mill effluents can, at some concentrations, reduce the effectiveness with which juvenile chinook salmon (*Oncorhynchus tshawytscha*) utilize food for growth (Figs. 11-22, 11-23). These same wastes can increase the food and the production of some organisms that the salmon utilize for food; but, through toxic and other effects, the wastes may decrease the availability of other fish food organisms. Thus, to learn how and at what concentrations kraft mill effluents may change the production of valuable species of fish in nature is no simple problem. We have made a beginning through studies of the effects of these wastes on the production of juvenile salmon in laboratory stream communities such as those we discussed in earlier sections of this chapter. As our reader will see, even these studies under relatively simple conditions are not easy to interpret. Future studies in more natural experimental streams like Berry Creek will be necessary before we can be sure how well what we have learned in the laboratory applies to nature.

Ellis (1968), in a spring 1967 experiment, found the curves relating production of juvenile chinook salmon to their biomass in laboratory stream communities to be typical in having maxima at intermediate biomasses (Fig. 17-17). But the curve of production for the salmon in the three streams receiving 1.5 percent by volume of untreated kraft mill

effluent was much lower than the curve for the salmon in the three streams not receiving effluent, and it had a much more steeply descending right limb. Food organisms of the salmon were more abundant in the streams receiving kraft mill effluent than in the other streams, because of the greater production of amphipods, which benefited from the introduction of the effluent. Nevertheless, except at the highest biomass, the salmon in the streams receiving effluent appear to have consumed only about as much food as the salmon in the streams not receiving the effluent (Fig. 17-18). The big differences in production, at the intermediate and low biomass levels, between streams receiving and streams not receiving effluent (Fig. 17-17) cannot, then, be attributed primarily to differences in food availability or consumption. But at a kraft mill effluent concentration of 1.5 percent—about 3 mg/l BOD—the effectiveness with which the salmon utilize for growth the food they consume is reduced (Fig. 11-22). We can attribute the great reduction in production occasioned by the presence of the effluent at the low and intermediate biomass levels primarily to direct effects on the ability of the fish to grow.

At the highest biomasses of salmon, food consumption in the control stream remained high, but it declined precipitously in the stream receiving kraft mill effluent (Fig.

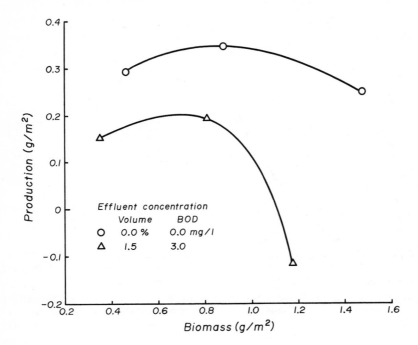

FIGURE 17–17 Relationships between the biomass of juvenile chinook salmon and their production in laboratory streams receiving untreated kraft mill effluent and in streams not receiving such effluent during a 17-day experiment conducted in the spring, 1967. After Ellis (1968).

17-18), even though this stream had food densities higher than any of the others. As the density of an animal population increases, social stress among its individuals is aggravated (Chapter 12). Observations on the social and feeding behavior of the juvenile salmon at the highest biomass suggest that the presence of kraft mill effluent somehow increased social stress and greatly reduced feeding, this resulting in the great decline of food consumption. It appears, then, that effects of the effluent on the feeding behavior of the fish were the primary cause of the precipitous decline in production at the high biomass (Fig. 17–17). Lower concentrations of either treated or untreated kraft mill effluent that have little or no direct effect on growth or behavior can increase the production of juvenile salmon by increasing the production and availability of their food organisms.

The effluents from many industrial pro-

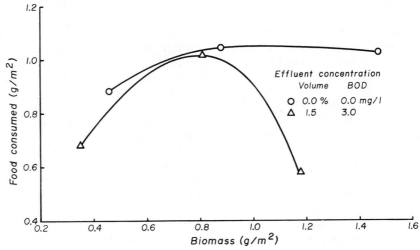

FIGURE 17–18 Relationships between the biomass of juvenile chinook salmon and their food consumption in laboratory streams receiving untreated kraft mill effluent and in streams not receiving such effluent during a 17-day experiment conducted in the spring, 1967. After Ellis (1968).

cesses, such as the kraft process, are very complex mixtures; this is at least in part responsible for their complex effects on stream biota. Other materials finding their ways into streams are not necessarily complex mixtures; nonetheless, their influences on stream communities can still be very complex. We used laboratory stream communities to begin to understand these influences of dieldrin, a chlorinated-hydrocarbon insecticide. Into laboratory streams containing sculpins, midge larvae, and algae, we continuously introduced dieldrin at concentrations of 0.06 and 0.6 parts per billion (ppb). Because of adsorption on organic materials present in the streams, the concentrations actually maintained in the water were probably lower than these. As our reader might expect, the densities of midge larvae in the streams having the highest concentration of dieldrin were greatly reduced (Fig. 17-19); this undoubtedly was in large part responsible for the reduction in the rate of growth of the sculpins in these streams. But we know that very low concentrations of dieldrin also can directly influence the ability of sculpins to utilize for growth the food they manage to consume (Fig. 11-21). Thus, we must suppose that the great reduction in sculpin production at the high dieldrin concentration (Fig. 17-20) was occasioned by direct effects on the growth of the sculpins as well as by effects on their food chain. It is of interest to note—by comparing the position of the ascending limb of the curve of production for sculpins in streams having the lowest concentration of dieldrin or none with the position of the descending limb of the curve for sculpins at the high concentration—the great reduction in the capacity of the streams to produce sculpins that was caused by these two effects of dieldrin (Fig. 17-20).

To write, as we did at the beginning of this section, that substantial changes in the environment of a species must almost inevitably lead either to an increase or to a decrease in its production is only to express the experience and beliefs of biologists with regard to such things. The effects of environmental change on production have actually been measured in relatively few instances. But we know that changes in environment affect the reproduction, survival, and growth of animals (Chapters 10, 11, 13, 16). Production, only the product of growth and biomass, too must then be affected. If we do not as yet know much about the relationships between environmental change and production, we must not be impatient; after all, years of study have taught us all too little about the effects of environmental change on reproduction and survival.

Production should be a useful parameter, once it is better understood. We have burdened our reader little with the rapidly growing literature, a literature mainly on fish. Chapman (1967) has reviewed this literature, and much of it contains quite valid estimates of production. But, as with

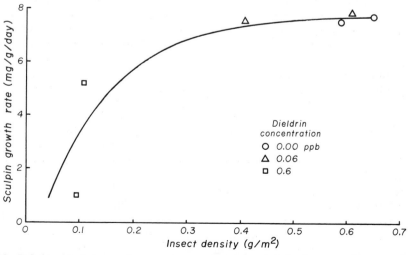

FIGURE 17-19 Relationship between the growth rate of sculpins and the density or biomass of the insects upon which they fed in laboratory streams into which different concentrations of dieldrin were introduced.

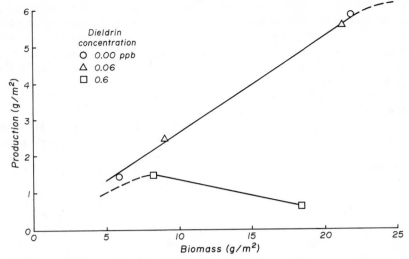

FIGURE 17-20 Relationships between the biomass of sculpins and their production in laboratory streams into which different concentrations of dieldrin were introduced.

any biological parameter, our interest is not so much in the values obtained, however valid, as in the conditions and causes of these values. In this respect, biology is like all science. Studies of production must become more concerned with conditions and causes. In our enthusiasm for a delightful new parameter, we would be well advised not to forget this.

THE UTILIZATION OF POPULATIONS

Man's utilization of a virgin forest, a population of birds, or a community of tide pool organisms may be entirely esthetic, this utilization requiring no removal by man of plants or animals from their populations. This esthetic capacity distinguishes man from the other animals of the earth. But, like all animals, man requires food and other resources, the utilization of which is dependent on his removing plants and animals from their populations. The amount man removes from a population is the *yield* of that population.

The *yield* of a population is dependent on social and economic factors as well as on biological factors. The *maximum equilibrium yield* is dependent on the capacity of an ecosystem to produce that population: it is the largest amount that can be continuously removed from a population under existing environmental conditions (Ricker, 1958, p. 20). Whatever the capacity of an ecosystem

to produce a particular species, the yield in a given period of time may approach zero or be much greater than the yield the population of that species can sustain indefinitely. A yield may be very low because man chooses for social and economic reasons either not to exploit a population or to exploit it very little. Or it may be very low because man through overexploitation has reduced the biomass and thus the production of the population to low levels. The yield may for a while be higher than the maximum equilibrium yield, but this is the overexploitation that leads to reduction in the biomass, the production, and, ultimately, the yield of a population. Not often has man shown the wisdom to exploit natural populations at the level of maximum equilibrium yield. Both the yield and the maximum equilibrium yield of a population may be low because the capacity of the ecosystem to produce that population is low. This may derive either from natural factors or from the effects of man's activities, his discharging of wastes being an all too frequent example.

A population is in *equilibrium* when increments to its biomass (stock) that result from reproduction (or recruitment) and production are balanced by decrements resulting from natural mortality and removal by man (yield): its biomass will not change, so long as there is no emigration or immigration. If reproduction and production increments are greater than decrements caused by natural mortality and removal by man, biomass will increase; if smaller, biomass will decrease.

This *biomass equilibrium* can occur at any level of biomass consistent with the capacity of an ecosystem to support a population: it need not be the *maximum maintainable biomass*. Likewise, an *equilibrium yield* exists for each level of biomass: it is the yield that would lead to neither an increase nor a decrease in biomass. An equilibrium yield need not be the *maximum equilibrium yield,* which can exist only at one level of biomass, just as maximum production of a given species in a particular environment can exist only at a certain biomass.

Ricker (1958, pp. 253–255) has developed an example, having some factual basis in the North Sea demersal fish stocks, that illustrates for a population increasing logistically the relationship between biomass and equilibrium yield (Fig. 17–21). This example will help us to explain how, through time, changes in biomass come about, and how these changes influence the equilibrium yield. In Chapter 16, we explained that the increase in number or biomass of a population through time often tends to follow an asymmetrical sigmoid curve. This curve can sometimes be approximated by the logistic equation, the rate of increase of the population decreasing as population size approaches an asymptotic value. This decline in the rate of increase is usually supposed to be caused by the detrimental effects that increasing number or biomass has on the reproduction and growth of individuals in the population.

The increase of biomass through time of the fish population in Ricker's (1958) example is shown by the logistic curve in Figure 17-21. From a very low biomass, the population first increases rapidly, because recruitment and production are much greater than mortality. As biomass increases further, however, losses through mortality begin to approach gains through recruitment and production, until, at the asymptotic biomass of 220,000 tons, losses and gains balance, and biomass can increase no further. Now, if at the point in time when biomass reaches 100,000 tons (level 1) man should begin to remove from the population 40,000 tons annually (the equilibrium yield for that biomass), the biomass would increase no further; neither would it decrease (Fig. 17-21). At a biomass of 60,000 tons (level 2), annual removal of the equilibrium yield of 32,000 tons would result in no further changes in biomass. No changes in biomass would accompany these removals because

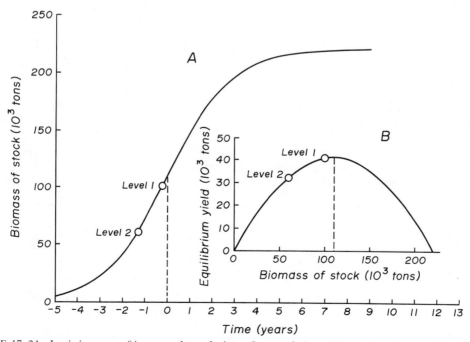

FIGURE 17-21 Logistic curve of increase through time of a population of fish in the absence of fishing (A), and the relationship between the biomass of the population and equilibrium yield (B), as explained in the text. After Ricker (1958).

recruitment and production would balance natural mortality and removal (yield).

The curve relating equilibrium yield to biomass in Ricker's (1958) example is shown in Figure 17-21. For good reason, it resembles the curves relating production to biomass that we considered in earlier sections of this chapter (Figs. 17-6, 17-7, 17-10). At very low biomasses, recruitment and production are low because few individuals are present. At very high biomasses, limits of food and space severely restrict recruitment and production; furthermore, natural mortality is high. And little is left over for man. At some intermediate level of biomass, the removal by man can be maximized and along with natural mortality still be balanced by recruitment and production (Fig. 17-21): this removal is the maximum equilibrium yield. Fisheries biologists have long supposed that a population of fish that has never been exploited has a potential equilibrium yield of zero at its existing biomass (Fig. 17-21): yield comes from replacing natural mortality with fishing mortality and from increased reproduction and production made possible by more effective utilization of food and space at lower biomasses occasioned by fishing.

Now, to our reader, this discussion of yield may seem a little too much, and he may well wonder why we have burdened him. There are really several good reasons. It is, after all, the yield of a population that usually most directly interests man. Public alarm over the effects of pollution on an aquatic resource becomes justifiably greatest when yields decline. But overexploitation and other factors also can cause declines in yield, and these should be understood, for there

may not always be causal relations between pollutional changes in the environment and declines in yield. We hope our elementary and inadequate explanation of the dynamics of yield will help our reader to appreciate the meaning of yield. This is an important aspect of the theory of exploitation (Beverton and Holt, 1957; Ricker, 1958), and it is important to all biologists and others interested in water pollution, not just to fishery and other resource biologists.

It should now be apparent to our reader that there is no constant relationship between yield and production, even for a particular species in a given environment. Production and equilibrium yield change in an understandable manner with changes in biomass. But yield itself depends not only on biological factors but also on social and economic factors determining the level at which man chooses to exploit a population. Even so, values for production and for the ratio of production to yield obtained in a few studies may be of some interest and are given in Table 17-5. We need to know more about the biological basis of production and yield before we can predict the effects of pollutional changes on the yield of aquatic resources to man.

Reductions in the biomass of populations can sometimes increase production and equilibrium yield, but severe reductions reduce both (Fig. 17-21). Environmental changes, pollutional or otherwise, may reduce biomass by increasing mortality or by decreasing reproduction or growth. Reproduction, mortality, and particularly production determine *equilibrium yield*. The relationships among these variables are complex. Man finally turns to parameters like *yield*,

TABLE 17-5   *Production* (P) *and ratios of production to biomass* (B) *and production to yield of different species of fish in lakes and streams studied by various investigators. Yield of salmon measured as migrant smolts. After* Chapman (1967).

WATER	SPECIES	P (g/m²/yr)	RATIOS		REFERENCE
			P/B	P/yield	
Cultus Lake	Sockeye salmon	5.9	2	2.1–3.6	Ricker and Foerster (1948)
Deer Creek	Coho salmon	9.0	2.5	2	Chapman (1965)
Horokiwi Stream	Brown trout	54.7	2	15	Allen (1951)
New York lakes	Brook trout	6–12	1–1.5	2	Hatch and Webster (1961)
Lawrence Creek	Brook trout	9.3–10.6	1.5–2.1	10–15	Hunt (1966)
Wyland Lake	Bluegill	9.1	1.2	2.9	Gerking (1962)

because his time and understanding are limited. He can put a value on yield, an economic value, and this helps him to make decisions as to the proper use of his resources. He should do this, for there are decisions to be made that cannot wait. But, hopefully, he will continue to seek more full understanding; and, again hopefully, until his understanding is greater, he will keep in mind the factors that may influence the parameters he uses.

SELECTED REFERENCES

Allen, K. R. 1950. The computation of production in fish populations. New Zealand Science Review 8:89.

———. 1951. The Horokiwi Stream: a study of a trout population. New Zealand Marine Department Fisheries Bulletin 10. [ix] + 238pp.

Beverton, R. J. H., and S. J. Holt. 1957. On the dynamics of exploited fish populations. Ministry of Agriculture, Fisheries and Food, London. Fishery Investigations Series 2, 19:3-533.

Brocksen, R. W., G. E. Davis, and C. E. Warren. 1968. Competition, food consumption, and production of sculpins and trout in laboratory stream communities. Journal of Wildlife Management 32:51-75.

———, ———, and ———. 1970. Analysis of trophic processes on the basis of density-dependent functions. Pages 468-498. *In* J. H. Steele (Editor), Marine Food Chains. University of California Press, Berkeley. In press.

Chapman, D. W. 1967. Production in fish populations. Pages 3-29. *In* S. D. Gerking (Editor), The Biological Basis of Freshwater Fish Production. Blackwell Scientific Publications, Oxford. xiv + 495pp.

———. 1968. Production. Pages 182-196. *In* W. E. Ricker (Editor), Methods for Assessment of Fish Production in Fresh Waters. International Biological Programme Handbook 3. Blackwell Scientific Publications, Oxford. xiii + 313pp.

Davis, G. E., and C. E. Warren. 1965. Trophic relations of a sculpin in laboratory stream communities. Journal of Wildlife Management 29:846-871.

——— and ———. 1968. Estimation of food consumption rates. Pages 204-225. *In* W. E. Ricker (Editor), Methods for Assessment of Fish Production in Fresh Waters. International Biological Programme Handbook 3. Blackwell Scientific Publications, Oxford. xiii + 313pp.

Ivlev, V. S. 1945. Biologicheskaya produktivnost' vodoemov. Uspekhi Sovremennoi Biologii 19:98-120. (Translated by W. E. Ricker, 1966. The biological productivity of waters. Journal of the Fisheries Research Board of Canada 23:1727-1759.)

Lindeman, R. L. 1942. The trophic-dynamic aspect of ecology. Ecology 23:399-418.

Neess, J., and R. C. Dugdale. 1959. Computation of production for populations of aquatic midge larvae. Ecology 40:425-430.

Ricker, W. E. 1946. Production and utilization of fish populations. Ecological Monographs 16:373-391.

———. 1958. Handbook of Computations for Biological Statistics of Fish Populations. Fisheries Research Board of Canada Bulletin 119. 300pp.

———, and R. E. Foerster. 1948. Computation of fish production. Bingham Oceanographic Collection Bulletin 11:173-211.

Warren, C. E., and G. E. Davis. 1967. Laboratory studies on the feeding, bioenergetics, and growth of fishes. Pages 175-214. *In* S. D. Gerking (Editor), The Biological Basis of Freshwater Fish Production. Blackwell Scientific Publications, Oxford. xiv + 495pp.

———, J. H. Wales, G. E. Davis, and P. Doudoroff. 1964. Trout production in an experimental stream enriched with sucrose. Journal of Wildlife Management 28:617-660.

PART VI

Community Ecology

Communities composed of populations of plants and animals represent the highest level of biological organization of which we are aware. They are dependent on the physical and chemical conditions and resources present at their locations, and together with these conditions and resources they form ecosystems. A persistent change in the conditions or in the resources of an ecosystem will lead directly to a change in the abundance of one or more populations. But because the populations of communities interact, such a persistent change will lead indirectly to changes in the abundance of many if not most of the populations in the community. Thus, a different community will develop at such a location. Changes in communities are a natural part of the long story of life and have led to the biological world of today. Very recently and very rapidly in this story, man has changed the conditions and resources of ecosystems, and inevitably there have followed great changes in communities. Many of these changes have been essential to man's existence on earth, but many have not been essential and have taken something of value from the earth and from man. This may involve even more than a loss of beauty. Man must grow in wisdom and care.

18 The Community as a Unit of Study

We are dealing with a series of increasingly abstract concepts. The individual idea is fairly concrete — examples of the individual can be seen, handled, preserved.... Populations can normally be defined without great difficulty. At the community level we have difficulty even in definition, both in defining the general concept and in defining particular communities. Yet natural communities are 'real' enough; a particular population lives, not in isolation, but in a network of relationships with other populations. The biosphere is not a random aggregation of individuals and populations but a series of distinct and differing patterns — forests, lakes, seas, grasslands, deserts — a series of differing communities.

MARSTON BATES, 1960, pp. 550-551

The community concept of modern ecology is one of the fruitful ideas contributed by biological science to modern civilization. Its importance is threefold. Through its numerous direct and indirect applications it is of value to such practical fields as agriculture, animal husbandry, wild life conservation, and medicine. Natural orderability, made clear by comprehension of the concept, is important to philosophical thought. Lastly, it is of especial importance to the professional ecologist.

ALLEE ET AL., 1949, p. 436

THE COMMUNITY CONCEPT

When plant and animal species have the opportunity to colonize a location where physical and chemical conditions are not too extreme and where energy and material resources are available, an assemblage of species will occur after a period of time. Complex interrelationships will develop among these species as well as between them and the conditions and resources of the location. If these conditions and resources are not too variable, this assemblage of plant and animal species will come to take on rather definite characteristics; it can then be considered a *community*. This biological community, being dependent on the conditions and resources of its location, may change if persistent changes in these occur. Such a community change is a reflection of changes in the plant and animal populations composing the community. Because of interdependencies, a change in conditions or resources leading to the disappearance of one species may lead to the disappearance of other species; and new species may replace them. Thus, a new assemblage of species can occur which may be distinguishable as a different community.

The concept of the biological community as a natural unit of biological organization cannot be traced to a single origin. The plant

281

community idea has been expressed in the writings of botanists for perhaps two centuries; in the writings of zoologists, the animal community idea has been present for nearly a century. But with such outstanding exceptions as Forbes (1887), not until early in this century did biologists begin to use in their work, thought, and writing a clear concept of biological communities of interacting populations of plants and animals. Clements and Shelford (1939), who contributed greatly to the early development of the biological community concept, trace its history through the writings of many outstanding biologists.

For the biological community to be considered a level of biological organization, particular communities must be so organized as to be distinct and more or less cohesive. Given similar primary physical and chemical conditions, different locations in the same region will usually be colonized mainly by the same species of plants and animals, and the communities that develop will be quite alike; but they will be different from communities developing at nearby locations having different primary conditions. Thus, the same type of community can develop at different locations and times, but it will be distinct from other types of communities because of differences in species composition. The cohesion of a particular community stems from the dependencies that develop among its populations of plants and animals and the conditions and resources of its location. These dependencies frequently involve energy and material resources, plants requiring solar energy and nutrients, herbivores requiring plants for food, and carnivores requiring animals for food. Plants may provide shelter for animals and animals may participate in the pollination of plants. Different species or groups have, then, different functions in the community; and a certain stability comes to the community through the regulatory interactions of these functions. The same functional groups are present in nearly all communities, but the species in these groups may be different. Distinctness of their species composition, dependencies among their species, functional groupings of these species, and some degree of stability together provide a basis for considering biological communities to be natural units of biological organization.

The cell, the individual organism, the population, and the community are different levels of biological organization. We can recognize the individual organism as a level of biological organization (Chapter 9) and also the population as a level of biological organization (Chapter 15), without expecting the same mechanisms and degrees of integration in both. Even though they may be functionally analogous, the nerve and hormonal communication systems within the individual are not the same as the sensory and genetic communication systems of the population (Chapter 15); neither is the population structurally or behaviorally as integrated as the individual. Likewise, we can recognize the community as a level of biological organization without expecting the same mechanisms and degrees of integration that we find at other levels of biological organization. Indeed, we should expect to find and we do find new mechanisms and new kinds of integration in the community. The community has self-regulatory properties deriving from population interactions that cannot be understood apart from the community as a whole.

The community, then, has certain functional properties that are analogous to functional properties of the individual organism and even to those of the cell (Table 18–1). Some biologists have followed Phillips (1935), Emerson (1939a), and others before them in thinking of the community as a *superorganism;* but there are those who object to this (Tansley, 1935). Alfred North Whitehead (1925), who was influenced by biological thought during the development of his *theory of organic mechanism,* provided a philosophical basis for the idea of superorganism:

The concrete enduring entities are organisms, so that the plan of the *whole* influences the very characters of the various subordinate organisms which enter into it. In the case of an animal, the mental states enter into the plan of the total organism and thus modify the plans of the successive subordinate organisms until the ultimate smallest organisms, such as electrons, are reached. Thus an electron within a living body is different from an electron outside it, by reason of the plan of the body. [p. 111]

The concept of an organism includes, therefore, the concept of the interaction of organisms. [p. 146]

For some, there may be conceptual value in thinking of communities as superor-

TABLE 18–1 *The community doctrine as compared to cell and multicellular organism doctrines.*
After Allee et al. (1949).

CELL	MULTICELLULAR ORGANISM	COMMUNITY
Composed of definitive protoplasms	Composed of definitive cells and tissues	Composed of definitive organisms and species
Has anatomy (cytological)	Has anatomy (tissues and organs)	Has anatomy (pyramid of numbers)
Has symmetry and gradients	Has symmetry and gradients	Has aspects of symmetry and gradients (stratification)
Has ontogeny (cell development)	Has ontogeny (embryology)	Has ontogeny (succession)
Has limitations of protoplasmic amounts (size, surface-volume ratio)	Has limitations of cell numbers (size, surface-volume ratio)	Has limitation of population numbers
Regeneration of parts	Regeneration of parts	Regeneration of parts
Division of labor between protoplasms	Division of labor between cells	Division of labor between organisms and species
Cycles of protoplasmic behavior	Cycles of cellular behavior	Cycles of organismic and species behavior
Self-sustaining organization (dynamic equilibrium)	Self-sustaining organization (dynamic equilibrium)	Self-sustaining organization (dynamic equilibrium)
Successful integration of whole determines survival of parts and repetition of parts	Successful integration of whole determines survival of parts and repetition of parts	Successful integration of whole determines survival of parts and repetition of parts
Homology of cytological parts	Homology of tissues and organs	Homology of phylogenetically related species in different communities
Senescence and rejuvenescence of cell	Senescence and rejuvenescence of organism	Senescence and rejuvenescence of community
Phylogeny of gene pattern	Phylogeny of cellular pattern	Phylogeny of species pattern
Selection of whole cell unit determines survival of gene pattern	Selection of whole organismic units determines survival of cell pattern	Selection of whole community determines species and organism pattern
Controls internal protoplasmic environment and establishes optima	Controls intercellular environment and establishes optima	Controls environment within community and establishes optima
Selects or rejects protoplasmic building materials	Selects or rejects tissue-building materials	Selects or rejects organisms (species) that harmonize or do not harmonize with community
Retrogressive evolution of cytological structure (chloroplasts)	Retrogressive evolution of tissue structure and of organs (eyes of cave fish)	Retrogressive evolution through species elimination

ganisms. For others, it may be enough to consider the community as merely another level of biological organization. However we choose to view the community, we should guard against the danger of our view becoming a conceptual block to study of the functional parts of communities, their individual organisms and populations. Only such studies can explain the existence of communities as natural units of biological organi-

zation. It is in part this danger that has caused Bodenheimer (1958) and others to warn against the blanket acceptance of the community concept.

We will be considering in some detail in the last section of this chapter the importance of the community concept. For now, we will only state that it has placed emphasis on the need to understand individuals and populations in their natural relations with others, and that it has directed much biological thought along fruitful lines of investigation.

THE ECOSYSTEM IDEA

The environment of an individual organism can be defined in useful ways (Chapter 7); defining the environment of a population is more difficult. In our analytical efforts, it is often important to be able to distinguish between the biological system we are studying and the environment controlling that system. But it is almost impossible to distinguish between a biological community and its environment. In addition to modifying the environment of its component parts—as do all biological systems—the community greatly modifies the conditions and resources of its location. The phytoplankton in a lake may reduce light penetration, temporarily deplete nutrients, and cause diurnal fluctuations of dissolved oxygen. In these and other ways, the phytoplankton modifies its own environment and the environment of all organisms in its community. Thus, more than other units of biological organization, the community becomes part of its own environment. Sometimes it is most useful to consider the community and its environment as composing a single integrated system. For these and other reasons, the idea of ecosystem has become important in ecology. An *ecosystem* can be defined as a biological community together with the physical and chemical conditions and the resources of its location.

Important ideas only infrequently first appear in definitive form. The title of Forbes' (1887) charming paper, "The Lake as a Microcosm," clearly suggests that he had the idea of ecosystem in mind, even though he gave less emphasis to physical and chemical conditions and resources than did later writers. This idea, given different names, appeared in the writings of several terrestrial and aquatic ecologists before Tansley (1935) proposed the term ecosystem and defined it in the manner now generally accepted. Lindeman's (1942) paper, "The Trophic-Dynamic Aspect of Ecology," stimulated more general use of the ecosystem idea, for he made apparent the need to consider the entire ecosystem when nutrient cycles and energy transfer in a community are studied.

Lakes often form well-defined ecosystems. Here the contribution of nutrient and other materials from the surrounding land over any moderate period of time is usually small in comparison to the materials introduced into the cycles within lakes over geologic time. And the amounts of materials leaving lakes are usually relatively small. Mainly requiring only the ever-necessary input of solar energy, a lake ecosystem is quite self-sustaining. Even solar radiation and climate are encompassed in the ecosystem as defined.

Not all ecosystems are so well defined as lakes. Where annually imported nutrient materials represent a large part of those available to the community, ecosystem boundaries are blurred, and the ecosystem idea is less useful. A given reach of a river, with materials imported from upstream and exported downstream, presents this problem. And when wastes are discharged into a water, the wastes—perhaps even the discharging industry or community—must be encompassed in the ecosystem. But even in these instances, the ecosystem idea retains much of its value, for it provides a conceptual framework for budgeting the energy and material economy of a system; imports and exports can be taken into account (Chapter 19).

THE NATURE AND STUDY OF COMMUNITIES

Most biologists consider the community as a level of biological organization, but their views on the extent and nature of community integration range widely. Some view the community as only a loose aggregation of plant and animal populations; at the other extreme, there are biologists who view the community as a superorganism (Table 18–1). Such divergent views in any science come in part from different conceptualizations of the problems involved and in part from insufficient information. All biologists can agree that more must be known to establish and explain the existence of communities as natural units of biological organization. The way in which we view the nature of communities has much to do with how we study them and what we learn. It is well to remember that were we to view communities in a different way and study them in a different way, we might learn different things that could change our view of the nature of communities.

The characteristics of communities are determined by the populations of different species that compose them. Many communities are distinctive in appearance, even on casual examination. Woodland communities and bullrush pond communities are examples of this. Frequently, however, large plants are not present to give communities distinctive appearances. Aquatic communities often lack large plants. Careful sampling of such an aquatic community and identification and enumeration of each of its species will usually reveal that it too has a distinctive species composition separating it from other aquatic communities.

In most communities, only a few species are represented by relatively large numbers of individuals, most species being represented by lesser numbers and some species by but a few individuals. The *diversity* of a community is sometimes expressed as the ratio of its total number of species to the logarithm of its total number of individuals. Diversity in tropical communities has usually been found to be high, these communities having large numbers of species, each represented by relatively few individuals. Temperate zone communities tend to be less diverse and arctic communities still less diverse. Several hypotheses have been advanced to explain this latitudinal gradient in species diversity (Pianka, 1966), but this is an exceedingly difficult problem that will not soon be solved. More is known about the causes of changes in the diversity of a single community when there are marked changes in physical, chemical, or biological conditions over a short period of time. The introduction of domestic or industrial wastes into an aquatic environment often results in reduction of community diversity, many species being eliminated or becoming rare and a few species becoming very abundant (Chapter 20). This is usually the result of most of the species being unable to tolerate the new conditions while individuals of the few tolerant species increase in number because of increased energy and material resources. Were the new conditions to exist over thousands of years, a new community having much greater diversity would probably evolve.

Vertical stratification tends to develop in communities and give them a certain spatial structure. In a dense forest or in a deep lake, photosynthesis may be restricted mainly to the upper stratum by light availability. Animals dependent for their food or shelter on living foliage or algae will be mainly associated with this stratum, as will be animals that depend on these animals for food. Organisms dependent on decomposing plant materials for food, or others dependent on these organisms, will occur in lower strata. In deep waters, differences in temperature or dissolved oxygen concentration can lead to community stratification.

The integrity of biological communities stems mainly from the nutritional interdependencies of the species composing them (Chapter 19). These interdependencies give communities a functional *trophic structure;* energy and material transfer and utilization within communities is sometimes called *community metabolism.* At the base of this structure we find the plants that are capable of utilizing light energy through photosynthesis; these are the *primary producers* upon which nearly all other life ultimately depends. *Herbivores* and *omnivores* utilize plants for food; and they, in turn, are utilized by *carnivores.* Finally, many kinds of *decomposers,* mainly microorganisms, break down the tissues of dead plants and animals, thus returning nutrients, other metabolites, and heat to the ecosystem. Particularly since publication of Lindeman's (1942) classic paper, many ecologists have devoted themselves to study of nutrient cycles and energy transfer in communities. Their progress in elucidating these systems has been slow, for the complexities are enormous. Yet the importance of trophic processes to the persistence of communities and to man himself demands that extensive effort to increase our understanding be continued.

Those of us who live in the temperate world are all aware of at least some of the periodic changes in communities. We watch for the first appearance of leaves in the spring, enjoy somewhat nostalgically their coming down in the fall, and observe the seasonal movements of birds. And during summer evenings, it is often mosquitoes that attract our attention. Seasonal periodicities are primarily under the control of light and temperature, though the organisms themselves may have intrinsic rhythms. Seasonal changes occur not only in terrestrial environments but also in aquatic ones, as in the case of spring blooms of phytoplankton. Daily rhythms in the metabolism and activities of plants and animals are also very much under the control of light and temperature. In the marine environment, daily and semimonthly tidal cycles are of great importance to life, particularly the lives of intertidal organisms whose activities must be synchronized with tidal changes.

Many kinds of communities progressively alter the characteristics of their locations and thus pave the way for establishment of different communities that are more suited to the new conditions. These, in their turn, may pave the way for still other communities. This is called *community succession*. The particular series of communities, or *sere*, that will develop through time at a particular location is determined by primary geological conditions, climate, and possible colonizing species present in the general region. A sere is composed of *seral communities,* the different communities present at different times during community succession. *Pioneer communities* are the first to become established in new or severely disrupted environments. These are followed by other seral communities until, in some cases, *climax communities* are formed. Climax communities tend to maintain their ecosystems in equilibria, and community succession proceeds no further.

A pond formed in sand dunes will first be colonized by a pioneer community whose plants are planktonic and benthic algae. Organic material accumulating in the pond will eventually provide the substrate necessary for the growth of submerged vascular plants and permit the establishment of their associated community. Later, around the periphery of the pond, emergent vegetation will begin to grow, and it will gradually encroach toward the middle as organic materials accumulate and the pond becomes shallower. The marsh community thus formed will over a long period of time so fill the pond that a meadow community can become established. Various terrestrial seral communities can then lead ultimately to the forest climax community characteristic of the region.

Community succession is not so clear or direct in most aquatic ecosystems as it is in a pond. Communities in small, rapidly flowing streams tend to remain in a pioneer seral stage. The continuous export of organic matter and the rigorous and seasonally fluctuating conditions in such a stream do not allow its community to alter directionally its ecosystem so as to develop conditions favoring later seral communities. Margalef (1960) has pointed out that different communities occurring at one time in different downstream reaches of a stream and river system are analogous to seral communities, the upstream communities exporting materials and changing conditions for those downstream. Communities in the lower reaches of river systems may represent later seral stages and may be more stable than those upstream; but they do not correspond well to terrestrial climax communities. Marine intertidal and coral communities resemble climax communities in some of their characteristics.

We must distinguish between community succession and the idea of *community evolution.* There is much evidence of community succession and some understanding of the processes by which one seral community follows another. How a particular seral community has over evolutionary time come to have the characteristics by which it is recognized is a much more difficult question. There is good evidence that plants and animals in interspecific systems simpler than the biological community have evolved together (Emerson, 1939b; Allee et al., 1949). Undoubtedly, species are sometimes replaced by others in communities (Table 18-1). Whether or not such replacements are quite permanent and favor the stability and persistence of the entire community is the important evolutionary question. Different kinds of communities, insofar as they are real units of biological organization, have somehow come into being, a kind of evolution. Evolutionary processes leading to different levels of biological organization may be different. In the absence of careful and long-term studies of communities, biologists have been reticent about discussing in detail the evolution of entire biological communities (Emerson, 1939a; Bates, 1960).

Some of the questions ecologists have raised concerning the nature of communities are very big indeed; it is difficult to envision just how they can be answered. Probably no community has been studied in detail and in its entirety over a sufficiently long period of time to establish either its constancy of species composition or its general stability. Such a study would be a monumental undertaking. Yet discussions of nutrient cycles and energy transfer, species diversity, and community evolution should at least have as their bases comparative information from many such studies. And communities as we observe and describe them are the outcome of many processes that need to be understood to establish and explain the existence of communities as natural units of biological organization. It is not enough to describe and discuss communities as a whole. Ecologists interested in the community level of organization would be well advised to take a page from the notebooks of physiologists who study the systems of the individual organism to explain its stability. Again it is as Bartholomew (1964, p. 8) has written: "each level finds its explanations of mechanism in the levels below."

WHY WE MUST STUDY COMMUNITIES

Some men must study communities as other men must climb mountains: their presence is a challenge. For still others, useful results are the goal. But to the scientist, the study of nature is more than a challenge, and it is more than useful, as Poincaré (1913, pp. 366–367) so eloquently explains:

The scientist does not study nature because it is useful; he studies it because he delights in it, and he delights in it because it is beautiful. If nature were not beautiful, it would not be worth knowing, and if nature were not worth knowing, life would not be worth living. Of course I do not here speak of that beauty which strikes the senses, the beauty of qualities and of appearances; not that I undervalue such beauty, far from it, but it has nothing to do with science; I mean that profounder beauty which comes from the harmonious order of the parts and which a pure intelligence can grasp.

Recognizing that the nature of man requires challenges and the satisfaction of his intellectual drives, we must also recognize other very compelling reasons for him to study and come to understand biological communities. Before the problems of environmental change became a common concern, Clements and Shelford (1939, p. 24) wrote:

At the most primitive levels, human families and societies are merely integral parts of the biome [community]. It was only with the advent of

agriculture and the control of the habitat by culture and especially of urbanization that man achieved such mastery of biome and habitat as to become an outstanding dominant of a new order. Such dominance, however, is chiefly the consequence of the development of steel and machinery. In pastoral areas, man perhaps is still to be reckoned as a constituent of the biome rather than the superdominant in it. Although ecology has advanced beyond the simple distinction of the natural and the artificial, it is evident that this still suggests an important difference in the reactions and coactions exerted by man at the various culture levels, a difference, however, that runs the entire gamut from influence to dominance and superdominance. Consequently, as suggested earlier, bio-ecology [community ecology] may at present concern itself chiefly with modern man in the role of coactor or reactor in the biome, leaving for sociology and related fields the development and structure of human communities per se. However, in basic studies of social processes and origins, bio-ecology must lay the foundation on which the superstructure of the other social sciences can be reared.

This was written in 1939. Since then atomic bombs were dropped on Hiroshima and Nagasaki, and the future of man for the first time came under a real cloud of doubt. And had atomic energy never been developed, there still would have come that other cloud: the accelerating growth of human population in the face of an ultimately limiting environment. Ecology and the social sciences probably cannot provide all the understanding necessary to remove these clouds; and were they able to do so, there would still be the question of human wisdom. But man must try.

Just as in science men face big questions and lesser questions, so also in managing his ecosystem man faces not only big problems but lesser problems too. In both science and environmental management, man can work on problems that can be solved; and, hopefully, this will contribute ultimately to solution of the bigger problems. The problems of increasing the production of valuable organisms, or decreasing the production of nuisance organisms, can be solved, in spite of increasing urbanization and industrialization and accompanying changes in terrestrial and aquatic environments. In a world where more food and less disease can help to better the condition of man and increase social stability, the solution of these problems cannot be demeaned. To understand the production of biological populations so as to be able to manage them, man must study the communities of which they are a part and upon which they depend for their existence. This can be put no better than it was by Stephen Forbes (1887, republished 1925, p. 537):

If one wishes to become acquainted with the black bass, for example, he will learn but little if he limits himself to that species. He must evidently study also the species upon which it depends for its existence, and the various conditions upon which *these* depend. He must likewise study the species with which it comes in competition, and the entire system of conditions affecting their prosperity; and by the time he has studied all these sufficiently he will find that he has run through the whole complicated mechanism of the aquatic life of the locality, both animal and vegetable, of which his species forms but a single element.

Man's activities, as well as natural phenomena, bring about changes in terrestrial and aquatic environments. Each plant and animal species has a particular range of environmental conditions within which it can adapt and be successful. Of the many species of plants and animals in a community, not all will be able to adapt to a change in environmental conditions, and the diversity and species composition of the community will change. Changes in diversity and species composition can be an early warning to man of environmental changes resulting from water or air pollution. Not all changes in biological communities signal danger or even conditions undesirable for man (Chapter 20). But man must continue to observe closely the composition of the communities of which he is a part, for he cannot afford to overlook the possibility that they are changing in ways that are not in his best interest.

Communities change with changes in the physical and chemical conditions and resources of their locations; on the other hand, the communities themselves change these conditions and resources. Organic materials introduced into rivers do not persist in their complex form but are used for maintenance and growth by the organisms of the aquatic community; water, carbon dioxide, minerals, other metabolites, and heat are the ultimate end products. Communities adapted to conditions of organic enrichment develop in flowing waters receiving organic wastes; these communities tend to return such waters to their original state. Biological treatment systems developed for domestic and industrial wastes depend on communities that are biologically adapted to particular wastes (Chapter 21). Not only to understand how communities in nature tend to maintain waters in their natural state, but also to understand and improve the biology of waste treatment systems, man must study nutrient cycling, energy transfer, and the species composition of communities adapted to maintain themselves in the presence of large quantities of organic materials and their metabolites.

SELECTED REFERENCES

Allee, W. C., A. E. Emerson, O. Park, T. Park, and K. P. Schmidt. 1949. Principles of Animal Ecology. W. B. Saunders Company, Philadelphia. xii + 837pp.

Bates, M. 1960. Ecology and evolution. Pages 547–568. *In* S. Tax (Editor), The Evolution of Life (Evolution after Darwin, Vol. 1). The University of Chicago Press, Chicago. viii + 629pp.

Clements, F. E., and V. E. Shelford. 1939. Bio-ecology. John Wiley & Sons, Inc., New York. vi + 425pp.

Emerson, A. E. 1939a. Social coordination and the superorganism. American Midland Naturalist 21: 182–209.

Forbes, S. A. 1887. The lake as a microcosm. Bulletin of the Peoria Scientific Association. (Reprinted 1925. Bulletin of the Illinois State Natural History Survey 15:537–550.)

Lindeman, R. L. 1942. The trophic-dynamic aspect of ecology. Ecology 23:399–418.

Phillips, J. 1935. Succession, development, the climax, and the complex organism: an analysis of concepts. Part 3. The complex organism: conclusion. Journal of Ecology 23:488–508.

Pianka, E. R. 1966. Latitudinal gradients in species diversity: a review of concepts. American Naturalist 100:33–46.

Tansley, A. G. 1935. The use and abuse of vegetational concepts and terms. Ecology 16:284–307.

19 Energy and Material Transfer

Although certain aspects of food relations have been known for centuries, many processes within ecosystems are still very incompletely understood. The basic process in trophic dynamics is the transfer of energy from one part of the ecosystem to another.
RAYMOND LINDEMAN, 1942, p. 400

The original statement of the trophic dynamic point of view (Lindeman, 1942) and many subsequent studies have regarded the matter without much reference to the kinds of organisms involved in each level. At the time such an abstraction led to a great advance in understanding, but now we must attempt, as always, to re-introduce what was discarded in the earlier effort to get some quantitative elementary understanding of the process.
G. EVELYN HUTCHINSON, 1963, p. 686

ENERGY, MATERIALS, AND LIFE

Life is an unstable state, the maintenance of which requires energy. And the form of life requires materials. The fundamental problem of all life, then, is the obtaining and utilizing of energy and materials. So far as man knows, neither energy nor material is created or destroyed, though their forms may change. But the price of change in form is the degradation of energy resources to heat evenly dispersed, a not generally usable energy source. The first and second laws of thermodynamics appear to apply, as indeed they must, to communities and ecosystems as well as to individual organisms (Chapter 11) and populations. But neither an individual organism nor an ecosystem is a closed system within which these laws operate. Into the individual organism must pass energy and materials, and from it leave materials and heat, because the concept of the individual organism does not include its environment. The same occurs in any ecosystem, even the earth, for the ecosystem earth is dependent on solar radiation and passes heat on into the void of space. Our accounts of energy and material transfer and transformation in ecosystems as in individual organisms must include import and export. Within an ecosystem, minerals may cycle indefinitely, but organic materials with their energy loads can undergo only a few transformations before available energy is degraded to heat; without the input of energy, the system would run down.

The energy and material resources of any ecosystem are ultimately limited. And in almost inscrutable ways, evolution has placed in each ecosystem occupants striving for each bit of space and nutrient. Where there are nutrients and light, plants of many kinds come to live and reproduce and fill those niches available to plants. Upon these depend herbivores and upon them carnivores, and upon all these depend decomposing microorganisms, until there are hardly space and nutrients for more. Though forever trying new ways, evolution has had millions of years to fill the niches of most ecosystems, and we must suppose that not many are left unfilled. So through its structure, physiology, behavior, and nutrient relationships, each kind of organism has a place in its ecosystem, though not usually a bountiful place and never an independent place. It is the very interdependence of the different kinds of organisms that is most characteristic of ecosystems. In energy and material relationships, we perhaps see this interdependence most clearly.

To understand the success or failure of populations of particular species, we must, among other things, know their energy and material relationships within the ecosystem in which they live. This may be reason enough to study energy and material transfer, for man is interested in the success of many different species. When he alters ecosystems, as when he discharges his wastes into waters, he changes energy and material relationships; in this way he changes the distributions and abundances of some of the species he values most. Other times, he causes the development of populations that, to him at least, are nuisances. And in so many other ways, man is dependent on the ecosystems of which he is a part. To change them much may be inviting trouble. To avoid undesirable changes in these ecosystems, he must come to understand their economy.

Biologists know this and study ecosystems for these reasons, but they study them for other reasons too. Each species evolves only within an ecosystem, where it takes its place and plays its role, sensitively adapted for its own needs and those of the community. Energy and material resources are involved, and apart from these the evolution of species cannot be understood. As to communities, their evolution, succession, diversity, and stability depend on energy and material relationships. Biologists cannot leave alone the intellectual problems these phenomena present.

So the energy and material relationships of organisms are important, important for many reasons. And important things man usually seeks to understand. How will he do this? What are the energy and material relationships of animals? How have they been studied, how are they to be studied, what can we hope to learn, and, after all, what is the possible value of anything we might learn?

We would be presumptuous to suppose that in these few pages, or even in many more, we could adequately answer these questions, but this chapter is our perhaps superficial attempt.

In Chapter 11, Bioenergetics and Growth, we considered the energy and material relationships of the individual organism. In Chapter 17, Production, we considered some of the relationships between the food resources and the production of populations. Biological communities are composed of individual organisms and their populations. We are in error, we are vastly in error, if we suppose that energy and material transfer in ecosystems can be really understood without our knowing their transfer in individuals and populations, because "each level finds its explanations of mechanism in the levels below" and because it finds "its significance in the levels above" (Bartholomew, 1964, p. 8). The community, or the ecosystem, is the level of biological organization above the population. But most communities are composed of unknown numbers of species and individuals, and these species are many. Can we ever hope to adequately measure energy and material transfer in most of the populations of a reasonably complex ecosystem? Probably not. How much, then, can we hope to understand of energy and material transfer in ecosystems? This is hard to say. We do not yet and probably never will know all the changes that go on in the human body. Yet medical science has demonstrated that we can do pretty well in managing the living body. Knowing these things, many biologists take the very hopeful view that understanding of whole ecosystems can come with very little study or understanding of their parts. We do not hold this view. Still, fine biologists like Margalef (1968) have on the basis of relatively few and simple measurements contributed to our thinking about ecosystems. The "black box" approach may often be employed in science, but it has real dangers. Apart from giving a false sense of satisfaction that impedes more penetrating investigations, it may confront us with another danger: the *phantom problem*. Max Planck (1949, pp. 52–53) describes, in terms meaningful to any scientist, how it is:

. . . an experience annoying as can be, to find after a long time spent in toil and effort, that the problem which has been preying on one's mind is totally incapable of any solution at all—either because there exists no indisputable method to unravel it, or because considered in the cold light of reason, it turns out to be absolutely void of all meaning—in other words, it is a *phantom problem*, and all that mental work and effort was expended on a mere nothing.

THE STUDY OF TROPHIC RELATIONS

Energy and Material Resources and the Food Habits of Animals

The study of trophic, or nutritional, relations did not begin with Raymond Lindeman (1942), as he well knew. To have been so successful over the past million or more years, man has had to observe and remember and teach trophic relations; this must have been important in his early evolution. To hunt, he had early to learn where animals feed. And, later, to grow crops he had to learn the best sites. Early naturalists, even before Aristotle, must have been intrigued with the feeding relations of animals and aware of their importance. Recent advances in the study of trophic relations have been based on no less than 5000 years of natural history, animal husbandry, and agronomy.

We seem to recall that, at some early but now forgotten stage of our education, we learned that a science in its development must become qualitative before it becomes quantitative. To the extent that this is so, it is the Achilles' heel of the study of trophic relations today. Seemingly quantitative procedures and thought too often are being pressed far beyond the still rudimentary qualitative knowledge. Somehow there have been published estimates and supposedly profound discussions of the efficiencies with which animals utilize their food resources, not only without sound measurements but even without real knowledge of the identity of the resources. This is scholasticism, and it would seem silly were it not for the time and energy wasted and the precedents set. Unfortunately, when we would be quantitative and theoretical, we find we do not know what the animals eat. The real study of trophic relations must begin with studies of food habits, as tedious and even mundane as these may seem.

But perhaps we should begin with plants, the basis of the economy of any reasonably extensive and complete ecosystem. The primary energy resource of *autotrophic plants*—those able to convert through photosynthesis solar energy into the chemical energy of plant protoplasm—is not difficult to identify: it is light. Not all of the light spectrum is utilized, and some plants, though primarily autotrophic, may make limited use of energy-rich compounds from their environment. Still, in the main, we can be satisfied that we know the important energy resource of autotrophic plants. For particular species, however, we are often lacking in knowledge of their essential nutrient materials, and this knowledge is a necessary part of our understanding of energy and material transfer in ecosystems.

There are autotrophic plants, bacteria in this case, that rather than being photosynthetic are *chemosynthetic*. These bacteria obtain from the degradation of mineral compounds the energy necessary to synthesize needed organic materials. But by far most of the kinds of bacteria, and the fungi, and all animals are *heterotrophic*: they must meet their energy needs through the degradation of organic materials, the products of life. It is the heterotrophic organisms, be they plants or animals, that most complicate the efforts of ecologists to analyze ecosystems. Those heterotrophic organisms, particularly microorganisms, that degrade the remains of plants and animals are sometimes known as *decomposers*. The tremendous importance of their returning mineral nutrients to the economy of ecosystems is widely recognized; the details of their ways of life are largely unknown.

But most men, including ecologists, seem to be more interested in animals than in bacteria, except perhaps for pathogens. And animals provide us with fascinating, important, and sometimes difficult problems to study. Animals, though all heterotrophic, are still very different in their nutritional requirements. Obviously some are *herbivores*, some *omnivores*, and some *carnivores*. Some feed on *detrital materials* and play the role of decomposers. Some are quite specialized in their feeding relations, others quite generalized. Most are sufficiently adaptable to take advantage of the opportunities of the day or the season to meet their needs. These trophic interdependencies are difficult to study, but they contribute to the stability of ecosystems, as Darnell (1968) has thought-fully discussed.

In these times of general awareness of the food problems of men over the earth, the ecological idea of *food chain* has become a part of the thinking of more people. It is a simple idea: some species of plant is eaten by some herbivorous species which in turn is consumed by a carnivore, this constituting a food chain. Ecologists use this expression frequently, because it has some conceptual value, and the idea focuses attention on certain problems of energy and material transfer. For example, animals that are predominantly herbivorous, even some populations of man, usually find more bountiful food resources than do predominantly carnivorous animals. Their food chain is shorter by one link, and less energy is lost as heat in the processes of transfer and transformation of energy and materials than would be if the food chain were longer. But as useful as this expression may be in the day to day communication of men, its conceptual value is quite limited. The pathways of energy and material transfer to few species can be represented by a simple chain. Rather, they must usually be represented by a tangled webbing, albeit a webbing of chain. The idea of *food web*, then, though not so nice for simple ecological thought, more nearly represents reality and the problems with which we must deal, and it does suggest interactions important to the stability of populations and communities.

Even a greatly simplified representation shows the feeding relations of herring (*Clupea harengus*) to be rather complex (Fig. 19-1), fortunately for its food organisms, for the herring, and for the animals including man that depend very much on the herring as a food resource. For, with declines in the abundance of one of its food organisms, the herring can compensate by shifting its feeding more to others, thus relieving this food organism and stabilizing the food resource and the production of herring. The food relations of young herring differ from those of the adult, and those of the adult shift from season to season and from place to place. Thus, the full analysis and representation of the feeding relations of herring would tax severely the biologist and artist alike, and our comprehension too. And were we to include in our representation all the organisms that depend upon herring as an energy and material resource, we would have pictured a major part of the food web of an important marine community. It would not be a simple picture. Further, our reader should remember, this is only the qualitative picture: we should not be too excited about the possibility of knowing the rates of transfer of energy and materials through its tangled webbing.

Occasionally, ecologists have attempted to work out in some detail major parts of food webs of natural communities; these audacious souls have earned our respect. Cummins, Coffman, and Roff (1966) obtained a

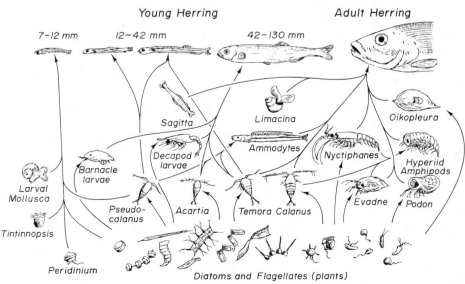

FIGURE 19-1   Food web leading to the production of herring. After Hardy (1959).

qualitative picture of a part of the food web of a riffle community in a small Pennsylvania stream. Their representation of this for only one time during the year (Fig. 19-2) begins to suggest the complexity with which we are dealing. Work like this is as valuable as it is rare. The complexity revealed should not discourage our attempts to find order. Rather, it should encourage us to seek the real order of nature, not order that is merely of conceptual convenience to man.

Another ambitious and energetic biologist, Darnell (1958, 1961), has pictured for us the results of his very nice study of the food habits of only the fish and larger molluscs and crustaceans of an estuarine community in Louisiana (Fig. 19-3). We may safely suppose that his conclusions are relevant to many communities other than the one he studied (Darnell, 1961, pp. 566-567):

1. The Lake Pontchartrain community is a broadly open system exchanging nutrients, producers, and consumers with adjacent fresh-water and salt-water areas as well as with neighboring marshes and swamps.

2. Consumers within the lake apparently depend in great measure upon primary production which takes place outside the lake. Hence, the estuarine community may be trophically unbalanced.

3. The most conspicuous single food item in the diets of the consumers of this community is organic detritus with its attendant bacteria.

4. Individual species do not appear to conform to specific trophic levels on the basis of the following considerations:

   a. Omnivory on the part of most, if not all, of the major consumer species,
   b. Nutritional opportunism among the consumers,
   c. Ontogenetic change in the food habits of the consumers,
   d. Importance of organic detritus in the nutrition of the consumer species,
   e. Complex nature of the origin of detritus.

5. An alternative model [Fig. 19-3] of the trophic relations of the community is presented which represents the observed food [of particular age groups] of each species in the form of a

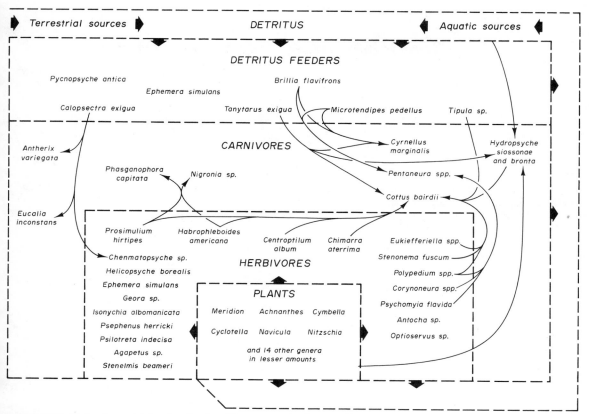

FIGURE 19-2 A partial food web of a riffle community in a small stream, Linesville Creek, Pennsylvania. After Cummins, Coffman, and Roff (1966).

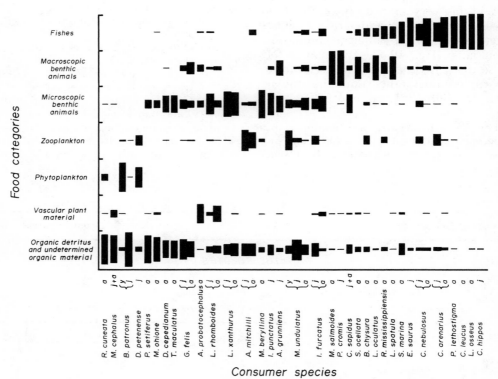

FIGURE 19-3   Utilization by young (y), juvenile (j), and adult (a) animals of different food resources in Lake Pontchartrain, a Louisiana estuary. The total height of the bars in each column represents 100 percent of the amount of food consumed by an age group of a particular species of animal. This total height is also the distance between the marks separating the food categories. After Darnell (1961).

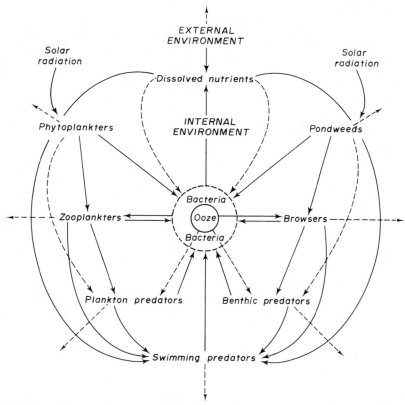

FIGURE 19-4   Major pathways of energy and material transfer in a senescent lake, Cedar Bog Lake, Minnesota. After Lindeman (1941a).

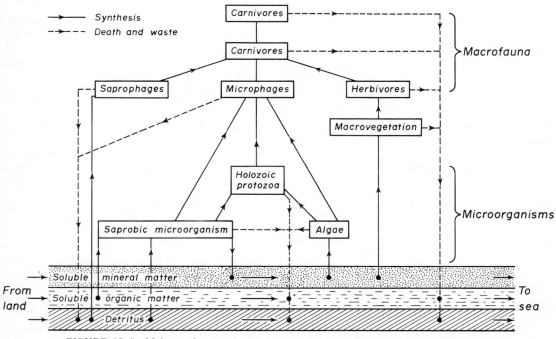

FIGURE 19-5   Major pathways of material transfer in a stream. After Hawkes (1962).

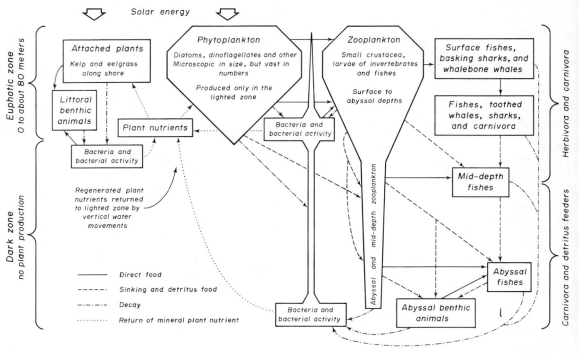

FIGURE 19-6   Major pathways of energy and material transfer in the sea. After Sverdrup, Johnson, and Fleming (1942).

spectrum and in which each type of food can be represented quantitatively and comparatively.

6. The really abundant consumer species of the Lake Pontchartrain community comprise 2 groups: those which feed heavily upon organic detritus and those which exhibit a broad range of food tolerance.

Over the years, ecologists have had fun by trying to diagram the major pathways of energy and material transfer in ecosystems of different kinds. If we will remember that such diagrams have not been intended as more than general and provisional representations, they can be helpful in putting our thoughts in order. They are an attempt to show in broad outline the pathways of entire ecosystems. But we often find outlines inadequate when we proceed to fill in with details, details of community food webs in this instance. Lindeman (1941a) prepared such a diagram for the senescent lake he studied (Fig. 19-4). Lakes and ponds usually represent better defined ecosystems than do streams or marine bodies of water: we do not have so much import and export of materials with which to deal. Hawkes' (1962) hypothetical representation of material transfer in a stream community (Fig. 19-5) well illustrates how much less applicable is the ecosystem concept to a given reach of

stream: here import and export dominate the picture of material transfer. Most marine ecosystems are neither so well defined as pond ones nor so poorly defined as those of streams. The fine general representation of a marine ecosystem (Fig. 19-6) prepared by Sverdrup, Johnson, and Fleming (1942), in addition to providing an outline of marine interrelationships, does for us two things: it indicates how far in understanding man has come to be able to prepare such an outline, and how much farther he has yet to go to fill in the details.

Lindeman's Trophic Dynamic Viewpoint

While this, his sixth completed paper, was in the press, Raymond Lindeman died after a long illness on 29 June, 1942, in his twenty-seventh year. . . . Knowing that one man's life at best is too short for intensive studies of more than a few localities, and before the manuscript was completed, that he might never return again to the field, he wanted others to think in the same terms as he had found so stimulating, and for them to collect material that would confirm, extend, or correct his theoretical conclusions. The present contribution does far more than this, as here for the first time, we have the interrelated dynamics of a biocoenosis presented in a form that is amenable to a productive abstract analysis. . . . it is to the present paper that we must turn as the major contribution of one of the most creative and generous minds yet to devote itself to ecological science.

Thus G. Evelyn Hutchinson sensitively wrote of the work of his young colleague (Lindeman, 1942, pp. 417–418) and touched on the aspirations and thinking that made his friend a fine scientist. Lindeman opened vistas for many ecologists, as he hoped and as Hutchinson anticipated.

At some phase of his intellectual development and study of Cedar Bog Lake in Minnesota, Raymond Lindeman (1941a, 1941b) attained that perspective distinguishing fine scientists everywhere. Perhaps the most important part of this perspective is the realization that from some of the myriad possible observations on any isolated system must come understanding of fundamental processes operating not only in that system but also in similar systems and even in systems not so similar. Trophic relations in a small lake are representative of the informational detail that tests the intellect of ecologists. Lindeman (1942) saw that common to all life was the need for energy and materials; more important, he saw that this common need suggested an approach to the analysis of ecosystems, the common denominator being energy. He knew that entire ecosystems, not just the biological communities or lesser parts of them, must be included in the most meaningful analyses; and he believed the rates of energy and material transfer, not existing amounts of material, to be most important in the functioning and history of ecosystems. But for the ecologist, even with a common denominator, there remain the many species with which to deal. Lindeman thought that functionally different species could be grouped into different trophic levels: autotrophic plants, herbivores, carnivores, and decomposers. Perhaps these ideas together would permit the most meaningful analysis of ecosystems.

Time has shown the value of Lindeman's holistic view, but his simplifying assumption of trophic levels has misled many ecologists. With his critical mind and the grace of a few more years, Raymond Lindeman, more than others, would have seen this; and from his passing ecology has suffered. Recognizing his magnificent contribution, we must focus attention on its weaknesses, as he would have done.

Lindeman (1942), drawing heavily on an unpublished manuscript of Hutchinson, suggested that organisms be assigned to particular *trophic levels* according to their energy and material resources: autotrophic plants constituting the first trophic level; herbivores, the second; carnivores consuming primarily herbivores, the third; and animals consuming primarily carnivores, the fourth. He considered the "productivity" of an entire trophic level to be its rate of energy intake; for trophic level n, this would be symbolized by λ_n. Fundamental to Lindeman's rationale for the examination of ecosystems was the idea of "progressive efficiencies":

$$E = \frac{\lambda_n}{\lambda_{n-1}} \times 100$$

where λ_n and λ_{n-1} are rates of energy intake by a given trophic level and the next lower level, autotrophic plants being the lowest level. Comparison of these efficiencies between different trophic levels might further understanding of the operation of ecosystems.

While terminology and the nature of measurements are not crucial to our argument, they can be a source of confusion, and perhaps we should digress briefly for clarification. Lindeman's "productivity" is not equivalent to production as we have defined it in Chapter 17. His concept of productivity applied to herbivores or carnivores is equivalent to food consumption, whereas we have defined *production* in terms of total tissue elaboration. We prefer to follow Ivlev (1945) in reserving the term *productivity* for the basic capacity of an ecosystem to produce a particular species, whatever the level of its production. Lindeman (1942) estimated energy intake of a trophic level by summing "turnover" (mean biomass × number of life cycles), respiration, predation, and decomposition; this he termed "corrected productivity." Allen (1951, p. 168) notes that "turnover" as determined by Lindeman would be an underestimate of total tissue elaboration. And Slobodkin (1962, p. 84), neglecting materials not assimilated, notes that energy intake would be the sum only of total tissue elaboration and respiration. Lindeman's measurement rationale, then, is in serious doubt, but this is trivial, unless we were more concerned with the values he obtained than with his argument. Let us return to the argument.

While commending the holistic approach of Lindeman and Hutchinson, Ivlev (1945, p. 105) early advanced perhaps the most crucial criticism of the trophic level concept when he wrote that it "does not, in all its endearing simplicity, correspond to the state of affairs in nature." Darnell (1961, p. 566) saw this:

Evidently (as suggested by Peterson and Jensen, 1911) the consumers exhibit a broad disregard for narrow trophic lines. In fact, the ability of a given species to utilize alternate foods (often from different 'trophic levels') appears to be one of the main buffering factors which tend to stabilize population levels in complex natural communities. To ignore the inherent diversity and the system of alternatives would seem to be overlooking the very essence of trophic integration involved in community balance. Actually, Lindeman (1942) cautioned that the higher trophic levels are not as distinctly recognizable as the lower levels, and in the present estuarine community, at least, they do not seem to be recognizable at all.

Some have suggested that, when the energy and material resources of particular species come clearly from more than one trophic level, the species involved be assigned proportionally to two or more trophic levels (Cummins, Coffman, and Roff, 1966; Darnell, 1968). When the necessary extent of such treatment of data is recognized, we wonder what is left of the trophic level concept, and how much such procedures can really contribute to our understanding of ecosystems.

Even were it possible to assign most of the organisms of a complex ecosystem to trophic levels, monumental efforts would be necessary to determine the rates of energy or material intake and assimilation, production, and respiration of only the energetically most important species. For no reasonably extensive and complex ecosystem has this even been seriously attempted; yet there are those who persist in writing about the energy intake and production of animal trophic levels and their attendant efficiencies. After an ambitious and imaginative study of the trophic structure and dynamics of Silver Springs, Florida, Odum (1957a, p. 100) had to conclude that "in spite of the efforts expended and the constancy and reproducibility of ecosystem a really satisfactory series of measurements of herbivore and carnivore production has not been obtained." It would

indeed be surprising if in this or in any other study of a large, complex ecosystem satisfactory measurements could ever be made on the energy or material intake and production of most of the species. In Chapter 17, we endeavored to explain how really difficult it is to obtain reliable estimates of the production of even one animal species, and how very tenuous must be any estimates of its food consumption. The trophic level concept changes none of this: its meaningful use requires real information, species by species.

Were we to know the food habits of all the animals in an ecosystem, were trophic levels really definable, and were we to know the energy and material transfer species by species, the grouping and summing of the values for the many organisms by trophic level might be the simplest way to treat this remarkable information. But would it be the most meaningful, or would it mask the truly important interactions that stabilize ecosystems? Surely with such information we would find much more powerful procedures for the functional analysis of ecosystems.

In retrospect, then, we believe that Lindeman's simplifying assumption of trophic levels has led some ecologists to be careless in studies of trophic relations. Without real information on the energy or material intake and production of even the important species in particular ecosystems, without even adequate information on food habits, these ecologists have advanced fanciful values for energy and material transfer between trophic levels, have computed associated efficiencies, and have entered into supposedly profound discussions. This is not science, and it is bad for biology.

But in the balance, Lindeman's emphasis on the ecosystem, his stressing of its dynamic aspects, and his suggestion of energy as a common denominator have encouraged ecologists to take more holistic views and to seek general relationships. He did not counsel that the search for principles be hurried or that it proceed without needed information. Beyond his important contribution to the ecological argument, Raymond Lindeman gave us a point of view; few are able to give so much.

Ivlev's Product of Interest Rationale

The death of Viktor Sergeevich Ivlev on December 3, 1964, ended the career of one of the world's most active experimental scientists in the fishery field. . . .

Whatever his surroundings, throughout this life Ivlev occupied himself principally with experimental studies on the physiology, nutrition, and behaviour of aquatic animals, both fishes and invertebrates. . . . his versatility became very widely recognized following the appearance in 1945 of the review article 'The biological productivity of waters.' . . . still his major contribution to a synthesis of ecological and physiological information concerning aquatic production. . . .

This William Ricker wrote in the foreword to his improved translation of what is probably the most useful single contribution to trophic ecology (Ivlev, 1945), a translation published "as a memorial to a talented scientist, who was also a most engaging personality." We never had the anticipated opportunity to enjoy Ivlev personally, but, through his many writings, he was perhaps our most important teacher, guiding our faltering steps through a field he knew so well, a field strewn with pitfalls. On at least two occasions, after having solved difficult trophic problems, we found we had not read our teacher well, for he had known the solutions. Raymond Lindeman (1942) gave us a star to check our course; Viktor Ivlev (1945) more nearly gave us a path to follow. What is this path, and how long and how firm may it be?

As we have noted, Ivlev (1945) did not accept the assumption of trophic levels with "all its endearing simplicity." He agreed with Hutchinson and Lindeman on the necessity of an ecosystem view and the value of dealing in dynamic and energetic terms. But he did not believe in the reality of trophic levels or in the possibility of obtaining the information necessary to evaluate quantitatively, or even qualitatively, all the trophic pathways of entire ecosystems. He believed that, if we are to progress far in understanding, we must focus our attention and energy on a *product of interest*. The product may be a single species, or it could be a few species depending on very nearly the same trophic pathways. The interest may be commercial, or it could be purely scientific. In any event, a particular product may be carnivorous, omnivorous, herbivorous, or even autotrophic. Its production would be explained by working out its principal trophic pathways through its food organisms and their food organisms back to the primary nutrients and solar energy. An aquatic

ecosystem has many possible products and its *productivity* for a particular product is its basic capacity to produce that product, regardless of the prevailing rate of production, which is much dependent on the biomass of the product (Chapter 17).

Ivlev knew that to explain the production of one animal, we must know not only its food habits but also the food habits of the animals along at least its most important trophic pathways. Organisms contributing little to production of the product of interest he believed could be safely neglected, thus reducing to some extent the difficulty of our task and narrowing our attention to the most important pathways. But, as Ivlev well knew, the remaining task of determining the rates of production and utilization along even these fewer pathways is not for the hurried or the weak to undertake, and it may never be accomplished for an animal with a complex food web.

The production rate of any product of interest would be explained by measuring at every step along each of its major trophic pathways back to solar energy not only the production rate of each species involved but also its rate of utilization by the organisms on the next step. For most carnivores, then, we would need to know not only the production rates of tens of species of organisms but also the efficiency with which each of these species cropped the production of every one of its important food species, and the efficiency with which it utilized for growth the food consumed. Determination of the efficiency with which an animal utilizes for growth the food it consumes is not the greatest problem: growth efficiencies can be approximated with bioenergetic studies such as Ivlev proposed and we outlined in Chapter 11. But the measurement of the production and utilization of more than a few animal species becomes a monumental undertaking (Chapter 17). And, if we were to follow Ivlev, the efficiency with which each consumer cropped each of its food species would be expressed as the ratio of its consumption of that species to the production of that species.

However, such a solution of the problem, by reason of its very great magnitude and difficulty, would bring to naught every attempt at a quantitative analysis of the production process. Consequently, the discovery of general laws and patterns that characterize the phenomenon must be one of our basic tasks in solving the productivity problem. My own efforts along these lines should not be overrated. . . . (Ivlev, 1945, p. 114)

With this humility, Ivlev, more than others, clearly saw the problem; and, even in this paper, he anticipated his later work on density-dependent relationships (Ivlev, 1947, 1961a, 1961b) that may help to explain production phenomena when we cannot measure all the rates of production and utilization along trophic pathways.

The path that Ivlev suggested for trophic analysis, then, is firm, but it is much too long, as he knew. The kind and quality of observations needed by biologists who would deal in trophic levels and entire ecosystems are no different than those needed by Ivlev; but, trophic levels or no trophic levels, these biologists must have information on hundreds rather than tens of species. Ivlev saw this and narrowed his approach to products of interest. But Ivlev knew that even this was too much, and he, unlike too many others, was willing to call a spade a spade and unwilling to deal in fantasy.

First, Ivlev (1945) contributed to the clarity of our thinking: he outlined the necessary complexity of any real analysis of rates of energy and material transfer through important parts of ecosystems. Second, he suggested useful ways of getting the information necessary at each step along any trophic pathway. He cannot be held accountable for the many steps and pathways nature has found useful. The production of a few products of interest may one day be explained by Ivlev's scheme, and maybe even the trophic relations of a simple ecosystem or two. But even when such explanations are beyond our means, Ivlev's approach will assist all of us in apportioning our investigative efforts on the basis of choice of products of interest and their most important energy resources.

The Question of Efficiency

Lindeman's (1942) paper has led many ecologists to a preoccupation with the question of efficiency. It is almost as though there were no other questions of trophic relations, or perhaps as though there were no other approach to the study of trophic relations. Ideas sometimes do this to men.

Each transformation of energy or material can be accomplished only by the expenditure of some energy, which then appears as heat of little value. Beginning with solar radiation and the energy of plant material formed through photosynthesis, each step through every trophic pathway must lose as heat some of the original energy supply until, finally, there is no more. Trophic pathways, then, can be only as long as the losses at their steps will permit. Lindeman saw this, and examination of the question of the possible lengths of food chains by estimation of progressive efficiencies—ratios of energy intakes of successive trophic levels—became a principal theme of his paper.

There then followed a rash of efficiencies. Everyone had to propose an efficiency of a new kind, the values of which would provide some marvelous insight, until not even a specialist could keep track of them. Odum (1959, p. 54) gives a fair sample of these; Koslovsky's (1968) list is shocking. Now, what is this all about?

An efficiency is the ratio of available energy or material coming out of a system to energy or material going into the system; it is neither more nor less than this. Such inputs and outputs exist at all levels of biological organization from the molecular level to the community or ecosystem. Theoretically, and only theoretically, efficiencies could be determined at any of these levels. We can talk and write about the efficiencies of energy and material transfer at all of these levels as much and as fruitlessly as we may wish. But we can determine efficiencies only under those few circumstances permitting measurement, *independent measurement*, of energy and material input and output.

Growth efficiencies of individual animals can be determined in the laboratory when we feed them known amounts of food and measure their growth (Chapter 11). Measurement of fecal wastes can make it possible to determine *assimilation efficiencies* under laboratory conditions. It is quite possible to hold populations of animals feeding on populations of food organisms in the laboratory and obtain fairly reliable and independent measurements of their production, their food intake, and the availability of their food. Slobodkin (1959, 1960, 1961, 1962) and his students (Richman, 1958; Armstrong, 1960) have succeeded in holding populations of *Daphnia* and other animals in the laboratory and obtaining estimates of their ecological and other efficiencies at different levels of food availability and different rates of removal of the animals produced. *Ecological efficiency* is defined as the ratio of the consumption of one population by others to its own consumption of food. Studies of individual animals and populations of animals in the laboratory are valuable, and their results may even be suggestive of possible efficiencies of energy and material transfer in nature, but no more than suggestive.

Meaningful estimates of the efficiencies of energy and material transfer between populations of animals in nature require the same *independence of measurement* of input and output at each step along any trophic pathway as do laboratory estimates of efficiency. It is perhaps possible, with sufficient interest and effort, to estimate the production of almost any population; but this is usually a difficult undertaking, even for fish (Chapter 17), and satisfactory estimates for invertebrates are few and far between. There are ways of approximating the food consumption of animals in nature (Davis and Warren, 1968), but the most reliable of these are based on the growth rates of the animals (Chapter 17), and hence there is no independence of the measurements needed for efficiency ratios. Such estimates of consumption may be of some value as order of magnitude approximations of food utilization in nature. But efficiency ratios based on them are no more than laboratory growth efficiencies, and we should not delude ourselves as to their meaning regarding the transfer of energy and materials in nature.

So much for populations in nature; what about the efficiencies of energy and material transfer between possible trophic levels? Trophic level *intake efficiencies, assimilation efficiencies, production efficiencies, cropping efficiencies,* and *ecological efficiencies* in almost endless combinations and varieties have been published and discussed. Except at the plant level, these have little or no scientific basis. To the extent they exist, trophic levels are only groups of populations of different species of organisms of the same trophic kind. And what we cannot do with one population, we cannot do with groups of populations. The measurements must still be made, species by species.

But beyond all this, if the efficiencies of energy and material transfer between populations and even between trophic levels in nature could be determined, would knowledge of these efficiencies be as valuable as is commonly supposed? Undoubtedly there would be some value in this knowledge, but insufficient to make it our goal. The efficiency with which an individual animal utilizes for growth the food it consumes is not a constant but is a function of its rate of food consumption (Chapter 11). And at different levels of population biomass, the proportions of the food a population consumes that are necessary for maintenance are greatly different, and, hence, the proportions available for growth are different (Chapter 17). Thus, the efficiencies of food utilization for individual growth and population production are in nature a function of population biomass. They can range from zero, for short periods of time, to rather high values. We know not what any determination of efficiency we might make really represents, other than that it is only one point on an efficiency curve.

Some ecologists recognize this and suggest that efficiencies are only meaningful when populations are in a "steady state," a sort of dynamic equilibrium (Slobodkin, 1960). Populations in nature may tend toward equilibrium densities, but they generally fluctuate greatly, sometimes violently (Chapter 16), and the physicist's steady state provides a very poor model. Presumably, changing efficiencies are reflecting the operation of mechanisms that tend to maintain populations in equilibria. High and low efficiencies may be as interesting as ones for populations at their mean densities. But with only a few determinations, painfully obtained, how are we to know which are which? Probably trophic levels are more stable than the populations they include. This should make their efficiencies, whatever they may be, more constant. But adjustment of populations provides much of the stability trophic levels, communities, or ecosystems may have; understanding of this stability can hardly come from the search for efficiency constants.

The methods by which we seek to obtain estimates of the efficiency of energy and material transfer between animal populations in nature are certainly not "indisputable"; indeed, they are hardly scientific methods at all. And the efficiency values we seek, although perhaps not "devoid of all meaning," have considerably less meaning than is commonly supposed. Is the question of efficiency dangerously close to being Max Planck's (1949) "phantom problem"?

Analysis by Means of Density-Dependent Functions

Nature insists and persists in being as she is, not always as we would have her be for the convenience either of our lives or of our science. When our scientific methods fail to represent or penetrate her complexity, we must be sufficiently adaptable to turn again to her for suggestions as to how best to proceed. The history of science shows that nature has not often failed us. We need not now look far: where nutrients are plentiful and light and water sufficient, plants usually are abundant; where plants are abundant, the animals that feed on them are generally abundant; and where herbivores are abundant we find that carnivores are abundant. Are not these abundances generally related? Do we need to know all the rates of energy and material transfer to obtain any understanding whatsoever? Perhaps we should examine this problem a little more closely.

To learn even in qualitative terms the energy and material pathways through a complex ecosystem, though perhaps possible, would require an investigation of formidable proportions. But to learn the rates of food consumption, assimilation, growth, and production of all the animals in a complex ecosystem—to learn the rates of energy and material transfer in its various trophic pathways—would be possible only in a dream. To measure the production of even one animal species is a difficult undertaking. We can make only the crudest approximation of its rate of food consumption from this information. If the species is an important carnivore feeding on only one or at most a few prey species, we may be sufficiently interested to undertake measurement of the production of each of these prey species. But if there are more prey species, as is so often the case, our interest is likely to wane and our scientific resources will usually be inadequate. This is the problem we face in seeking to understand the production of even one product of interest through acquisition of knowledge of the production of each of its prey species. The problem of so explaining the production of all the or-

ganisms of a complex ecosystem is this multiplied a thousand times. The history of the trophic-dynamic aspect of ecology, with all the interest and support this approach to ecology has received since publication of Lindeman's (1942) inspirational paper, gives no assurance that the presently popular pathways of scientific endeavor will lead to the understanding we seek.

If they will not, is there an alternative approach to the explanation of the production of a single product of interest? And could such an alternative approach be extended to encompass major parts of ecosystems? An alternative approach should not replace present approaches where they are profitable; it should complement and strengthen them. We need not search far for such an approach: it lies in the mainstream of biological thought. It derives from the idea that density-dependent processes are important throughout ecological systems. Indeed, Lindeman's idea that it is the rate of transfer of energy and materials rather than biomass that is important, taken to ridiculous extremes, has led to a departure in trophic studies from the historically and operationally important concept of *density dependence*.

Since publication of Lindeman's (1942) paper, those who have studied trophic relations have largely ignored if not belittled the importance of density or biomass, except as it has been needed to estimate production. It is indeed curious that they should have done so, particularly during this period of time, for the publication of papers by Volterra (1927, 1931), Lotka (1925, 1932), Nicholson (1933), and Smith (1935) led to great emphasis in ecological thought on the importance of density-dependent phenomena in population regulation (Chaper 16). Earlier, Howard and Fiske (1911) viewed population regulation much as did Nicholson. But we must go back in time much further: density-dependent phenomena were implicit in Malthus' (1798) population essay, which led Verhulst (1838) to formulate the logistic equation, and which was important to Darwin's (1859) conceptualization of natural selection. Implicit if not explicit in nearly all population studies is the idea that the nutritional bases of populations are among the most important density-dependent factors influencing them. Have most of the recent studies of trophic relations somehow left the mainstream of ecological thought and thereby been weakened?

Ivlev (1945, 1961a) was aware of the importance of density in trophic processes. In his mathematical formulation for the rate of food consumption of animals, he included variables for the density and the distribution of food. Later, Ivlev (1961b) developed energy budgets—including food consumption, respiration, and growth—for a zooplankton-eating fish as functions of the density of its food organisms. The significance of this contribution has gone largely unappreciated. And so has another of his contributions: in 1947 he published mathematical formulations of the growth and production of fish as functions of their biomass. Drawing on our own experience (Davis and Warren, 1965; Warren and Davis, 1967; Brocksen, Davis, and Warren, 1968, 1970), we have explained the relation between the production and biomass of animals on the basis of relations between their growth rates and the densities of their food organisms and between their biomasses and those of their food (Chapter 17). And, going further, we have suggested that density-dependent relations exist not only between any two steps along a trophic pathway but run from a product of interest back to primary energy and material resources (Brocksen, Davis, and Warren, 1970).

The main argument of this section has already been presented in Chapter 17, Production, in the section entitled Density Dependence in Trophic Processes. There we explained the curvilinear relationship between the production of a population and its biomass; we also explained the relationships between consumer growth rate and food density and between food density and consumer biomass that lead to production-biomass curves. We went so far as to present mathematical formulations of production based on these density-dependent relationships. Transfer of energy and materials in an ecosystem depends on their utilization by populations and the resulting production of these populations. Analysis of energy and material transfer in ecosystems, at least for animals, must be based on knowledge of the food utilization and production of the individual species constituting the ecosystems. There is no shorter way. The methods needed for the analysis of energy and material transfer in ecosystems are the methods needed for the analysis of population production. For animals, these were explained in Chapter 17; they need not be re-explained here. But there we took this analysis back only to the immediate food resource of a population. Can such an analytical approach be useful in explaining the steps back along trophic pathways to primary energy and material resources? We think so. Here we will only suggest how fruitful may be the analysis of density-dependent trophic relations throughout ecosystems.

The relationship between the biomass of a consumer and the density or biomass of its food will have a negative slope under some conditions and will have a positive slope under other conditions. When the basic capacity of an ecosystem to produce the food of a consumer remains constant through time, we can often expect the relationship between consumer biomass and food density to have a negative slope. Such an inverse relation will also be found when we compare consumer biomasses and food densities in different ecosystems having the same basic capacity to produce the food. But if we graph the relationship between consumer biomass and food density in an ecosystem having a changing basic capacity to produce the food, then we will usually find the curve to have a positive slope. And if we graph the relationship between consumer biomass and food density for ecosystems that differ in their capacity to produce the food, we will often find the curve defining the relationship also to have a positive slope. For sockeye salmon (*Oncorhynchus nerka*) and their zooplankton food organisms, we discussed in Chapter 17 the inverse relations that exist within lake systems and the direct relations that exist between lake systems having different basic productivity for zooplankton (Fig. 17-13). And there we explained why we believe a lake having very high basic productivity may have a relationship between consumer biomass and food density that is less negative in slope than a similar relationship for a lake having less capacity to produce food. We will present evidence that these two kinds of relationships—the inverse ones when capacity to produce food is constant, and the direct ones when capacity to produce food changes—may be very general, not only between a consumer and its immediate food resource but also between that food resource and its energy and material resources, and so on back to primary energy and material resources. Perhaps, then, we can generally represent the relationship between biomass levels at any two successive steps along a trophic pathway as we have done in Figure 19-7. Such relationships can be expected to be inverse when basic productivity is nearly constant, whether a single ecosystem or different ecosystems of similar productivity are involved. The relationships can be expected to be direct when the basic productivity of an ecosystem is changing through time or when ecosystems of different productivities are being compared.

At each step along any trophic pathway in which we wish to relate the growth and production of a consumer to the availability of its energy and material resource, there are at least three distinct advantages in dealing with the density or biomass of the resource rather than with its production rate. The first of these advantages is that, if the food resource comprises many species, it is usually relatively simple to determine their individual and combined biomasses, whereas it is extremely difficult if not impossible to deter-

mine their individual and combined production rates. The second advantage is that biomass is more likely to indicate the immediate availability of a particular species or of a total food resource to a consumer than is the production of a species or a food resource. This is because the production of a food species or resource passes into many trophic pathways and is not all available to a product of interest, whereas their biomass, existing as it does at a point in time, is an expression of food opportunity at that time, even if the opportunity is enjoyed by competing species. The third advantage is a conceptual one, but no less important. If the food species consume each other, their production rates are not additive, for some of the same energy and material is involved in the production of the different species. Their biomasses, however, are clearly additive, for they represent only the total amount of material of the same or different kinds existing at any moment and place. Some ecologists err in attempting to deal with the total production of, say, the benthos of an aquatic system when organisms within the benthos are consuming each other. Such a total represents the repeated summing of the same energy

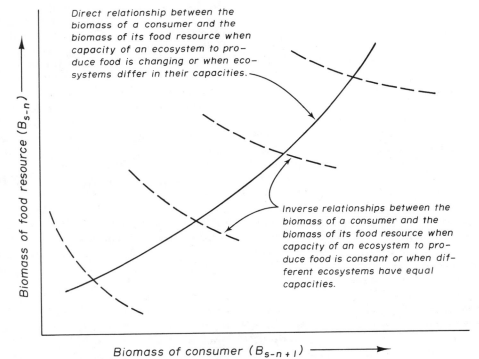

FIGURE 19-7   Theoretical relationships between the biomass of a consumer and the density or biomass of its food resource. After Brocksen, Davis, and Warren (1970).

and materials and is conceptually and methodologically useless. There is no similar difficulty in conceptualizing or using total biomass measurements of the benthos, even if total biomass is composed of organisms of different trophic kinds.

In Chapter 17, we presented evidence that the relationship between the biomass of juvenile sockeye salmon and the biomass of their zooplankton food organisms within a lake system was inverse, but that when lake systems of different basic productivity were compared the relationship was direct (Fig. 17-13). The main argument in Chapter 17 pertained, however, to carnivores consuming herbivores in systems having constant or similar basic capacities to produce herbivores. Thus, we have to this point in our book dealt mainly with the inverse relationships that exist between the biomasses of carnivores and herbivores within systems having constant basic productivity or between systems having similar basic productivity. We have dealt little with the direct relationships that must exist within systems having changing basic productivity or between systems having different productivity; neither have we dealt with the successive steps leading back along a trophic pathway from a product of interest to primary energy and material resources. We must do this now, and perhaps we should begin with laboratory experiments designed to elucidate these very relationships.

With six laboratory streams, we attempted to develop six ecosystems having different basic capacities to produce benthic algae and thus to support herbivorous insects and carnivorous fish (Brocksen and Warren, MS.). To do this, we exposed each of three pairs of streams to different light intensities and provided two current velocities at each light intensity. Light intensity most affected algal production, but its effect was modified somewhat by current velocity. We obtained, then, laboratory stream ecosystems of differing productive capacity, which permitted us to examine density-dependent relationships between steps from a product of interest, cutthroat trout (*Salmo clarki*), back along its trophic pathway to light energy. Because we worked only with six streams, and these generally had different productive capacities, we could not well define the inverse relationships we have shown to exist between biomasses of a consumer and its

food organisms when ecosystems having similar basic productivity are compared. This would have required several streams at each level of productive capacity, more streams than we had available for these experiments.

Figure 19-8 is our attempt to represent graphically—for one of several confirming experiments—not only the demonstrated direct relationships between biomass levels at successive trophic steps in ecosystems of differing productive capacity, but also the inverse ones that would usually have occurred had we enough streams of the same productive capacity available to simultaneously demonstrate these. Higher levels of light energy permitted the development of higher plant biomasses (Fig. 19-8A), and those streams having the highest light levels and plant biomasses also maintained the highest insect biomasses (Fig. 19-8B). But, if herbivorous insects were to consume appreciable amounts of plant material, increasing insect biomass at any given light level would lead to a decline in plant biomass (Fig. 19-8B). Then, plant biomass would be not only a function of light intensity but also a function of insect biomass. Similarly goes the argument for the next trophic step. The high insect biomasses accompanying high plant biomasses, with the associated insect production, permitted the maintenance of high trout biomasses (Fig. 19-8C). But since trout may consume sufficient of their food resource to decrease its density, any increase of trout biomass in streams having a given capacity to produce insects might be expected to decrease insect biomass (Fig. 19-8C). Insect biomass, then, is not only a function of plant biomass but also a function of trout biomass, if trout consume sufficient insects. Neither we nor other animals cropped the trout in the streams, but had such cropping occurred, these same phenomena could be expected at the next trophic step (Fig. 19-8D). Generally, then, we are only saying that the biomass of organisms on any trophic step not only is a function of the biomass of their food resource but also is a function of the biomass of their consumers, if the consumers remove a large proportion of these organisms. By dealing with both the food resource and the consumer of the organisms on any trophic step, we no longer need to limit our argument to just the inverse or just the direct kind of relationship, as we shall see.

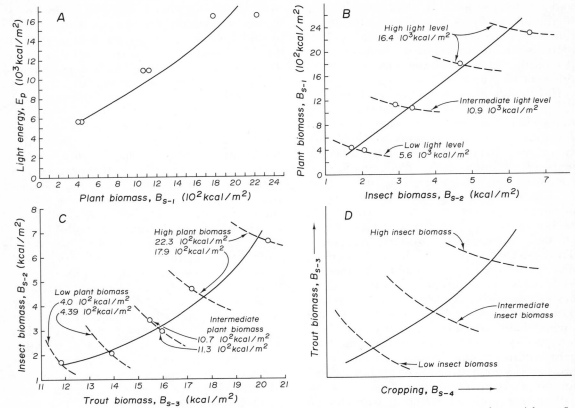

FIGURE 19-8 Density relationships between successive steps along the pathway of energy and material transfer leading from light energy to the production of cutthroat trout in laboratory stream communities during a spring 1968 experiment. Only the solid lines defining relations between systems having different productive capacities are reasonably well defined. The broken lines only suggest the probable inverse relations that would exist at each level of productive capacity were more streams tested. After Brocksen and Warren (MS.).

In Chapter 17, we developed equations for the production of a consumer as functions of its biomass and the biomass of its food resource. But these equations were restricted to ecosystems having constant or similar productive capacity for the food resource, because they included no functions of the energy and materials utilized by that food resource. Now, by incorporating functions reflecting the basic capacity of ecosystems to produce the food resource of a product of interest, we are in a position to write equations for the production of a product of interest that are not restricted to the condition of constant or similar basic productivity of the ecosystems (Brocksen and Warren, MS.). These equations, if their functions can be defined as we believe will sometimes be possible, will be useful not only when basic productive capacity is constant or similar but also when it is changing or different. Thus, these equations are much more gen-

eral and much more useful than the earlier equations, particularly in an environment ever changing, often because of man. And because they incorporate successive steps along trophic pathways, these equations permit more of an ecosystem view, encompassing important parts of ecosystems from products of interest back to primary energy and material resources. But before presenting and explaining such production equations, we must first see how the densities or biomasses of energy and material resources at successive steps along trophic pathways can be related.

Figure 19-8 will help our reader to conceptualize just how the biomass of organisms on any trophic step may be related to the density of their energy and material resources and to the biomass of their consumers, for it is just such relationships that this figure illustrates. If consumption by herbivores has little or no effect on plant biomass, then, as suggested in Figure 19-8A, we can represent plant biomass by the equation:

$$B_{s-1} = f_1 (E_p, M_p) \qquad (1a)$$

where

B_{s-1} = biomass of plants involved in a particular trophic pathway,

E_p = light energy, and

M_p = mineral nutrients of plants.

If, on the other hand, herbivores have an appreciable effect on plant biomass, as indicated in Figure 19-8B, we must in addition include in our expression for plant biomass some function of herbivore biomass. Then, our equation for plant biomass becomes:

$$B_{s-1} = f_{11}(E_p, M_p; B_{s-2}) \qquad (1b)$$

where

B_{s-2} = biomass of consumers on next trophic step, mainly herbivores but could include some carnivores.

Now if carnivores on step 3 of any trophic pathway have little effect on the biomass of organisms on step 2, we can represent the biomass on step 2 as follows:

$$B_{s-2} = f_2(B_{s-1}) \qquad (2a)$$

But if carnivores do have an important effect, as is generally the case, and as we have represented by the broken lines in Figure 19–8C, then we must include the biomass of these carnivores in our equation:

$$B_{s-2} = f_{12}(B_{s-1}, B_{s-3}) \qquad (2b)$$

where

B_{s-3} = biomass of all carnivores influencing the biomass of organisms on the previous step of a trophic pathway leading to a product of interest.

When cropping by man or other animals has little effect on the biomass of carnivores on step 3, then this biomass may be simply represented:

$$B_{s-3} = f_3(B_{s-2}) \qquad (3a)$$

Cropping may, however, have an important effect (Fig. 19-8D); and the animals involved, a fourth trophic step, must somehow be represented in our equation for the biomass of carnivores on the third step:

$$B_{s-3} = f_{13}(B_{s-2}, B_{s-4}) \qquad (3b)$$

where

B_{s-4} = biomass or some other appropriate representation of the intensity of cropping by animals on a fourth trophic step.

Having explained how the biomasses of organisms on successive steps along a trophic pathway leading to a product of interest might be related, we are now in a position to formulate production equations for any product of interest as functions of these biomasses or primary energy and material resources. The product of interest need not be and, indeed, usually will not be all of the species on a particular trophic step. Competitors of our product of interest cannot be ignored in production equations for this product. To formulate such production equations, we will always need to know one additional relationship: the relationship between the growth rate of the product of interest and the density of its food resource.

Now the production of any product of interest on any step along a trophic pathway, say product "a" on step 3, can be defined:

$$P_{s-3a} = G_{s-3a} \times B_{s-3a} \qquad (4)$$

where

P_{s-3a} = production of product a on trophic step 3,

G_{s-3a} = growth rate of product a, and

B_{s-3a} = biomass of only product a on trophic step 3.

In Chapter 11 and again in Chapter 17, we showed that the growth rate of carnivorous fish is often a function of the density of their food resource. Trout were the only important carnivores on trophic step 3 in the experiment we are here using as an example (Fig. 19-8), and we will take them to be our product of interest. In Figure 19-9, we show the relationship between the growth rate of these trout and the biomass of their food organisms, the insects on trophic step 2. The growth rate of our trout can then be simply represented as a function of the biomass of their prey:

$$G_{s-3a} = f_4(B_{s-2}) \qquad (5)$$

By substituting $f_4(B_{s-2})$ from equation 5 for G_{s-3a} in equation 4, we obtain:

$$P_{s-3a} = f_4(B_{s-2}) \times B_{s-3a} \qquad (6)$$

an equation for the production of an animal as a function of its own biomass and the biomass of its food resource.

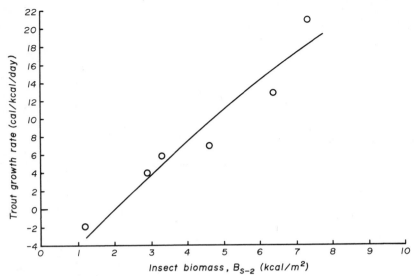

FIGURE 19-9   Relationship between the growth rate of cutthroat trout and the density or biomass of the insects upon which they fed in laboratory streams having different productive capacities during a spring 1968 experiment. Data of Robert W. Brocksen, Oregon State University.

But as we explained earlier, the biomass of organisms on one trophic step may often be related to the biomasses of organisms on the previous and on the succeeding steps. This, then, makes possible an array of production equations, equations involving functions of energy and material resource densities at all steps back along a trophic pathway from a product of interest to primary energy and material resources. As simply as possible, we will develop the alternative equations that are appropriate for including each of the successive trophic steps, because these equations represent an important advance in our ability to relate the production of an organism to the basic capacity of its ecosystem to maintain that production.

The biomass of organisms on step 2 of a trophic pathway may or may not be much influenced by the consumers—trout in our example—on step 3. If not, we can, by substituting $f_2(B_{s-1})$ from equation 2a for B_{s-2} in equation 6, obtain and equation for the production of our product of interest in terms of plant biomass (B_{s-1}) and its own biomass (B_{s-3a}):

$$P_{s-3a} = f_4\ [f_2(B_{s-1})] \times B_{s-3a} \qquad (7a)$$

But if all the carnivores, including our product of interest, on step 3 (B_{s-3}) do have an appreciable influence on the biomass of their food resource on step 2, then we must sub-

stitute $f_{12}(B_{s-1}, B_{s-3})$ from equation 2b for B_{s-2} in equation 6 to obtain our production equation:

$$P_{s-3a} = f_4\ [f_{12}(B_{s-1}, B_{s-3})] \times B_{s-3a} \qquad (7b)$$

This equation includes terms not only for plant biomass and the biomass of our product of interest but also for the biomass of all carnivore species on step 3.

In seeking production equations in terms of primary energy and material resources for a product of interest that is three steps along a trophic pathway, we find the number of possible sets of conditions, and hence the number of alternative equations, to increase. Perhaps our reader will bear with us a bit longer. If the organisms on step 2 (B_{s-2}) of a trophic pathway have little or no influence on plant biomass (B_{s-1}), and if the carnivores on step 3 (B_{s-3}) have little influence on the biomass on step 2, we can substitute $f_1\ (E_p, M_p)$ from equation 1a for B_{s-1} in equation 7a and obtain a production equation for a product of interest in terms of primary energy and material resources and its own biomass:

$$P_{s-3a} = f_4\ \{f_2[f_1(E_p, M_p)]\} \times B_{s-3a} \qquad (8a)$$

But if the carnivores do influence the biomass of their food resource and the herbivores do not influence plant biomass, then we must substitute $f_1(E_p, M_p)$ from equation

1a for B_{s-1} in equation 7b to obtain our production equation:

$$P_{s-3a} = f_4 \{f_{12}[f_1(E_p, M_p), B_{s-3}]\} \times B_{s-3a} \quad (8b)$$

Finally, we need two production equations for the alternative possibilities when the herbivores on step 2 (B_{s-2}) do influence plant biomass (B_{s-1}). Here, if the carnivores on step 3 (B_{s-3}) do not influence appreciably the biomass of organisms on step 2, we can substitute $f_{11}(E_p, M_p; B_{s-2})$ from equation 1b for B_{s-1} in equation 7a to obtain:

$$P_{s-3a} = f_4 \{f_2[f_{11}(E_p, M_p; B_{s-2})]\} \times B_{s-3a} \quad (9a)$$

But if the carnivores do influence the biomass of their food resource, then we must substitute $f_{11}(E_p, M_p; B_{s-2})$ from equation 1b for B_{s-1} in equation 7b to obtain our production equation involving primary energy and material resources:

$$P_{s-3a} = f_4 \{f_{12}[f_{11}(E_p, M_p; B_{s-2}), B_{s-3}]\} \times B_{s-3a} \quad (9b)$$

Now, in other chapters of this book we have asked our reader to follow the rationale of various mathematical representations of biological phenomena, and we hope that he has tried and benefited. But even our most patient reader may feel that being expected to follow in detail the equations we have just presented is a bit too much, and we would be inclined to agree with him; we are not sure

we even expect this. Why then, he may fairly ask, have we troubled him with these equations? For this reason: the production of an animal is a function not only of its own biomass but also of the productive capacity of its ecosystem; productive capacity is a changing thing, often because of man's activities; there can be no understanding of production without this being taken into account. These equations, as complex as they may at first seem, are remarkably simple representations that may be useful in relating the production of animals that interest man to the productive capacities of their ecosystems. Some of our readers may one day pursue the lines of investigation these equations suggest; for our other readers, we hope that casual reading of these equations will suggest how the production of animals is ultimately very much dependent on primary energy and material resources, and that there may be ways of representing this dependence.

The production of an animal may be almost any value, depending on the biomass of the animal and the productive capacity of its ecosystem. In Chapter 17, we showed this for juvenile sockeye salmon (Fig. 17-14). In the laboratory stream experiment we have been using as an example, we had insufficient streams to develop trout production-biomass curves for each level of productivity. But in Figure 19-10, we have attempted to

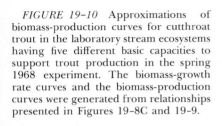

FIGURE 19-10 Approximations of biomass-production curves for cutthroat trout in the laboratory stream ecosystems having five different basic capacities to support trout production in the spring 1968 experiment. The biomass-growth rate curves and the biomass-production curves were generated from relationships presented in Figures 19-8C and 19-9.

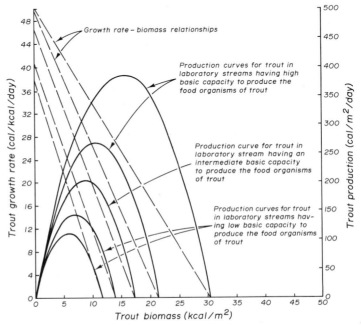

show that higher levels of production can be maintained at higher biomasses as the basic productivity of ecosystems increases: we must understand that we are dealing with systems of production-biomass curves. In nature, we will usually be unable to define these curves; nevertheless, we will fail to appreciate the meaning and the basis of the production values we do obtain unless we place them in this conceptual framework. When man changes an ecosystem, he moves its species onto higher or lower production-biomass curves.

As these equations become more generally known, some years of examination of existing data and collection of more appropriate data will be necessary before it can be ascertained just how generally in nature such density-dependent relations can be usefully defined. Until that time, they provide a much needed complementary view of trophic relations. It has not been advantageous that the conceptualization of trophic processes only in dynamic terms has dominated their study to the near exclusion of examination of their density-dependent phenomena: both views are needed. T. C. Chamberlin (1897, p. 841), scholar, founder and editor of the *Journal of Geology*, and perhaps the leading geologist of his time, saw more clearly than others how "a premature explanation passes first into a tentative theory, then into an adopted theory, and lastly into a ruling theory," which no science can afford. The assumption of trophic levels and the dynamic viewpoint of trophic relations together have come dangerously near to being a ruling theory. Chamberlin's "Studies for Students: The Method of Multiple Working Hypotheses" could have been written for those of us who are students of trophic relations. We can profitably consider his thoughts carefully. His antidote for the evolution of ruling theories was the advancement and testing of alternative hypotheses. And the method of multiple hypotheses, in addition to counteracting this poison, has another great advantage: few complex phenomena can ultimately be attributed to a single cause. When we limit our viewpoint, we limit our power of understanding: "while a single working hypothesis may lead investigation very effectively along a given line, it may in that very fact invite the neglect of other lines equally important" (Chamberlin, 1897, p. 846).

CHANGES IN ENERGY AND MATERIAL RESOURCES AND THE RESPONSES OF POPULATIONS AND COMMUNITIES

The success of populations and the nature of communities are utterly dependent on their resources of energy and materials. Change these and some populations will decline or disappear while others will appear or increase, and their communities will so change. It is understanding of this that we seek in studies of trophic relations. Such understanding is not only of biological interest. It is also of great importance in the management of natural resources and in preventing the activities of man from damaging the productivity of ecosystems for the plants and animals he values.

Now almost any change in the environment will influence directly the success of some species: a change in temperature, salinity, dissolved oxygen, or mineral nutrients. But because each species has interrelationships with many other species — particularly trophic relationships — an environmental change directly influencing one or a few species will indirectly influence a great many more, perhaps most if not all of the species of the community. Thus, nearly any environmental change is likely to change the energy and material resources of species throughout a community. Changes in temperature or salinity, or the introduction of toxic substances due to activities of man, can do this; studies of trophic relations are often pertinent to the analysis of their effects. But here we should perhaps primarily consider changes in trophic relations resulting from the introduction into ecosystems of additional usable energy and materials. This is often one of the outcomes of man's activities and often must be considered pollution.

An increase of solar radiation reaching an ecosystem — such as results when we remove by logging the forest canopy from a stream — changes the energy input. When plant mineral nutrients from many sources enter natural waters, materials are added to aquatic ecosystems. Suffice it to say there are ways of measuring the light energy and minerals that become the primary energy and material resources of aquatic communities. These permit photosynthesis and the production of plants, which become the major energy and material resource that develops within ecosystems. Organic materials developing within an ecosystem are sometimes termed *autochthonous*, a terrible word, yet one sometimes used by ecologists. But animals often depend on organic materials formed outside rather than within their ecosystems. Leaves falling into a stream to be utilized for food by animals, or to be degraded by bacteria and fungi, are in this category; so are organic wastes introduced into natural waters by man. Such materials of origin foreign to an ecosystem may be considered *allochthonous*, another terrible word with which our reader could be confronted. Whatever names ecologists may choose for these two categories of energy and material resources, the distinction between the categories is important to understanding of the trophic processes of ecosystems in which man has an interest.

To this point in our text, we have written little about *primary production*, the production of plant material through photosynthesis, and we do not intend to write much about this now. Our most difficult problems of analysis of trophic relations are not with photosynthetic plants but with animals, bacteria, and other heterotrophic organisms. Photosynthetic plants do constitute a rather discrete trophic level, with their common energy resource in the sun. Thus, summing the production rates of different species of these plants presents no major conceptual or analytical difficulty. And, at least in aquatic ecosystems, there are practical procedures for approximating the total production rate of this trophic level. In fact, it is only this total rate and not the individual production rates of the species that can usually be determined in studying aquatic ecosystems. Photosynthesis is the common trophic process of photosynthetic plants, and the total prevailing rate of photosynthesis in many aquatic communities can be estimated.

There are several methods by which the photosynthetic rate of algae and submerged vascular plants can be estimated; these have been summarized by Odum (1959) and by Strickland (1960), among others. Perhaps the oldest, simplest, and still most reliable method for estimating the rate of photosynthesis of plankton algae is the *light bottle–dark bottle* technique (Maucha, 1924; Putter, 1924; Gaarder and Gran, 1927). Its principle has been applied to estimation of the production of benthic algae (McIntire et al., 1964).

Using carbon dioxide, other mineral nutrients, and radiant energy, colored plants through the process of photosynthesis manufacture a photosynthate of organic materials. In this process, oxygen is freed in direct proportion to the photosynthate formed. But simultaneously, oxygen is utilized by the plants and other organisms in their respiration. Thus, oxygen accumulation is less by the amount of this utilization. The rationale, then, of the *light bottle–dark bottle* method is that the sum of oxygen evolution measured in the light and oxygen utilization measured in the dark is equivalent to the total formation of photosynthate in the light. Any application of this rationale, of course, involves obvious and not so obvious assumptions, which for our purposes here we need not consider. We need only note that reasonable approximations of the total rates of photosynthesis in aquatic environments can usually be obtained.

There is, however, one major difficulty awaiting satisfactory solution. The total rate at which photosynthate is formed by plants includes not only organic materials accumulating as growth and available to other organisms but also organic materials being utilized by the plants for their own metabolic needs. Measurements of plant production based on oxygen evolution and utilization include both these categories of organic materials and are considered to be estimates of *gross production*. *Net production* of plants is considered to be their gross production less their own utilization of organic materials through respiration. It is this net production of plants that passes to other trophic kinds or leads to an increase in plant biomass. But, unfortunately, it is usually impossible to distinguish between the respiration of plants and the respiration of animals and other heterotrophic organisms in a community or in the experimental apparatus being employed. Thus, there is no really indisputable and generally useful method of estimating the net production of plants in nature. Still, knowledge even of their rate of gross production is of considerable interest.

According to the aquatic ecosystem, its kind, its level of radiant energy, available plant nutrients, and existing plant biomass, as well as temperature and other factors, rates of gross primary production range widely. Odum (1959) has brought together estimates of gross primary production that investigators have obtained for different ecosystems under various conditions (Table 19-1). From such a table, we can only surmise the energy, material, and other conditions that led to the production rates found in the different ecosystems, but that the productivities of these systems for plants are vastly different is obvious. Such differences in basic productivity profoundly influence not only the plants but also all the animals, bacteria, and other heterotrophic organisms that depend ultimately on plants making available the energy of solar radiation. To understand changes in energy and material resources and the responses of populations and communities, it is well to begin with beginnings.

Gross primary production represents the total formation within an ecosystem of the organic materials upon which plants and animals depend for their maintenance and growth. Although it is not usually possible to measure the respiration of particular species of plants and animals in natural aquatic systems, it is often possible to measure the total respiration of a plant and animal community. This is of considerable interest, for it gives us an indication of the rate at which organic materials are being metabolically degraded within the community. For the whole community, then, the relationship between gross primary production and community respiration can give us some indication as to whether the rate at which organic material is being formed within the ecosystem is greater or is less than the rate at which it is being degraded. If the ratio of gross primary production to community respiration is greater than one, we must suppose production of organic material is more than enough to balance degradation for metabolic purposes; if it is less than one, production

TABLE 19-1   Gross primary production in different aquatic ecosystems as measured by various investigators. After Odum (1959).

Ecosystem	Production (gm/m²/day)	Reference
Averages for long periods (6 to 12 months):		
Infertile open ocean, Sargasso Sea	0.5	Riley (1957)
Shallow, inshore waters, Long Island Sound; year average	3.2	Riley (1956)
Texas estuaries, Laguna Madre	4.4	H. T. Odum (unpublished)
Clear, deep (oligotrophic) lake, Wisconsin	0.7	Juday (1940)
Shallow (eutrophic) lake, Japan	2.1	Hogetsu and Ichimura (1954)
Bog lake, Cedar Bog Lake, Minnesota (phytoplankton only)	0.3	Lindeman (1942)
Lake Erie, winter	1.0	Verduin (1956)
Lake Erie, summer	9.0	Verduin (1956)
Silver Springs, Florida	17.5	Odum (1957a)
Coral reefs, average three Pacific reefs	18.2	Kohn and Helfrich (1957)
Values obtained for short favorable periods:		
Fertilized pond, N.C., May	5.0	Hoskin (unpublished)
Pond with treated sewage wastes, Denmark, July	9.0	Steeman Nielsen (1955)
Pond with untreated wastes, South Dakota, summer	27	Bartsch and Allum (1957)
Silver Springs, Florida, May	35	Odum (1957a)
Turbid river, suspended clay, N.C., summer	1.7	Hoskin (unpublished)
Polluted stream, Indiana, summer	57	Odum (1957b)
Estuaries, Texas	23	H. T. Odum (unpublished)
Marine turtle-grass flats, Florida, August	34	Odum (1957b)
Mass algae culture, extra CO_2 added	43	Tamiya (1957)

within the ecosystem must be insufficient to support existing metabolism. Perhaps we can think of a community as a whole as being *autotrophic,* or capable of self-support, if plant production exceeds all respiration, and as being *heterotrophic* if the reverse is true. This use of the concepts of autotrophy and heterotrophy is different than when they are applied to particular species, because communities in either category have both kinds of organisms. Still, if we do not let this special use of terminology confuse us, it does call attention to certain important characteristics of communities as a whole.

Figure 19–11 suggests possible coordinates of gross primary production and community respiration for communities of different kinds. Communities positioned along the 45 degree diagonal have production values equal to utilization and could be capable of self support. Communities positioned above the diagonal—those having a ratio of gross primary production to community respiration greater than one—are producing organic matter in excess of its metabolic degradation. Such communities are either accumulating or exporting organic material. A eutrophic lake is usually accumulating organic material; a river into which organic materials have been introduced but which

has reached its mesosaprobic zone is probably exporting downstream most of the excess production of organic materials. In communities positioned below the 45 degree diagonal, gross primary production is less than the degradation of organic materials. These communities either are mainly utilizing previously stored organic materials, as in the case of humic lakes, or are depending on organic matter being introduced, as in the polysaprobic zone of a river receiving large amounts of organic wastes (Fig. 19–11). To properly reflect the overall energy and material budget of a community, these coordinates must be based on information obtained over annual cycles of production and respiration. Even then, though we obtain a useful picture of the community as a whole, we must not delude ourselves into thinking that we have obtained any great understanding of mechanisms. For this we must go back to the populations involved.

Now we have written as though communities having gross primary production in excess of community respiration were capable of "self-support." This is not, of course, strictly correct. Biological life on the earth itself requires continual input of solar radiation, and so it is for each of the earth's ecosystems and communities. We sometimes view ecosystems as being self-sufficient by including in their definition the input of energy from the sun. Perhaps this is useful in the ecosystem concept, but we do not so define biological communities: strictly, they are never self-supporting. We must always consider the variable input of solar radiation. In the real world, we cannot often evaluate the changes that occur in the gross primary production of communities when there comes much change in the passage of solar radiation to plant life. Intrigued with the possibilities, biologists find ways of examining them, and often in doing this they retreat to the laboratory. We and our associates have done this many times, particularly in examining the trophic relations of stream communities. As we discussed in earlier sections of this chapter, laboratory stream communities have made this possible. The results of our endeavors have sometimes been instructive. McIntire and Phinney (1965) investigated the relationships between radiant energy and gross primary production when the necessary carbon dioxide was available in different concentra-

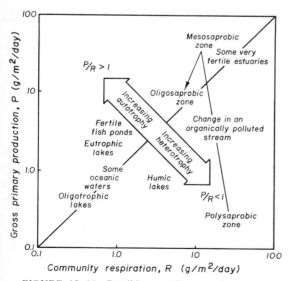

FIGURE 19–11 Possible coordinates of gross primary production and community respiration for different ecosystems, some tending toward autotrophy, others tending toward heterotrophy. After Odum (1956).

tions. With increasing light intensity, the photosynthetic rate of benthic algae increased rapidly until available carbon dioxide began to be limiting, and, then, much further increase in photosynthesis could not occur (Fig. 19-12). At higher concentrations of carbon dioxide, greater amounts of light energy could be utilized effectively. Relationships of this kind between available energy and materials and the production of organisms are to be generally expected; it is, however, helpful to know the form of the relationships and the limiting levels of energy and materials, even in simple laboratory systems. For in nature too, these are the beginnings, the beginnings of changes in biological communities with all their plants and animals.

It would be nice if, from their original entry into ecosystems, we could follow the transfer of energy and materials in quantitative terms through even the most important trophic pathways. Numerous investigators have attempted to do this by following the ambitious format used by Odum (1957a) in his studies of Silver Springs, Florida. But only the results of the very detailed study conducted by Teal (1957) on a tiny spring remain at all convincing. As we have endeav-

ored to make clear in earlier sections of this chapter, we are unable to obtain the measurements of energy and material intake, assimilation, respiration, growth, and production necessary to make real the information presented in diagrams of energy and material transfer for even moderately complex and extensive ecosystems. Very intensive studies of small and relatively simple ecosystems may sometimes yield sufficient information to permit tentative quantitative diagramming of some of the important pathways of energy and material transfer, but even then the representations are open to serious question. Such are the results of Teal's (1957) work on Root Spring, and such are our own results on laboratory stream communities. Still, perhaps these studies give us some insight into changes that may take place in the communities of more complex and important ecosystems when their energy and material resources are changed.

Even for the simple communities we have studied in laboratory streams (Davis and Warren, 1965), the approximations necessary render quite disputable such a fairly complete diagram of energy and material transfer as Figure 19-13. Light energy, gross plant production, community respiration,

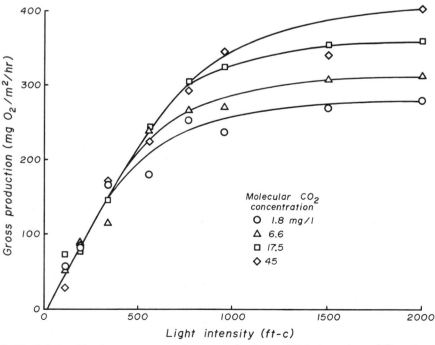

FIGURE 19-12   Relationships between gross primary production and light intensity at different concentrations of carbon dioxide in laboratory stream communities. After McIntire and Phinney (1965).

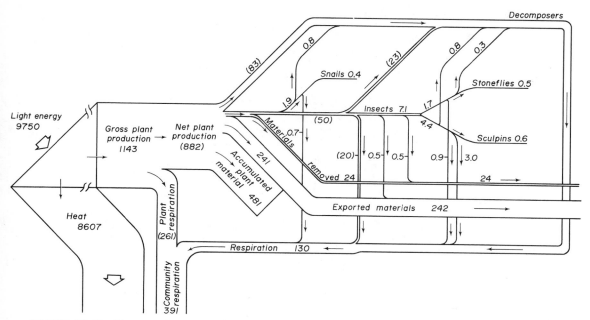

FIGURE 19–13 Diagram of energy and material transfer in a laboratory stream community. All values are given in kilocalories per square meter for a 74-day period. Two-thirds of community respiration was arbitrarily attributed to plant respiration and one-third to all other respiration. After Davis and Warren (1965).

and accumulation and export of plant material were satisfactorily measured, as were sculpin, stonefly, and snail production rates. By accounting the major fates of herbivorous food organisms, mainly insect larvae, we were able to obtain reasonable estimates of their production. But plant respiration was arbitrarily taken to be two-thirds of community respiration, animal respiration was estimated from studies of metabolic rates in respirometers, and decomposer respiration was assumed to be the remainder. Estimates of food consumption rates were based on growth rates of the animals, and then everything was made to balance. This is hardly a rigorous analysis of community energetics, but it is about as good as any that has ever been made. We should not get too excited about the possibility of this approach to the analysis of natural ecosystems, but when and where reasonably satisfactory measurements can be made, perhaps they will have something to tell us.

In another laboratory stream experiment (Brocksen, Davis, and Warren, 1968), we exposed one stream to twice the light energy supplied to another; in consequence, the gross plant production in the stream receiving the most light was about double that in the other stream (Fig. 19-14). In the stream

receiving the lesser light energy, much less plant material was thus made available as food for the snails and for the insects that were food for trout. And in this stream, both snails and trout lost weight; their production values were negative. Considerable production of snails and trout occurred in the other stream, even though their biomasses were more than twice those in the stream having the low levels of light and plant production (Fig. 19-14). The ultimate capacity of all ecosystems to produce animals that may be of interest to man depends very much on their basic energy and material resources.

When man introduces quite low concentrations of organic materials into natural waters, he is sometimes surprised by the very profound changes that come about in the aquatic communities. Were he to appreciate fully the magnitude of the energy contribution that even low concentrations of energy-rich materials represent, his surprise would probably be less. For many years, we enriched some sections of our experimental stream, Berry Creek, by the continuous introduction of sucrose so as to maintain a concentration of about 4 mg/l (Warren et al., 1964). Superficially, at least, this was not a great change in water quality, not one causing appreciable decreases in dissolved oxy-

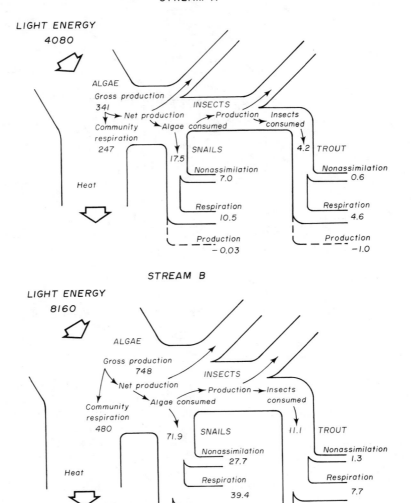

FIGURE 19–14 Increases in gross primary production, snail production, and cutthroat trout production occasioned by exposing one laboratory stream community to twice as much light energy as another. All values are given in kilocalories per square meter for an 80-day period. After Brocksen, Davis, and Warren (1968).

gen, for example. Still, over a year, this represented an energy input of 130,000 kcal/m² to the riffle of an enriched section, more than twice the energy of solar radiation reaching that riffle, which was shaded by a forest canopy (Fig. 19-15). This energy and material resource, sucrose, along with necessary mineral nutrients present in the stream, led to the production of a lush growth of the bacterium *Sphaerotilus natans* blanketing the streambed. We did not succeed in measuring the actual production rates of *Sphaerotilus* and of most of the organisms utilizing it for food. But measure-

ments we were able to make are indicative of the great change the sucrose brought to the community of this small stream. For example, the production of snails, which previously had to depend on algae and leaves for food, was increased from 40 to 330 kcal/m² when *Sphaerotilus* became available as an energy and material resource (Fig. 19-15). The biomass of herbivorous insects was more than doubled, and that of the carnivorous insects was quadrupled. And the production of cutthroat trout, depending on these insects for food, was increased many times (Fig. 19-15).

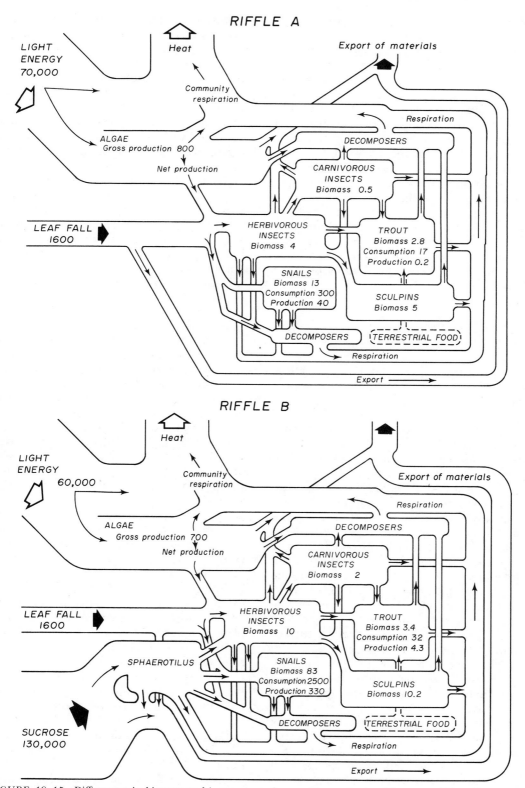

FIGURE 19-15 Differences in biomass and in energy and material transfer in riffle communities of Berry Creek Experimental Stream, Oregon, occasioned by continuous enrichment with sucrose. All values are in kilocalories per square meter per year.

Not always can we hope that organic enrichment of streams will have such an apparently beneficial effect as increasing trout production. Even low levels of enrichment can over long periods of time so change the character of natural waters as to render them unsuitable for species that man values. And, of course, higher levels can in much shorter periods of time lead to changes disastrous to species and communities man treasures. But we have made our point sufficiently clear: man must watch closely any changes he brings to the energy and material resources of aquatic communities; inevitably these will lead to changes in the abundance of some species and so to changes in the communities; sometimes man may view these changes as beneficial, but more often he has not.

TROPHIC STUDIES AND WATER POLLUTION BIOLOGY

Perhaps now we can attempt a realistic evaluation of the contribution trophic studies can make to water pollution biology. In this we will endeavor to be clear. Then, if we are wrong, we will at least be clearly wrong. Should there by now remain any doubt in our reader's mind, the first thing we must make clear is our position on fundamental knowledge of biology, however fundamental it may be. The success of man's management of the living resources of the earth will in large part be dependent on his understanding of their biology, but their management cannot await anything like full understanding. Thus, we must often pursue research that we know will fall far short of this understanding. In some reasonable balance, then, research directed toward immediate goals and research directed toward fundamental understanding must proceed; both will contribute to water pollution biology. So when we take a skeptical view of the possible contributions some approaches to the study of trophic relations can make to water pollution control, our reason is not that we believe these approaches to be too fundamental for mundane questions; the best of biology is needed to answer many of these questions. Our skeptical view stems from our conviction that some of these approaches are simply poor science. This man cannot afford, whatever may be his objectives.

Most water pollution problems involve, in one way or another, changes in the pathways by which aquatic populations obtain energy and materials, changes that lead to differences in the success of these populations and in the composition of aquatic communities. Some of these problems originate from changes in the energy and material resources of aquatic communities, as when organic substances or mineral nutrients are introduced into streams or lakes. But changes in trophic pathways may follow other kinds of changes in aquatic ecosystems: changes in temperature, or the introduction of toxic substances. The importance of energy and material resources to the success of populations and the composition of communities demands that man, in his own best interest, gain understanding of how his activities may influence trophic pathways and the consequences of such influences. The need for this understanding is so imperative that available scientific resources should support the soundest relevant science. History warns us of the danger in supposing that some science is more sound or is more relevant to particular problems than is other science. Yet, keeping our minds as open as is humanly possible, we must recognize that not all approaches are scientifically tenable or are likely to be relevant to water pollution control. We cannot pursue all kinds of trophic studies at the same rate; let us favor the most sound and, when water pollution control is our objective, the most relevant.

The ecosystem idea encourages a broad view of the community-environment complex. Such a holistic view is important in all biological studies, including those directed toward the solution of water pollution problems. It should be retained, but without loss of regard for the necessity of understanding wholes through knowledge of the functioning of their parts. We have warned our reader, at other points in this book, that a holism not giving due regard to parts is intellectually sterile. As we study the trophic problems resulting from water pollution, let us keep these considerations in mind; they help to retain the soundness of science. But often it is frustration by complexity, complexity such as we see in trophic relations, that drives scientists to holistic views that give little regard to the functioning of parts. Our problem is to understand the functioning of parts as they lead to the behavior of wholes. Sometimes our interest may be in a part; but, then, we cannot neglect the whole, for the part is only a part and would be different away from the whole: "This arises from the intimate character of the relation of whole to part" (Whitehead, 1925, p. 207). Our work must begin with careful definitions of problems as we see them and proceed through careful evaluations of the material and intellectual resources we can command in seeking their solution. Then we must choose those measurements we can reasonably expect to make, measurements of the part of interest, of the most important interacting parts, and of the whole, measurements that will aid us most in our search for understanding.

When man introduces energy or materials into aquatic ecosystems, there nearly always will come some changes in trophic pathways. If the changes be such that there follows a loss of value of the water involved, our problem is likely to be judged one of water pollution. For the ecologist, whether or not a change be judged pollutional, the problem is much the same: the need to evaluate the effect of the change on the aquatic populations and their community. And here it is not only possible but it is both practical and necessary to begin at the beginning of the trophic scheme. What were and what have become the main energy and material resources of the aquatic community involved? Sometimes we are too late in our interest to determine without doubt the conditions that existed before man changed them; but often, after evaluation of existing conditions, we can make reasonable judgments about prior ones. At whatever time our efforts begin, available solar radiation should be determined, as should amounts and sources of mineral nutrients. Great changes are often occasioned in the trophic pathways of aquatic ecosystems when man introduces energy-rich organic materials, even in amounts that may seem, superficially at least, to be small. It is important that the sources, character, and quantity of these materials be known, because they are likely to be major contributors to the economy of the ecosystems. And other contributions to this economy, whatever their source, should be evaluated, if they are likely to be significant. The energy and material resources of a community, including those introduced by man, will ultimately determine the fate of its populations; our understanding must begin with knowledge of these resources.

Some idea of the overall economy of an aquatic ecosystem can come from comparison of the gross photosynthesis of its plants to the total respiration of all its organisms. As trophic studies go, measurements of the gross production of plants and the community respiration in aquatic ecosystems are usually relatively simple. And, for the effort involved, the broad picture of plant production and energy and material utilization obtained is often sufficiently rewarding. When man introduces large amounts of organic materials into aquatic ecosystems, the tendency for community respiration to exceed plant production gives us some indication of the total impact of these materials. If the impact is great, the changes that occur are likely to be ones man does not wish. But these changes we will need to identify, for scientific as well as practical reasons.

It is here that we should, for both scientific and practical reasons, become more concerned with parts, those parts that interest us most, hopefully the ones upon which we can obtain meaningful measurements. We need not now go over again the reasons why we should give our attention to one or at most a few *products of interest*, rather than diffuse it hopelessly over entire trophic levels, even if there be such. Often, man is most concerned about but one or a few endangered species in polluted waters. Since we can never properly study all species, why not devote our attention to these? If our interest and scientific resources do not fail us, we can probably obtain reasonably adequate estimates of their production rates, something we cannot expect to accomplish for many species. But the production of any population is made possible by its energy and material resources, and changes in production may be expected to follow changes in these resources. Production is a variable phenomenon, and, if its study is to be worthwhile, it must be related to the conditions making it possible, particularly the energy and material resources.

For our product of interest, then, if it is an animal, we must learn what it eats, what kinds of organisms appear in the stomachs of the individuals of its population. Studies of food habits are tedious, but they can usually be accomplished at reasonable cost, and they are absolutely necessary if we are to define the pathways by which our product of interest receives its energy and materials. Further, if our animal is a carnivore, we should determine the kinds of food upon which its principal prey species depend. Thus we will know the major trophic pathways leading from this animal back to primary energy and material resources. But this is only a qualitative picture: it can suggest reasons for changes in the production of an animal; it cannot relate the level of production of an animal to the energy and material resources determining this level. Can we ever hope to do this?

We have already devoted much of this chapter to explaining why we believe the production of an animal cannot usually be related to its energy and material resources on the basis of the production rates of all the species constituting these resources. Very simply, there are three reasons: the task of measuring the production rates of many species becomes too great; the production of any food species passes into many trophic pathways and is not all available to a given consumer; and the production rates of food organisms of different trophic kinds are not additive. The density or biomass of a food resource as a measure of its immediate availability to a consumer does not present these conceptual and analytical difficulties. Perhaps not always but often the growth rate of a consumer can be defined as a function of the biomass of its food resource. Then, the production of the consumer can be formulated as a function of its biomass and the density of its food organisms. There is evidence that the biomass of organisms on any step along a trophic pathway may often be a function of the biomass of their food organisms and the biomass of their consumers. In a previous section of this chapter, we burdened our reader with a lengthy explanation of this phenomenon because, when it occurs, we have a practical rationale for relating changes in the production of a product of interest to changes in the primary energy and material resources of its ecosystem. We come back, then, to the kind of change man often induces in aquatic ecosystems, a change in their primary energy and material resources, the kind of change we must understand in our efforts to control water pollution.

There has come an increased public awareness of the destructiveness of water pollution to human material, esthetic, and ethical values, and public support is now making scientific endeavors in this area more feasible and respectable. But science, like other human social activities, is terribly subject to fads. Biologists should not become interested in water pollution solely because such interest is now more popular. Many water pollution problems are fundamentally problems of trophic relations. Studies of trophic relations, when sound, are quite worthwhile, whether or not they are directed toward the solution of practical problems. But over the past 25 years, studies of trophic relations have become faddish in ecology. And biologists, those who are interested in water pollution and those who are not, should not clamber thoughtlessly aboard scientific bandwagons, trophic or other. Their responsibilities to themselves, to their science, and to society require that they carefully evaluate the scientific merit of each available approach and, if their objectives are practical, the relevance of each approach.

SELECTED REFERENCES

Brocksen, R. W., G. E. Davis, and C. E. Warren. 1968. Competition, food consumption, and production of sculpins and trout in laboratory stream communities. Journal of Wildlife Management 32:51-75.

———, ———, and ———. 1970. Analysis of trophic processes on the basis of density-dependent functions. Pages 468-498. *In* J. H. Steele (Editor), Marine Food Chains. University of California Press, Berkeley. In press.

———, and C. E. Warren. Density dependence of trophic processes in laboratory stream communities. Manuscript.

Chamberlin, T. C. 1897. The method of multiple working hypotheses. Journal of Geology 5:837-848.

Cummins, K. W., W. P. Coffman, and P. A. Roff. 1966. Trophic relationships in a small woodland stream. Verhandlung der Internationalen Vereinigung für theoretische und angewandte Limnologie 16:627-637.

Darnell, R. M. 1958. Food habits of fishes and larger invertebrates of Lake Pontchartrain, Louisiana, an estuarine community. Publications of the Institute of Marine Science (University of Texas) 5:353-416.

———. 1961. Trophic spectrum of an estuarine community, based on studies of Lake Pontchartrain, Louisiana. Ecology 42:553-568.

———. 1968. Animal nutrition in relation to secondary production. American Zoologist 8:83-93.

Davis, G. E., and C. E. Warren. 1965. Trophic relations of a sculpin in laboratory stream communities. Journal of Wildlife Management 29:846-871.

Hutchinson, G. E. 1963. The prospect before us. Pages 683-690. *In* D. G. Frey (Editor), Limnology in North

America. The University of Wisconsin Press, Madison. [xviii] + 734pp.

Ivlev, V. S. 1945. Biologicheskaya produktivnost' vodoemov. Uspekhi Sovremennoi Biologii 19:98-120. (Translated by W. E. Ricker, 1966. The biological productivity of waters. Journal of the Fisheries Research Board of Canada 23:1727-1759.)

_____. 1961a. Experimental Ecology of the Feeding of Fishes. (Translated by D. Scott.) Yale University Press, New Haven. viii + 302pp.

_____. 1961b. Ob utilizatsii pishchi rybami-planktofagami. Trudy Sevastopoloskoi Biologicheskoi Stantsii 14:188-201. (Translated by W. E. Ricker, 1963. On the utilization of food by plankton-eating fishes. Fisheries Research Board of Canada Translation Series 447. 15pp. + [ii].)

Lindeman, R. L. 1941a. Seasonal food-cycle dynamics in a senescent lake. American Midland Naturalist 26:636-673.

_____. 1941b. The developmental history of Cedar Creek Bog, Minnesota. American Midland Naturalist 25:101-112.

_____. 1942. The trophic-dynamic aspect of ecology. Ecology 23:399-418.

McIntire, C. D., R. L. Garrison, H. K. Phinney, and C. E. Warren. 1964. Primary production in laboratory streams. Limnology and Oceanography 9:92-102.

Margalef, R. 1968. Perspectives in Ecological Theory. University of Chicago Press, Chicago. viii + 111pp.

Odum, E. P. 1959. Fundamentals of Ecology. 2nd ed., in collaboration with H. T. Odum. W. B. Saunders Company, Philadelphia. xvii + 546pp.

Odum, H. T. 1956. Primary production in flowing waters. Limnology and Oceanography 1:102-117.

_____. 1957a. Trophic structure and productivity of Silver Springs, Florida. Ecological Monographs 27:[55]-112.

Slobodkin, L. B. 1960. Ecological energy relationships at the population level. American Naturalist 94:213-236.

_____. 1962. Energy in animal ecology. Pages 69-101. *In* J. B. Cragg (Editor), Advances in Ecological Research. Vol. 1. Academic Press, New York. xi + 203pp.

Strickland, J. D. H. 1960. Measuring the production of marine phytoplankton. Fisheries Research Board of Canada Bulletin 122. viii + 172pp.

Teal, J. M. 1957. Community metabolism in a temperate cold spring. Ecological Monographs 27:[283]-302.

Warren, C. E., J. H. Wales, G. E. Davis, and P. Doudoroff. 1964. Trout production in an experimental stream enriched with sucrose. Journal of Wildlife Management 28:617-660.

20 Biological Indices

Now for the native gifts of various soils,
What powers hath each, what hue, what natural bent
For yielding increase. First your stubborn lands
And churlish hill-sides, where are thorny fields
Of meagre marl and gravel, these delight
In long-lived olive-groves to Pallas dear.
Take for a sign the plenteous growth hard by
Of oleaster, and the fields strewn wide
With woodland berries.

VIRGIL, 30 B.C., p. 57

It may be then that a practical answer to the problems of river pollution in this country would be the classification of rivers according to the uses by which each would best serve the human community. If this were done, the stream-bed communities, associated with the conditions required in each river, could be more clearly defined and thereby be of even greater value in assessing pollution.

H. A. HAWKES, 1962, p. 422

ENVIRONMENTAL CHANGE AND POLLUTION

For reasons we believe to be very good, *we have not titled* this chapter "Biological Indices of Pollution." The problem of measuring environmental change and the problem of defining pollution are confused when we speak of biological indices of pollution. Almost any reasonably persistent change in an environment will bring about changes in the populations and in the community there. Suitably described, these biological changes can be very useful *biological indices* of the environmental change that caused them. Throughout this book we have endeavored to explain—and indeed all of biology teaches—how changes in environments bring about changes in individual organisms, in their populations, and in their communities. Since the environment at a location in large part determines what organisms are able to inhabit that location, the inhabitants should be biological indices of environmental changes. This is an ancient, a proven, and a very useful idea, as we shall see. Changes in populations and communities most certainly indicate environmental changes. Some environmental changes man has chosen and will continue to choose to consider pollution. But many other environmental changes man has not and will not consider pollution. Biologists, with biological indices as one of their tools, are well equipped to detect the changes that occur in environments and to interpret for society the biological significance of these changes; but it is for society to determine which changes are to be classified as pollution and must be prevented. The biological index idea will be most useful if the problem of measuring environmental change and the problem of defining pollution are, insofar as possible, kept distinct.

Long before Virgil wrote his beautiful

Georgics—certainly one of the earliest agricultural extension bulletins—to commend rural life to the returning Roman legionnaires of Augustus and to instruct them in the culture of the soil, man used biological indices of his environment: biological indices of the seasons, soils, waters, and habitats of the earth. How helpful they must have been to tell him when and where to hunt and fish, to gather food, to graze his flocks, to plant his crops, and to carry on many other activities of his life. Indeed, we must suppose that many animals other than man learn to use biological indices to guide their search for food and perhaps other of their activities.

During the eighteenth and nineteenth centuries, there was considerable interest among biologists in means of predicting the time of occurrence of certain seasonal events. The stage of development of plants, responding in an accumulative way to temperature, was found to be one useful biological aid to prognostication of these events. From the early part of our century, the use of plant species to indicate various soil types and plant associations has greatly interested botanists. Petersen (1918) used individual species of invertebrate animals as biological indices of the presence of particular marine communities. Freshwater biologists since the beginning of this century have been concerned about the development of pollutional conditions, and they have stressed the value of many different ways of using particular species or communities of plants and animals as biological indices of these conditions. Often the waters that these biologists studied were badly polluted, no matter how one might wish to define pollution, and the plants and animals they found to be characteristic of these waters came to be known as "biological indices of pollution." But the waters these biologists studied should, in our opinion, have been considered polluted because their condition was bad for man, not simply because their communities had changed, though sometimes community changes in themselves are bad for man. Not all changes in biological communities are evidence of pollution. Thus, we believe it better to separate the idea of biological indices of environmental change and the idea of pollution than to join them in the expression "biological indices of pollution."

With all their understanding of a forever-changing nature—changing with the dynamics of populations, with the evolution of species, and with the succession of communities—biologists as a group still remain deeply concerned about the changes man's activities have caused in the natural communities of the earth. They recognize that most of the changes occasioned by agriculture, for example, are not only beneficial but essential to man's existence. But they find it hard to accept changes, such as those caused by waste disposal, that are not likely to be beneficial and that often seem unnecessary. Biologists are often inclined to view any such change as intrinsically bad, however small the change. In advancing this view, they seem to forget that the very presence of man on the earth requires some adjustment in natural communities and that even highly treated waste products, when discharged into waters, bring about some change, however small, in aquatic communities.

Thus, Dr. Ruth Patrick (1953, p. 33) expressed the deep feelings of many biologists when she wrote:

It is, then, the changes which have occurred too recently for such evolution to take place which bring about a definite reduction in number of species. Such changes are usually due to the activities of man and are defined as pollution. Thus by this definition pollution is defined as any thing which brings about a reduction in the diversity of aquatic life and eventually destroys the balance of life in a stream.

Were Dr. Patrick's definition of pollution to be accepted by society, any reduction in community diversity would be pollution, and biological indices of this would be direct evidence of pollution; they would be "biological indices of pollution," however small the changes they might indicate. But the feelings of many biologists notwithstanding, society has not been able to define pollution in this manner. We endeavored to explain this in Chapter 2, and there we attempted to define water pollution in a way that might be socially acceptable: any impairment of the suitability of water for any of its beneficial uses, actual or potential, by man-caused changes in the quality of the water. Some changes in aquatic communities are consistent with man's beneficial uses of water; some are not. Biological indices are useful in monitoring changes, whether or not society determines the changes to be pollution.

When waters are classified as to their beneficial uses, when the water quality characteristics necessary for these uses are determined, and when the biological communities associated with waters of particular quality are known, biological indices can be even more useful for detecting shifts in water quality that, in endangering beneficial uses, are appropriately considered pollution. This is the sense of the quotation from Hawkes (1962) that heads this chapter.

Before discussing biological indices further, we will first attempt to describe the kinds of changes that occur in aquatic communities when wastes are introduced into rivers, lakes, and estuaries. Then, we hope to be able to explain some of the ways in which information on species and communities has been used to develop biological indices of environmental conditions. Only after this can we really consider the usefulness of biological indices.

CHANGES IN AQUATIC ECOSYSTEMS

Zones in Rivers Receiving Sewage

When sewage is introduced into a river, there begin sequences of events in time and distance of flow that lead to different environmental conditions and different aquatic communities in successive reaches of the river, until natural processes eventually return the river to something like its original condition. Other putrescible organic wastes, such as those from food processing industries, may cause similar but not identical changes in rivers, for the composition and the decomposition products of these wastes differ somewhat from those of sewage. Very early in the history of water pollution biology, Kolkwitz and Marsson (1908, 1909) — on the basis of their extensive studies of rivers and streams receiving sewage and other putrescible organic wastes in Germany — proposed their *saprobic system* of zones of organic enrichment and their classification of many plant and animal species according to the zones in which these species survived. The successive reaches of a river below a point of waste introduction were designated as *polysaprobic, alpha-* and then *beta-mesosaprobic,* and finally *oligosaprobic,* on the basis of their physical and chemical

characteristics as well as the number and kinds of organisms that inhabited them. Because of the historic importance of the *saprobic system,* because later systems have been its derivatives, and because its originators can best explain to our reader what they had in mind, we quote, in translation and at some length, the original definitions of the saprobic zones of Kolkwitz and Marsson (1909; translation published 1967, p. 86).

I — The zone of polysaprobia is characterized chemically by a certain degree of wealth of high-molecular, putrescible and organic substances (protein components and carbohydrates) which enter the collectors in the directly putrescible sewage from cities and agricultural, industrial and other enterprises. A decrease of oxygen content of the water accompanied by reduction manifestations, formation of hydrogen sulphide in the mud and an increase of carbon dioxide often are the chemical sequels of this.

Organisms generally occur in great numbers but with a certain monotony; especially Schizomycetes and (usually bacteriophage) colorless flagellates are frequent. The bacteria developed in standard nutrient gelatine may exceed 1 million per ccm water. Organisms with high oxygen requirements obviously are generally completely absent. Fishes usually avoid remaining in this zone for any length of time.

Plant organisms preferring hydrogen sulphide in this zone may recur in H_2S sources in the oligo-saprobiotic zone.

II — The zone of the mesosaprobia is divided into a alpha- and/or beta-saprobiotic section. It generally succeeds the polysaprobiotic zone. In the alpha-section which adjoins the former, self-purification takes place still rather aggressively as already mentioned but with the simultaneous occurrence of oxidation manifestations — in contrast to zone I — which are conditioned in part by the oxygen production from chlorophyllous plants.

The protein components contained in the water are probably already decomposed down to asparagin, leucine, glycocoll, etc., which results in a qualitative difference from zone I.

In the beta-section, the decomposition products already approach mineralization. Normal, generally nitrate-containing effluents from the trickle fields are most properly included in this zone.

All organisms of the mesosaprobiotic zone usually are resistant to minor action by sewage and its decomposition products. Notable is among other factors the content of the zone of Diatomaceae, Schizophyceae and many Chlorophyceae and some higher plant organisms. Higher and lower animal organisms are also found in a great number of individuals and varieties.

III — The zone of the oligosaprobia is the domain of (practically) pure water. If it was preceded by a self-purification process locally or chronologically, it succeeds the mesosaprobiotic zone and then represents the termination of mineralization. However, we here include also lakes in which the water does not undergo a mineralization process properly speaking. The oxygen content of the water may often remain permanently close to the saturation limit and occasionally even exceed the latter (as a function of the air dissolved in the water). The content of organic nitrogen usually does not exceed 1 mg/lit. The water is generally transparent to a considerable depth, except at times of abundant plant growth. The number of bacteria developed on standard nutrient gelatine is generally low. In contrast to the polysaprobiotic zone, it amounts to only a few hundreds and infrequently thousands of bacteria per ccm water.

Kolkwitz and Marsson (1908, 1909) published long lists of the species of plants and animals they believed to be associated with each of these zones, the species being classified as polysaprobic, alpha- or beta-mesosaprobic, or oligosaprobic. And from the publication of these papers, the idea of "biological indices of pollution" received its early impetus.

Patterned after the saprobic system, other systems of zoning organically enriched rivers have been proposed (Fig. 20-1). These later systems are conceptually similar but differ in nomenclature and in delimitation of zones. European workers have made the most use of these ideas and have followed the *saprobic*

system, whereas Americans have tended to favor the system proposed by Whipple, Fair, and Whipple (1927). But difficulties inherent in any more or less arbitrary classification of environments so dynamic as those of rivers, and difficulties in properly assigning species to these environments, plague those who use these systems. Their rigid application has fallen into disfavor with some European biologists and most American biologists (Hynes, 1960; Bartsch and Ingram, 1966). Nevertheless, those who proposed and those who have used these systems have found them of value in ordering their thoughts on the complex interrelationships in rivers receiving organic wastes. Few are the biologists interested in water pollution whose thinking has not been influenced by these zonal classifications of rivers.

With increasing volume or biochemical oxygen demand (BOD) of sewage introduced, the depression in dissolved oxygen content of a river below the point of introduction becomes greater, and greater time and distance of flow downstream are necessary before stream reaeration processes return the dissolved oxygen to near air-saturation levels (Fig. 20-2). The *oxygen sag curve* can be formulated by taking into account BOD loading, reaeration, and time of flow (Phelps, 1944), this making it possible to predict—but only very approximately—the ef-

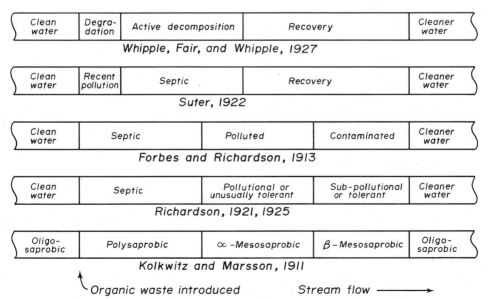

FIGURE 20-1 Systems various investigators have suggested for classifying the zones of rivers receiving sewage. After Whipple, Fair, and Whipple (1927).

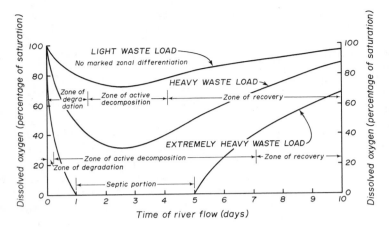

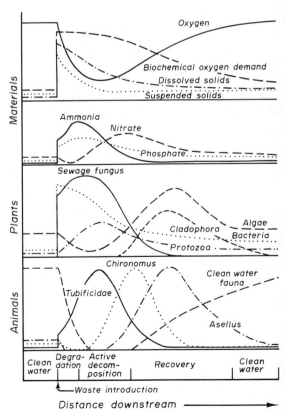

FIGURE 20-2 Oxygen sag curves and zonal classifications for the Ohio River below Pittsburgh when different amounts of organic wastes were being introduced. After Whipple, Fair, and Whipple (1927).

fects of waste discharges on the oxygen resources of a river. The magnitude and downstream extent of oxygen depression change with changes in waste load, and thus the zonal classification of the river changes, if it is based on water quality or plankton organisms (Fig. 20-2). But organisms living on the bottom and having longer life cycles than plankton organisms cannot reflect the patterns of short-term changes in water quality. This is one of the difficulties inherent in use of the various systems of zonal classification of rivers.

Some of the more important successional changes occurring in a river when sewage is introduced are shown in Figure 20-3. As Kolkwitz and Marsson knew, bacteria and protozoa predominate in the zone where oxygen becomes severely limiting. Farther downstream, these decline, and algae—benefiting from increases in mineral nutrients—first become very abundant and then decline. Where oxygen concentrations are very low, the number of species of larger invertebrates is greatly reduced, but tubificid worms, midge larvae, and other forms able to tolerate conditions in this zone become numerically very abundant, for much food and little interspecific competition exist here. As the river returns to its natural condition, the number of species able to live there increases, and strong numerical predominance of particular species tends to disappear.

Not often are the wastes introduced into larger streams and rivers of only one kind, and the points of waste introduction are usually numerous. Thus, not often can we expect to find typical examples of change in rivers receiving sewage from only one

source. The Upper Trent in England receives some toxic wastes as well as sewage via Fowlea Brook, but the results of a fine study reported by Butcher (1946) suggest that the effects of the sewage predominate (Table 20-1). In the Upper Trent below

FIGURE 20-3 Typical changes in the water quality and the plant and animal populations of a river as it passes through various zones following the introduction of sewage. After Hynes (1960).

TABLE 20-1 *The influences of sewage on the water quality, fauna, and flora of the River Trent below the entry of Fowlea Brook. After Butcher (1946).*

Station	Miles below Fowlea Brook	Water Quality					Animals Collected									Algae per Square Millimeter					
		Oxygen (per. sat.)	Ammonia N (mg/l)	BOD-4 hr (mg/l)	Tar Acids (mg/l)	Cyanide as CN (mg/l)	Tubificidae	Chironomidae (red)	Asellus	Leeches	Gammarus	Molluscs	Caddis	Ephemeridae	Sewage Fungus	Stigeoclonium	Nitzschia palea	Gomphonema parvulum	Chamaesiphon spp.	Cocconeis placentula	Total Algae
Milton	Above	107	0.2	3.6			4			5	11	11	24	7	Absent	30				820	2140
Trentham	2	0	9.1	13.0	1.25	0.22	7								Domin.	190	130	16			270
Strongford	3	31	8.2	10.6	1.0		No collection								Domin.	1620	680	133	16		1200
Tittensor	5	19	14.0	10.3	0.7		50	7							Domin.		2380	600			5040
Stone	8	34	11.5	8.3	0.4		12	9							Com.	15,320	5250	3390			24,130
Aston	10	41	12.4	6.4	0.2		30	24							Com.	1930	620	690			3266
Sandon	13.5	40	11.5	7.0	0.04		25	7	6						Mod.	320	250	660	20		1440
Ingestre	17	44	11.5	6.8			2		36	3					Mod.	370	100	3000	330		4420
Wolseley	22	64	5.6	4.3			1		54	7	1	3			Rare	150		1950	830		4900
Rugeley	24	69	6.6	5.2					75	14	3	1			Absent	3460	620	1950	2050	1930	15,800
Handsacre	27.5	70	3.6	4.9			2		3	5	38	24	6		Absent	30	1720	9170	1570	1330	13,170
Yoxall	30	87	2.4	4.6			2		1	4	45		26		Absent	320	8000	4200	2020	3500	17,160
Wichnor	35	77	0.9	4.6					11	7	2		27		Absent	860	150	1280	200	1480	5350

where Fowlea Brook enters, we see the disappearance of sensitive forms and the early increase in numbers of tubificids and midge larvae. As conditions improve downstream, these more tolerant forms decline in number, and the more sensitive crustaceans, molluscs, caddisflies, and mayflies reappear. These changes in the abundance of animals as well as changes in the abundance of the benthic algae and sewage fungus well exemplify the typical effects of sewage on stream organisms (Fig. 20-3).

It would be negligent for an American biologist to leave a discussion of the zones of rivers receiving organic wastes without mentioning the monumental study of the Illinois River conducted by Richardson and his mentor Forbes. Richardson (1928) appended a list of their important publications to his final summary of this study, "a study of 1,308 biological collections made by the Natural History Survey between 1913 and 1925, inclusive, over a 225-mile stretch of the Illinois River and in its connected bottom-land lakes" (1928, p. 469). Our allotted space will permit no proper summary of this important contribution. Suffice it to say that Richardson was able to define the zones of organic pollution and their animal associations in this complex river-lake system having many waste outfalls. And even today, his thoughts on the concept of biological indices remain instructive, pertinent, and fresh.

Changes in Stream Communities Caused by Toxic Substances

The composition of river and stream communities is affected very differently when the wastes introduced are primarily of a toxic rather than a putrescible organic nature. As we have explained, domestic sewage—in addition to leading to depression of oxygen concentration and to other conditions intolerable for some species—provides a new and abundant resource of energy and materials for those species tolerating the conditions that develop in receiving waters. Thus, while the number of species may be severely reduced by domestic sewage, some few species become exceedingly abundant. Unlike sewage, most toxic substances directly affect nearly all forms of life unfavorably. For only a few kinds of bacteria do toxic

substances provide an important energy resource. Species differ in their tolerance for the same toxic substance, but at some level most toxic substances can destroy any form of life. When highly toxic materials are introduced into a river providing inadequate dilution, their effect is to eliminate most life. As the concentration of these materials is gradually reduced downstream—by dilution with increased streamflow, by adsorption on bottom materials, and by such biological activity and accumulation as may occur—the most tolerant species begin to reappear, and the reappearance of these species is followed by the successive downstream reappearance of other species, according to their tolerance. But because toxic substances afford direct benefit to few if any organisms, great increases in the numbers of individuals of particular species are not usually found. Numbers increase only gradually downstream as the toxic substances are diluted. This is the simple and usual effect of the direct action of toxic wastes.

But in ecological systems, indirect actions may be as important as direct ones, and occasionally some species do become more abundant as a result of indirect actions of toxic wastes. Though the energy and material resources of a stream may not be appreciably increased by toxic substances, the elimination of sensitive herbivores may permit more resistant algae to become very abundant; the elimination of some predators may permit their more resistant prey to increase; or a resistant species may be favored by the elimination of its competitors. Still, these are more the expected perturbations than they are dominant characteristics of the effects of toxic substances on stream communities.

Studies in England on the River Churnet and the River Dove into which it flows illustrate well the effects of toxic wastes on stream communities (Butcher, 1946). The Churnet received wastes from the Leek Sewage Works and the Leekbrook Dye Works some 7 miles above the point of entry of wastes from a copper works at Froghall (Table 20-2). After exhibiting the typical community changes of a stream receiving putrescible organic matter, the Churnet was well on its way to recovery from the effects of the organic wastes by the time it reached the point of introduction of the toxic wastes from the copper works.

TABLE 20-2 *The influence of toxic copper waste, entering the River Churnet at Froghall, on the fauna and flora of the River Churnet and the River Dove into which it flows. After* **Butcher** (1946).

Station	Miles below Leek Sew. Works	Oxygen (per. sat.)	Ammonia N (mg/l)	BOD-4 hr (mg/l)	Copper (mg/l)	Tubificidae	Chironomidae (red)	Chironomidae (small green)	Asellus	Leeches	Molluscs	Gammarus	Caddis	Ephemeridae	Total Animals	No. of Species	Sewage Fungus	Stigeoclonium	Nitzschia palea	Gomphonema parvulum	Chamaesiphon sp.	Cocconeis	Chlorococcum	Achnanthes affinis	Total Algae
Above Leek	Above	83	0.05	4.3		16		355		4	13	187	16	4	634	27	Absent	53			36	348			700
Below sewage works		67	1.1	6.2		20	23	50	1	43	14	12			234	14	V.rare	59	59	41	212	70			1200
Below dye works		45	1.1	10.2		1045	72	1	1						1120	6	Dom.	70	322	29	17	13			490
Cheddleton Bridge	2	35	0.6	4.2		>5000	238								>10,000	2	Dom.	196		143		3			1430
Basford Green	3	36	0.6	6.5		>5000	264								>10,000	2	Dom.	1199	2533	673					7062
Above Froghall	7	49	0.9	4.8	Nil	2752	2400		61	24	16				5248	11	Rare	206	41	71	80	16			1081
Below Froghall copper works	7.5	52	0.9	4.8	1.0				Nothing								V.rare						38	84	141
Oakamoor	10.5	63	0.6	3.3	1.2				Nothing								Absent						21	144	211
Alton	12.5	78	0.6	3.3	1.2				Nothing								Absent	2840			2598		2208	4670	33,300
Rocester	18.5	82	0.2	3.6	0.6				Nothing								Absent	52,016					3005	665	58,400
Below Churnet Mouth	18.8	95	0.02	2.2	0.12			16					1		17	2	Absent	1400		16,500	4400		9400	2802	27,600
Doveridge	21	91	0.01	2.3				20					4	8	22	4	Absent								
Sudbury	27	99	0.03	1.8				13					8		24	6	Absent	850	700	15,400	1800		19,500	142	55,100
Scropton	31	101	0.02	2.4	0.12			11					5	1	18	7	Absent								
Monks Bridge	38	111	0.02	2.5				57		24			12	31	108	10	Absent	930	11,500	9750	2300		6200		53,300
R. Dove above Churnet Mouth	–	100	0.07	2.0	Nil			313		1	1		33	59	478	30	Absent		100	100	1800				2800

Water Quality: Miles below Leek Sew. Works, Oxygen (per. sat.), Ammonia N (mg/l), BOD-4 hr (mg/l), Copper (mg/l).
Animals per 0.1 Square Meter: Tubificidae, Chironomidae (red), Chironomidae (small green), Asellus, Leeches, Molluscs, Gammarus, Caddis, Ephemeridae, Total Animals, No. of Species, Sewage Fungus.
Algae per Square Millimeter: Stigeoclonium, Nitzschia palea, Gomphonema parvulum, Chamaesiphon sp., Cocconeis, Chlorococcum, Achnanthes affinis, Total Algae.

From the entry of the effluents from the first copper works the fauna was completely exterminated and for over 10 miles there was not a sign of a single animal; at first algae were extremely rare and even sewage fungus was absent while between 1 to 2 pts/million of copper were found in the water. Five miles further down the algae had increased very greatly and the principal species were four green algae and three or four diatoms only rarely seen in other parts of the river. (Butcher, 1946, p. 95)

Greater dilution of the copper occurred where the Churnet flows into the Dove; thus all animals were not eliminated from the Dove, though their species were reduced from 30 above to 2 below the confluence (Table 20-2). And perhaps because of elimination of the animals, the algae became more abundant. With the downstream decrease of copper in the Dove, both the number of animal species and the numbers of individuals in these species gradually increased. But even 19 miles below the confluence of the two streams, the persistence of copper had not permitted the fauna of the Dove to regain its natural variety and abundance.

Lake Eutrophication

And the fish that was in the river died; and the river stank, and the Egyptians could not drink of the water of the river; and there was blood throughout all the land of Egypt. (Exodus 7:21)

Thus was described the first plague of Egypt, in which the forces of nature aided Moses in making the waters not just of the Nile but throughout the land unfit for life. The "blood" in the water was most probably a heavy bloom of blue-green algae, imparting to the water the red color, the foul smell, and the noxious quality. *Oscillatoria rubescens* and other species of blue-green algae have often appeared in lakes that have reached a stage of eutrophication in which severe nuisance conditions may develop, frequently the result of civilization bringing a twentieth-century plague.

Lake *eutrophication* is the increase over time of the nitrogen, phosphorus, and other nutrients necessary for the production of algae and other aquatic plants. It is an important part of the successional process by which lakes pass from their original nutrient-poor or oligotrophic condition, through a more productive *eutrophic* condition, to become shallower with sedimentation and smaller with the encroachment of marsh, until, over geologic ages, lake basins are returned to land and terrestrial life.

The relative importance of physical processes, as compared to biological processes, in this *hydrarch succession* depends on the geological origin of a lake, its morphology, its tributaries, and the surrounding terrain. Sedimentation may originally be mainly mineral, but in some lakes, as eutrophication proceeds, the death and deposition of plant and animal life become most important. To all lakes, some nutrients come from the surrounding soils, and some soils have more to contribute than others. Agriculture, with its use of fertilizers, can greatly hurry the process of eutrophication. But it is the discharge of organic wastes into lakes—wastes which even after secondary treatment contain great amounts of plant nutrients—that has most often reduced the time of lake eutrophication from geologic ages to less than the life span of man.

The eutrophication of lakes by sewage has occurred in many places throughout the world, but perhaps nowhere has this process been so closely observed from its earliest stages as in some of the beautiful Alpine lakes of Europe, of which the story of the *Zürichsee*, or Lake Zürich, in Switzerland may be most notable. The histories of this and other lakes suffering such eutrophication have been traced by Hasler (1947). Following apparent shifts in the diatom flora, the first eruption of the blue-green alga *Oscillatoria rubescens* in Lake Zürich occurred in 1896. As this lake continued to be enriched with sewage, different diatom species became the most abundant, and the blooms of *Oscillatoria rubescens* became greater and more frequent, until in 1936 it formed a scum with the smell of fish oil over most of the lake, and the stream leaving the lake flowed red-brown in color. Thus, the pattern of lake eutrophication with sewage was announced: a shift in the diatom and algal flora from nutrient-thrifty species to those able to dominate at high nutrient concentrations, a shift leading to the ascendancy of blue-green forms, and blooms coming in greater magnitude and frequency. And with the increasing rain of decomposing organic materials to the bottoms of lakes, anaerobic conditions may develop in the deeper waters, thus favoring the return of nutrients to the waters above to compound the nutrient problem. As the environment of the once oligotrophic Lake Zürich changed, populations of salmonids and whitefish declined and were replaced by minnows and other forms more tolerant of the new conditions.

Small and medium-sized lakes have most often been the ones to reach the stage of sewage-caused eutrophication where nuisance conditions develop: the Madison Lakes in Wisconsin (Lackey and Sawyer, 1945; Lackey, 1945) and Lake Washington (Edmondson, 1968). And when they do, the local press and a concerned public get into the act. It takes no careful limnological study to detect the change when this stage is reached, for taste and odor problems, scum on the beaches, and loss of valued fish populations are biological indices of the change that are obvious to all. But very large bodies of water may be excessively fertilized by the huge metropolises of today. When this happens, as in the case of Lake Erie, a national literary magazine, *The Saturday Review* (October 23, 1965), is capable of expressing indignation:

Lake Erie's water is the life blood that feeds the industrial complexes of Detroit, Toledo, Cleveland, Erie, Buffalo, and intermediate points.

Without it, the economy of the lake basin, with its more than 25,000,000 people, would wither and die. (Seltzer, 1965, p. 36)

But along with the billions of gallons of water these social-industrial giants use and return to Lake Erie each day, go their enriching waste products.

Collapse of the cisco (*Coregonus artedii*) fishery of Lake Erie in 1925 was attributed by many to the introduction of wastes from the Detroit and Toledo area at the western end of the lake, but one must be careful in making such an assessment, for overfishing and other factors may have been involved. Studies conducted during 1929 and 1930 (Wright and Tidd, 1933) did indicate that this lake, already moderately rich in nutrients, was evidencing in its phytoplankton some of the effects of enrichment by wastes in limited areas of its western part. Sensitive mayflies were being replaced by tubificid worms in the bottom fauna over perhaps 100 square miles influenced by rivers draining the most populated areas. Even this large area was only about 8 percent of the western part of the lake, but changes were clearly apparent in Lake Erie at this relatively early date.

Over the years of the twentieth century,

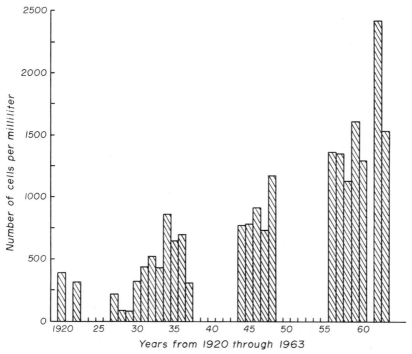

FIGURE 20-4 Mean phytoplankton abundance in the central area of Lake Erie during the years from 1920 through 1963 for which complete records are available. After Davis (1964).

there has been an apparent increase in the concentration of major ions, nitrogen, and phosphorus in the waters of Lake Erie (Beeton, 1961). Davis (1964) brought together available data on the phytoplankton of the central area of the lake. Since 1920, there has been an uneven but marked increase in the average number of cells per milliliter (Fig. 20-4). Lake Erie, like many larger lakes, appears to have been originally characterized by algal blooms in the spring and fall and low algal densities during the winter and summer. But through the years the magnitude and duration of the blooms have so increased that over most of the year phytoplankton density now remains high (Fig. 20-5). And with the changes in magnitude and duration of the blooms have come changes in the species of diatoms and algae dominating these blooms. Here again we

find the ascendancy of blue-green algae — the troublemakers *Oscillatoria, Anabaena,* and *Microcystis* — and various species of green algae have entered the picture.

In 1953, during a period of thermal stratification, the deeper waters of western Lake Erie became severely depleted of oxygen, and nymphs of the mayfly *Hexagenia* suffered a heavy mortality (Britt, 1955), a situation that continues to recur with increasing severity. Since 1929, the bottom fauna has exhibited the changes typically caused by excessive organic enrichment (Table 20-3): species intolerant of developing conditions decrease while the tolerant ones, like tubificid worms (Oligochaeta) and midge larvae (Chironomidae), become very abundant. Following the 1925 collapse of the fishery on the cisco — a species living in oligotrophic lakes — there have been other changes in the

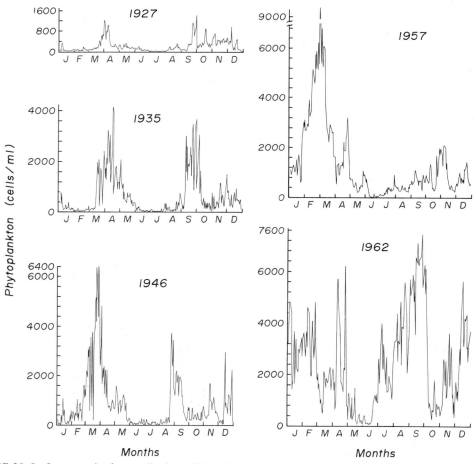

FIGURE 20–5 Increases in the amplitude and duration of phytoplankton blooms occurring in the central area of Lake Erie as eutrophication caused by organic wastes continued during the years from 1927 through 1962. After Davis (1964).

TABLE 20–3 Influence of eutrophication on the bottom fauna of western Lake Erie near South Bass Island, 1929, 1930, and 1958. Animals in number per square meter. After Beeton (1961).

ANIMALS	1929 STATION 158 (20 samples)	1930 STATION 158 (16 samples)	1958 STATION 2 (40 samples)	1958 STATION 3 (7 samples)
Hirudinea	6	4	37	27
Oligochaetes	6	3	559	635
Hexagenia	312	515	49	44
Chironomidae	81	22	257	280
Gastropods	12	24	7	0
Fingernail clams	16	8	55	37
Trichoptera	20	1	3	0

fish populations of Lake Erie, more tolerant but less desired species replacing the more valued ones, until the fisheries now have but little of their original value. It is as though, with changes in the setting and the lake character, we were rereading the story of the *Zürichsee*.

Community Changes in an Estuary

In a remarkably extensive and very nice study of the great estuary known as San Francisco Bay, into which flow the Sacramento River and the San Joaquin River, Filice (1954, 1958, 1959) has shown that the changes caused by sewage and by toxic wastes in estuarine communities are as distinctive as we have come to expect them to be in freshwater communities: sewage reducing the number of species but greatly increasing the numbers of individuals of tolerant forms; toxic wastes reducing not only the number of species but also the total number of individuals. Filice examined the composition of benthic animal communities at 460 stations along about 50 miles of the southern shore of the upper part of this estuary. Nearly 50 waste outfalls empty into this reach; some of his stations were at these outfalls, others were marginal to them, and still others were quite removed (Filice, 1954). Of the 460 stations, 312 were either not influenced by any waste discharges or were influenced only by domestic wastes or only by industrial wastes; these 312 stations are the ones that here concern us.

Filice (1954, 1959) classified these stations into five categories according to their kind and extent of contamination by wastes: *maximum domestic pollution* occurred at outfalls from domestic installations or food processing plants; *marginal domestic pollution* occurred in regions somewhat downstream from the domestic outfalls; *maximum industrial pollution* occurred at outfalls whose effluents were toxic and not metabolizable by bottom invertebrates; *marginal industrial pollution* occurred where these toxic wastes had been somewhat further diluted; and *normal stations* occurred where contaminated water did not usually flow. The total number of individuals of each species found in all samples collected from the stations in each of these categories of contamination is given in Table 20-4 by species and category. The numbers enclosed in the parentheses are the numbers of the various species that would have been expected to be found at stations in the various categories if the species were neither favorably nor unfavorably influenced by conditions at those stations. That is to say, if the numerical distribution of a species among the five contamination categories were random, the numbers in the parentheses, which have been adjusted for differences in sample size, would be expected to occur. Thus, deviations of the observed numbers from those in parentheses indicate the extent of favorable or unfavorable effect of the different conditions on the various species.

Whereas 39 species were found at stations Filice considered to be uninfluenced by wastes, only 5 were found at stations nearest domestic waste outfalls (Table 20-4). Sta-

TABLE 20–4   Influence of domestic and industrial wastes on fauna of San Francisco Bay. Numbers of individual animals collected according to species at stations not influenced by wastes, at stations at and marginal to domestic waste outfalls, and at stations at and marginal to industrial waste outfalls. Numbers in parentheses are numbers of each species that would have been expected at the various stations had these species not been influenced, either favorably or unfavorably, by either the presence or absence of wastes. After Filice (1959).

Species	Total Specimens	Normal Stations	Stations Influenced by Domestic Wastes		Stations Influenced by Industrial Wastes	
			At Outfalls	Marginal to Outfalls	At Outfalls	Marginal to Outfalls
Gemma gemma	8335	957 (4221)	0 (329)	7374 (2456)	3 (251)	1 (1077)
Mya arenaria	724	276 (367)	0 (29)	446 (213)	1 (22)	1 (94)
Rhithropanopeus harrissii	98	34 (50)	6 (4)	47 (29)	1 (3)	10 (13)
Polydora uncata	687	199 (348)	8 (27)	470 (202)	0 (21)	10 (89)
Nereis succinea	427	192 (216)	5 (17)	178 (126)	0 (13)	52 (55)
Nassarius obsoletus	129	29 (65)	2 (5)	97 (38)	0 (4)	1 (17)
Capitella capitata	7	4 (3)	3 (0)	0 (2)	0 (0)	0 (1)
Balanus improvisus	2828	512 (1432)	0 (112)	2115 (834)	0 (85)	201 (365)
Macoma inconspicua	268	28 (136)	0 (11)	231 (79)	0 (8)	9 (35)
Corophium spinicorne	190	102 (96)	0 (7)	81 (56)	0 (6)	7 (25)
Corbicula fluminea	67	43 (34)	0 (3)	18 (20)	0 (2)	6 (9)
Pisidium occidentale	31	24 (16)	0 (1)	3 (9)	0 (1)	4 (4)
Nereis lighti	9	4 (5)	0 (0)	1 (3)	0 (0)	4 (1)
Balanus crenatus	97	16 (49)	0 (4)	81 (29)	0 (3)	0 (12)
Glycinde armigera	90	53 (46)	0 (3)	37 (26)	0 (3)	0 (12)
Macoma inquinata	87	10 (44)	0 (3)	77 (26)	0 (3)	0 (11)
Nephtys caecoides	79	76 (40)	0 (3)	3 (23)	0 (2)	0 (10)
Mytilus edulis	51	3 (26)	0 (2)	48 (15)	0 (1)	0 (7)
Ampelisca milleri	49	40 (25)	0 (2)	9 (14)	0 (1)	0 (6)
Tapes semidecussata	41	20 (21)	0 (2)	21 (12)	0 (1)	0 (5)
Modiolus demissus	28	15 (14)	0 (1)	13 (8)	0 (1)	0 (4)

tions marginal to domestic waste outfalls yielded 30 species, this indicating much less reduction in the number of species as these wastes became more diluted. Thus, in the areas where domestic wastes were the most concentrated, very few species could survive, and none of these were present in large numbers. But in the surrounding areas where the wastes were more diluted, many but not all species found conditions more or less suitable. And some species, perhaps able to take advantage of a new energy and material resource, became quite numerous, much more numerous than would be expected if these species were unaffected by the waste. Among these were the clams *Gemma gemma, Mya arenaria,* and *Macoma inconspicua,* the polychaete worm *Polydora uncata,* and the barnacle *Balanus improvisus.*

The industrial wastes believed to be toxic, on the other hand, appeared to have eliminated all but three species in the areas where the wastes were most concentrated, and these species were present only in very low numbers (Table 20-4). Where these wastes were more dilute in the immediately surrounding areas, 12 species were found; but—with the exception of *Nereis lighti* which Filice explained as a sampling artifact—none of these species appeared to be favored by

TABLE 20–4 *Continued.*

Species	Total Specimens	Normal Stations		Stations Influenced by Domestic Wastes				Stations Influenced by Industrial Wastes			
				At Outfalls		Marginal to Outfalls		At Outfalls		Marginal to Outfalls	
Polydora ligni	26	6	(13)	0	(1)	20	(8)	0	(1)	0	(3)
Eteone californica	22	0	(11)	0	(1)	22	(6)	0	(1)	0	(3)
Conopeum commensale	21	7	(11)	0	(1)	14	(6)	0	(1)	0	(3)
Odostomia fetella	17	5	(9)	0	(1)	12	(5)	0	(0.5)	0	(2)
Ostrea lurida	7	6	(3)	0	(0)	1	(2)	0	(0)	0	(1)
Stylochus franciscorum	7	0	(3)	0	(0)	7	(2)	0	(0)	0	(1)
Streblospio benedicti	5	4	(2)	0	(0)	1	(1)	0	(0)	0	(1)
Heteromastus filiformis	5	2	(2)	0	(0)	3	(1)	0	(0)	0	(1)
Crepidula convexa	5	0	(2)	0	(0)	5	(1)	0	(0)	0	(1)
Gnorimosphaeroma oregonensis	1	0	(0.5)	0	(0)	1	(0)	0	(0)	0	(0)
Parapleustes pugettensis	17	17	(9)	0	(1)	0	(5)	0	(0.5)	0	(2)
Nerinides acuta	15	15	(8)	0	(1)	0	(4)	0	(0.4)	0	(2)
Polydora sp. a	12	12	(6)	0	(0.5)	0	(3)	0	(0.4)	0	(2)
Polydora brachycephala	9	9	(5)	0	(0)	0	(3)	0	(0)	0	(1)
Synidotea laticauda	5	5	(2)	0	(0)	0	(1)	0	(0)	0	(1)
Alcyonidium polyoum	3	3	(1)	0	(0)	0	(1)	0	(0)	0	(0.4)
Macoma nasuta	2	2	(1)	0	(0)	0	(1)	0	(0)	0	(0)
Polydora sp. b	1	1	(0.5)	0	(0)	0	(0)	0	(0)	0	(0)
Pontharpinia sp.	1	1	(0.5)	0	(0)	0	(0)	0	(0)	0	(0)
Hesperonöe complanata	1	1	(0.5)	0	(0)	0	(0)	0	(0)	0	(0)
Clinocardium nuttalia	1	1	(0.5)	0	(0)	0	(0)	0	(0)	0	(0)
Pectinaria californiensis	1	1	(0.5)	0	(0)	0	(0)	0	(0)	0	(0)
Total species		39		5		30		3		12	
Total bottom dredged (liters)	923.86	467.86		36.47		272.32		27.83		119.38	

the presence of industrial wastes, and most of them were found in much lower numbers than would have been expected in the absence of any effect of the wastes.

Now in the case of an estuary, we see effects of putrescible organic wastes and toxic wastes on the composition of communities that are very reminiscent of those we saw in freshwater communities. Domestic wastes at high concentrations may eliminate most species of animals; but as these wastes become more dilute, the numerical ascendancy of some species is phenomenal. The effects of toxic industrial wastes are different. These wastes eliminate most species and favor few if any, even after they have become more dilute.

BIOLOGICAL INDICES OF ENVIRONMENTAL CHANGE

Species, Communities, and Samples

There are, then, good reasons for the distributions and abundances of individual species to change with changes in their environments, as we have attempted to explain throughout this book. Changes in the distributions and abundances of individual species bring changes in the plant and animal assemblages we call communities. We have every reason to accept the ideas that changes through time and space in the distributions and abundances of particular species and in the composition of communities are caused by persistent environmental changes and that these biological changes can be used as biological indices of environmental changes. In the preceding sections, we gave examples of changes in species numbers and changes in communities that occur when the environments in different ecosystems change. Now we can tell something of how changes in numbers of individual species, in the composition of communities, and in community diversity have been used as biological indices of environmental change. But before doing this, we should perhaps say just a little about the different groups of organisms various biologists have favored and about the whole problem of sampling, a matter so integral and influential in our efforts and results.

Although we will refer in this section to *community composition* and *community diversity*, these usages are not strictly correct, because nearly always in studies of this sort only some naturally or conveniently grouped part of the entire community is examined. The grouping may include only the plankton a net can capture, or only the macroscopic benthic invertebrates, or only the benthic algae. Bacteria, fungi, and protozoa are not usually included, even though the importance of their roles in the community can hardly be overstated. When microorganisms are studied, other components of the community are often excluded. Then, reference may be made to the "protozoan community" or to the "algal community," a very different usage of the community concept (Chapter 18). Realizing the difficulty of incorporating into most studies all the plant and animal elements of communities, we can understand this practice. In following this practice, we are doing the possible and taking advantage of the methods of community analysis that have proved useful even in studying parts of communities. But it is well for us to remember that there remains much possibility for misunderstanding, because of what is left undone.

Biologists are human; being human, they are very inclined to believe that their particular specialities and knowledge are not only useful but perhaps the most useful for treating certain ills of the world. Thus, especially in the literature of water pollution biology, some biologists have argued, not always very convincingly, that the groups of organisms in which they specialize are most useful as biological indices of environmental change. This is understandable, but it is also silly. Any group of organisms, properly studied, can be useful as a biological index of change; and those specialists who know most about particular groups are likely to be the ones who study them properly. The results of such studies not only have biological index value but contribute toward understanding of changes in communities. Who is to say, without reference to particular circumstances, which kinds of studies are most valuable?

For reasons of convenience, or because of the information to be derived, organisms of a particular habitat are often favored in studies of biological indices of environmental change. If we are to learn something of the history of a particular location on a river, the organisms that have lived at that location, the benthic organisms, will be most useful; and those with the longest lives will be indicative of conditions extending the furthest back in time. But if we wish to learn something of the history of a particular mass of water as it moves down a river, studies of the plankton organisms living and moving in that mass, like studies of its chemistry, will be most useful. In lakes or in estuaries, as in the sea, the problems are somewhat different; sometimes studies of plankton and sometimes studies of benthic organisms will be most convenient or useful. But most often, as in rivers, the information gained from study of planktonic and benthic organisms is complementary, the greatest value deriving not from the study of one or another group but from study of the community within which they interact.

To describe the distribution and abundance of species and the composition of their communities through space and time presents a sampling problem. The distribution of living things on the earth, either in space or in time, is not random; and some kinds are rare and some are common. Where we sample, when we sample, and how much we sample may profoundly influence the conclusions we reach on the basis of our observations. Is a species absent or only rare? Have we taken enough samples to properly reflect the composition of a community? Common sense, for which there is no substitute, is a wonderful guide. But these important questions are difficult. Along with the study of communities, there have developed sampling theory and statistical usages to which we must turn when our conclusions are highly dependent on sampling procedures.

Indicator Organisms

The *indicator organism* concept in its purest form, a form often discussed and little used, is simply that the presence of a particular species is indicative of the existence of certain environmental conditions that interest man, whereas its absence is indicative of the absence of those conditions. The continued persistence of a species at a location is most assuredly evidence that conditions there are suitable for its existence; but its absence does not always mean unsuitable conditions, because there may have been no individuals to colonize the location. But only some of the conditions that make up the environment of any species are usually of interest to man, and most of the remaining ones, though necessary for the species, are unknown. For an individual species to be useful as an indicator organism, it must have a rather narrow range of suitable environmental conditions that are known and of interest to man. Few are the species that satisfy these qualifications, and those that do are likely to be rare in their numbers of individuals, not a very promising state of affairs for application of the indicator organism concept in its purest form.

When we consider how variable nature usually is in space and time, even in particular habitats, we must conclude that most abundant species cannot have very narrow ranges of suitable environmental conditions. And when we seek indicator organisms that may be useful in assessing pollution, we must remember that nearly all the species on earth evolved before man created pollutional conditions. Admittedly, those organisms that now appear to be favored by certain pollutional conditions must have evolved in environments having some of these conditions in common. But the fact that they can become more abundant in the new than in the old environments attests to the breadth, not the narrowness, of their ranges of suitable environmental conditions. The natal environments are still around, as are various

other suitable environments; and tubificid worms, red midge larvae, *Oscillatoria,* and other organisms apparently favored by pollutional conditions are in many of these environments. Thus, neither the presence nor the absence of this or that species alone can be taken as very reliable evidence of the existence of the particular range of conditions man chooses to consider pollutional. The question of abundance of the particular species needs consideration, as does the abundance of other species with which it is associated.

Kolkwitz and Marsson (1908, 1909) attempted to classify many plants and animals according to the ranges of conditions in rivers receiving putrescible organic wastes within which each species appeared to be most favored: "Such a classification presupposes that the respective organisms are uniquely dependent, within relatively narrow limits, on the chemical composition of the water for their distribution and development in situ" (1908; translation published 1967, p. 47). As we have already indicated, these organisms were classified as being *polysaprobic, alpha-* or *beta-mesosaprobic,* or *oligosaprobic.* Hynes (1960) has summarized many of the criticisms of this system, perhaps the most crucial being that few organisms are "uniquely dependent, within relatively narrow limits," on environmental conditions. Still, probably most of the organisms that Kolkwitz and Marsson placed in particular categories do tend to be favored by conditions these categories represent; it is not seemly to quibble too much over the details of a conceptual advance from which our work continues to benefit. And Kolkwitz and Marsson never intended that their classification be used to permit application of the indicator organism concept in its purest or simplest form: "As we have stressed frequently, the main emphasis in the evaluation of the waters should in general not be laid on the individual organisms but on the biocenoses [communities] whose particularities cannot here be described in detail" (1909; translation published 1967, p. 86).

Richardson (1928), following—as so many others since have done—the general theme of the Kolkwitz and Marsson concept, classified the benthic invertebrates of the Illinois River according to their tolerance of conditions developing from the introduction of sewage and other putrescible organic wastes. The principal categories of his classification were *pollutional group, sub-pollutional group,* and *cleaner-water species.* He did, however, separate the sub-pollutional group into three subcategories, and he provided additional categories for organisms having air-breathing structures and for organisms favored by rapid current. Though Richardson went to considerable length to classify organisms according to the environmental conditions most favoring them, he himself had no illusions about the value of the indicator organism concept. His evaluation of this concept is as pertinent and sound today as it was then (1928, p. 410):

Various extensive published lists, as well as questions frequently asked by workers newly interested, seem to imply, to say the least, an overconfidence in the simplicity and efficacy of the use of a few or many so-called index organisms in the determination of degrees of stream pollution. Less frequently are we asked to name those kinds, particularly of small bottom organisms, which are likely to be most useful for that purpose. This question often appears to be a definite reflection of the fact that various published lists, including our own, are much too long to be useful to the uninitiated worker without a good deal of explanation; as both it and other variations of inquiry seem to result from an impression that biological determinations of the extent of injury by sewage or other waste can be made by a more or less rule-of-thumb mechanical method, the practice of which calls for little by way of preliminary knowledge, except the names and identity of the species. As a matter of fact, the number of small-bottom-dwelling species of the fresh waters of our distribution area that can be safely regarded as having even a fairly dependable individual index value in the present connection is surprisingly small; and even those few have been found in Illinois to be reliable as index species only when used with the greatest caution, and when checking with other indicators.

If this is the reasonable evaluation of the use of individual species as indicator organisms that we believe it to be, why then did Kolkwitz and Marsson, Richardson, and others who followed them go to such great lengths in their attempts to classify organisms according to the environments that seem to favor them? For this reason: such information as we can obtain on the requirements of particular species helps us to understand why, in the face of environmental change, some species disappear—this leading to a reduction in the number of species—and other species come to have more individuals. These were the two kinds of changes in communities we saw—in previous sections of this chapter—to occur when putrescible organic materials were introduced into rivers, lakes, and estuaries. Changes in the number of species and in number of individual organisms are themselves evidence of environmental change. We will be discussing in the next two parts of this section the use of community composition and community diversity as biological indices of environmental change. The best community diversity indices so far developed cannot take into account valuable information on the environmental requirements of the species involved, and, as good as these indices are, they are the poorer for this lack. Analyses of community composition are difficult and often subjective, but they can be most valuable when there is knowledge of the environmental requirements of the species composing the communities. Thus, the much maligned classifications that have followed Kolkwitz and Marsson provide a valuable base in autecology we would well not overlook in our search for better biological indices.

Community Composition

Early in their search for organisms that would typify particular communities or be indicative of environmental conditions, biologists learned that the common species, the ones that appeared most often in their samples, were usually successful over too wide a range of environmental conditions to be indicative of particular conditions of interest to man. Species having narrower ranges of suitable environmental conditions were often found to be rare, and their occurrence in samples, even when they were present, was doubtful. Knowledge of the relative abundances of species, either common or rare, was found to be more useful in the development of biological indices than was knowledge only of their presence or absence. As knowledge of the relative abundances of many species in a community and knowledge of their environmental requirements were incorporated into the evaluation of environmental conditions, such evaluations became more dependable. Community composition, then, has come to be accepted as being much more reliable than particular indicator organisms for evaluating environmental conditions, pollutional or otherwise.

The use of community composition as evidence of environmental change in space or time requires, first, identification and enumeration of species and individuals at different locations or times and, second, some means of analyzing the data and presenting the findings. Hynes (1960, p. 163), with reference to the saprobic systems and other formal methods of analysis we will consider in this section, has justifiably argued:

In conclusion I would stress that in my view it is a great mistake to try to evolve formal methods such as those quoted above. In nature little is simple and straightforward, and a rigid system can lead only to rigidity of thought and approach. Each river or stream and each effluent is different, so the pattern of pollution varies from place to place. But although the pattern varies the phenomenon is nonetheless detectable, as has I hope been made clear in previous chapters. If numerical data are collected and tabulated, or drawn as histograms, the effects of pollution are clearly shown even when it is very slight. There is neither need of, nor advantage in, a formal classification into zones, which in any event are not clearly defined, nor is anything to be gained by elaborate graphical methods.

Leaving out organisms whose occurrence "was too spasmodic to illustrate anything" or which were scarce, Hynes (1960, p. 83) prepared a table (Table 20-5) to show the effects of very small amounts of ammonia and cyanide on the composition of a stream community at different locations during different years. From such a table, most of us can draw reasonably valid conclusions regarding the effects of the toxic substances on the composition of the community. But only a fine biologist like Hynes, having experience with the communities involved, can reduce the great volume of species and numerical data to a table as "simple" as this without biasing the findings or losing too much valuable information. Yet for the mind to comprehend the meaning of great masses of data, such reductions are necessary. And biologists, some of whose methods we will be considering, have sought ways of making this information comprehensible not only to themselves but also to others, a very important matter. But almost any method of data reduction results in information loss, the very sophisticated diversity indices based on information theory being no exception. We must guard against such losses. Tables of species and numbers can contain all the basic information, and knowledge of the biology of the species can be used in their interpretation. To the extent that such tables can be objectively analyzed and comprehended, they are certainly to be recommended.

On the basis of her extensive studies of streams in the Conestoga Basin in Pennsylvania, Dr. Ruth Patrick (1949, 1950) proposed a "biological measure of stream conditions." Plants and animals were classified into seven groups, those organisms considered to respond similarly to stream conditions being placed in the same group, even though they were of very different kinds having different biologies (Table 20-6). Histograms were prepared in which the height of the bar for each group represented the number of species in that group at a particular station as a percentage of the number of species in that group found at nine typical "healthy" stations (Fig. 20-6). If any single species occurred in exceptionally large numbers at a station, the bar for its group was doubled in width. Stations were classified as *healthy, semi-healthy, polluted,* or *very polluted,* according to an arbitrary scheme (Table 20-6). Methods such as this can probably reveal rather marked changes in community composition, but the arbitrary establishment of degrees of pollution without reference to social values is not very helpful, as we have explained. Probably a more crucial question, however, is whether or not this is a biologically sound and efficient use of valuable data on the effects of environmental change on community composition. This classification of organisms into the seven groups does not specify the stream conditions, pollutional or other, to which the organisms in each group are supposed to respond similarly. As we have explained, putrescible organic wastes and toxic wastes can affect the same species as well as different species very differently. Different systems of classifying organisms into groups would be necessary for different kinds of wastes. Further, all protozoa, which were placed in one group, or all insects and crustacea, which were placed in another group, do not respond similarly to the same environmental change, as we have seen. This kind of lumping of data on organisms of different kinds having different requirements can only result in the obscuring of community changes and the loss of valuable information.

Wurtz (1955), recognizing the analytical procedures used by Patrick (1949, 1950) to be biologically unsound, reanalyzed data

TABLE 20–5   *Numbers of animals collected in April 1955, 1956, 1957, at different stations above and below the point of discharge of exceedingly small amounts of ammonia and cyanide into a river. Particular plants are indicated as being absent (O), present (P), common (C), or abundant (A). Mixing of toxic effluent was complete by station E. Flow pattern of effluent was such that station C was not affected and station D was affected maximally. After* Hynes (1960).

ORGANISM	YEAR	STATION AND DISTANCE IN YARDS FROM OUTFALL									
		A 2200 Above	B 50 Above	C 100 Below	D 100 Below	E 600 Below	F 700 Below	G 950 Below	H 1200 Below	I 1600 Below	J 1900 Below
Flatworm											
Polycelis felina	1955	5	1	19	6	24	123	38	21	26	7
	1956	1	2	8	0	1	4	1	0	1	5
	1957	1	2	0	0	3	2	13	1	1	4
Mayflies											
Baetis rhodani	1955	56	44	67	20	131	56	28	25	32	20
	1956	61	18	36	2	6	0	1	0	2	0
	1957	329	347	475	9	43	40	51	42	63	58
Rhithrogena semicolorata	1955	267	263	214	31	375	62	53	119	54	47
	1956	149	88	253	2	6	3	11	14	2	54
	1957	102	169	239	4	47	8	2	8	2	2
Stoneflies											
Amphinemura sulcicollis	1955	62	119	60	23	48	45	13	24	39	27
	1956	228	175	166	11	34	12	48	15	23	70
	1957	130	267	185	42	131	24	38	19	15	8
Leuctra spp.	1955	22	27	16	0	2	1	0	3	3	2
	1956	4	6	8	0	0	0	0	0	2	0
	1957	89	57	18	1	4	0	1	1	0	0
Isoperla grammatica	1955	17	30	10	5	14	8	1	3	1	1
	1956	69	105	132	23	13	3	18	6	6	8
	1957	43	41	29	2	10	0	0	0	0	0
Caddisflies											
Rhyacophila dorsalis	1955	4	5	2	3	5	9	3	3	5	5
	1956	12	6	5	1	2	1	0	0	0	0
	1957	13	19	8	10	7	7	5	9	15	10
Hydropsyche instabilis	1955	0	0	0	0	0	0	0	1	0	0
	1956	3	9	6	1	9	9	7	1	8	11
	1957	39	55	22	19	27	15	40	21	40	36

from the Conestoga Basin and presented these data in a manner remedying some of the difficulties. He classified organisms that were not part of the plankton into four groups: *burrowing, sessile, foraging,* and *pelagic.* He further classified these organisms into *tolerant* and *nontolerant groups;* this is to some extent possible with regard to conditions created by putrescible organic wastes, but it is not really appropriate for toxic wastes, a distinction Wurtz failed to recognize. According to his scheme, the data are presented in histograms in which each bar represents the number of species leading one of the four ways of life at a location, as a percentage of

the total number of species at the location (Fig. 20–7). The absolute numbers of tolerant and nontolerant species in each group are shown by numerals above or below its bar. Wurtz considered a community to be typical of clean water if more than 50 percent of the species were in the nontolerant category. This system, then, does allow utilization of information on the tolerance of species; in fact, it requires such information. This is one of its difficulties, because such information is rarely available for all species, and few species are clearly tolerant or nontolerant. The investigator is given no flexibility in the use of such infor-

TABLE 20–5   *Continued.*

ORGANISM	YEAR	STATION AND DISTANCE IN YARDS FROM OUTFALL									
		A 2200 Above	B 50 Above	C 100 Below	D 100 Below	E 600 Below	F 700 Below	G 950 Below	H 1200 Below	I 1600 Below	J 1900 Below
Sericostoma personatum	1955	3	4	1	1	1	12	2	5	6	1
	1956	1	4	14	1	5	3	3	0	1	5
	1957	3	2	4	1	1	1	4	0	2	3
Beetle											
Esolus parallelopipedus	1955	53	94	100	2	23	13	2	17	8	23
	1956	11	6	38	0	4	4	2	2	7	7
	1957	40	55	12	0	3	11	14	11	10	12
Biting midges											
Ceratopogonidae	1955	4	0	0	2	1	3	0	0	0	0
	1956	26	11	38	14	10	14	12	12	21	18
	1957	2	13	3	23	41	20	21	13	15	10
Nonbiting midges											
Orthocladiinae	1955	109	76	153	66	51	38	25	104	20	85
	1956	70	163	100	37	40	69	90	103	10	44
	1957	37	48	25	0	23	23	38	9	10	7
Tanypodinae	1955	0	0	1	1	0	1	1	1	1	1
	1956	0	1	3	1	2	4	0	1	6	2
	1957	2	1	1	6	7	9	6	6	2	9
Alga											
Lemanea	1955	C	C	C	P	P	C	C	C	C	C
	1956	A	A	A	0	P	P	C	C	A	C
	1957	A	A	A	0	P	C	C	A	A	A
River moss											
Hypnum palustre	1955	C	P	P	C	C	A	A	A	A	A
	1956	P	C	C	C	A	A	A	A	A	A
	1957	C	P	C	P	C	A	A	A	A	A
Total number of animals in sample	1955	654	802	698	288	278	452	223	405	285	325
	1956	680	653	931	136	187	246	230	164	133	304
	1957	976	1158	1105	336	394	187	255	154	191	184

mation as he may have available. And this system permits no exploitation of valuable information on relative abundance, information that not always but often is obtained in the course of biological investigations.

Biologists like Patrick and Wurtz are to be commended for their search for means of reporting complex biological data in forms readily usable by engineers and other nonbiologists, an important consideration in water pollution control. Some, like Beck (1955), have developed simple numerical indices that can easily be included along with other water quality data in reports and regulations. On the basis of his considerable experience

as a biologist with the Florida State Board of Health, Beck (1954, 1955) proposed the use of a "biotic index":

Biotic index = 2(n Class I) + (n Class II)

where n represents the number of macroscopic invertebrate species either in Class I (tolerant of no appreciable organic pollution) or in Class II (tolerant of moderate organic pollution but not of conditions nearly anaerobic). More weight in the computation of the index was given to Class I organisms, because they are the least tolerant of pollutional conditions. A stream nearing septic conditions will have a biotic

TABLE 20–6 Classification of various organisms into groups and method of determining stream condition based on these groups proposed as a "biological measure of stream conditions" by Dr. Ruth Patrick (1949, 1950).

CLASSIFICATION OF ORGANISMS INTO GROUPS

Group	Organisms
1	The blue-green algae, certain green algae, and certain rotifers
2	Oligochaetes, leeches, and pulmonate snails
3	Protozoa
4	Diatoms, red algae, and most green algae
5	All rotifers not in Group 1, clams, prosobranch snails, and tricladid worms
6	All insects and crustacea
7	All fish

METHOD OF DETERMINING STREAM CONDITION

Stream Condition	Results
Healthy	Groups 4, 6, and 7 each contain more than 50 percent of number of species found in that group at 9 typical "healthy" stations.
Semihealthy	(a) Either or both Groups 6 and 7 less than 50 percent, and Group 1 or 2 less than 100 percent, or (b) Either Group 6 or 7 less than 50 percent, and Groups 1, 2, and 4 100 percent or more, or Group 4 contains exceptionally large number of individuals.
Polluted	(a) Either or both Groups 6 and 7 are absent, and Groups 1 and 2 50 percent or more, or (b) Groups 6 and 7 both present but less than 50 percent and Groups 1 and 2 100 percent or more.
Very polluted	(a) Group 6 and 7 both absent and Group 4 less than 50 percent, or (b) Either Group 6 or 7 is present and Group 1 or 2 less than 50 percent.

index value of zero, whereas Beck found streams receiving moderate amounts of organic waste to have values from 1 to 6 and streams receiving little or no waste to have values usually over 10. This system was useful for Beck in his work in Florida. In judging any system, this is perhaps the most important consideration; biologists and other investigators adopt the procedures that their own experience has shown to be effective. Beck's biotic index takes into account neither the relative abundance of individuals of different species nor the presence of most of the species in the community. But time does not always permit intensive investigations. We must be careful to evaluate biological index systems within the framework of the objectives that they have been designed to accomplish.

Thus, we see that biologists have struggled with the problems of analysis and representation of data on the effects of pollutional changes on community composition. When data on the species and numbers of individuals present in communities are available, it is a shame to reduce these data by procedures resulting in the loss of information. The use of tables, such as Hynes (1960) suggests, need not result in the loss of information; and available biological knowledge can be exploited in their interpretation. If it is necessary or desirable to reduce such information to very simple terms, diversity indices based on information theory are far more efficient than other procedures. Such indices should complement tabular analysis. But the intensive investigations these two procedures require are not always possible or even necessary for practical purposes. Experienced biologists know that very limited sampling and simple means of analysis and representation can often be very useful in studies of aquatic environments.

Community Diversity

One must wonder at the great *diversity*, or variety, of kinds of life that exists in some communities: tropical rain forests, coral reefs, even communities in brooks. But there are some communities that exhibit *monotony*, or very little diversity, in their composition, either because they have very few species or because one or two species are very abundant and other species are very rare. Perhaps in most communities there are a few species that are quite abundant, many species that are quite rare, and many other species that are neither very abundant nor very rare. Such communities are neither very diverse nor very monotonous; they are intermediate in diversity. Theoretically, *maximum diversity* exists when there are a great many species, each represented by one individual, and *minimum diversity* exists when all individuals belong to the same species;

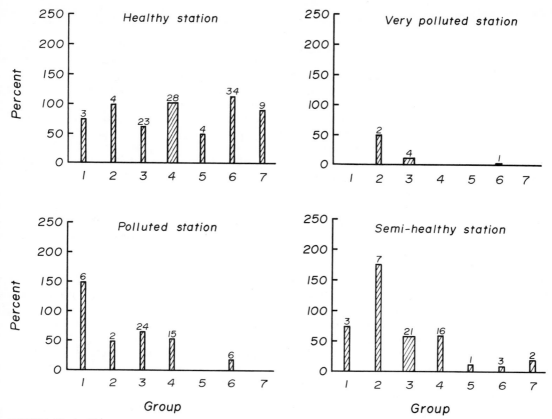

FIGURE 20-6 Changes in community composition of Little Muddy Creek, Conestoga Basin, Pennsylvania, resulting mainly from introduction of decomposable organic wastes, as shown by Dr. Ruth Patrick's "biological measure of stream conditions." Height of the histogram bar for each group represents the number of species in that group at a particular station as a percentage of the number of species in that group found at nine typical "healthy" stations. A bar having double width indicates that at least one species in the group occurred in unusually high numbers. Numerals above bars indicate number of species in groups found at particular stations. After Patrick (1950).

FIGURE 20-7 Changes in community composition of Lititz Run, Conestoga Basin, Pennsylvania, resulting from introduction of decomposable organic wastes, stream recovery, and finally dilution with waters of Conestoga Creek. Organisms are classified as being tolerant or nontolerant of pollutional conditions and as burrowing (B), sessile (S), foraging (F), or pelagic (P). Each bar represents the number of species in a particular category as a percentage of the total number of species found at a station. The number of species in each category is given by the numeral above or below its bar. After Wurtz (1955).

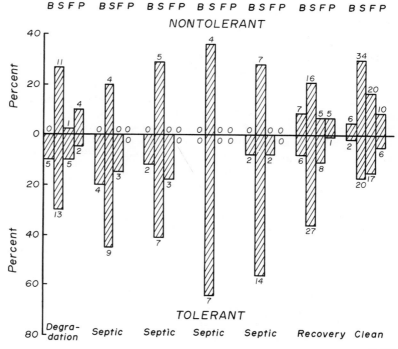

neither of these conditions can, of course, exist in any natural community.

Community diversity is a wondrous thing to most of us. To biologists it is more: a fascinating and difficult problem to solve. Biologists have a very poor understanding of the factors that permit great diversity in some communities and little in others. Stability of physical and chemical conditions, high capacity for primary production, large size—these are factors in ecosystems often believed to lead to greater diversity. But substrate morphology, competition, predation, and evolutionary time are probably also important in various ways. There are many theories (Pianka, 1966), none of which alone can explain diversity. One thing does seem clear: a marked environmental change, without the passage of sufficient time for the evolution of a new community to occur, is likely to lead to a reduction in the diversity of the community at the affected location. Our reader cannot help but recall, for we have tried his patience with re-emphasis, that the introduction of putrescible organic wastes into rivers, lakes, and estuaries leads to a reduction in the number of species and the numerical ascendancy of a few species. This is a reduction in community diversity caused by an environmental change. If we can find an efficient way of measuring community diversity, we should have a numerical value that could be one useful biological index of environmental change.

Wilhm and Dorris (1968), in a brief paper of considerable importance to biologists interested in water pollution control, evaluated two general types of *community diversity indices* on the following bases: their utilization of the distribution of individuals among species, their relative independence of sample size, and their being dimensionless. Wilhm and Dorris then illustrated the use of the more efficient kind of index, the one based on information theory, as a biological index of environmental change caused by waste discharges.

Margalef (1951) proposed an index to community diversity that can be computed from the formula

$$\overline{d} = \frac{(s - 1)}{\log_e n}$$

where s is the number of species and n is the total number of individuals in a community. As pointed out by Wilhm and Dorris, the three hypothetical communities shown in

Table 20-7 would exhibit very different levels of diversity—A appearing most diverse and C appearing most monotonous—yet diversity values obtained from $(s-1)/\log_e n$ would be the same for the three communities. Relative abundance of different species is an important characteristic of the structure of a community and inevitably influences our idea of its diversity. Valuable information on relative abundance, if available, is wasted when diversity indices that do not incorporate this information are used. Wilhm and Dorris further concluded that diversity values computed with formulas of this kind are not sufficiently independent of sample size. Finally, were we to compute diversity indices with total biomass rather than with total number of individuals—as might sometimes be useful—we would find the diversity values to vary with our choice of weight unit, if we used formulations of this kind; they are not dimensionless.

Margalef (1956) was perhaps the first to propose that methods derived from *information theory* be used in obtaining diversity indices. In this approach, diversity is equated with the uncertainty that exists as to the species of an individual we might select at random from a community. Obviously, the more species in a community and the more nearly equal their abundances, the greater would be the uncertainty as to the species of the individual we might select. Since diversity is equated with this uncertainty, the greater the uncertainty, the greater the diversity. Information content is a measure of uncertainty, and thus it is a possible measure of diversity. Patten (1962) used such an index of diversity in his study of the phytoplankton of Raritan Bay. Brillouin

TABLE 20-7   Three hypothetical communities having the same number of species (s) and total number of individuals (n) that yield the same diversity index, $\overline{d} = (s - 1)/\log_e n$, even though their distributions of individuals (n_i) among species differ greatly. After Wilhm and Dorris (1968).

| COM-MUNITY | INDIVIDUALS IN SPECIES i (n_i) | | | | | TOTAL INDI-VIDUALS | TOTAL SPECIES |
	n_1	n_2	n_3	n_4	n_5	n	s
A	20	20	20	20	20	100	5
B	40	30	15	10	5	100	5
C	96	1	1	1	1	100	5

(1962) has considered the scientific application of information theory and the mathematics involved. We are poorly qualified to discuss this matter; even were we better qualified, the purpose of our presentation here and the limits of space would not encourage us to do so. The explanation given by Wilhm and Dorris (1968, p. 478) seems sufficient for our purpose:

The formula given by Brillouin (1960) [1962] as a measure of diversity (or information) per individual is,

$$\overline{H} = (1/N) \left(\log N! - \overset{s}{\underset{1}{\Sigma}} \log N_i!\right)$$

where N is the number of individuals in s species and N_i is the number of individuals in the i'th species [and logarithms are interpreted to base 2]. Assuming reasonably large values of N and N_i, the logarithms of the Γ functions may be approximated by Stirling's formula to yield

$$\overline{H} = -\Sigma(N_i/N) \log_2(N_i/N).$$

The population ratio (N_i/N) is estimated from sample values (n_i/n) to yield the equation

$$\overline{d} = -\Sigma(n_i/n) \log_2(n_i/n).$$

A theoretical maximum diversity, $\overline{d}_{max}$, and a theoretical minimum diversity, $\overline{d}_{min}$, can be calculated.

$$\overline{d}_{max} = (1/n) [\log_2 n! - s \log_2(n/s)!]$$
$$\overline{d}_{min} = (1/n) \{\log_2 n! - \log_2 [n - (s-1)]!\}$$

Then the position of d between these extremes can be calculated by the redundancy expression, r.

$$r = \frac{\overline{d}_{max} - \overline{d}}{\overline{d}_{max} - \overline{d}_{min}}$$

Redundancy is an expression of the dominance of one or more species and is inversely proportional to the wealth of species. Thus, indexes derived from information theory permit an expression not only of the compositional richness of a mixed-species aggregation of organisms, $\overline{d}$, but also of the dominance of one or more species, r.

The diversity index $\overline{d}$, based on informa-

tion theory, is different for each of the three hypothetical communities given in Table 20-7; it does take into account the relative abundances of the different species, something not done by Margalef's original index (1951). Furthermore, it is relatively independent of sample size. Diversity indices, based on information theory, calculated from increasing numbers of pooled samples from three different habitats and from a sample board are given in Table 20-8. For each habitat, the value in the first column represents the diversity index $\overline{d}$ determined from the first sample, while the value in the tenth column represents this index for all ten samples pooled. These diversity values reach 95 percent of their asymptotic values by the first sample from the spring, the third sample from the meadow, the fourth sample from the stream, and the fifth sample from the board. Further increase in the number of samples taken would undoubtedly have revealed more rare species, but the contribution of rare species to the value of this diversity index is quite small. Thus, values of this index are relatively independent of sample size, a very important consideration. Finally, should it be desirable to consider diversity in terms of biomass rather than in terms of numbers of individuals, the values of the diversity index based on information theory will not depend on the choice of weight unit, for these values will have no dimensions.

Wilhm and Dorris (1968) computed diversity indices, based on information theory, for communities above and below the points of introduction of different kinds of wastes into fresh and marine waters (Table 20-9). As we have come to expect, community diversity first decreased below the points of waste introduction, and then, with time and distance

TABLE 20-8 *Diversity indices,* $\overline{d} = -\Sigma(n_i/n)\log_2(n_i/n)$, *computed for three communities and for samples from a sample board that illustrate the relative independence of indices from sample size when index based on information theory is used. Values have been computed from pooled samples increasing from 1 to 10. After* Wilhm and Dorris (1968).

HABITAT	ORGANISMS	NUMBER OF SAMPLES POOLED									
		1	2	3	4	5	6	7	8	9	10
Meadow	Insects	2.37	3.61	3.98	4.02	4.07	4.08	4.16	4.16	4.17	4.16
Stream	Macroscopic	2.34	2.98	3.10	3.15	3.16	3.20	3.27	3.30	3.28	3.28
Spring	Macroscopic	1.34	1.39	1.40	1.37	1.35	1.38	1.36	1.37	1.38	1.40
Sample board		2.18	2.21	2.35	2.63	2.77	2.80	2.87	2.85	2.89	2.86

TABLE 20–9   *Changes in diversity indices,* $\bar{d} = -\Sigma(n_i/n)\log_2(n_i/n)$, *based on information theory, with distance of flow below points of waste discharge into natural and other waters. After* Wilhm and Dorris (1968).

| | | DIVERSITY INDEX ($\bar{d}$) | | |
| | | Above Outfall | Near Outfall | Downstream |
WATER	WASTE			
Skeleton Creek	Domestic and oil refinery		0.84	1.59	3.44
Skeleton Creek	Domestic and oil refinery	3.75	0.94	2.43	3.80
Otter Creek	Oil field brines	3.36	1.58		3.84
Refinery Ponds	Oil refinery		0.98	2.79	3.17
Keystone Reservoir	Dissolved solids		0.55		3.01
Alamitos Bay	Oil field brines		1.49	2.50	
Alamitos Bay	Oil field brines		1.44	2.70	
Alamitos Bay	Storm sewer		1.45	2.81	

of flow, it gradually returned to levels typical for the particular waters. The diversity values for the communities of Alamitos Bay are of especial interest, because Reisch and Winter (1954), who made the observations, concluded from their subjective analysis of the species present and their relative abundances that the wastes involved had little if any effect on the biological communities. The diversity indices computed by Wilhm and Dorris clearly indicate that there were effects that were not revealed by subjective analysis of the tabulated data. This is important: it is evidence that, as irreplaceable as are good methods of tabular analysis, the unaided human mind cannot extract from complex tables all evidence of changes in community composition; good diversity indices are also necessary.

In 1605, in his great *Advancement of Learning*, Sir Francis Bacon (1605, republished 1952, p. 40) wrote: "Doth any give the reason, why some things in nature are so common, and in so great mass, and others so rare, and in so small quantity?" Later by 350 years, Professor Hutchinson (1959) wrote on "Why are there so many kinds of animals?" Before Bacon, men were aware of diversity, and hopefully, long after Hutchinson, men will be interested in the manifestations of diversity and studying its causes. But a diversity index is only a manifestation; alone it provides no framework for causal explanation. It is a useful and powerful tool for describing certain community phenomena; the explanation of these phenomena requires other knowledge. It can be a very useful biological index of community or

environmental change. It should not replace the use of other knowledge about the biology and environmental requirements of the species contributing to diversity. Tables of species and numbers, as subjective as their analysis may be, still permit use of knowledge about individual species in their interpretation. There is no single path to understanding.

THE USEFULNESS OF BIOLOGICAL INDICES

From the early chapters of this book, we have endeavored to explain how changes in environmental conditions lead to changes in the distribution and abundance of particular species and, thus, to changes in the composition of communities. The explanation of these biological changes may lie at any level of biological organization: cellular, individual organism, population, or community. Changes in the distribution and abundance of a particular species may result from direct effects of a primary environmental change on its survival and reproduction or from indirect effects following changes in the distribution and abundance of other species with which it is associated. Either way, a succession of events leads to changes in community composition. In this chapter, we have seen that the waste products of civilization bring about changes in river, lake, and estuarine environments that lead to changes in the distribution and abundance of individual species and, hence, to alterations in communities. We have reviewed the origins of the biological index concept. And we have described some of the ways in which biologists have applied this concept to the detection of environmental changes caused by the introduction of wastes into natural waters. When a society is confronted with an important social problem, many of the citizens of that society seek to contribute their special knowledges to its solution. The biological index concept is one of the contributions biologists have made to the solution of the social problem of pollution. Though this concept has dominated biological thought relevant to water pollution, biology has, in addition, much more to offer, as we hope this book has made clear. A biological index is but a tool, albeit a useful tool. If biologists have sometimes been overly enthusiastic in their claims for biological indices—as we pointed out, perhaps too critically, some time ago (Doudoroff and Warren, 1957)—this is understandable. Given a fascinating idea having some relevance to an important social problem, concerned people will want to contribute to the betterment of society. But the social and biological complexity of pollution problems cannot be cut by a single tool, no matter how useful. In the future, biological indices will have a role, but a role more proportional to their usefulness as but one of many possible biological approaches to the solution of water pollution problems.

Changes in biological indices are evidence of environmental changes. The idea of index suggests that the biological indices themselves provide no direct evidence that uses of the waters involved are endangered. If *Oscillatoria* or *Sphaerotilus*—organisms which, because of their very nature, create water pollution problems—is the biological index, then, of course, the index and the pollutional effect are the same. But, generally, changes in aquatic communities need not affect uses of waters, even fisheries. Other water quality characteristics may be involved in decreasing the value of particular uses. For this reason, other water quality parameters are usually measured. This aids in the interpretation of the biological index information; it provides additional information as to whether or not water uses are being endangered; and it provides environmental information to increase our understanding of community changes. We should not expect too much of information collected primarily for index purposes.

Community composition and diversity can be extremely sensitive biological indices of environmental change. Rather slight changes in environmental conditions, if persistent, can lead to changes in community composition and diversity. But are we to view all environmental changes that lead to changes in biological indices as pollution to be controlled? We cannot usefully hold this view, because man cannot possibly persist on the earth without influencing the communities of which he is really a part. In this regard, he is not different from other organisms. McKee (1967, pp. 259-260) has stated this very well:

To some people, pollution means any departure from a condition of pristine purity. There is, however, no true base line of pristine purity, for the parameters of natural water quality have been changing inexorably since prehistoric times as a result of natural phenomena. The purists sometimes fail to recognize that the propagation, proliferation, and prosperity of each life form on this planet exert an influence on the environment. Man is no exception. The development of his civilization, economy, and well-being must inevitably have an impact on the environment. . . .

The mere fact that parameters of water quality are changed as a result of man's activities does not in itself constitute pollution. There must be evidence that these changes are adverse to one or more beneficial uses of water, and such adversity must be unreasonable, i.e., more than trivial or superficial.

Society has not chosen and cannot choose to consider all environmental changes as pollution to be controlled (Chapters 2 and 22). Biological indices can detect a range of environmental changes from very slight ones, which man cannot prevent, to very great ones, which man, for his own good, must prevent. How, then, are we to determine which of the environmental changes that biological indices detect are to be viewed as pollution to control?

To answer this question, we must consider the beneficial uses of the particular waters involved. It is only when we consider the uses society chooses to make of particular waters that we are able to relate biological indices of environmental change to that range of environmental changes to be considered as pollution that should be controlled. Once a water has been classified as to the uses and level of each use society deems most beneficial for man, it becomes possible to establish the water quality standards necessary to protect these uses (Chapter 3). Then the aquatic community that persists in the range of water quality conditions permitting these uses can be described. These uses may not require that no effluents of any kind be introduced into the water. Should waste introduction consistent with desired uses be permitted, the aquatic community developing would be different than the one persisting when no waste introductions occurred. Resulting changes in biological indices in this range could not be considered evidence that pollution, as we have defined it, was occurring. But should gradual increases in waste introductions occur over the years, further changes in biological indices would occur; these changes could well be warning that further degradation of water quality might endanger the desired uses of the water. Thus, not all changes in environmental conditions can be considered pollution, and not all changes in biological indices can be taken as evidence that pollutional changes in the environment are occurring. But when a water has been classified for particular uses, and when a known aquatic community persists under water quality conditions permitting those uses, changes in biological indices resulting from degradation of water quality may well be evidence of pollution. This is what Hawkes (1962) meant in the quotation with which we began this chapter.

We would hope that man will choose to keep a few of the waters of the earth in essentially their primitive condition. In these waters, any change in biological communities would be evidence of pollution. In other waters, we probably must accept changes that are consistent with and necessarily follow those uses wisely judged by society to be in man's best interest. Rivers and lakes of recreational or water supply value may reasonably be permitted to change to some point, but the appearance of *Oscillatoria rubescens* may well be evidence that any change has gone too far. Properly used, biological indices can give us early warnings. Our con-

ceptualization of the relationships between society and water pollution as well as our knowledge of the responses of communities to environmental change determine the usefulness of biological indices.

SELECTED REFERENCES

Butcher, R. W. 1946. The biological detection of pollution. Journal of the Institute of Sewage Purification 1946(2):92-97.

Doudoroff, P., and C. E. Warren. 1957. Biological indices of water pollution, with special reference to fish populations. Pages 144-163. *In* C. M. Tarzwell (Editor), Biological Problems in Water Pollution. Transactions of the 1956 Seminar. R. A. Taft Sanitary Engineering Center, U.S. Department of Health, Education, and Welfare. 272pp.

Filice, R. P. 1959. The effect of wastes on the distribution of bottom invertebrates in the San Francisco Bay estuary. Wasmann Journal of Biology 17:1-17.

Hasler, A. D. 1947. Eutrophication of lakes by domestic drainage. Ecology 28:383-395.

Hawkes, H. A. 1962. Biological aspects of river pollution. Pages 311-432. *In* L. Klein, River Pollution. 2. Causes and Effects. Butterworth, London. xiv+456pp.

Hynes, H. B. N. 1960. The Biology of Polluted Waters. Liverpool University Press, Liverpool. xiv+202pp.

Kolkwitz, R., and M. Marsson. 1908. Ökologie der pflanzlichen Saprobien. Berichte der Deutschen Botanischen Gesellschaft 26a:505-519. (Translated 1967. Ecology of plant saprobia. Pages 47-52. *In* L. E. Keup, W. M. Ingram, and K. M. Mackenthun (Editors), Biology of Water Pollution. Federal Water Pollution Control Administration, U.S. Department of the Interior. iv+290pp.)

———. 1909. Ökologie der tierischen Saprobien. Beiträge zur Lehre von der biologischen Gewasserbeurteilung. Internationale Revue der Gesamten Hydrobiologie und Hydrogeographie 2:126-152. (Translated 1967. Ecology of animal saprobia. Pages 85-95. *In* L. E. Keup, W. M. Ingram, and K. M. Mackenthun (Editors), Biology of Water Pollution. Federal Water Pollution Control Administration, U.S. Department of the Interior. ix+290pp.)

Patrick, Ruth. 1950. Biological measure of stream conditions. Sewage and Industrial Wastes 22:926-938.

Richardson, R. E. 1928. The bottom fauna of the middle Illinois River, 1913-1925: its distribution, abundance, valuation, and index value in the study of stream pollution. Bulletin of the Illinois State Natural History Survey 17:387-475.

Wilhm, J. L., and T. C. Dorris. 1968. Biological parameters for water quality criteria. Bioscience 18:477-481.

21 Biological Waste Treatment

Dead matter which ferments and putrifies is not obedient, at any rate inclusively, to forces of a nature purely physical or chemical. It is life which rules over the work of death and the dissolution of animal and vegetable matter. This constant return to the atmospheric air and to the mineral kingdom of the constituents which vegetables and animals have borrowed from them is an act related to the development and multiplication of organized beings.

LOUIS PASTEUR

The treatment of waste waters in biological oxidation plants may be regarded as the environmental control of the activity of populations of the necessary organisms.

H. A. HAWKES, 1963, p. vii

WATER POLLUTION CONTROL, WASTE TREATMENT, AND BIOLOGY

The objective of water pollution control is to minimize harmful effects occasioned in natural waters by introduction of the waste products of man's activities. This is also the objective of waste treatment. Physical, chemical, and biological principles and processes are applied in waste treatment to attain this objective. Because waste treatment is such an integral part of water pollution control, and because biology is such an integral part of waste treatment, no book pretending to consider biology and water pollution control can ignore waste treatment. This chapter, then, is our attempt, however cursory, to explain some of the biology of waste treatment.

Man finds no new processes in his exploration of natural phenomena; he only becomes aware of processes which extend back nearly to the beginning of life. Creative as he is, man can combine old processes and forms in new ways, but we have no reason to believe the components of his creations to be new. Like their physical and chemical coun-

terparts, the biological processes involved in waste treatment occur nearly everywhere in nature on the earth; most particularly, they occur in the waters of the earth. Before man ever introduced wastes into these waters, all the groups of organisms presently involved in waste degradation in natural waters were present. And the organisms involved in engineered systems of biological waste treatment are the not-too-distant progeny of organisms degrading those wastes that reach natural waters. A biological waste treatment plant is a very specialized ecosystem. It is man's attempt to compress in space and time degradation processes that he does not wish to extend far in these dimensions over the earth. He has accomplished this by rather remarkable engineering involving biological systems, which we intend to sketch. To do this, we need not introduce our reader to additional biological ideas. At this late point in a book on biology, we would have failed were this necessary. It would be tedious to indicate the earlier chapters pertaining to this one, for most of them contain relevant biological ideas.

To those of us who are biologists unspecialized in the application of biology to waste treatment, and to all who would learn something of the biology of waste treatment, Hawkes (1963), with his book *The Ecology of Waste Water Treatment*, has rendered a service. His own contributions, as well as those of others, to a unified theory of the biology of waste treatment are apparent in this and other of his writings (Hawkes, 1961).

HYDROLYSIS, OXIDATION, AND REDUCTION

Hydrolysis, oxidation, and reduction are degradation processes upon which all biological waste treatment depends and, indeed, upon which all life depends. The synthesis of new protoplasm is also involved in waste treatment, but hydrolytic splitting of molecules to permit passage of their components across cell membranes and oxidation and reduction to provide energy are necessary for this synthesis. As we shall see, organisms of many different kinds are involved. Through the individual and community activities of these organisms, organic wastes are converted to carbon dioxide, minerals, water, heat, new protoplasm, and stable organic materials. The synthesis of new protoplasm is essential in the operation of biological waste treatment systems, but these must be operated so as to yield no more new protoplasm than necessary, because this too requires disposal.

Hydrolysis is the process by which organic and inorganic compounds are split into fragments by the addition of water. Hydrolyses are spontaneous reactions in which small amounts of energy are liberated, but, in the absence of catalysts, they usually proceed very slowly. Hydrolytic enzymes, or *hydrolases*, are involved in the biological degradation of carbohydrates, fats, and proteins. Hydrolysis is often a necessary first step in such biological degradation, for these compounds are usually too large to pass across cell membranes. In animals, including protozoa that enclose their food before digestion begins, hydrolysis is internal. Bacteria, however, secrete into their external medium hydrolases that break their nutrients into fragments that can be absorbed. Large polymeric molecules are usually hydrolyzed in several steps, each step being catalyzed by a specific enzyme. Glycogen and starch, which are polymeric carbohydrates, are hydrolyzed

into monosaccharides, proteins into amino acids, and fats into alcohols and fatty acids.

Oxidation and *reduction* are extremely important in chemical reactions, including the biochemical ones involved in the degradation of organic matter. Oxidation was formerly viewed either as the addition of oxygen to a substance or the removal of hydrogen, whereas reduction was viewed as the reverse of either process. But many reactions now viewed as oxidations or reductions involve neither oxygen nor hydrogen, and the earlier idea has been extended to include any change in the number of electrons of a substance. Now, a substance losing one or more electrons is said to be *oxidized*, and a substance gaining one or more electrons is said to be *reduced*. A substance accepting electrons thus tends to act as an *oxidizing agent* and one yielding electrons tends to behave as a *reducing agent*. In any reaction in which one substance is oxidized, another must be reduced. Biological oxidation and reduction reactions are catalyzed by enzymes that are usually specific for the particular organisms and reaction steps involved.

Oxidation-reduction processes may proceed either in the presence of free oxygen, *aerobically*, or in its absence, *anaerobically*. Organisms adapted to live only in the presence of free oxygen are called *aerobic organisms*, whereas those adapted to live only in its absence are termed *anaerobic organisms*. *Facultative anaerobes* are able to maintain themselves under either of these conditions. In streams receiving organic wastes, either aerobic or anaerobic conditions may exist, and organisms adapted to one or the other condition bring about the degradation of these wastes. In aerobic waste treatment systems, such as trickling filters and activated sludge plants, aerobic organisms bring about the hydrolysis, oxidation, and reduction of waste substances. In sludge digestion, a process used to stabilize organic solids, anaerobic organisms carry out these processes in the absence of oxygen. Except for organisms using light energy, oxidation-reduction processes provide the energy organisms need to maintain themselves; and, along with hydrolysis, these processes make available the materials as well as the energy necessary for growth. Of the myriad microorganisms present in streams and waste treatment facilities, each kind has its own niche and may obtain its energy from only one step in a long degradation sequence. Chemosynthetic bacteria obtain their energy from inorganic compounds, whereas heterotrophic organisms obtain energy from organic compounds.

The complete aerobic oxidation of an organic compound, which may involve several stages requiring the activities of different microorganisms, yields carbon dioxide and water, as in the case of cane sugar, the carbohydrate sucrose:

$$C_{12}H_{22}O_{11} + 12O_2 \rightarrow 12CO_2 + 11H_2O + Energy$$

If, as in proteins, nitrogen is present in the molecule, it will be converted to ammonia and eventually, in the presence of certain bacteria, to nitrate, as we shall see. Any sulfur or phosphorus present will eventually be oxidized to sulfate or phosphate.

The anaerobic oxidation of carbohydrates—including such resistant ones as cellulose—fats, proteins, and most organic substances leads to the production of methane, carbon dioxide, and reduced forms of nitrogen (ammonia) and sulfur (hydrogen sulfide), when these elements are present. Certain species of anaerobic bacteria are responsible for methane formation. Though the exact steps are not clearly understood and are probably not the same for different substrates, the overall reaction for methane production from propionic acid, a fatty acid, can be represented:

$$4CH_3CH_2COOH + 2H_2O \rightarrow$$
$$7CH_4 + 5CO_2 + Energy$$

Decomposition of the nitrogenous products of protein breakdown leads to the production of ammonia, whether mediated by aerobic or by anaerobic bacteria. But, under aerobic conditions, if phosphates are present and inhibiting levels of organic materials or toxic substances do not exist, oxidation of the ammonia can proceed. The first step, the conversion of ammonia to nitrite, is accomplished by bacteria in the genera *Nitrosococcus* and *Nitrosomonas*:

$$2NH_3 + 3O_2 \rightarrow 2HNO_2 + 2H_2O + Energy$$

The nitrite can then be oxidized to nitrate by bacteria of the genus *Nitrobacter:*

$$2HNO_2 + O_2 \rightarrow 2HNO_3 + Energy$$

These, then, are chemosynthetic bacteria, obtaining their energy from the oxidation of inorganic compounds. Similarly, sulfur bacteria of the genus *Beggiatoa* obtain energy by the aerobic oxidation of sulfides to free sulfur:

$$2H_2S + O_2 \rightarrow 2H_2O + 2S + Energy$$

Under anaerobic conditions, when nitrates are absent and organic material is present, other bacteria are able to reduce sulfates to hydrogen sulfide, the organic matter serving as the reducing agent, providing the hydrogen, and itself being oxidized to carbon dioxide:

$$H_2SO_4 + 8H \rightarrow H_2S + 4H_2O + Energy$$

The foul smell of organically polluted waters when they are septic is often due to hydrogen sulfide, a compound which is very toxic to aquatic life.

SELECTION AND DESIGN OF TREATMENT FACILITIES

Primary, Secondary, and Tertiary Treatment

It is usually technically feasible to return the waters involved in the transport of domestic and industrial wastes to almost any desired level of purity. The desired or required level of purity of treated waste waters should be determined by possible dilution in receiving waters and the uses that are to be made of those waters. But in practice, at the present stage of development of waste treatment technology, economic considerations usually limit the treatment of wastes to levels of purity below those desirable and technically attainable. In the selection of waste treatment facilities, then, there occurs some sort of adjustment between desired economies and desired levels of purity of treated waste waters.

Waste treatment is usually accomplished in a series of more or less discrete stages, the level of treatment attained being roughly correlated with the number of stages in the overall process. These stages are sometimes conveniently grouped under the headings primary treatment, secondary treatment, and tertiary or advanced treatment. Where dilution in receiving waters is great or when effluent requirements are lax for other reasons, only primary treatment may be practiced. More often, now, secondary treatment in addition is required. And, with increasing use of our waters, we can expect that tertiary treatment of the effluents of

TABLE 21–1   *Extent of purification of raw sewage by various treatment processes. After* Klein (1966).

PROCESS	APPROXIMATE PERCENTAGE OF REDUCTION		
	5-day BOD	Suspended solids	Bacteria
Plain sedimentation	30–40	40–75	25–75
Septic tank	25–65	40–75	40–75
Chemical precipitation	60–75	70–90	40–80
Sedimentation + contact beds	50–75	70–80	50–80
Sedimentation + trickling filters (low-rate)	80–90*	80–90*	90–95
Sedimentation + activated sludge	85–95*	85–95*	90–98
Sedimentation + intermittent slow sand filters	90–95*	85–95*	95–98
Oxidation ponds	75–95†	‡	>99.9

*Even higher reductions are sometimes attained.
†BOD after filtration.
‡Suspended solids may be high because of presence of algae.

secondary processes will become more common. The design of primary and secondary waste treatment facilities is well covered in sanitary engineering texts (Imhoff and Fair, 1956; Clark and Viessman, 1965; Fair, Geyer, and Okun, 1968).

Primary treatment involves the removal of most of the floating and settleable solids from wastes by mechanical means. Wastes are usually first passed through trash racks consisting of steel bars spaced at intervals of 2 to 4 inches. Sand and coarser mineral materials may then be removed in grit chambers. These two treatment steps protect pumps and other equipment from damage. Next, the wastes are sometimes passed through fine screens that are self-cleaning. More often, they are passed directly into skimming devices and settling tanks that may be either separate or combined units. The removal of settleable solids in *primary settling tanks* yields *primary sludge*, the disposal of which we will consider later. More than half the materials suspended in sewage can be removed through primary treatment.

Secondary treatment involves the removal of most of the colloidal and dissolved organic materials present in wastes. This is usually accomplished under aerobic conditions by biological oxidative decomposition and the production of biological growths that are removed in secondary settling tanks as *secondary sludge*. The primary and secondary sludges are usually then decomposed in anaerobic sludge digestion tanks. Trickling

filters and activated sludge tanks are now the devices most often used to maintain the aerobic conditions and the intimate contact between waste waters and organisms necessary for biological removal of waste materials. Some idea of the possible extent of purification of sewage by different combinations of primary and secondary treatment can be gained from Table 21-1.

Tertiary treatment is any additional processing of secondary treatment effluents undertaken to further improve their quality. Tertiary treatment may be used to further reduce almost any undesirable constituent of wastes: suspended solids, BOD, nitrogen, phosphorus, toxic substances, or bacteria. Numerous methods have been explored (Klein, 1966, pp. 151-161), and some of these as well as new ones will be more commonly used in the future.

Trickling Filters

In the development of primary treatment facilities, the application of the principles of physics through sound engineering design will usually suffice. The design of no successful facility—primary, secondary, or tertiary—can be contrary to these principles and sound engineering; but in the design of secondary treatment facilities employing biological oxidative decomposition and removal of colloidal and dissolved organic materials from wastes, biological principles must also be considered. Physical and chemical characteristics of the environment required by the biological community that is to accomplish the waste treatment set limits on the design of secondary waste treatment facilities. The nature of the waste to be treated and the biological community desired will determine these limits. In a way, then, we are designing an ecosystem, a microcosm we hope to control rigorously for our own purposes. In such an ecosystem, we must be able to control waste or nutrient delivery, assure sufficient oxygen, provide for intimate contact between wastes and organisms over necessary periods of time, and manipulate in other ways the kinds and abundances of organisms developing. Experience more than biological theory has guided waste treatment efforts, but experience has been a good teacher; these efforts have been quite successful. Still, it is not too much to hope that the rapidly developing theory of biological waste treatment (McCabe and Eckenfelder, 1956) will permit improvements in waste treatment processes.

Trickling filters, or *bacteria beds* as they are sometimes called, are essentially beds of stone or other material provided both with apparatus for distributing the waste over the surface and with a drainage system for ventilation and effluent removal. These beds are usually retained within masonry or concrete walls and are 6 to 8 feet in depth, but they may be as shallow as 3 feet or as deep as 12 feet. Very often, the beds are circular in shape, this making distribution of the waste over the surface possible by means of simple rotary devices powered by jet action or motors (Fig. 21-1). When rectangular beds are employed, the wastes may be distributed from fixed sprinklers or from traveling bridges. In whatever manner the wastes are delivered to the beds, they are applied to any one area of a bed only intermittently, usually between two and six times per hour. This increases ventilation and the all-important delivery of oxygen to the biological community. *Loading* of a filter may be expressed in terms of waste volume or BOD per unit time in relation either to the surface area or to the volume of the bed. No single means of

FIGURE 21-1 A picture of a trickling filter with a section removed so as to show construction details. Photograph courtesy of Link-Belt/FMC.

SINGLE STAGE TRICKLING FILTERS

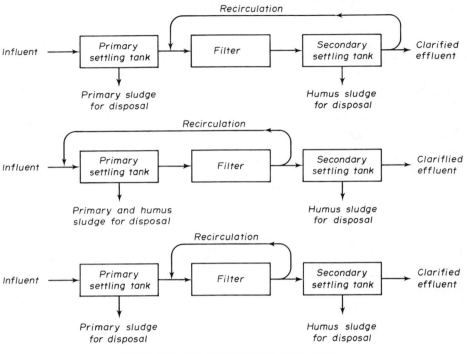

TWO STAGE TRICKLING FILTERS

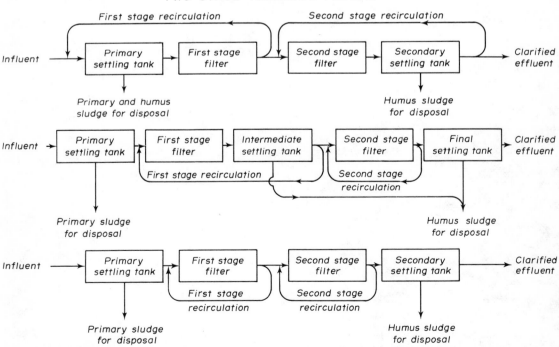

FIGURE 21-2   Flow diagrams for single stage and two stage trickling filters having different patterns of flow recirculation. After Clark and Viessman (1965).

describing loading provides all the information necessary to control conditions in a trickling filter.

The bed itself functions to provide a very large attachment surface for the biological community and to permit the free flow of waste and air in contact with the community. Rock of various kinds is usually employed, but other materials may be satisfactory. The primary requirements are that the material resist crushing and chemical disintegration and that the particles be as nearly spherical and uniform in size as possible. This will prevent packing of the bed material and maximize the interstitial space. The particle size selected for a particular facility should be determined on the basis of the strength and volume of the waste to be treated, for these will influence accumulation and flushing of the biological film. The ideal size for a given waste should provide the maximum surface area consistent with adequate distribution of waste and air and removal of solids. The most suitable particle diameter for different wastes will be usually between 1 and 4 inches.

Trickling filters are provided with underdrains for removal of effluent and for ventilation (Fig. 21-1). These drains may consist of grooves in the floor, loosely fitted tiles, or a false floor constructed of special tile. As a result of differences between air temperatures within and outside of the bed, air is drawn through the drains, as in a chimney.

After primary treatment, the wastes are passed through the trickling filters, whose effluents may be returned to primary sedimentation basins to settle the humus sludge produced — generally an inefficient practice — or the sludge may be settled in secondary sedimentation basins. Many different schemes of treatment by means of trickling filters are employed (Fig. 21-2), the scheme selected for a particular waste usually depending on its characteristics and volume. Recirculation of the effluent through a trickling filter has many advantages, including stabilization of the character and volume of the waste water flowing through the filter. Higher rates of application are maintained in this way, thus making bed utilization more uniform and improving flushing action. Sometimes two filters are operated in series, and sometimes their sequence is periodically altered. The mode of operation of a trickling filter system has profound effects on the ecology of the bed, and better control of biological growth and waste treatment is possible where recirculation and other means of regulation are available to the plant operator.

When secondary treatment of wastes is necessary, either the trickling filter or the activated sludge process is usually employed. The *choice of process* depends upon the characteristics and volume of the waste to be treated, the proximity of large population centers and available land, the climate, and economic considerations. Trickling filters require more land area than activated sludge plants, may cause odor and fly nuisances, and are more costly to construct. On the other hand, the trickling filter process is usually less sensitive than the activated sludge process to changes in effluent character and volume and to many toxic substances. Thus, once installed, trickling filters are generally less costly to operate. The activated sludge process responds much more quickly to control measures, and, when very large volumes of waste water are to be treated and trained operators can be employed, it is the process most often adopted. Activated sludge processes have been developed for treating wastes that present difficult treatment problems, an example being pulp and paper mill effluents having high temperatures and high and variable pH (Gehm, 1956).

Activated Sludge Facilities

The biological community bringing about the degradation of organic wastes in the *activated sludge process* is maintained in suspension in the liquid either by diffused air or by mechanical agitators in aeration tanks (Fig. 21-3). Here, the biological community along with adsorbed materials forms a flocculent mass, the *activated sludge*. The vigorous mixing of waste, sludge, and air accomplished insures delivery of nutrient materials and oxygen to the community.

Continuous flow of the wastes through the aeration tanks is maintained. The tanks themselves may be long channels — 10 to 15 feet deep, 10 to 30 feet wide, and 100 to 400 feet long — or they may be more nearly square or circular in shape. Long tanks are designed to permit mainly transverse mix-

FIGURE 21-3   Secondary and tertiary sewage treatment plant of the South Tahoe Public Utility District in 1965. Since this photograph was taken, the capacity of this plant has been increased from 2.5 to 7.5 million gallons per day and even more advanced waste treatment has been provided. This has been necessary to prevent eutrophication of Lake Tahoe, high in the Sierra Nevada Mountains lying between California and Nevada. Attracted by this beautiful lake, vacationers and permanent residents have increased in number many times in recent years. The plant, as shown before expansion, provided for primary sedimentation (building in center right), activated sludge treatment (lower right), secondary sedimentation (circular pond), sludge digestion (large circular tanks), and tertiary treatment (building in upper right). Photograph courtesy of Cornell, Howland, Hayes, and Merryfield, Engineers and Planners, Corvallis, Oregon.

CONVENTIONAL

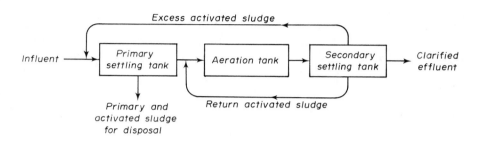

STEP AERATION

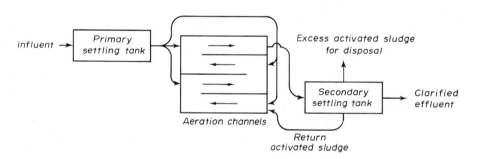

REAERATION

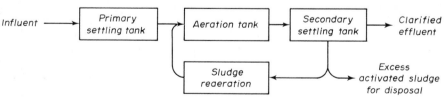

FIGURE 21-4   Flow diagrams showing different methods of operating activated sludge treatment facilities. After Imhoff and Fair (1956).

ing, whereas square or circular tanks are designed for complete mixing.

When diffused air is used for mixing and aeration, the air from compressors is introduced into the tanks in a manner designed to achieve the pattern of mixing desired. Mechanical mixing and aeration, when employed, are accomplished by rapidly rotating paddle wheels, brushes, or propellers. If biological oxidation is to be unimpeded, anaerobic conditions should not be permitted to develop even at the influent end of a tank, and the dissolved oxygen concentration should be no lower than 2 mg/l at the outlet. If a highly nitrified effluent is desired, oxygen concentrations should be 3 to 6 mg/l at the outlet of a channel or, if complete mixing is practiced, throughout the aeration tank.

In a conventional activated sludge plant, the waste, after primary settling, is passed through the aeration tank (Fig. 21-4). Passage time depends on plant design and operation and is generally from four to eight hours. The aeration tank effluent then goes to a secondary settling tank, where the sludge is removed from the final effluent. Most of the settled sludge is returned to the aeration tank, the remainder being removed for further processing and disposal along with sludge from the primary settling tank. Recycling of activated sludge is essential to the process, because any given volume of waste water can produce only about 5 percent of the sludge required for its treatment. The return flow of sludge and water is generally about 20 percent of the flow of the influent waste water. The "age" of the activated sludge in the aeration tank is controlled to three or four days by regulating the return flow. As we shall see, the age of the sludge is important in determining its biological oxidation activity.

There are many variations of the conventional activated sludge process. In one of these, the returning sludge is aerated for about two hours before being reintroduced into the waste aeration tank (Fig. 21-4). This permits further oxidation of adsorbed and absorbed materials and can reduce the time of treatment of wastes in the aeration tank, because initial removal of waste materials is by adsorption and absorption, processes requiring less time than complete oxidation. Step aeration—more properly, step loading—is employed in some treatment

plants (Fig. 21-4). Here, wastes are added to the aeration tanks at two or more points along their lengths, thus equalizing the load on the oxygen resource and the sludge activity. Sometimes, aeration tanks are operated in series. As with trickling filter processes, combinations of these basic treatment plant designs are employed in almost bewildering variety.

Sludge Digestion Facilities

Sludge from primary settling tanks, humus sludge from trickling filters, and activated sludge require disposal, a very difficult problem because of their large volume and putrescible nature. Most of their volume is due to water, which accounts for 94 to 99 percent of their weight. Removal of water is generally a necessary step in sludge disposal. A sludge having a moisture content of 94 percent by weight occupies only one third the volume of a sludge having a 98 percent moisture content, if both contain the same weight of dry sludge. Sludges may be dried on beds providing for removal of some of the water by filtration. But their putrescible nature can still cause problems that can be avoided if the sludges are stabilized by treatment before their drying and ultimate disposal on land.

Decomposition of sludges is usually accomplished under anaerobic conditions in *sludge digestion tanks* by communities of microorganisms. Primary and secondary sludge sedimentation and sludge digestion may be combined in two-storied structures such as Imhoff tanks, where sedimentation takes place in chambers separated from digestion chambers located below them. Where large volumes of wastes must be processed, separate sludge sedimentation and sludge digestion tanks are usually employed. Anaerobic digestion requires months at temperatures below 60 F, so digestion tanks are usually heated to a temperature range of 80 to 95 F, where communities of *mesophilic* microorganisms can degrade the sludge in periods of 20 to 30 days. More rarely, temperatures from 110 to 130 F are maintained, a range in which *thermophilic* microorganisms can digest sludge in periods of five to ten days. A lower quality of liquid effluent and odor and process control problems have discouraged use of the thermophilic process.

Neither humus sludge nor activated sludge alone supports enough biological activity for adequate anaerobic treatment, but when they are combined with primary sludge, digestion progresses well; this is the usual practice. The products of anaerobic sludge digestion are the digested sludge, a supernatant liquor, and gases. The digested sludge has only about one third its original volume and is a rather inoffensive and very humus-like material, quite unlike the "humus sludge" produced by trickling filters. The supernatant liquor retains objectionable characteristics and is usually recycled through secondary treatment. The principal gas produced is methane, which has high fuel value, and some of this is usually burned to heat the digesters. Anaerobic sludge digesters, then, usually incorporate features for sludge introduction and mixing, maintenance of anaerobic conditions, heating, removal of digested sludge and supernatant liquor, and collection of gases.

Lagoons

In waste treatment processes, a variety of different kinds of lagoons may be used for waste storage, equalization of waste flow and quality, percolation, sedimentation, aerobic or anaerobic degradation, or some combination of these. Lagooning may be employed as the sole treatment process or in combination with other processes. It is usually practiced only in areas of low human population, where large acreages of land are available and the possibilities of public hazard and nuisance are reduced.

Lagoons may be used to store wastes for release during periods of high stream flow when adequate dilution is possible. Unless lagoons are carefully sealed, sufficient percolation may occur to contaminate ground waters with toxic substances or pathogenic organisms. Where these hazards do not exist and sandy soils are present, percolation may be used as a means of waste disposal.

Lagoons are sometimes used primarily for sedimentation, either with or without chemical precipitation. But even when the lagoons are employed primarily for aerobic or anaerobic waste degradation, sedimentation occurs. If lagoons are to be maintained for aerobic biological oxidation, primary

sedimentation—either in tanks or in ponds provided specifically for that purpose—is advisable. Oxidation ponds should be no deeper than 5 feet if aerobic conditions are generally to be maintained, unless mechanical aeration is practiced. Biological oxidation is accomplished mainly by bacteria. Algae, through photosynthesis, then utilize the carbon dioxide and other plant nutrients produced and also provide additional oxygen for the oxidative processes. The sludges produced are degraded by facultative anaerobes, including bacteria, protozoa, insects, and worms. Because a large part of the degraded waste materials is used in algal synthesis, the effluents of oxidation ponds usually have a high content of reasonably stable organic material. Nevertheless, this algal material will decompose in the streams into which it is introduced, and economical means of removing the algae from oxidation pond effluents are being sought. Oxidation ponds are most effective in areas and seasons having considerable sunlight.

BIOLOGY AND OPERATION OF TRICKLING FILTERS

Trickling Filter Communities

Even a cursory examination of trickling filters or activated sludge systems in operation is sufficient to convince one that these are very specialized ecosystems that man has made possible for waste treatment. Were we to examine these ecosystems more closely, we would find that the communities inhabiting them are also specialized, as we should expect. They differ from naturally occurring communities to the extent that their ecosystems have been made to differ from natural ones. And from one trickling filter or one activated sludge system to another, communities differ as environmental conditions differ, even if opportunities for colonization by plants and animals are the same. Hundreds of species of plants and animals aid man in the degradation of his wastes, and their interrelationships are innumerable and poorly understood. There will come no Alexander to cut for us this Gordian knot, but we need learn no new biological concepts to gradually solve the problems it presents. Whatever the ecosystems, communities, populations, and species, there operate the same biological processes that we have devoted this book to explain.

The nature of the biological community that develops in a particular trickling filter depends on environmental conditions there, the opportunities for plant and animal colonization, and the succession of plants and animals made possible by their activities. Environmental conditions are determined primarily by the design and operation of a filter and by the nature of the wastes being treated, but the activities of the organisms themselves are also important in determining these conditions. The size of bed particles and the manner of waste application will largely determine the delivery of oxygen and nutrients to the community. We intend, in this brief survey of biological waste treatment, to distinguish little between sewage treatment and industrial waste treatment, for the biological processes are similar. But differences in wastes have profound effects on the communities that depend upon them. Sewage usually contains in adequate amounts the nutrients upon which most communities depend, and concentrations of potential toxicants are not often harmfully high. Industrial wastes, on the other hand, are often low in such essential nutrients as nitrogen and phosphorus, and they may be of a very toxic nature. Still, communities able to degrade industrial wastes, including some of their toxic components, can develop.

Trickling filters may be *colonized* (Chapter 18) by organisms present in the wastes, by spores carried through the air, by insect reproduction, or by intentional introduction of organisms through the efforts of sewage plant operators. Of the many species reaching trickling filters, which become successful colonizers depends on existing environmental conditions. Aquatic bacteria, fungi, algae, protozoa, and some higher organisms are present in most sewages, and from among these usually come the initial colonizers. Industrial wastes often do not have a full complement of the organisms necessary for their degradation, and initial "seeding" of a trickling filter with unsettled effluents from other plants treating similar wastes may be desirable. Colonization by organisms dependent for their food on the *biological film* of microorganisms covering the rocks of the bed must await some development of this film. Protozoa, aquatic worms, and aquatic and semi-aquatic insects graze upon this film. As would be expected, the grazing animals in trickling filters come from surrounding aquatic and semi-aquatic environments (Lloyd, 1944). Crisp and Lloyd (1954) have compared the environments of mud flats and trickling filters to determine why so few of the mud-flat species are successful in colonizing trickling filters. Food resources and physical conditions necessary for survival and reproduction appear to be the determining factors, unless the wastes are quite toxic.

Succession, as we explained in Chapter 18, is change that occurs in community composition resulting from the activities of some species paving the way for colonization by other species. This occurs when a new trickling filter is put into operation. Here it is sometimes referred to as *maturation,* the trickling filter not reaching its operational potential until community succession has led to a community well suited to degrade the waste being treated. In occurring through time, this is an example of the classic concept of succession. In a more general sense, succession through space can perhaps be considered to occur in flowing waters (Chapter 18), where upstream communities modify the waters in such a manner as to make possible the existence of different downstream communities. Succession in this sense too may be considered to occur in trickling filters as wastes flow down through their beds (Hawkes, 1963).

In the case of sewage, for example, we might expect the activity of heterotrophic bacteria to be greatest in the upper levels of trickling filters and the activity of autotrophic bacteria, such as the nitrifying ones, to be greatest in the lower levels; there is some evidence of this (Mills, 1945; Barritt, 1933). Protozoa usually associated with poorly degraded wastes tend to be more abundant toward the surfaces of beds, whereas those associated with highly degraded wastes dominate the lower levels (Barker, 1946). Stratification of kinds of larger animals could result from differences in their food habits or from differences in physical and chemical conditions at different levels in trickling filters.

Natural selection of strains of bacteria genetically capable of tolerating and degrading certain toxic wastes (Chapter 7) may occur along with colonization and succession during the maturation of a trickling filter;

under operational conditions, however, these biological phenomena can probably never be distinguished. Suffice it to say that, by whatever biological processes, there often develop trickling filter communities not only capable of degrading organic materials in the presence of toxic substances but also capable of degrading substances which themselves are highly toxic. There are known to be bacteria able to oxidize phenols, thiocyanates, thiosulfates, and cyanides present in gas production wastes (Happold and Key, 1932).

The community of a trickling filter, like any other community, is composed of many species, each occupying a *niche* (Chapter 7), each requiring of its environment certain physical and chemical conditions and nutrient resources, and each contributing nutrients and other necessary conditions to the environments of other species. The matrix of the community covering the rocks of a trickling filter is composed of zoogloeal bacteria and fungi forming the glutinous biological film. This film provides the homes of the other organisms, and their nutrient materials derive from the activities and cells of its microorganisms as well as from the inflowing wastes. We earlier distinguished between *heterotrophic* and *autotrophic* organisms, the former obtaining their energy from organic compounds, the latter from solar radiation or from inorganic substances. We should perhaps further classify heterotrophic organisms according to their energy and material sources, though we hesitate to do so. We dislike excessive terminology, it is difficult to avoid ambiguity, and relatively few organisms are discriminating in their nutritional habits. Hawkes (1963, p. 35) has suggested that animals preying or grazing on living organisms be considered *holozoic* organisms and that plants and animals utilizing dead and decaying matter be considered *saprobic* organisms. He would further separate the saprobic organisms into *saprozoic* ones, mainly animals that ingest particulate materials, and *saprophytic* ones, such as bacteria and fungi, which absorb organic materials that are in solution in their environment.

We do not believe it worthwhile to encumber this brief consideration of biological waste treatment with long lists of the plant and animal species that have been found in trickling filters. By now, our reader must realize that the species composition of trickling filter communities will vary according to region, climate, wastes being treated, and treatment plant operation. If we are to understand particular waste treatment processes, it is important that we know the species composition of the communities degrading the wastes. But this information alone, without knowledge of the environmental conditions and the functional roles of the species involved, is of little value. In this regard, the study of waste treatment communities differs not in the least from the study of other communities. Many fine biologists, knowing this, have endeavored to relate the species composition of trickling filter communities to environmental conditions. Hawkes (1963) has reviewed the work of these biologists, many of whom provide extensive lists of species to which our reader can refer.

Different species in different communities play similar or the same functional roles (Chapter 18). Identification of these roles, then, permits some consideration of the functioning of communities without precise identification of the species involved, though there are dangers in proceeding too far without adequate identification. Nevertheless, it will be convenient for our purposes to consider groups of organisms having common functional roles in trickling filter communities. The primary energy source of trickling filter communities is the organic matter suspended and dissolved in the wastes being treated. In this, they differ from most other communities, which depend primarily on the energy of solar radiation made available through plant photosynthesis. Though algae grow on the surfaces of trickling filters where light is present, photosynthesis is not essential to the communities of trickling filters or to their operation. Bacteria and fungi, utilizing complex and degraded organic materials, provide the basis for the economy of trickling filter communities. Upon their cells feed the animals, mainly protozoa, rotifers, nematodes, oligochaete worms, and insects. The animals also utilize some suspended and dissolved waste materials, and the bacteria and fungi degrade not only the waste materials but also the cells of their own kinds and the bodies of animals. Thus, the energy and materials present in a waste as it flows downward through a trickling filter are utilized by

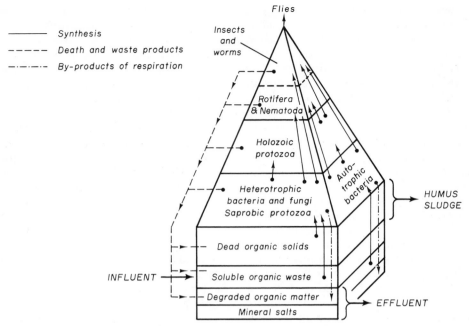

FIGURE 21-5 Main pathways of energy and material transfer in treatment of organic wastes by means of trickling filters. Respiration at each level converts some of energy in materials to heat. After Hawkes (1963).

successive groups of organisms in their respiration and growth, until most of the waste is transformed to heat, carbon dioxide, water, mineral salts, and the cellular material exported as humus sludge. The many pathways of energy and material transfer are intricate, not well known, and can be only poorly represented (Fig. 21-5).

*Energy and Material Transfer
in Trickling Filters*

Analysis of biological waste treatment processes is essentially a problem in community energetics (Chapter 19). Trickling filters are operated to oxidize and in other ways remove from wastes those materials whose biochemical oxygen demand (BOD) or toxicity makes their introduction into natural waters undesirable. Soluble materials able to pass across cell membranes are absorbed directly by the organisms of the biological film. Other soluble materials and colloidal and larger particles are adsorbed onto the film where hydrolysis by extracellular enzymes makes their absorption possible. Some of the particulate waste matter is directly ingested by protozoa and other animals. Then, through the respiration and growth of suc-

cessive groups of organisms, the energy-rich materials present in wastes are utilized and reutilized until they are either largely oxidized or retained in the cells of organisms exported from the filter as humus sludge. Here is a problem in community energetics, a problem requiring analysis of pathways of energy and material transfer.

But any meaningful community energetics must begin with the bioenergetics of individual organisms and populations, the nutrient materials they obtain, their respiration, and their growth. We have already discussed in considerable detail the bioenergetics and growth of individual animals (Chapter 11) and the food consumption and production of populations (Chapter 17); the ideas we explained can be applied directly to the animals in trickling filter communities. These ideas can be applied with little modification to the analysis of nutrient consumption and production of populations of bacteria and fungi, for all heterotrophic organisms are confronted with the problems of obtaining energy-rich materials and utilizing them in respiration and growth. Indeed, our bioenergetic equation (Chapter 11) was derived from an equation originally used to describe the bioenergetics of cultures of fungi (Terroine and Wurmser, 1922; Ivlev,

1945). It has been shown that four species of fungi common in sewage treatment plants utilize less nutrient material in respiration and more in growth than do zoogloeal bacteria (Water Pollution Research Board, 1956); thus, their *growth efficiency* is higher than that of the bacteria. Hawkes (1963, p. 47), however, has pointed out that maximum growth efficiency is not necessarily desirable in waste treatment. High growth efficiency is desirable only when the sludge produced has economic value, not when it presents a problem of disposal.

In microorganisms, growth and production are more intimately related to reproduction than they are in multicellular organisms. The reason for this is simple: cell division in microorganisms leads to more organisms, not larger organisms. Increases in the biomass of a microorganism population result, then, mainly from reproduction; in multicellular animals, growth in the size of individual organisms and their reproduction are quite distinct processes, either of which can lead to population biomass increases. The balance between reproduction and death,

population dynamics (Chapter 16), determines the rate of change of biomass in microorganism populations more nearly than it does in populations of larger organisms. *Population growth*, then, in reference to microorganisms, can simultaneously and correctly signify both numerical and biomass increase; only the units of measurement would differ.

The pattern of growth of microorganism populations through time tends to be sigmoid, as it is with larger organisms (Chapter 16). Cultures of microorganisms are sometimes considered to pass through different growth phases (Fig. 21-6). Depletion of nutrients—sometimes accumulation of toxic metabolites—finally leads to a cessation of growth. After nutrients are decreased to starvation levels, respiration of an *endogenous* sort occurs, and this utilization by the organisms of their own cell materials leads to a decrease in their number and biomass (Fig. 21-6). In the biological treatment of wastes, the growth of populations of microorganisms is usually maintained between the last part of the logarithmic phase and the beginning of the endogenous phase.

FIGURE 21-6 Microorganism population densities, growth rates, and phases of growth. After Hawkes (1961) and Monod (1949).

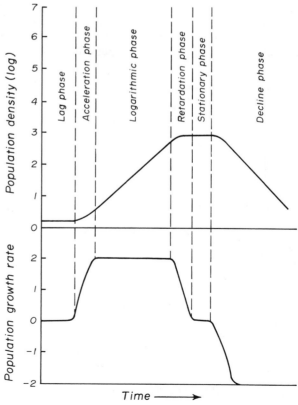

The growth rate of microorganism populations not only is a function of time, or their age, but also is a function of nutrient or waste concentration. Hawkes (1961) has suggested a hypothetical relationship between microorganism growth rate and waste concentration (Fig. 21-7) that is very reminiscent of relationships between the growth rates of animals and the densities of their food organisms (Chapter 11). Nutrients limit the growth of microorganisms only when nutrient concentrations are lower than some relatively high level (Fig. 21-7). At still lower concentrations, nutrient availability will be only sufficient for the microorganisms to *maintain* their biomass; and at concentrations lower than this, endogenous respiration and a decrease in biomass will occur. In biological waste treatment, waste concentrations are usually maintained between the high levels that are limiting and the maintenance levels (Fig. 21-7). But nutrient concentrations change as a waste passes down through a trickling filter; this is its purpose. In consequence, the growth rates of heterotrophic microorganisms tend to be highest in the upper levels of the trickling filter bed, where concentrations of complex organic compounds are highest. The growth rates of the autotrophic microorganisms tend to be highest in the lower levels of the bed, where their nutrient materials have become more available. Recirculation of ef-

fluents is often practiced in part to make waste concentrations and the growth of microorganisms more uniform throughout trickling filters. *Intraspecific* and *interspecific* competition for nutrients can obviously limit the growth of populations of microorganisms.

Through growth and production, then, populations of bacteria and fungi retain in their cells materials synthesized from waste products. But some part of the waste products, probably less than half, is oxidized during this process, the principal end products being carbon dioxide, water, and heat. These bacteria and fungi become the energy and material sources of successive generations of their own and related kinds. And all kinds become the food of protozoa, rotifers, nematodes, oligochaete worms, and insects. With each reutilization of materials deriving from the original waste products, additional material is oxidized, until the cells and bodies of organisms exported from trickling filters as humus sludge represent but a small part of the energy and materials originally present in the wastes.

The amount of humus sludge produced by conventional trickling filters, on a per capita basis, may be as little as one half the amount of sludge produced by the activated sludge process, and the energy content of humus sludge per unit of dry weight may be as little as two thirds that of activated sludge

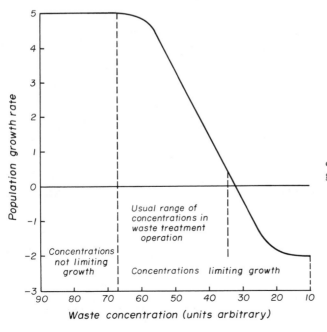

FIGURE 21-7 Relationship between waste concentration and population growth rate of microorganisms. After Hawkes (1961).

(Imhoff and Fair, 1956, pp. 180–181). This is partially because in trickling filters the production of bacteria and fungi is utilized not only by protozoa but by higher organisms as well, whereas in activated sludge, utilization and reutilization of materials are dependent primarily on bacteria and protozoa. Thus, in trickling filters, larger grazing animals help man to dispose of the sludge.

These larger animals, in survival, reproduction, and growth, must conform to the same biological laws as all other animals, laws that are not really different for bacteria and fungi. In our earlier discussion of population dynamics (Chapter 16), we considered the factors that operate to control the distribution and abundance of animals—operate to control their numbers. Our reader may remember that these factors are sometimes classified as being either *density-dependent* or *density-independent*. Hawkes (1963, p. 43) believes that density-dependent processes, particularly competition for food, are important in controlling the numbers of multicellular animals in trickling filters but that the course of events may be altered by changes in temperature or flow.

There have been fine ecological studies on the oligochaete worm and larval fly populations of trickling filters. When only one dominant species is present, intraspecific competition for food may be important in limiting its numbers (Hawkes and Jenkins, 1951; Hawkes, 1952). When only a few species are present, perhaps because of the presence of toxic substances, these species often attain higher numbers than they would were there more species (Lloyd, 1945); this could be explained by reduction in interspecific competition for food (Hawkes, 1963, p. 90). But Andrewartha and Birch (1954, pp. 604–612), de-emphasizing competition as they do, interpret studies by Lloyd (1943), Reynoldson (1947a, 1947b, 1948), and Lloyd, Graham, and Reynoldson (1940) as evidence of other kinds of interactions between species. The feeding activities of some species may cause plants and associated animals to be dislodged from filters; predation of some species on the eggs and larvae of their own and other kinds occurs. We would not entirely discredit any of these explanations. It would be strange indeed if biological processes of one kind always operated to the exclusion of others. But some things are clear: trickling filters

operate as they do because of the presence of multicellular organisms. In consuming bacteria and fungi, the multicellular organisms limit the densities of the microorganisms, thus tending to keep the microorganisms in the logarithmic phase of their growth and tending to prevent the accumulation of excessive biological film. And utilization of the bacteria and fungi by higher organisms results in a further release to the atmosphere of carbon dioxide and heat.

Operational Control of Trickling Filters

The objective of operational control of a waste treatment process is to produce an effluent having desired characteristics. In biological waste treatment, the environment, the activities, and the biomasses of the populations and communities of organisms responsible for waste degradation must be controlled, if treatment objectives are to be realized. The populations and communities of trickling filters are controlled mainly through maintenance of appropriate environmental conditions. In the activated sludge process, in addition to maintenance of a suitable environment, direct control of the activity and biomass of the organisms is exerted by recirculating different amounts of the sludge.

For a waste of particular characteristics and volume, control of a trickling filter environment depends mainly on the manner of waste introduction. Different methods of surface application, effluent recirculation, or use of filters in series give treatment plant operators some control over the environments, communities, and final effluents of trickling filters. Distribution and rate of flow of wastes through trickling filters as well as waste nutrient concentrations can be controlled by these methods. Appropriate use of these methods can also help to maintain uniform and adequate delivery of oxygen and nutrients to the biological film by preventing plugging of filter beds with excessive community biomasses. But nutrients essential in the environments of most microorganisms are deficient in some industrial wastes; if trickling filters are to be employed in their treatment, these wastes may need to be mixed with sewage or commercial fertilizers. Toxic substances and temperature are two important environmental factors over which the plant operator will usually have little control. Rudolfs et al. (1950) have reviewed some of the literature concerned with the problem of toxic interference with waste treatment processes. Seasonal changes in temperature can have profound influences on trickling filter operation and efficiency, partially because of their effects on the balance between film growth and animal grazing (Hawkes, 1963).

Ideally, a trickling filter should be operated so as to maintain an optimal amount and distribution of biological film. The film need not be thick to function properly, so the problem usually becomes one of excessive accumulation restricting waste flow and air circulation. In controlling this film, the populations of grazing animals assist the treatment plant operator. Hawkes (1963, pp. 82-86) has noted, however, the disagreement as to whether it is this grazing or the scouring action of the liquid that is primarily responsible for film control. He suggests that grazing may be relatively more important in film control when trickling filters are operated in the conventional manner than when they are operated at very high flow rates resulting in greater scouring action.

Recirculation of a considerable portion of trickling filter effluent (Fig. 21-2) offers several interrelated advantages in control. More constant hydraulic and waste concentration loading, higher flow rates, and lower waste concentrations are achieved in the filter. As a result of the last two factors, heterotrophic activity and the biological film are distributed much more evenly down through the bed, and excessive accumulation of film in the surface layers is prevented (Hawkes, 1961).

Where two filters are operated in *series* (Fig. 21-2), there exists the possibility of periodically reversing the sequence in the series. This, too, provides a means of controlling growth of the biological film (Hawkes, 1961). The first filter in a series receives the waste at a higher concentration and the film may accumulate. Then, when this filter is used as the second one in the series, it receives effluent of a lowered waste concentration. This concentration is insufficient to maintain the high film growth and accumulation possible at the higher waste concentration, and endogenous respiration leads to a decline in film biomass (Fig. 21-7). Thus, accumulated film, first on one filter and then on the other, is reduced.

For the same volume and strength of waste applied to a trickling filter, there is yet another way of controlling accumulation of biological film. The speed of rotation of the waste distribution arms can be varied. Decreasing rotation speed increases the amount of waste applied at one time to a given bed location but decreases the frequency of application to that area, the *frequency of dosing*. If application at each location occurs every minute, there is little change with time in the waste concentration at particular bed depths. But if application is at longer intervals, say every 15 minutes, waste concentration at a particular depth declines markedly during the intervals between applications (Hawkes, 1961). During each interval, declining nutrient concentrations lead to declining microorganism growth rates and may lead to endogenous respiration (Fig. 21-7), thus preventing excessive film accumulation. At very low and very high frequencies of application, the efficiency of waste treatment declines, but high efficiencies can be obtained within the range of frequencies effective in film control.

Control of the growth of microorganisms of the biological film controls, of course, the food supply of the grazing animals. In the last analysis, this may be the best means of controlling the abundance of flies (*Anisopus, Culicoides, Psychoda*) that emerge from filters to become a public nuisance. Insecticides have sometimes been useful in controlling these insects, and when such control permits oligochaete worms to become more abundant, excessive accumulation of the biological film may still be prevented. But the use of insecticides has disadvantages, including a possibly toxic effluent, and, once initiated, it must be continuous to be effective, because the original fauna can quickly become re-established. Trickling filters are complex biological systems whose effectiveness depends on complex ecological equilibria that are perhaps better maintained by nutritional than toxic materials.

BIOLOGY AND OPERATION OF ACTIVATED SLUDGE SYSTEMS

Activated Sludge Communities

The environment of an activated sludge system is a very specialized one, even more so than a trickling filter; we cannot imagine a natural ecosystem with a similar environment. Fantastically rich in dissolved and suspended organic matter, violently agitated to keep the flocculent sludge in suspension and in contact with this matter, maintained aerobic by the introduction of great amounts of oxygen, regulated by continuous reintroduction of sludge, the activated sludge system has no natural counterpart. An environment of almost unbelievable opportunity for some species of microorganisms, it is most inhospitable for larger aquatic organisms, most of which could not withstand the buffeting.

It is not at all surprising, then, that activated sludge systems have been colonized almost entirely by various species of bacteria, fungi, protozoa, and the small but multicellular rotifers and nematode worms. Only very rarely do small crustaceans, insects, and oligochaete worms become established; insufficient light usually prevents the development of algae. Colonizers may come in with the waste, particularly with sewage, so that a sludge of high activity can develop. But organisms capable of degrading some industrial wastes may be of rare occurrence or not even present in those wastes; then, "seeding" of the activated sludge system with sludge from plants treating similar wastes or even sewage can hurry establishment of the treatment process. Bacteria, not fungi, are primarily responsible for degradation of wastes in the activated sludge process. Operational problems are usually associated with any ascendence of fungi. The composition of the bacterial flora will depend on the nature of the waste. The composition of the protozoan fauna too may depend on the waste, but it depends perhaps more on the stage of community succession permitted by sludge recirculation.

In a properly operating activated sludge system, most of the bacteria, protozoa, and other organisms are gathered together in flocculent masses, the *activated sludge* itself. For each kind of organism, according to its needs, the activated sludge floc provides mechanical substrate, adsorption surfaces for nutrient materials, or food organisms; it is more than just a place to live. The initial biological floc has generally been assumed to result from the activities of *zoogloeal bacteria*, which produce a polysaccharide slime agglutinating into masses their cells, those of other organisms, adsorbed colloids, and all manner of biological debris. There is now some

evidence that bacteria, regardless of their slime-producing properties, become a part of a floc when they lack sufficient repulsive surface charge to prevent contact and when, once held together by colloidal surface forces, their energy of motility is too low for them to separate (McKinney, 1956). Whatever the initial formation mechanisms, the floc becomes a mass of cells, cell products, and adsorbed materials having structure and function essential in the activated sludge process.

An effective activated sludge floc does not arise immediately after aeration of a waste, but only after a *community succession*, the initial stages of which possess little or no floc. Different bacterial floras as well as protozoan faunas characterize the stages of this succession. The total organic content of the sludge-liquid mixture is low in the initial stages, when little sludge has been produced to accumulate by recirculation. It has been suggested that during sewage treatment nonproteolytic bacteria predominate until sludge accumulation provides the nutritional substrate for predominance of proteolytic ones (Allen, 1944). Unless purification has proceeded to rather advanced stages and a high oxygen concentration is maintained, the activity of nitrifying bacteria is usually not great. Initially, amoeboid and flagellated protozoans able to utilize dissolved and particulate waste materials predominate (Fig. 21–8). Later, as bacteria increase, flagellates and free-swimming ciliates able to utilize bacteria for food increase. As floc formation progresses, crawling and attached ciliates come to predominate, and these may be accompanied by rotifers and nematodes. This succession has been carefully described (Barker, 1949), as have some of the causal competitive interactions of the species (McKinney and Gram, 1956). If little sludge is recirculated, the succession will remain in its early stages; if much is recirculated, as is the usual practice, succession will progress to stages characterized by attached ciliates, stages producing rather well oxidized effluents (Fig. 21–8).

Proper design and operation of treatment plant facilities can help to provide the environmental conditions necessary for the development of biological communities able to degrade not only sewage but also industrial wastes presenting difficult treatment problems (McCabe and Eckenfelder, 1956). These communities are composed of dif-

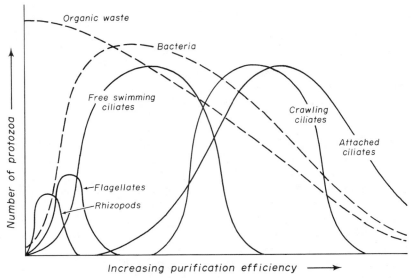

FIGURE 21-8 Hypothetical curves showing succession of protozoa in relation to degree of purification of organic wastes and the densities of bacteria present in the activated sludge process. After Hawkes (1963).

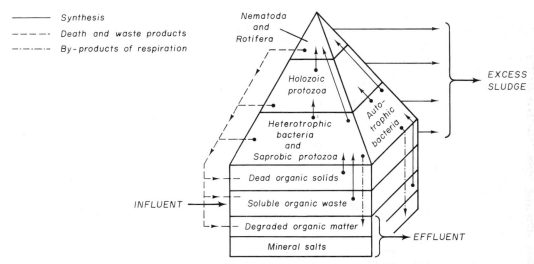

FIGURE 21-9 Main pathways of energy and material transfer in treatment of organic wastes by means of the activated sludge process. Respiration at each level converts some of energy in materials to heat. After Hawkes (1963).

ferent species occupying different *niches* (Chapter 7) in the waste treatment ecosystems. Niche is a concept sometimes used to cover the needs and roles of a species in an ecosystem. Particular wastes and particular ways of operating activated sludge systems provide, at some point in time, the needs of certain species that form the community functioning to degrade the wastes. Changes in plant operation and changes caused by the organisms themselves lead to new environmental conditions permitting the development of other communities. And, of course, different wastes lead to the development of different environmental conditions and communities. Waste treatment practice being what it is, few of the environmental conditions, the species, and their roles are usually known. Were they known, we could hardly present them here, for it would be a long and complex presentation. Some general understanding of the activated sludge process can come from consideration of the roles of various groups of organisms involved (Fig. 21-9). But we must not take only general understanding and representation too seriously. Were we to know more, efforts at simple diagrammatic representation would most certainly be frustrated.

Energy and Material Transfer in Activated Sludge Systems

The patterns and magnitudes of energy and material transfer in activated sludge systems depend primarily on the stages of community succession maintained through sludge recirculation. Activated sludge systems, under good operating conditions, are usually characterized by ciliate protozoa and a sludge having good settling qualities. In conventional activated sludge processes, the amount of sludge maintained in the system is generally several times the amount of materials present in the incoming waste and is determined by the amount of sludge recirculated. The "age" of the sludge increases with increasing recirculation; this matter we will return to in the next section. With increasing sludge age, there is time for community succession to pass to later stages.

Initially, rapid removal of suspended and dissolved materials present in wastes occurs through their adsorption and absorption by

the gelatinous matrix and bacteria of the floc. Colloidal and other suspended materials are rapidly adsorbed onto the floc, where most are hydrolized by extracellular enzymes of the bacteria, thus making their absorption possible. Soluble materials in the waste, when of appropriate molecular structure and size, are absorbed directly by the organisms, without initial extracellular degradation. "The extent to which the waste is accumulated, either on the surface of the floc or as storage products within the cell, depends upon the relative rates of adsorption, absorption and oxidation in relation to the concentration of the waste" (Hawkes, 1963, p. 128).

The bioenergetics of individual organisms and the dynamics of their populations are as important in determining community composition and patterns and magnitudes of energy and material transfer in activated sludge systems as they are in trickling filter systems. But these principles need little redevelopment here, for they are similar in both kinds of systems. Populations of bacteria pass through similar growth phases, whether in activated sludge, trickling filters, or most other environments (Fig. 21-6). And in activated sludge just as in trickling filters, the growth rates of bacterial populations may be high, low, or even negative, as the availability of nutrient materials changes (Fig. 21-7).

The microbial populations of most activated sludge systems appear to be maintained between the latter part of the logarithmic phase and the early part of the endogenous phase of their growth. The synthesis of cell material from waste nutrients by bacteria in the logarithmic phase of growth is quite efficient, relatively little of the nutrients being converted through respiration into carbon dioxide, water, and heat. This, of course, leads to a greater production of sludge per unit of waste than would occur if the populations were maintained closer to the endogenous phase of their growth, where most of the waste materials would be oxidized and little new sludge production would result (McKinney, 1957). Since sludge disposal is always a problem, it may be advantageous to maintain the bacterial populations near their endogenous growth phase, so that little sludge and a highly oxidized and relatively well nitrified effluent are produced. But, longer waste retention time, higher rates of sludge recirculation, and more oxygen must be provided; the economics of plant operation again becomes involved.

Even though something is known of the feeding habits of protozoa in activated sludge systems, the importance of their roles in determining the nature of the sludge and the liquid effluent produced has not really been evaluated (Hawkes, 1963, pp. 72-74). This is not surprising, because determination of energy and material utilization by protozoa in such systems would be exceedingly difficult. Except perhaps in early stages of succession (Fig. 21-8), the proportion of waste materials utilized directly by protozoa appears to be small (McKinney and Gram, 1956). We must turn to their feeding on bacteria to suggest the possibly most important functions of protozoa in activated sludge systems operating in the later stages of community succession. Protozoa that feed on bacteria dispersed in the liquid medium may reduce both the turbidity and the BOD of the effluent. Other protozoa, feeding on the bacteria of the floc, may control these bacterial populations, thus tending to keep them in the logarithmic phase of their growth. And protozoa, in utilizing for their own respiration the bacteria, decomposing cells, and other materials present in the floc, increase the amount of organic matter degraded to carbon dioxide, water, and heat; thus, they aid in the disposal of sludge. When present in significant numbers, rotifers and nematodes, adding another step to the food chain, contribute in the same manner to further waste degradation.

It is not at all unusual, then, with conventional activated sludge systems operated in conjunction with primary treatment, to obtain reductions of BOD, suspended solids, and bacteria in the liquid effluent in excess of 90 percent of the amounts present in the untreated wastes (Table 21-1). Perhaps 50 percent of the organic wastes introduced into the aeration tanks are biologically oxidized, the products being mainly carbon dioxide, water, and heat. The remainder is largely incorporated into the activated sludge, which must be removed by sedimentation to obtain a liquid effluent having the desired reduction in content of organic materials.

Operational Control of Activated Sludge Systems

Activated sludge systems are usually considered to be more sensitive than those of trickling filters: more sensitive to changes in the volume, strength, temperature, pH, or toxicity of the wastes being treated. Perhaps an activated sludge system is more sensitive because the entire community in its turbulent ecosystem is suddenly exposed to any change. In a trickling filter, the upper strata of the community may absorb any shock, thus protecting the lower strata so that they may function almost normally. Perhaps the activated sludge process is more sensitive mainly because the successional stage of the activated sludge is so important in determining treatment efficiency and is so dependent on waste concentration and retention time. Perhaps the main reason is the greater simplicity of the activated sludge community, for ecologists know that simple communities often fluctuate violently in the face of environmental change. Most probably, at one time and place or another, any one or all of the suggested causes may be responsible for the sensitivity of the activated sludge process to environmental change.

Operational control of the activated sludge ecosystem—its environment and community—is designed to obtain an effluent of the desired characteristics. Difficulty of obtaining this control is sometimes mentioned as a disadvantage of the activated sludge process. But the wide adoption of this method of treatment, both for sewage and industrial wastes, is evidence that successful operational control is not only possible but is very generally accomplished. Through maintenance of proper nutrient balances, waste-floc contact, dissolved oxygen levels, and sludge age, activated sludge systems can be operated to produce liquid effluents of the desired characteristics.

Essential to biological oxidation in an activated sludge system is the maintenance of an aerobic environment, oxygen concentrations of from 1 to 2 mg/l usually being maintained throughout most of the aeration tank. Automatic control of aeration may one day be more generally practiced. The development of reliable polarographic instruments for continuous monitoring of dissolved oxygen concentration has made this practical. Such instruments can be coupled to variable aeration devices to maintain oxygen concentrations in aeration tanks at desired levels.

But control of oxygen, whether or not automatic, is only one aspect of the management of an activated sludge ecosystem. Control of the stage of community succession, by sludge recirculation, is also important. Biological indices (Chapter 20) of successional stages may here prove of value to the plant operator. As we noted in a previous section, the efficiency of BOD removal and the clarity of the effluent tend to increase through the successional stages typified by different kinds of protozoa (Fig. 21-8). Thus, amoeboid and flagellated protozoa are generally indicators of early and quite inefficient stages, free-swimming ciliates of intermediate and somewhat more efficient stages, and attached ciliates of later and quite efficient stages, which are usually maintained in activated sludge systems (McKinney and Gram, 1956). But, as Hawkes (1963, p. 136) has warned, each kind of waste, and indeed each treatment plant, may have its own characteristic species; the use of organisms as indicators of plant operation should be based on experience. There are other uses of microscopic examination of activated sludge. The development of a low density floc having poor settling characteristics and causing a poorly clarified effluent is known as *bulking*. The overloading of a plant, insufficient aeration, and toxic substances appear to be among its initial causes. Whatever may be the causal sequence, filamentous bacteria — such as *Sphaerotilus* — and fungi are frequently associated with bulking, and ascendence of these organisms may be a useful indication of impending difficulty.

Perhaps the single most important factor determining the successional stage of an activated sludge community is the average *sludge age*. One way in which this can be approximated is by determining the ratio of dry weight of activated sludge in the system either to the dry weight of suspended solids entering per day or to daily BOD load, preferably BOD load if the wastes are mainly in solution (Imhoff and Fair, 1956). In conventional activated sludge systems, the optimal sludge age appears to be about three or four days. *High rate systems*, which are effective for treating some wastes, usually are maintained so as to have sludge ages between 0.2 and 0.5 day.

There is one other type of control parameter that we must mention before concluding our discussion of the activated sludge process. The *sludge volume index* is one measure employed to determine the settling qualities of sludge. It is generally defined as the volume in milliliters of floc settling in 30 minutes for each gram dry weight of suspended solids in the effluent containing floc and liquid. A good sludge will have an index of 50 to 100 ml/g, whereas a poor sludge, one with bulking characteristics, may have an index of 200 ml/g or higher (Imhoff and Fair, 1956). Since the production of a clarified effluent of low BOD by the activated sludge process requires the removal of nearly all the activated sludge, the settling characteristics of the sludge are very important.

BIOLOGICAL WASTE TREATMENT PERSPECTIVES

We must now bring this chapter on biological waste treatment to a close. Like this book itself, it has become longer than we intended, and we dare not try our reader's patience further. We had hoped to include sections on the biology and operation of anaerobic sludge digestion systems and oxidation ponds, the former being of great importance and the latter of importance in some areas. Hopefully, however, our outline of the application of hydrolysis, oxidation, reduction, and synthesis in biological waste treatment processes will help our reader to obtain additional information from symposia and reviews on anaerobic digestion (McCabe and Eckenfelder, 1958) and oxidation ponds (Klein, 1966). We have discussed trickling filter and activated sludge ecosystems, and, after all, sludge digestion systems and oxidation ponds are only other ecosystems used in waste treatment, though they do have important differences.

Now, in our few remaining paragraphs, we would like to attempt a little broader and perhaps longer view of biology and waste treatment. In Chapters 1 and 4, we sketched the early history of biological waste treatment, and with the present chapter we have brought our history to the present. Waste treatment is an intensely practical problem requiring practical solutions, which began to appear early in its history as a result of the dedicated efforts of a small number of investigators, mainly engineers and chemists. From the beginning there have been biologists interested in waste treatment, but never very many. This field of human endeavor has been more dependent on experience than on sound theory, perhaps understandably so where biological systems have been involved. This should not be taken as depreciation of perhaps our greatest teacher, experience. But rapid progress in a technical field most often depends on the evolution of sound guiding theory from experience.

In recent decades, there has been brought into being a very respectable body of theory of waste treatment (McCabe and Eckenfelder, 1956, 1958), theory to which biologists are beginning to contribute significantly (Hawkes, 1963), theory that will undoubtedly guide our future progress. Indeed, those of us who have viewed this technical and scientific field from the outside might well wonder what our possible contributions could be. It would be a mistake to underestimate them, though neither would we want to be too hopefully naive. Man has not been particularly successful in anticipating either his successes or his failures. The waste treatment problems facing him are immense and offer nearly unlimited scope for scientific endeavor. Although we may err in anticipating future problems and their solutions, it would be hazardous not to make an attempt. The theory and practice of waste treatment will advance more in the next decade than in the last three decades. Biologists are needed in this work; it offers them not only a chance to help their fellow man but also a chance for fascinating studies of individual organisms, populations, and communities.

Accepting that our problem of disposing of all kinds of wastes—domestic, agricultural, and industrial—will increase with increasing population, we probably can also accept that we will find ways of treating these wastes, thus lessening their harmful effects on our environment. But there is an awesome wastage of organic and inorganic materials, available energy, and heat in the generally practiced treatment of wastes, a wastage so eloquently pointed out by Hynes (1960). There have been attempts, some successful, to put to good use the wastes of man's activities. Yet, as reasonable and thrifty as such use may seem, the economics of waste utilization has generally been discouraging. And the wastage goes on and perhaps will go on for a long time in the future. Still, there may be higher economic laws that man cannot long afford to disobey: the economics of energy and material transfer in his ecosystem, the earth. Biologists, particularly ecologists, are sensitive to these laws; they can contribute by further exploring them and teaching their findings. But this is not enough, for if it is found that the wastage cannot be afforded, then the ways of utilization must be found. Here, too, biologists have an important contribution to make. In biological waste treatment as it has been generally practiced, hydrolysis, oxidation, and reduction must be maximized, the synthesis of energy-rich end products minimized. But the biological synthesis of useful end products is one route to waste utilization, a route in the realm of the biologist.

SELECTED REFERENCES

Barker, A. N. 1946. The ecology and function of Protozoa in sewage purification. Annals of Applied Biology 33:314–325.

Clark, J. W., and W. Viessman, Jr. 1965. Water Supply and Pollution Control. International Textbook Company, Scranton, Pennsylvania. xiv + 575pp.

Fair, G. M., J. C. Geyer, and D. A. Okun. 1968. Water and Wastewater Engineering. Vol. 2. Water Purification and Wastewater Treatment and Disposal. John Wiley & Sons, Inc., New York.

Hawkes, H. A. 1961. An ecological approach to some bacteria bed problems. Journal of the Institute of Sewage Purification 1961 (2): 105–132.

———, 1963. The Ecology of Waste Water Treatment. Pergamon Press Ltd., Oxford. ix + 203pp.

Imhoff, K., and G. M. Fair. 1956. Sewage Treatment. 2nd ed. John Wiley & Sons, Inc., New York. vi + 338pp.

Keup, L. E., W. M. Ingram, and K. M. Mackenthun (Editors). 1967. Biology of Water Pollution: A Collection of Selected Papers on Stream Pollution, Waste Water, and Water Treatment. Federal Water Pollution Control Administration, U.S. Department of the Interior. iv + 290pp.

Klein, L. 1966. River Pollution. 3. Control. Butterworth, Washington. xv + 484pp.

McCabe, J., and W. W. Eckenfelder, Jr. (Editors). 1956. Biological Treatment of Sewage and Industrial Wastes. Vol. 1, Aerobic Oxidation. Reinhold Publishing Corporation, New York. vii + 393pp.

———, and ——— (Editors). 1958. Biological Treatment of Sewage and Industrial Wastes. Vol. 2, Anaerobic Digestion and Solids-Liquid Separation. Reinhold Publishing Corporation, New York. vi + 330pp.

McGauhey, P. H. 1968. Engineering Management of Water Quality. McGraw-Hill Book Company, New York. [vii] + 295pp.

PART VII

Conclusion

Each kind of organism modifies to some extent its environment and the environment of other organisms. It takes nutrients from its environment and returns heat and waste materials. The bodies and waste products of one kind of organism become the energy and material resources for other kinds. Most kinds of organisms alter the environment in some ways that favor their own persistence and that of other organisms and alter the environment in other ways that are harmful to themselves and others. Man does all these things, necessarily so, but in altering the environment he seems to be going too far. The inertia of his cultural development in general and his technological development in particular appears to be carrying him along in ways that many believe to be not only wrong but dangerous. It is one thing to believe or even to know that ways are wrong or dangerous, but it is another thing altogether to know what should be done, to know how to do it, and to have the wisdom to act. To provide for himself now and into the distant future, man will need to alter the environment. Along with beneficial changes will come changes that are destructive to all kinds of life including man. Long ago man recognized the benefits of environmental manipulation; now he is recognizing the dangers. If he can soon learn to distinguish between the kinds of changes that are truly essential and cannot be avoided and the kinds of changes that are not essential and must be avoided, there is yet time for him to come to live in stable harmony with his own kind and with other life on earth. Man has the capacity to determine what he should be doing and how it can be done. The question is really more one of his wisdom.

22 *Ecological Change: Its Acceptability and Evaluation*

It is through his ideas that man has shifted from being just one more species in a biological community into becoming a sort of geological force, altering the whole surface of the planet and affecting in one way or another the lives of all other organisms.

MARSTON BATES, 1962, p. 9

The greatest social problem facing man today is the ecological one of harmonious adjustment to the ecosystems of which he is a part. Ecologists must provide technical answers, and they must come to fill an increasingly important role in formulating national and international policies. We are in nowise suggesting that it is possible or even desirable to maintain the ecosphere in a primeval state; what we are suggesting is that it well behooves man to learn to predict the effects of his activities on the ecosphere and to govern those activities accordingly.

FRANK BLAIR, 1964, p. 18

ENVIRONMENTAL CHANGE AND MAN

Water pollution is ultimately an environmental problem of man; and man will determine, however wisely, what pollutional changes are acceptable. We have devoted most of our book to discussion of ways to increase our understanding of the environmental requirements of aquatic organisms; but the objective of this book is to relate biology to water pollution control. In concluding, then, we must somehow relate the environmental requirements of aquatic organisms to the environmental requirements of man. We are not here primarily concerned with relating man's physical and chemical requirements to those of other organisms, though, through studies of other organisms, much has been learned about the possible effects of environmental factors on

man. Our concern is with the roles that aquatic organisms and water play in man's environment and with how man can use his knowledge of these organisms in managing his environment to best satisfy his needs.

This viewpoint, of course, leaves self-centered man managing the resources, the plants, and the animals of the earth in his own best interest, perhaps without regard for other life on earth. It is appalling to some that man, a recent development in the history of this planet, should have come to view it and its life as being here to satisfy his needs. While recognizing the moral and philosophical validity of this concern, we must also recognize that man has become the dominant animal on earth; and, hopefully, he will remain so. The peoples of the earth will need to utilize intensively and wisely its resources to fulfill their needs and aspirations. Man will be wise to encourage great diversity of plant and animal life wherever possible, but he cannot do this in agricultural and urban areas. Still, in his own interest, man must learn to recognize the importance of all life in seeking his own best role in nature.

What do we mean by man's environment? This is not an easy concept. It is difficult to distinguish clearly between an organism and its environment, particularly in the case of man. As we explained in Chapter 7, the environment of the individual organism is most easily conceptualized: the sum of the phenomena influencing its development, metabolism, and activities. It is more difficult to distinguish between a population and its environment, and still more difficult to distinguish between a biological community and its environment, for populations and communities are parts of their own environments. The difficulty in distinguishing between organisms and their environments lies in the fact that they form interacting complexes, organism-environment systems, in which each influences and is influenced by the other. The ecosystem concept is helpful, ecosystems being the interacting systems formed by biological communities and the conditions and resources of their locations. It is easier to think of man as a part of an ecosystem than it is to conceptualize the environment of man.

Conceptualization of the environment of man is far more difficult than is conceptualization of the environment of other organisms, for the nature of man and his cultural evolution have further blurred the distinction between organism and environment. Is the adaptive clothing of a man part of the man or part of his environment? And the ideas man generates, are they part of him or part of his environment? Certainly they influence him profoundly. Marston Bates (1962, p. 8) writes of man's *conceptual environment*. It is sometimes helpful, even necessary, in our efforts to understand the responses of man to his environment to attempt to distinguish between man and factors in his environment. But in doing so, we must remember that man and his environment form an interacting system; human populations are parts of particular ecosystems which together form the ecosystem earth. Moreover, it will be well to remember that, although man has learned to control his environment more than have other or-

ganisms, the changes he brings about in his ecosystem initiate multitudinous other changes over which he has no control. Along with his valiant attempts to control his environment, man will need to adapt in meaningful ways, if the future ecosystem earth is to be stable and desirable for man.

Man has changed the earth; he has changed its lakes, rivers, and estuaries, sometimes necessarily, sometimes needlessly. Some have argued that our waters should receive no wastes, that we should maintain them in their natural state or return them to this condition as soon as possible. Superficially, this is an appealing position to take. But those who do so fail to recognize that most uses of waters change them from their natural state. Insofar as such changes decrease the value of waters for possible uses, the changes can be considered pollutional. Man must use the waters of the earth, most of them intensively. It is desirable, even necessary, that man maintain some waters in essentially their natural state; this would permit only scientific and very limited recreational use. But it is misleading to argue that man can or should maintain in their natural state most of the waters of the earth.

In the beginning, primitive human societies were no doubt well adapted to their environment and changed it relatively little; a few such societies still persist in isolated areas of the world. As man learned to use fire, to domesticate animals, and to plant crops, the changes he wrought in his ecosystem became greater. Improvements in agriculture and transporation and the use of fossil fuels made the Industrial Revolution possible, and the changes man caused became profound. Now, with the Scientific Revolution, the rapidity of change is a cause for alarm, when we consider that the greatest changes have occurred in the last few hundred years of man's history. Most of the changes appear to be beneficial for man. There has been a kind of adaptation, a cultural adaptation, but one that differs from the more biological adaptation of primitive societies. This new kind of adaptation is not everywhere sound; it confronts us with new dangers. Still, if we could, few of us would return to the adaptation of primitive societies. And, of course, population growth has made this impossible.

Man's use of the earth's resources and the changes he brings about in his ecosystem cannot be considered apart from his present-ly existing populations and their perhaps inevitable growth. To do so is to make use-less abstractions. Furthermore, we cannot consider man's use of any element of his environment—soil, water, or air—apart from the others and the human populations in-volved; the very nature of an ecosystem is the interdependency of its components. The most profound biological problem of which man is aware is his own uncontrolled popu-lation growth on an ultimately limited earth. The problem differs in different parts of the world, but it may confront all peoples. Along with controlling his populations, man will need to utilize intensively the earth's re-sources; and there will come many changes in its ecosystems. But intensive use must be wise use; resources cannot be wasted or spoiled. Though intensive use will bring many changes in the soil, water, and air of the earth, these cannot be spoiled for use and reuse; though waters will be changed, they cannot be changed in ways diminishing their value.

The question is not whether there are going to be changes in man's environment but whether man can adapt to these changes. And, if he can, what is the price of adapta-tion? Primitive peoples, still living on the earth in isolated areas, have demonstrated that through biological, social, and cultural adaptation, "mankind can survive, retain en-dearing traits, and create independent and viable cultures without technology, even un-der the harshest circumstances" (Dubos, 1965, p. 254). Other men appear to have adapted to seemingly undesirable conditions of urban life. In writing of the people of the mill town of Leigh, in Lancashire, England, where the dreary surroundings are obscured only by smog, Dubos (p. 273) goes on to say:

They have remained highly active and produc-tive physically, biologically, and intellectually. Their expectancy of life is not very different from that of people of the same stock and economic status living in uncontaminated areas, nor is their birthrate smaller.

Neither the people of Leigh nor the people of the Amazon Valley would survive long if their environments were exchanged; but with time, man's potential for biological, so-cial, and cultural adaptation permits him to become biologically successful under as-tonishingly different circumstances. Dwellers in our modern cities seem to have adapted to

the agitations and tensions there. Neverthe-less, while we should not underestimate the adaptability of man, we must not depend on its dimly understood processes to insure man's future.

Moreover, whatever the adaptive poten-tials of man may be, adaptation has its costs, not all of which are apparent. Adaptation can best be evaluated historically, for it is generally a process requiring long periods of time. Adaptational changes to environmental conditions now confronting man may, over long periods of time, be excessively costly in biological, social, or cultural terms. And even if man can adapt biologically, socially, and culturally to survive, reproduce, and func-tion under greatly changed environmental conditions, such biological success must not be man's sole objective.

The biological view of adaptation is inadequate for human life because neither survival of the body, nor of the species, nor fitness to the condi-tions of the present, suffice to encompass the richness of man's nature. The uniqueness of man comes from the fact that he does not live only in the present; he still carries the past in his body and in his mind, and he is concerned with the future. (Dubos, 1965, p. 279)

The adaptation of primitive societies still existing will not be possible much longer, however tragic may be the loss of these so-cieties. Modern man does not and cannot seek such adaptation. The problem of human population growth in the face of limited resources will not permit this. Thus, man must learn enough and be wise enough to plan his adaptation. He must learn how his ecosystem will change in response to his activities. He must learn to manage his social institutions for humanity's ultimate good. Hopefully, he will be wise enough to know what is ultimately good. And, hopefully, he will learn enough and be wise enough to retain some of the natural beauty of the earth and its waters, however intensive their use may need to become.

THE ACCEPTABILITY OF ENVIRONMENTAL CHANGES

What environmental changes are accept-able is a social question that must ultimately be answered by each society; the wisdom of the answers will depend on each society's values and its foresight of the biological, social, and cultural consequences of environmental changes. Hopefully, the pathways societies will follow will be selected in the main not for short-term benefits to a few peoples but for the ultimate good of mankind. Man's knowledge and his ways of making choices and putting them into effect are most imperfect. Still, on the basis of imperfect knowledge, he must make choices and endeavor to follow them. He cannot do otherwise.

We are confronted with the problem of acquiring knowledge of the dangers and benefits, both biological and social, of possible environmental changes. But for man, biological and social dangers and benefits are usually inseparable; thus, acquisition of this knowledge is exceedingly difficult. Few changes are clearly either injurious or beneficial.

We must endeavor to acquire knowledge of environmental changes hazardous to man's health or to his resources; here our concern is water resources. Gross pollution of our waters is clearly an immediate hazard to many aquatic resources, a hazard that can be documented biologically with relative ease. Lesser environmental changes may also be immediately detrimental to aquatic life; and insofar as they produce measurable changes in the species involved, they can often be evaluated. But changes not likely to be harmful in the near future may in the long run be destructive of aquatic environments. The ultimate biological effects of increasing river temperatures, eutrophication of lakes, and accumulation of biologically stable pesticides or radioactive materials are difficult to predict. Yet their effects may be more destructive and more permanent than those of obviously hazardous gross pollution, which often can soon be eliminated. Biologists seeking to evaluate hazards to aquatic resources face much the same problem as those evaluating other hazards to the well-being of man. René Dubos (1965, p. 220) has written:

Public concern with the dangers of environmental pollution has led many different branches of government to encourage new lines of scientific research and to enact various control programs. Despite these efforts, however, it is clear from the proceedings of the many national and international conferences held during recent years that knowledge concerning the medical significance of the pollution problem is still extremely scant. In most cases, knowledge hardly goes beyond the description of a few *immediate* pathological responses. Hardly anything is known of the *delayed* effects of pollutants on human life, even though they probably constitute the most important threats to health in the long run.

In determining the acceptability of particular environmental changes, societies should have information on the probable social and cultural consequences of these changes. Social scientists seeking to provide this information are confronted with difficulties similar to those natural scientists face; the addition of the human dimension, if anything, makes their studies more complex. Changes in food resources can obviously have profound social and cultural effects, as can increased crowding of peoples into urban areas. Environmental pollution, through its long-term effects on the health of peoples, may bring about undesirable social and cultural changes. These changes, however, will usually be more difficult to evaluate than those resulting from food shortages or crowding. And the needs of man, as we might wish him to be, go beyond food and space and bodily health, perhaps to needs for diversity of landscape and activities, and needs to be creative and to enjoy beauty. Dubos (1965, p. 279) was concerned about such needs when he wrote:

Man is so adaptable that he could survive and multiply in underground shelters, even though his regimented subterranean existence left him unaware of the robin's song in the spring, the whirl of dead leaves in the fall, and the moods of the wind—even though indeed all his ethical and esthetic values should wither. It is disheartening to learn that today in the United States schools are being built underground, with the justification that the rooms are easier to clean and the children's attention not distracted by the outdoors!

We must hope that social scientists will find ways of determining the social and cultural consequences not only of biological deficiencies but also of esthetic deficiencies in human existence.

For evaluation of the human ecosystem, no single set of social parameters can be relied upon, be it psychological, sociological, or economic. With regard to economics, a social science in which the methodology for studying environmental problems is advancing rapidly, Boulding (1967, p. 53) writes:

It may be, of course, that for the present generation or two this is simply a problem in economics. We have to manipulate the rewards of the system so that pollution is not rewarded. It is a problem, however, which may easily go beyond economics. . . .

The multidimensional nature of man and his interactions with all the circumstances of his ecosystem must not be forgotten in our attempts to determine the possible consequences of environmental change.

The acceptability of a particular environmental change is a social choice or decision; hopefully, it will be wise and enlightened with understanding of the possible biological, social, and cultural consequences of change. We use the term social choice loosely here, for societies rarely choose between clear alternatives; they move along indistinct paths and make frequent corrections in their course toward ill-defined goals. The mechanisms societies use in determining their courses toward their goals vary vastly. Democratic governments, though ponderous and perhaps inefficient, appear to be best for most people most of the time. But they depend on an enlightened citizenry making its desires known. Scientists, educators, and governmental administrators are faced with the formidable task of informing their people of the possible consequences of environmental changes, so that the participation of people in determining the course of their civilization is enlightened. And the desires of even an enlightened people, except as imperfectly represented by their decisions on the few questions voted upon, are difficult to assess. Yet, resource administrators must assess public desires, and such assessment should include understanding of the processes by which attitudes are formed and choices made. For attaining this understanding, White (1966, p. 127) proposes research that:

. . .will throw light on how decisions in truth are made, on how the professional's own preferences figure in the proposed solutions, on what he thinks the citizen prefers, on what the citizen, given a genuine choice, does prefer, and on how all of these may shift with the circumstances and experience surrounding the choice.

Man must find ways of choosing between environmental alternatives; he cannot depend on chance to lead to desirable circumstances, not with the rate at which he is changing his environment.

Having determined its goals, however imperfectly, a society must endeavor to manage and adapt to its environment so as to favor attainment of these goals. Such attainment may, however, in various degrees be prevented by man's inability to manage his ecosystem to the extent necessary or by the constraints his own institutions place on his possible actions. There are limits to the extent to which man can manage his ecosystem; beyond these he must adapt, as Bates (1962, p. 13) has written:

We have come to feel that we are so far apart from the rest of nature that we have but to command. Yet, that old aphorism of Francis Bacon still applies—you cannot command nature except by obeying her.

And Gaffney (1966) has observed that existing social institutions may prevent the application of otherwise excellent solutions to environmental problems, as can existing technology. Fortunately or unfortunately, the evolution of social institutions may be much slower than the evolution of technology.

BIOLOGICAL EVALUATION OF CHANGE

When the physical and chemical conditions of ecosystems are changed beyond their normal ranges, unusual changes, however small, can be expected to occur in the individual organisms, population, and communities of these ecosystems. Sufficient study can disclose these changes. Much will one day be known about both usual and unusual changes in aquatic and other ecosystems; this will increase man's ability to manage and to adapt to his environment. But such full understanding of his environment is an objective man cannot soon attain; and, before it is attained, he must make continually the most enlightened decisions possible regarding his environment. Profound understanding of his ecosystem will one day provide man with the best possible basis for prediction of the effects of environmental changes; some of our investigative effort must be directed toward this end. Nevertheless, we must focus much of our effort on problems for which solutions are now needed. We must seek to make the best possible use of our limited efforts and knowledge in resolving these problems.

Man's interest in particular waters and kinds of waters and in their fauna and flora along with his knowledge of the kinds of pollutional changes occurring provides a basis for selection of critical problems and promising approaches. There will be some waters that for scientific or esthetic reasons a society may wish to retain in essentially their natural state. In other waters, a society may wish to insure that there will be no decrease in the production of valuable species or increase in the production of nuisance species. There will be waters in which changes in the production of species will be permitted so long as all the associated benefits and costs are acceptable. Other waters may be so valuable for industrial or agricultural use that changes short of those leading to elimination of aquatic life may be acceptable. Such a classification of waters according to man's interest in them provides one means of relating social desires to the problem of biological evaluation of environmental change; it also aids in the selection of the biological approaches likely to be most effective in solving particular problems.

Since human knowledge is limited, man cannot know all he might wish about controlling his activities. Accordingly, he generalizes from existing knowledge until experience confirms or leads to correction of his ideas. This is so in our daily life; it is so in science. We cannot study in detail every instance of pollution, every water, or every species. We must apply knowledge obtained about some species in certain waters under particular conditions to similar species in other waters whose conditions appear nearly the same. A system of water use classification and water quality standards is helpful in this; but it represents only a formalization of the way biologists, without the aid of such a system, use their knowledge in evaluating pollutional changes.

Conditions Permitting Natural Communities

We have already noted that societies may choose to keep some waters in essentially their natural or primitive state for scientific or limited recreational use. There remain but few large lakes, rivers, and estuaries in this condition, although there are many smaller waters not much influenced by man

as yet. Man's needs will probably permit very few large waters to be retained in their natural state; perhaps more small waters can be so maintained. At this time in history, man's past and present activities have made it impossible to determine what primitive conditions were in most of our larger freshwaters and in many of our smaller ones.

Virtually any but the most restricted uses of waters will bring about some changes in their physical and chemical characteristics. These in turn lead to changes in the biological communities of these waters, changes which may be gross or which may be hardly detectable, according to the magnitude and the biological importance of the physical and chemical changes. With sufficient effort, biologists can describe the composition of aquatic communities—the species of both macroscopic and microscopic plants and animals that are present and their relative abundances. Such detailed descriptions of aquatic communities have rarely if ever been made. But the methodology by and large exists, and the importance of having detailed descriptions of aquatic communities under different conditions warrants the expenditures necessary to obtain them. When such descriptions of communities are made, as increasingly in the future they will be, detailed studies of existing physical and chemical conditions should also be made. Only in this way can we begin to know the conditions necessary for the existence of particular communities and kinds of communities.

The composition and structure of any aquatic community is determined by the species historically present at its location, by the opportunities for colonization by other species, by the responses of individual organisms and their populations to existing physical, chemical, and biological conditions, and by the interactions of all of these. Knowledge of these conditions and processes would undoubtedly be the best possible basis for understanding how particular communities persist under certain sets of conditions. But the hundreds of species present in most aquatic communities and the complexities of their interactions will probably never permit man to predict by this method the changes that will occur in communities with particular environmental changes. He cannot determine even the effects of the most important pollutional changes on the survival, reproduction, growth, and behavior of each of the hundreds of species in an ecosystem, let alone their interactions. Thus, to predict the environmental conditions necessary for the persistence of particular aquatic communities, we must depend primarily on knowledge of the composition of communities existing under known conditions and knowledge of community changes that have occurred when conditions have changed in known ways.

Wherever waters yet persist in essentially their primitive states, we should at the earliest moment possible obtain detailed descriptions of their biological communities and the conditions under which these communities are existing. Should societies desire to maintain any of these waters in their primitive states, they may do so by permitting no uses of the waters that lead to any change in the composition or structure of their communities. Studies should be continued on waters that have been described in their primitive state, even when they are to be utilized more intensively. The results of these studies are likely to be of value in the future management of the waters involved; and, perhaps even more important, information on the changes occurring in their communities with changing conditions can be invaluable in the management of other waters.

Biological indices of environmental change, which we discussed in Chapter 20, are basically descriptions of different components of aquatic communities. None of these are so detailed or so complete as the descriptions of aquatic communities we are here recommending; none include all the components of aquatic communities, their bacteria, fungi, and algae, their other plants, and their animals. Some have claimed various biological indices to be very sensitive to environmental changes of all kinds. No single component of an aquatic community is likely to be indicative of all possible changes that might influence its other components. Therefore no particular biological index can be considered as valuable as detailed and complete descriptions of aquatic communities under different known conditions. Nevertheless, if future studies can establish useful relationships between environmental conditions, particular biological indices, and the composition of aquatic communities as a whole, these indices will take on new value. With reference to water quality control, the greatest value of biological in-

dices, as well as of complete descriptions of aquatic communities, is in those instances in which a society wishes to protect the composition of aquatic communities from change. This is especially so where primitive conditions are to be maintained. No other biological approach to the evaluation of problems of this sort is more direct or more practical.

Conditions Not Changing Production of Populations of Interest

There will be waters that societies either cannot or do not choose to maintain in their primitive state but in which either the maintenance of unrestricted production of valuable populations or the prevention of increased production of nuisance organisms is desired. Changes in community composition and structure may not result in or be accompanied by either decreases in the production of populations of value or increases in the production of nuisance organisms. Thus, the most direct and effective approach to evaluation of conditions that may change the production of populations of interest is to study the production of these populations and the conditions leading to this production. In doing so, it may be necessary to study food and other organisms that may influence the production of the organisms of direct interest, but emphasis in these studies should not be placed on evaluation of changes in community composition or structure.

Not only pollutional changes but also changes in natural conditions can influence the production of populations (Chapter 17). Only in recent years have measurements of the production of fish populations been made; in few instances will information be available on production before environments were influenced by the activities of man. For these and other reasons, periodic measurements of the production of fish populations will not alone provide a basis for predicting the effects of environmental change. They should be accompanied by other studies in nature and in the laboratory.

Studies of reproductive success and of the survival and movements of the age groups of fish populations in relation to environmental conditions in nature should be conducted along with studies of growth rates and biomasses of these groups (Chapters 16 and

17). The food habits of individuals in the different age groups in relation to the availability of their food resources should also be studied. To be able to relate the production of populations to environmental conditions, we must understand how these conditions influence the availability of the food resources of populations; some studies should be directed to this end (Chapter 19). Studies like these should be made at different locations and times; from them will come knowledge of value in predicting the effects of environmental changes on the production of animal populations.

In the laboratory, the effects of particular environmental factors on the survival, development, growth, and behavior of fish or other aquatic organisms can be studied (Chapters 10, 11, 12, and 13). Such studies coupled with studies of fish production in laboratory ecosystems (Chapter 17) can be useful not only in explaining changes in production observed in nature but also in predicting the effects anticipated environmental changes may have on fish production.

The relative simplicity of laboratory experiments, whether on individual organisms, populations, or communities, means that their conditions differ in important ways from conditions in complex natural ecosystems. The results of laboratory experiments need to be interpreted with great care. Small experimental streams (Warren et al., 1964) and lakes (Hasler, 1964) where natural conditions can to some extent be controlled provide a promising means of studying the effects of pollutional and other conditions on the production of populations of interest. These studies have some of the advantages of laboratory experiments and can provide a safer basis for predicting the probable course of events in larger ecosystems.

Nuisance populations that societies will choose to control in some waters, with the exception of biting insects, will perhaps most often be autotrophic or heterotrophic plants. Increases in the production of such populations will often result from human activities, leading to increases in plant nutrients. Adequate methods have been developed for measuring the production of autotrophic plants; generally useful methods for measuring the production of heterotrophic plants are yet to be made available. Measurements of the photosynthetic rate of algae and other autotrophic plants in relation to environmental conditions in nature can provide some understanding of the causes of particular problems. But the complex interactions of plants with environmental conditions — primarily light, temperature, and nutrients — make full understanding of the causes of changes in production difficult to obtain in nature alone.

Determination of the specific nutrient requirements of particular species of autotrophic and heterotrophic plants can best be accomplished through laboratory studies of pure cultures. Here, under different light and temperature conditions, the combinations and concentrations of nutrients that favor the reproduction and growth of particular species can be determined. This information together with information on the production of these species under different known environmental conditions in nature can be exceedingly useful in explaining and predicting the effects of changes in nutrients resulting from the activities of man.

But the interactions of plants not only with the physical and chemical conditions of their environment but also with other plants and animals are exceedingly complex. Such interactions are impossible to unravel in nature alone, and few of these interactions exist in cultures in the laboratory. For this reason, laboratory ecosystems and experimental streams and lakes should be utilized in the study of nuisance problems. Into these ecosystems, under some control, different wastes and nutrients can be introduced to determine what effects they may have on the production of nuisance populations under relatively natural conditions.

The determination of conditions influencing or likely to influence the production of valuable or nuisance populations is a most difficult matter requiring the expenditure of considerable human and material resources. Moreover, this approach to the solution of biological problems is in its infancy and has a limited, though developing, methodology. Still, the complexity and importance of many of our water pollution problems now demand that expenditures in their solution be great; valuable resources are involved. Where our concern is the production of particular populations, we must begin the difficult task of evaluating the determining conditions; other biological approaches, however developed or interesting, will not provide the needed information.

Conditions Leading to Changes Having
Acceptable Benefit-Cost Balances

In some waters, a society may choose not to maintain conditions permitting maximum production of valuable populations or only minimum production of nuisance organisms. Here, a society may choose to maintain conditions permitting levels of production of valuable or nuisance populations that provide some desired balance between all the benefits and all the costs associated with so maintaining these waters. Evaluation of conditions that will lead to such a balance is the most difficult general problem facing biologists and social scientists interested in resource management. In solving this problem, biologists must do more than determine conditions unlikely to lead to change in the production of valuable and nuisance populations; they must also determine how much change will occur in these populations with each change in environmental conditions. For their part, social scientists must attempt to evaluate all the benefits and costs associated with maintaining each possible set of environmental conditions, benefits and costs not in economic terms alone but in terms of man's ultimate biological and social welfare as well. It is for societies to choose, we hope wisely, what balance they wish to maintain between all these benefits and costs.

When biologists have reached the point of being able to predict the changes which will occur in the production of particular populations with changes in environmental conditions, they will have attained a level of understanding that all scientists seek. The value of such understanding goes beyond the establishment of desired benefit-cost balances. Efforts to obtain information necessary to establish these balances are not wasted even when they fall short of this goal, for they lie in the mainstream of ecological thought. The biological approaches necessary to relate changes in production of aquatic populations to changes in environmental conditions are the same as those needed to determine conditions that will not decrease the production of valuable populations or increase the production of nuisance populations, which we discussed in the preceding section. But wider ranges of environmental conditions must be studied in the laboratory, in experimental streams and lakes, and in uncontrolled nature.

Social scientists, in attempting to evaluate all the benefits and costs associated with maintaining particular sets of environmental conditions, are faced with the problem of finding dimensions not only for economic but for biological, social, ethical, and esthetic conditions as well. Even with dimensions for these biological and social values, there would yet remain the need for common denominators to provide means for balancing the different kinds of benefits and costs.

There has been, in recent years, considerable interest in the application of economic theory and method to solution of natural resource and environmental health problems. Economists have developed ways of using economic value as the dimension and the common denominator for determining benefit-cost balances. They have made progress in finding ways of evaluating even recreational resources (Clawson, 1959; Brown, Singh, and Castle, 1965); and, hopefully, they will find ways of determining the expenditures that societies are willing to make in order to retain other kinds of esthetic experiences. Ethical values are undoubtedly beyond the pale of economics. There are, of course, ecological, technological, institutional, and other constraints on any system of environmental management that man might wish to use. In at least one respect, this is very fortunate, for most of our ethical and at least some of our esthetic values are protected by legal, religious, traditional, and other social constraints on man's activities.

Such biological and social evaluations of all the benefits and costs associated with maintaining the condition of the environment at different levels would provide the ultimate basis for wise selection of ways to manage our environment. But, in any particular period of time, societies will have to choose the apparently most desirable balance between benefits and costs on the basis of the best evaluations then available.

Conditions Permitting Residual
Populations of Some Desirable Species

There may be waters whose value for domestic, agricultural, industrial, power, or navigational use is so great that a society would choose to maintain them only in such condition as to prevent gross nuisance problems and to permit residual populations of some desirable though hardy aquatic species. Some lakes and the lower parts of some large rivers in densely populated, highly industrialized areas are likely to be in this category. The accumulated effects of all the uses such waters undergo, even though the effect of each use is small, may so alter them as to prevent the persistence of good populations of their native species. Increases in temperature, dissolved solids, and plant nutrients are difficult if not impossible to prevent in waters like these and can render them unsuitable for the production of valuable species or can increase the production of nuisance organisms. Further, these waters are subject to many other changes resulting from man's activities. Nevertheless, some of these waters can be maintained in a sufficiently attractive condition to encourage limited recreational use; this a society may choose to do.

The biological approaches most useful for determining conditions suitable for populations able to support limited recreational use will depend on local circumstances. Recreational fishing can sometimes be provided by stocking fish to be caught directly by anglers. Conditions not lethal for these fish, but still not suitable for their growth and reproduction, could make this sort of fishery possible. Limited growth and reproduction of some species may be possible in other waters and, perhaps supplemented with stocking, could provide for some recreational use. Perhaps the most effective means of determining conditions permitting such use would be laboratory studies coupled with monitoring of the physical and chemical conditions and the persistence of the desired species in nature. In the laboratory, emphasis should be placed on determining the effects of critical environmental variables on the survival, reproduction, and growth of the species to be maintained. Conditions not directly lethal to fish species should permit at least restricted production of some organisms that would be suitable food for the fish.

Anadromous fish, migrating to or from spawning and rearing areas in the upper reaches of river systems, frequently must pass through downstream waters that have been greatly altered by man's activities. Our society will usually be very anxious to make such passage of salmon and other valuable species possible. For this, we will most often be concerned with providing conditions suitable for survival and movement, not for embryonic development or growth; laboratory and field studies should be directed toward determining such conditions.

Secondary treatment and chlorination of the wastes of man and his animals can make waters receiving these wastes safe for recreational uses. If, however, these waters are unattractive because of odors or floating materials, they will discourage recreational use. Such conditions will not usually develop where treatment of wastes is sufficient to insure dissolved oxygen levels permitting the persistence of valuable fish populations. Nevertheless, very low levels of dissolved organic materials can stimulate nuisance growths of bacteria; and small increases in other plant nutrients can cause nuisance blooms of algae. When nuisance growths of bacteria and other plants occur in waters such as we are considering in this section, solutions for this problem are likely to be difficult to find. Even when sound evaluations of the biological problem can be made by procedures we have discussed in previous sections, the technological problem of control of such nuisances can be very great. Satisfactory solutions to the problems arising from multiple use of waters will not always be found.

MONITORING OF CHANGE

With increased utilization of their ultimately limited water resources, many societies will undertake, if they have not already done so, programs of monitoring their waters for changes in physical, chemical, or biological characteristics. Such monitoring programs, not only for heavily utilized waters but also for waters not yet subject to such utilization, have three important purposes: to provide information on natural and man-caused changes in waters in order to better manage them; to determine whether or not standards established for particular waters are being met; and to determine whether or not standards set for particular waters are providing the protection desired for the uses they were established to maintain.

Lack of information on the characteristics of particular lakes, rivers, and estuaries has been a great hindrance to the formulation of plans for their further protection and use. California and other states established sampling stations for the collection of water quality data before the federal government, in cooperation with local, state, and other agencies, established in 1957 the National Water Quality Network. These extensive sampling programs will continue to provide information on many waters in the United States, information that can be used as baselines in setting standards, whether these be set to maintain existing conditions or to return them to some prior level.

Most societies ultimately will establish various sorts of water quality standards in order to maintain the desired uses of particular waters. Without continual monitoring of the characteristics of these waters, it would be impossible to determine whether existing or additional uses of the waters are consistent with established standards. Water quality standards presumably will be established and revised periodically on the basis of the best available information regarding the water quality needs of the water uses to be protected. Whether or not particular standards are providing the desired levels of protection for these uses will ultimately be determined best by experience. Continual monitoring of the physical and chemical conditions of particular waters and periodic studies of the levels of use actually possible under these conditions can provide a basis for continuing evaluation of the adequacies of standards.

The system of water use classification and water quality standards adopted by a society as well as local circumstances will to a considerable extent determine the nature and extent of the monitoring to be done. The volume and quality of waste waters being discharged, the physical and chemical characteristics of receiving waters, and the biological organisms in these waters may be monitored. When the purpose of the monitoring is to determine whether or not particular uses of the water are receiving the protection desired, the monitoring could involve measurements of the production of valuable organisms or measurements of the success of other uses. More frequently, however, monitoring programs—whose purpose is usually to follow routinely the course of events—will utilize simple, relatively inexpensive measurements; intensive investigations will most often be used only in the development of particular case histories.

Even though we will be primarily interested in maintaining the quality of receiving waters at certain levels, it is usually important to have continuous records of the volume and quality of waste waters being discharged from particular sources. Such records are of value both to the discharger and to the responsible regulatory agency in determining whether or not requirements set for a particular waste source are being met. Further, records of all discharges into a receiving water can be used in combination with receiving water quality and flow data to determine whether requirements set for the various waste discharges are in total adequate to maintain established standards. The volume of all the important waste discharges should be continuously recorded. The quality measurements that should be made for particular wastes will depend on the nature of the wastes and on the uses and standards established for the receiving water. Where thermal pollution is a concern, effluent temperatures should be monitored; where salinity is a problem, the dissolved solids content of wastes should be continuously measured. If the problem is one of oxygen depletion in the receiving water, the biochemical oxygen demand of the waste should be monitored. Sometimes particular components of a waste are known to be responsible for its toxicity to aquatic organisms; when practical methods for measuring these components are available, they should be used routinely. In other instances, the complexity of toxic wastes or the lack of practical analytical methods may make acute toxicity bioassays the most reliable method of monitoring to protect aquatic life.

Most water quality standards will be established in terms of physical and chemical characteristics necessary to protect particular uses. Such standards can be defined more precisely than can most biological standards. Furthermore, measurement of physical and chemical characteristics is usually simpler and more reproducible than measurement of biological characteristics. The development of instruments that can continuously record some of the important physical and chemical characteristics of water has greatly reduced the expense and increased the reliability of monitoring these characteristics.

Biological characteristics of waters are sometimes incorporated into water quality standards. Among these, the coliform index and acute toxicity are perhaps most general.

Counts of organisms having nuisance potential and even measurements of community composition may one day be found useful as standards. When biological parameters are specified in standards, their routine measurement in monitoring programs can be expected. But more complex biological measurements will usually be reserved for periodic studies designed to determine whether or not physical and chemical standards are insuring the desired protection for aquatic life.

SELECTED REFERENCES

Bates, M. 1962. The human environment. (Horace M. Albright Conservation Lectureship 2.) University of California, School of Forestry, Berkeley. 22pp.

Brown, W. G., A. Singh, and E. N. Castle, 1965. Net economic value of the Oregon salmon–steelhead sport fishery. Journal of Wildlife Management 29:266–279.

Dubos, R. 1965. Man Adapting. Yale University Press, New Haven. xxii + 527pp.

Hasler, A. D. 1964. Experimental limnology. Bioscience 14(7):36–38.

Jarrett, H. (Editor). 1966. Environmental Quality in a Growing Economy. Published for Resources for the Future, Inc., by The Johns Hopkins Press, Baltimore. xv + 173pp.

McKee, J. E. 1967. Parameters of marine pollution — an overall evaluation. Pages 259-266. *In* T. A. Olson and F. J. Burgess (Editors), Pollution and Marine Ecology. Interscience Publishers (Wiley), New York. xvi + 364pp.

Roslansky, J. D. (Editor). 1967. The Control of Environment. (1966 Nobel Conference discussion.) North-Holland Publishing Company, Amsterdam. xi + 112pp.

Warren, C. E., J. H. Wales, G. E. Davis, and P. Doudoroff. 1964. Trout production in an experimental stream enriched with sucrose. Journal of Wildlife Management 28:617–660.

23 The Roles of Water Pollution Biologists

To begin, we should recognize that biologists have played a relatively minor role in water resources planning, and in the development of water and waste treatment. . . . the number of biologists was never large, in no way comparable to that of the various sanitary engineering groups who were involved more persistently.

CHARLES RENN, 1960, p. 157

You, the nation's biologists, must play a key role in lifting us out of our ignorance. You must provide sources of information which will give guidance as to the better procedures to follow in the future. You should go after this task with new effort. You must tell us what is needed, what it will cost, and how long the research might take. You should also recommend interim courses of action based on present knowledge.

STUART UDALL, 1964, p. 18

THE CHANGING ROLES OF BIOLOGISTS

The involvement and roles of biologists in the solution of problems associated with water pollution have changed with increased awareness of the nature and extent of these problems. Viewed initially as being primarily a hazard to public health, pollution problems were attacked principally by microbiologists and the medical and engineering professions. Even when there was concern for valuable aquatic resources, the efforts of but few ecologists were enlisted. Here and there, pioneering contributions were made by outstanding biologists, but not many others were interested in water pollution biology. As the full cost and complexity of water pollution has become apparent, so has the need for biologists. Today, increasing numbers of biologists are working on problems associated with water pollution; but in the years ahead, more biologists will be needed in im-

portant roles: aiding in the development of public policy; providing understanding of biological processes; developing biological approaches to the solution of difficult problems; and applying biology to water pollution control.

Present methods of waste treatment were developed in the late nineteenth and early twentieth centuries by engineers and chemists with but little help from biologists. Only in more recent years, as a few biologists have been engaged in waste treatment studies, has there come real understanding of the biological processes that make these methods so remarkably effective. But even today, much must be learned of the species and roles of organisms taking part in these processes, if biological waste treatment methods are to be utilized with maximum effectiveness. In the United States, many of the engineers and chemists pioneering the development of waste treatment methods were influenced by William T. Sedgwick, a biologist whose interest in teaching and in waste treatment and water supply was important in making clear the need for biology. Biological knowledge was early utilized in identifying the causes of many of the taste and odor problems in water supplies, microorganisms often being responsible.

A few biologists in the early twentieth century studied the effects organic wastes have on the composition of biological communities of rivers. Especially noteworthy among these were Kolkwitz and Marsson (1908, 1909) in Europe and Forbes and Richardson (1913, 1919) in the United States. Their studies were the forerunners of many investigations of changes in rivers resulting from waste introduction; and even today, many biologists interested in water pollution are conducting similar studies.

Perhaps publication by M. M. Ellis of his "Detection and Measurement of Stream Pollution" in 1937 can be taken as signifying a change in one of the roles of biologists from merely describing the effects of water pollution to determining for purposes of prediction the conditions under which aquatic organisms, particularly fish, might be expected to be successful. On the basis of his studies of dissolved oxygen concentrations and fish populations in many rivers and streams, Ellis concluded that an abundant and mixed fauna of warm-water fish can occur in waters in which the dissolved oxygen concentration does not fall below 5 mg/l. This value subsequently became probably the most widely adopted water quality standard for the protection of fish. Ellis (1937) also employed acute toxicity bioassays in evaluating the danger of toxic substances to fish. He was not the first to use bioassays for this purpose, and considerable improvement and standardization of methods came after his time. To many, information on the acute toxicity of wastes had appeared to be primarily useful for interpreting physical and chemical analyses; but there gradually came an awareness that, in controlling waste discharges for the protection of aquatic life, toxicity information can best be used more directly.

As it became apparent that biologists must attempt to predict the effects wastes will have on the success of aquatic populations, there came wider acceptance of the view that not only lethal effects but also effects on the reproduction, growth, and movements of organisms must be known. There were early studies such as those of Shelford and Allee (1913) and Shelford (1917) on the responses of fish to gradients of gases and toxic substances; but until the past decade or two, the principal attention of biologists interested in water pollution was directed to studies of stream communities and to the acute toxicity of pollutants. Now, biologists are investigating most of the problems of life in aquatic environments. In the United States, this has been made possible since about 1950 by increased involvement of the federal government in water pollution investigation and control. Not only have the programs of the federal agencies been broadened tremendously, but research grants to universities have made it possible for biologists having many different backgrounds and interests to apply new approaches to the solution of water quality problems. Thus, today, biologists are investigating problems ranging from improvement of water supplies to waste treatment, and from factors affecting individual species to overall changes due to eutrophication of aquatic environments.

As the extent and complexity of the biological problems of water pollution have become known, and as the resources for the solution of these problems have become available, increasing numbers of biologists have assumed more and more important roles in the management of water resources. This has occurred not only in governmental agencies and in universities, but also in metropolitan sewage districts, in industry, and in independent consulting firms. Where initially biological studies were usually an adjunct to necessary chemical and physical studies, biological studies are now recognized as basic to most water quality control programs. With this recognition, the involvement of biologists in planning and administering water quality control programs has increased.

The roles of the different organizations involved in water quality control—governmental, academic, metropolitan, industrial, and consulting—differ considerably; and it follows that the roles of their professional people, biologists and others, differ. Although there is considerable overlap in function, personnel of governmental agencies and universities are primarily responsible for the research necessary; governmental agencies are responsible for seeing that receiving water quality is maintained;

and industrial and metropolitan organizations are responsible for the control of the quality of their effluents. Wherever biologists are employed, they will be involved in water quality research or management or both. The functions of organizations and their personnel are not easily separated or defined. But we will endeavor to discuss the roles of biologists as they pertain to formulation of water pollution control policies, to increasing understanding of the effects of water quality changes, to development of approaches to pollution problems, and to more routine application of biological knowledge.

WATER POLLUTION CONTROL POLICIES AND BIOLOGICAL KNOWLEDGE

Water pollution control policies are but a part of the overall water and other resource management policy of a society. The state of scientific and technical knowledge influences these broad policies, and particular policies involve scientific and technical problems; but it is not the professional responsibility of natural scientists, social scientists, or engineers to establish these policies. Water resource management policy develops in societies as a result of the actions of men through their various social institutions, and it is influenced by geographic, historic, cultural, governmental, economic, and other social factors.

In a democratic society, it is ultimately the desires of the people that determine policies of resource utilization. These desires are only imperfectly represented through democratic institutions, and they are not always consistent with each other or with resource realities. Further, patterns of resource utilization and management change slowly and with difficulty; established patterns place constraints on the ways even democratic societies are able to manage their resources for the public good. Administrators, in attempting to interpret and carry out public intentions, have these severe limitations placed upon their efforts. Hopefully, it is the public that through its legislative bodies determines the nature of public resource policies; but administrators must elucidate these policies and make them effective. Theoretically, then, administrators are in-

volved secondarily in policy formulation. These administrators have various professional and other backgrounds that, presumably, are helpful in their part of policy formulation; but the professional competence of administrators ideally should be used only in the elucidation and carrying out of policies whose first expression was a reflection of the desires of the people.

Even though water pollution control policy should have its origin in legislative bodies and is not primarily a scientific or engineering matter, scientific, engineering, and other knowledge is necessary to make such policy effective. In the past, administration at the policy level has been mainly in the hands of the public health and engineering professions. The social and biological complexity of water pollution problems now makes it imperative that more social and biological scientists be enlisted in this level of administration.

Water pollution control policies to be effective must take into account biological and social realities as well as technological possibilities. Biologists, then, must provide the relevant biological knowledge. Of the alternative paths from which a society must select its course, biologists must endeavor to provide an evaluation of the biological outcome of each. Only on the basis of such biological and social evaluations can a society hope to choose the best course and formulate effective policies.

Public policy must be continually scrutinized and periodically re-evaluated; knowledge and conditions change. Water pollution biologists must continually examine water quality control policies in the light of biological knowledge to determine whether or not these policies are consistent with reality. And the effective implementation of existing or new policies may require the acquisition of new biological knowledge, which biologists must identify and pursue. These biologists must, then, attempt to fully understand public policy and how it is being implemented, a difficult endeavor when policies are vague and their implementation diffuse.

Viable public policy in resource matters must take into account the broad range of human experience and activities that will be influenced by that policy. For biologists to be able to understand and evaluate such policy, they need considerably more knowledge of resource economics, political science, law, business management, technology, and even the humanities than they are likely to obtain in their formal education, which, like for other specialities, is narrow. Any policy represents a balancing of interests, and its final evaluation must take this into account. A policy that might seem less than desirable from a strictly biological viewpoint may be seen to be the best policy possible under existing conditions, when social and technological factors as well as biological ones are considered. To be effective at the policy level, biologists must prepare themselves to give intelligent consideration to the broad range of human experience and activities affected by policy. Their education cannot end on Commencement Day.

This preparation will increase the effectiveness of water pollution biologists in another of their roles pertaining to policy: the education of others on the biological aspects of water pollution. An enlightened public is essential to the development of sound policy, and the public is most likely to accept biological viewpoints from biologists who are not oblivious of social realities. And resource administrators facing problems of perplexing complexity will more readily accept biological advice from those aware of this complexity. Biologists working in industry can there greatly increase understanding of the importance and complexity of the biological problems associated with water pollution. They will be more effective in doing this if they have some appreciation of the economic and other pressures on top management.

UNDERSTANDING OF EFFECTS OF WATER QUALITY CHANGES

For man to be able to plan his activities intelligently, he must have some understanding of the changes they will bring about in his environment and of the effects these changes may have on him. This is particularly so where his primary resources — water, air, and soil — may be changed. A river, a lake, or an estuary with its biological community is an extremely complex, highly integrated system. Changes in water quality may bring about changes in the physiology and behavior of individual organisms, changes in populations, and changes in com-

munities; these in turn may lead to other changes in water quality. Full understanding of change and the processes that lead toward stability in such complex systems is an ideal toward which scientists work; it is an ideal that most recognize may never be fully attained. Yet each step toward its attainment leaves man better able to cope with his environment.

As we are faced with grave problems of pollution for which solutions are desperately needed now, it is understandable that many people are impatient with time-consuming, costly studies that promise no immediate solutions for these problems, even though these studies may one day provide the understanding that man needs to better manage his environment. Unquestionably, much emphasis must be given to research that promises soon to yield information of importance in solving critical problems. And we must be sure we are making the best possible use of information already available, because without some corrective action, even if this is not the ideal corrective action, our problems will grow worse. We cannot wait for full understanding and ideal solutions. But whenever man bases his actions on poor understanding, his chances of erring are great. Nevertheless, he must risk this when the dangers of inaction are even greater. At the same time, man must work toward full understanding of his environment, if he is to provide for his future. Clearly, then, both research that promises soon to yield useful information and research that can ultimately provide necessary understanding must be supported to the extent of our society's ability; a nice balance is needed, a balance that is difficult to maintain.

Research that is directed toward obtaining full understanding of the effects of water quality changes on aquatic organisms can perhaps most often be pursued best in a university atmosphere when biologists there are interested in water quality problems; but many governmental agencies and some industries are now making such research possible within their own programs. The interests and abilities of the biologist and the opportunity he is given to pursue his investigations will in large part determine the extent of his contribution to understanding of the effects of water quality changes. The extent of his possible contribution may, however, be limited by the general level of knowledge in his particular field of biology, for only infrequently can scientists expand the limits of knowledge much beyond existing levels. Thus, the possible contribution of a biologist interested in water quality problems depends very much on the work of earlier biologists who may have had little or no interest in water quality but who contributed to the body of pertinent biological knowledge.

Understanding of the effects of water quality changes on aquatic organisms will require knowledge of suborganismic, organismic, population, and community levels of organization of life, and it will require synthesis of this knowledge so that life's interactions may be comprehended. Biochemists, physiologists, pharmacologists, behaviorists, and ecologists interested in the effects of water quality changes on aquatic organisms will need to be enlisted in the pursuit of this knowledge. Knowledge in each of these areas of biological science has become far too extensive for any one biologist to be fully competent in all of them. Yet, knowledge of the effects of water quality changes on any single level of organization of life is really of little value unless it can be related to knowledge of the effects on the other levels. Hopefully, there will be some biologists who are sufficiently competent in all biological disciplines to bring together knowledge of the different levels of organization and to define the interrelationships which exist.

When we set out to study the effects of an unusual water quality change on a biological system, we frequently find that little is known of the responses of that system to variations of environmental conditions within the normal range. Basic biological studies are often needed before we can study the effects of a particular water quality change. A biologist interested in the effects of an environmental change on a complex biological system that is poorly understood cannot usually simultaneously develop the necessary basic knowledge and study the effects of the change. This consideration led to our earlier comments that the state of biological knowledge will in part determine the contribution of a biologist and that an individual scientist cannot usually extend the limits of knowledge much beyond existing levels. Further, the same consideration often makes it more efficient to gain reasonable

understanding of a biological system before it is subjected to wide ranges of many different environmental factors. If we do not do this, much of our effort may be wasted in measuring responses that we cannot hope to understand. Studies of the effects of environmental factors on systems that are not understood will usually need to be repeated.

Environmental factors that have not been studied do not present totally new biological problems. The mechanisms by which aquatic organisms adapt to their environment and the changes in biological systems that lead to their breakdown are not different for every possible change in the aquatic environment. Biological studies of the effects of water quality changes should be so conducted as to be useful in explaining how biochemical and physiological responses influence the behavior, survival, reproduction, and growth of individual organisms, and how these in turn influence the success of populations and the stability of communities. The results of such studies not only will be the most useful in solving particular water quality problems, but they also will greatly facilitate later studies directed toward the solution of other problems. Those biologists whose role is to increase understanding of the effects of changes in aquatic environments must avoid superficial studies, if they are to be productive in their role.

DEVELOPMENT OF BIOLOGICAL APPROACHES

Resource management biologists working for governmental, industrial, metropolitan, or consulting organizations need approaches and techniques that can be used to find solutions for particular water pollution problems. These biologists may frequently lack the background, the time, or the facilities needed to develop new approaches and techniques that can help them in their work. Even when the necessary knowledge of biological systems is available, rationales and methods for its use in the solution of pollution problems may need to be developed. This task will usually fall to biologists who can devote a considerable portion of their time to research. Frequently, however, the need for new approaches may first become apparent to biologists who daily must face practical problems of water pollution control.

The approaches biologists develop make use of knowledge of the various levels of biological organization. The most promising approaches are those that are biologically sound, that are practical, and that will yield information in terms that are most readily usable in the solution of water pollution problems. Because pollution problems are ultimately social in nature, the most valuable biological information will be that which can be related to economic, esthetic, or other social values.

Biological approaches and techniques that may have value in the solution of water pollution problems are many, but few have been carefully examined and tested. Relatively few biologists have been interested and have worked in the field of water pollution; and the various biological specializations are poorly represented among them. General appreciation of the complexity and importance of the biological problems associated with water pollution has increased in recent years; with this has come increased support for biological programs. The number of biologists interested in these problems and the specialities they represent are increasing; in the years ahead, many new approaches and techniques will be examined and tested. Some will be found valuable, and their application will hasten the solution of many difficult water pollution problems.

Of the many possible biological approaches to problems, those selected for development and application will to a considerable extent depend on the nature of the problems and on regulatory policies. But with time, problems and policies change, and new approaches are needed. While developing approaches to meet today's problems, biologists should be giving some attention to the development of approaches likely to become useful with changing conditions, for such development can require considerable periods of time. Many different biological approaches will be developed; many different biological approaches will be needed. A single approach cannot be expected to provide all the biological information needed in dealing with even one complex problem. Biologists should be careful not to oversell particular approaches that they may develop. When obviously valuable approaches are found, this may be difficult

to avoid. But overselling can lead to poorly planned control programs, wasted effort, and disappointment. Lack of confidence in the value of biological programs may follow.

APPLICATION OF BIOLOGICAL KNOWLEDGE

Toward the attainment of the water quality objectives of a society, it is the responsibility of resource management biologists to apply biological knowledge, whether they are employed by governmental, industrial, metropolitan, or consulting organizations. For solving problems, these biologists must make the best possible use of existing information or develop necessary information when this is readily possible. Rarely will existing and readily obtainable information be entirely adequate for positive judgment as to the best course of action. Nevertheless, decisions must be made and actions taken, even when some would wish for more information, information that could require many years to obtain. Existing problems need solutions now, even though the solutions be imperfect; agricultural, industrial, and urban development is continually increasing the magnitude of these problems. Not all biologists are suited for resource management work, either by temperament or by training. The research biologist is temperamentally inclined to be dissatisfied with the state of knowledge; he is trained to devote his energy to extending its limits. It may be hard for him to make important decisions influencing the course of human events on the basis of information he is likely to judge insufficient. The management biologist too may recognize the inadequacy of available information, but he must be able to make such decisions effectively without being too uneasy about the state of knowledge.

A management biologist will be helped by broad biological training; he is likely to need reasonable understanding of the physiology and ecology of both aquatic plants and animals. Particularly, he will need great familiarity with the water quality requirements of organisms. But if his recommendations are usually to be accepted and useful, they will need to be based not only on a comprehension of the biological aspects of pollution problems but also on a fair comprehension of their social and technological aspects. A broad education is difficult to obtain, but study of the social sciences and engineering will benefit the individual biologist and increase his value to his organization and society.

The roles of management biologists employed by different kinds of organizations will differ to some extent, but, in a general sense, these differences are more apparent than real. Management biologists must use their knowledge to reduce conflicts in water use. Those working for governmental agencies may be involved more in the application of regulatory authority; those in industrial or metropolitan organizations more with helping their organizations to avoid creating problems. But both roles should lead to reduction of conflicts. The role of the consulting biologist too is to reduce problems; when organizations cannot afford biologists on their staffs, consulting biologists can make exceedingly important contributions.

It is important that biologists employed by governmental, industrial, and metropolitan organizations work together in the interest of all. The objectives of regulatory agencies and of those responsible for discharging wastes into receiving waters differ and sometimes seem unreconcilable; but too often these groups are assumed to be perpetual antagonists. All groups in a society have a common interest in water quality. Differences in viewpoint are very often more over means than ends, except perhaps when extreme viewpoints are held, viewpoints that are not likely to prevail. Biologists have a common language and rather common viewpoints. Even when the attitudes of their employing organizations differ, biologists face many of the same problems and approach them in similar ways. Their patient efforts can do much to reduce the areas of conflict existing between agencies that must regulate in the public interest and those that must dispose of wastes but whose activities are essential to the public good.

One of the most important roles of management biologists is to assist in the development of water quality standards providing desired levels of protection for valuable aquatic life or preventing the occurrence of nuisance problems. On the basis of available information on the water quality requirements of organisms, it is the responsibility of biologists to recommend the sets of

standards necessary. This responsibility may fall most directly on biologists employed by regulatory agencies, but it is important that all biologists with pertinent knowledge contribute, for information cannot be overlooked if established standards are to be the best possible at any given time.

Management biologists must continually study the biological populations and the physical and chemical conditions of waters affected by waste discharges. Only on the basis of such studies can it be determined whether or not established standards are adequate to provide the desired levels of protection for aquatic resources. Generally applied water quality standards cannot be expected to be equally effective under all circumstances.

Through their studies, biologists must be able to recommend the treatment of wastes that will be necessary to provide the protection desired for aquatic populations. In this, biologists employed by industries can be helpful to regulatory agencies, for they have ready access to information on the industrial processes involved. They can be particularly helpful to industrial managers in recommending sites for new industry where difficult problems will not develop and in recommending ways of reducing problems where industry is located.

Conditions leading to nuisance problems will often not be prevented by generally applied water quality standards. Special effluent and receiving water standards may be necessary at particular locations to prevent the occurrence of these problems. Fairly extensive studies will frequently be required to determine the factors responsible for particular problems and to find solutions for them. Often, these studies will be assigned to management biologists. They will need to draw on basic knowledge of the biological processes involved and use approaches other biologists have developed for the solution of such problems.

Most of the waste treatment processes used to prevent or alleviate problems in receiving waters are dependent on the activities of organisms. These processes have been used with reasonable effectiveness for a long time, even though the activities of all the organisms are not fully understood. Basic and developmental studies, for which research biologists will mainly be responsible, should eventually result in increased understanding and more effective use of these processes. But even on the basis of present understanding, the control and operation of waste treatment plants dependent upon biological processes can be made more effective if biological as well as engineering and chemical knowledge is continuously applied.

THE FUTURE OF WATER POLLUTION BIOLOGY

The intensity of use of nearly all the waters of the earth will increase, and most uses of natural waters inevitably will change them. Even when the intensity of use has reached the point at which man must treat nearly all his wastes, changes that interfere with some uses will still occur and pollution problems will persist. Most of the changes man has wrought in his environment have been beneficial to him, and increased use of the earth's waters will be to his benefit. Although there will be undesirable changes in waters that he may be unable to prevent and must accept, he cannot afford needlessly to reduce the value of waters for any of their uses. There is no need to recount the practical reasons for this; but we must never forget the ethical and esthetic ones. In his finest image, man cannot use wastefully the resources of the earth or destroy their beauty. Biologists are trained to have a keen sense of the importance and beauty of the diversity and of the energy and material economy of ecological systems. They have a distinctive contribution to make in helping all men to appreciate this importance and beauty.

The best biological thought has not been directed to the solution of water pollution problems. Many fine scientists hesitate to undertake research that has direct application to solution of man's problems; they believe that this would hinder their pursuit of the understanding they seek. Others, with no hesitancy to undertake applied research, have viewed the opportunities for biologists in the field of water pollution to be limited. Charles E. Renn (1960, p. 158), himself a professor of sanitary engineering at The Johns Hopkins University, has said: "Most biologists, and probably the most able and aggressive, will not concern themselves with it at all. It will not attract them."

Increased public awareness of the importance and complexity of the biological problems associated with water pollution has, however, changed the outlook for biologists who may have wondered about opportunities in this field. And society can no longer afford to have but few of its finest biologists willing to work on problems directly related to man's problems of environmental change. Man is the most interesting organism in the ecological systems of the earth. For the biologist, he is in many ways the most difficult organism to study. But the physiological, behavioral, population, and community relationships of man provide not only some of the most difficult but also some of the most fascinating biological problems of which we are aware. It is for society to make the opportunities for research directly related to man's environmental problems so inviting that more of our finest biologists will choose to engage in this work. These biologists will view opportunity not solely or even primarily in terms of monetary remuneration but also in terms of the chances they are to be given to pursue, as they believe best, problems that fascinate them.

The traditions and approaches of biology have barely begun to be used in solving the environmental problems of man, except perhaps in medicine and agriculture. This is particularly so with respect to contamination of our water, air, and soil. Great biologists of the past and present have given biology a grand tradition and powerful approaches for solving problems of the individual organism, its population, and its community; herein lies the strength of biology. It is for biologists now and in the future to apply this strength to the solution of man's water and other problems.

Many of the problems associated with water pollution are amenable to solution through application of approaches that biologists have developed. Throughout this book we have dealt with these problems and with approaches of value in their solution; there is no need to recount them now. From broad approaches, specific approaches will in many cases need to be developed. This is one of the large initial tasks awaiting biologists.

In the years ahead, a great deal more support must be given to research directed toward increasing understanding of biological processes in the aquatic environment. We must come to understand the fundamental processes that lead to the production of valuable or nuisance organisms and how changes in water quality affect these processes. For this we will need much knowledge of the water quality requirements of organisms, of how changes in individual organisms affect their populations, and of the relationships between populations and communities. This is a big order; it cannot be less, for if man is to manage his aquatic resources, he must be able to predict the outcome of his activities.

Prediction of the course of events in complex biological systems is at best uncertain; the most reliable predictions are based on understanding. Much of our research must be directed toward providing the understanding necessary to make reliable predictions of the effects of water quality changes man's present and future activities may cause. It was lack of such understanding that caused such frenetic activity when the seriousness of pesticide, eutrophication, and temperature problems became apparent. We cannot always anticipate new technological developments; but we can learn the needs of aquatic life, so that when these developments arrive, their probable effects on life can soon be evaluated.

When man can predict reliably, he can plan his activities intelligently. Broad programs of water resource management are needed, programs that take into account man's future needs, material and esthetic. General and particular programs of water quality control are being developed and represent considerable advance in water resource management. But water quality in the future may depend as much on patterns of land use and urban and industrial development as on regulation of existing uses and development. Planning for water quality control must reach beyond the waters, to where and how we live and work.

Water quality conditions in most industrial nations of the world necessitate that present management programs be directed primarily toward alleviating existing problems and perhaps secondarily toward preventing future ones. But to truly manage aquatic resources, we must seek ways to enhance their production, not merely to prevent their destruction. Hynes (1960, pp. 170–174) has made an eloquent plea for such a positive attitude in our approach to the problems of waste disposal; we have not been overly imaginative in finding ways to benefit from the materials and energy in our wastes.

Water pollution biology is but a part of the environmental biology of man, the dominant organism in his ecosystem. For man to be successful in a humanly meaningful way, he must continually improve his relationships in this ecosystem. All biologists can contribute to the understanding that will make this possible. Water pollution biologists are directly engaged in this effort. The future of water pollution biology will depend on these biologists and, perhaps much more, on the young biologists who will be entering this work, on their abilities, training, imagination, and dedication, and on the support they receive from society, to which all this is really very important.

SELECTED REFERENCES

Hynes, H. B. N. 1960. The Biology of Polluted Waters. Liverpool University Press, Liverpool. xiv + 202pp.

Renn, C. E. 1960. Biological aspects of industrial waste problems. Pages 156–159. *In* C. M. Tarzwell (Editor), Biological Problems in Water Pollution. Transactions of the Second Seminar, 1959. R. A. Taft Sanitary Engineering Center, U.S. Department of Health, Education and Welfare. 285pp.

Tarzwell, C. M. 1963. Sanitational limnology. Pages 653–666. *In* D. G. Frey (Editor), Limnology in North America. University of Wisconsin Press, Madison. xvii + 734pp.

Udall, S. L. 1964. A message for biologists. Bioscience 14(11):17–18.

LITERATURE CITED

Adolph, E. F. 1964. Perspectives of adaptation: some general properties. Pages 27-35. *In* D. B. Dill, E. F. Adolph, and C. G. Wilber (Editors), Adaptation to the Environment. (Handbook of Physiology, Section 4.) American Physiological Society, Washington, D.C. ix + 1056pp.

Allee, W. C., A. E. Emerson, O. Park, T. Park, and K. D. Schmidt. 1949. Principles of Animal Ecology. W. B. Saunders Company, Philadelphia. xii + 837pp.

Allen, K. R. 1950. The computation of production in fish populations. New Zealand Science Review 8:89.

———. 1951. The Horokiwi Stream: a study of a trout population. New Zealand Marine Department Fisheries Bulletin 10. [ix] + 238pp.

Allen, L. A. 1944. The bacteriology of activated sludge. Journal of Hygiene 43:424-431.

American Public Health Association et al. 1960. Standard Methods for the Examination of Water and Wastewater Including Bottom Sediments and Sludges. 11th ed. American Public Health Association, Inc., New York. xxi + 626pp.

Anderson, B. G. 1944. The toxicity thresholds of various substances found in industrial wastes as determined by the use of *Daphnia magna*. Sewage Works Journal 16:1156-1165.

———. 1946. The toxicity thresholds of various sodium salts determined by the use of Daphnia magna. Sewage Works Journal 18:82-87.

Anderson, J. M. 1968. Effect of sublethal DDT on the lateral line of brook trout, *Salvelinus fontinalis*. Journal of the Fisheries Research Board of Canada 25:2677-2682.

———, and Margaret R. Peterson. 1969. DDT: Sublethal effects on brook trout nervous system. Science 164:440-441.

Andrewartha, H. G., and L. C. Birch. 1954. The Distribution and Abundance of Animals. The University of Chicago Press, Chicago. xv + 782pp.

Aquatic Life Advisory Committee, Ohio River Valley Water Sanitation Commission. 1955. Aquatic life water quality criteria: first progress report. Sewage and Industrial Wastes 27:321-331.

Ardern, E., and W. T. Lockett. 1914. Experiments on the oxidation of sewage without the aid of filters. Parts I and II. Journal of the Society of Chemical Industry 33:523-536; 1122-1124.

Aristotle. c. 336 B.C. Metaphysics. (Translated by W. D. Ross.) Reprinted 1952. Pages 495-626. *In* The Works of Aristotle, Vol. I. Great Books of the Western World, Vol. 8. Encyclopaedia Brittanica, Inc., Chicago. vii + 726pp.

———. c. 336 B.C. History of Animals. (Translated by D. W. Thompson.) Reprinted 1952. Pages 3-158. *In* The Works of Aristotle, Vol. II. Great Books of the Western World, Vol. 9. Encyclopaedia Brittanica, Inc., Chicago. v + 699pp.

Armstrong, J. T., Jr. 1960. The dynamics of *Daphnia pulex* populations and of *Dugesia tigrina* populations as modified by immigration. Ph.D. thesis. University of Michigan, Ann Arbor. 102pp.

Averett, R. C. 1969. Influence of temperature on energy and material utilization by juvenile coho salmon. Ph.D. thesis. Oregon State University, Corvallis. 74pp.

Bacon, F. 1605. Advancement of Learning. Reprinted 1952. Pages 1–101. *In* Francis Bacon. Great Books of the Western World, Vol. 30. Encyclopaedia Brittanica, Inc., Chicago. vi + 214pp.

Baerends, G. P. 1957. Behavior: the ethological analysis of fish behavior. Pages 229–269. *In* Margaret E. Brown (Editor), The Physiology of Fishes. Vol. 2, Behavior. Academic Press Inc., New York. xi + 526pp.

————, and J. M. Baerends-van Roon. 1950. An introduction to the study of the ethology of cichlid fishes. Behaviour (Supplement 1):1-242.

Banks, H. O. 1961. Priorities for water use. Pages 153-167. *In* Proceedings of the National Conference on Water Pollution, 1960. U.S. Department of Health, Education and Welfare, Washington, D.C. x + 607pp.

Baranov, F. I. 1918. [On the question of the biological basis of fisheries.] Nauchnyĭ Issledovatel'skiĭ Ikhtiologischeskiĭ Institut Izvestiia 1:[81]–128. (Translation deposited in International Fisheries Commission collection, 1938.)

Barker, A. N. 1946. The ecology and function of Protozoa in sewage purification. Annals of Applied Biology 33:314-325.

————. 1949. Some microbiological aspects of sewage purification. Journal of the Institute of Sewage Purification 1949(1):7-22.

Barnett, S. A. 1963. The Rat: A Study in Behaviour. Aldine Publishing Company, Chicago. xvi + 288pp.

Barrington, E. J. W. 1957. The alimentary canal and digestion. Pages 109-171. *In* Margaret E. Brown (Editor), The Physiology of Fishes. Vol. 1, Metabolism. Academic Press Inc., New York. xiii + 447pp.

Barritt, N. W. 1933. The nitrification process in soils and biological filters. Annals of Applied Biology 20:165-184 + plate.

Bartholomew, G. A. 1964. The roles of physiology and behaviour in the maintenance of homeostasis in the desert environment. Pages 7-29. *In* Homeostasis and Feedback Mechanisms. (Symposia of the Society for Experimental Biology 18.) Academic Press Inc., New York. viii + 460pp.

Bartsch, A. F., and M. O. Allum. 1957. Biological factors in treatment of raw sewage in artificial ponds. Limnology and Oceanography 2:77-84.

————, and W. M. Ingram. 1966. Biological analysis of water pollution in North America. Verhandlungen Internationale Vereinigung für Theoretische und Angewandte Limnologie 16:786-800.

Bates, M. 1960. Ecology and evolution. Pages 547-568. *In* S. Tax (Editor), The Evolution of Life. (Evolution after Darwin, Vol. 1.) The University of Chicago Press, Chicago. viii + 629pp.

————. 1962. The human environment. (Horace M. Albright Conservation Lectureship 2.) University of California, School of Forestry, Berkeley. 22pp.

Beamish, F. W. H. 1964a. Respiration of fishes with special emphasis on standard oxygen consumption. III. Influence of oxygen. Canadian Journal of Zoology 42:355-366.

————. 1964b. Respiration of fishes with special emphasis on standard oxygen consumption. II. Influence of weight and temperature on respiration of several species. Canadian Journal of Zoology 42:177-188.

Beck, W. M. 1954. Studies in stream pollution biology. I. A simplified ecological classification of organisms. Quarterly Journal of the Florida Academy of Sciences 17:211-227.

————. 1955. Suggested method for reporting biotic data. Sewage and Industrial Wastes 27:1193-1197.

Beeton, A. M. 1961. Environmental changes in Lake Erie. Transactions of the American Fisheries Society 90:153-159.

Belding, D. L. 1934. The spawning habits of the Atlantic salmon. Transactions of the American Fisheries Society 64:211-216.

Bernard, C. 1865. An Introduction to the Study of Experimental Medicine. (Republished 1949. Translated by H. C. Greene. Henry Schuman. xix + 226pp.

Bertalanffy, L. von. 1938. A quantitative theory of organic growth (Inquiries on growth laws. II). Human Biology 10:181-213.

Beverton, R. J. H., and S. J. Holt. 1957. On the dynamics of exploited fish populations. Ministry of Agriculture, Fisheries and Food, London. Fishery Investigations Series 2, 19:3-533.

Birch, L. C. 1948. The intrinsic rate of natural increase of an insect population. Journal of Animal Ecology 17:15-26.

Black, E. C. 1940. The transport of oxygen by the blood of freshwater fish. Biological Bulletin 79:215-229.

———. 1951. Respiration in fishes. Pages 91-111. *In* Some Aspects of the Physiology of Fish. University of Toronto Biological Series 59. Ontario Fisheries Research Laboratory Publication 71. 111pp.

———, and L. Irving. 1938. The effect of hemolysis upon the affinity of fish blood for oxygen. Journal of Cellular and Comparative Physiology 12:255-262.

———, D. Kirkpatrick, and H. H. Tucker. 1966. Oxygen dissociation curves of the blood of brook trout (*Salvelinus fontinalis*) acclimated to summer and winter temperatures. Journal of the Fisheries Research Board of Canada 23:1-13.

Black, Virginia S. 1957. Excretion and osmoregulation. Pages 163-205. *In* Margaret E. Brown (Editor), The Physiology of Fishes. Vol. 1, Metabolism. Academic Press Inc., New York. xiii + 447pp.

Blair, W. F. 1964. The case for ecology. Bioscience 14(7):17-19.

Bodenheimer, F. S. 1958. Animal Ecology To-Day. Monographiae Biologicae Vol. 6. Uitgeverij Dr. W. Junk, Den Haag. 276pp.

Boulding, K. E. 1967. The prospects of economic abundance. Pages [39]-57. *In* J. D. Roslansky (Editor), The Control of Environment. (1966 Nobel Conference discussion.) North-Holland Publishing Company, Amsterdam. xi + 112pp.

Breder, C. M., Jr. 1936. The reproductive habits of the North American sunfishes (family Centrarchidae). Zoologica 21:1-48 + 7 plates.

Brett, J. R. 1944. Some lethal temperature relations of Algonquin Park fishes. University of Toronto Studies Biological Series 52. Ontario Fisheries Research Laboratory Publication 63. 49pp.

———. 1946. Rate of gain of heat-tolerance in goldfish (*Carassius auratus*). University of Toronto Studies Biological Series 53. Ontario Fisheries Research Laboratory Publication 64:[5]-28.

———. 1952. Temperature tolerance in young Pacific salmon, Genus Oncorhynchus. Journal of the Fisheries Research Board of Canada 9:265-323.

———. 1956. Some principles in the thermal requirements of fishes. Quarterly Review of Biology 31:75-87.

———. 1957. The sense organs: the eye. Pages 121-154. *In* Margaret E. Brown (Editor), The Physiology of Fishes. Vol. 2, Behavior. Academic Press Inc., New York. xi + 526pp.

———. 1964. The respiratory metabolism and swimming performance of young sockeye salmon. Journal of the Fisheries Research Board of Canada 21:1183-1226.

———, and C. Groot. 1963. Some aspects of olfactory and visual responses in Pacific salmon. Journal of the Fisheries Research Board of Canada 20:287-303.

———, and D. MacKinnon. 1954. Some aspects of olfactory perception in migrating adult coho and spring salmon. Journal of the Fisheries Research Board of Canada 11:310-318.

Briggs, J. C. 1953. The behavior and reproduction of salmonid fishes in a small coastal stream. California Department of Fish and Game, Fish Bulletin 94. 62pp.

Brillouin, L. 1962. Science and Information Theory. 2nd ed. Academic Press Inc., New York. xvii + 347pp.

Britt, N. W. 1955. Stratification in western Lake Erie in summer of 1953: effects on the Hexagenia (Ephemeroptera) population. Ecology 36:239-244.

Brocksen, R. W. 1966. Influence of competition on the food consumption and production of animals in laboratory stream communities. M.S. thesis. Oregon State University, Corvallis. 82pp.

———. 1969. Density dependent relationships in trophic processes of simplified stream communities. Ph.D. thesis. Oregon State University, Corvallis. 83pp.

———, G. E. Davis, and C. E. Warren. 1968. Competition, food consumption, and production of sculpins and trout in laboratory stream communities. Journal of Wildlife Management 32:51-75.

———, ———, and ———. 1970. Analysis of trophic processes on the basis of density-dependent functions. Pages 468-498. *In* J. H. Steele (Editor), Marine Food Chains. University of California Press, Berkeley. In press.

———, and C. E. Warren. Density dependence of trophic processes in laboratory stream communities. Manuscript.

Brockway, D. L. 1963. Some effects of sub-lethal levels of pentachlorophenol and cyanide on the physiology and behavior of a cichlid fish *Cichlasoma bimaculatum*(Linnaeus). M.S. thesis. Oregon State University, Corvallis. 56pp.

Brodie, B. B., and R. P. Maickel. 1962. Comparative biochemistry of drug metabolism. Pages 299-324. *In* B. B. Brodie and E. G. Erdös (Editors), Metabolic Factors Controlling Duration of Drug Action. (First International Pharmacological Meeting, Vol. 6.) The Macmillan Company, New York. xviii + 330pp.

Brody, S. 1927. Growth and development with special reference to domestic animals. III. Growth rates, their evaluation and significance. Missouri Agricultural Experiment Station Research Bulletin 97. 70pp.

———. 1945. Bioenergetics and Growth. Reinhold Publishing Corporation, New York. xii + 1023pp.

Brown, Margaret E. 1946. The growth of brown trout (*Salmo trutta* Linn.) II. The growth of two-year-old trout at a constant temperature of 11.5°C. Journal of Experimental Biology 22:130-144.

——— (Editor). 1957a. The Physiology of Fishes. Vol. 1, Metabolism. Academic Press Inc., New York. xiii + 447pp.

——— (Editor). 1957b. The Physiology of Fishes. Vol. 2, Behavior. Academic Press Inc., New York. xi + 526pp.

———. 1957c. Experimental studies on growth. Pages 361-400. *In* Margaret E. Brown (Editor), The Physiology of Fishes. Vol. 1, Metabolism. Academic Press Inc., New York. xiii + 447pp.

Brown, V. M. 1968. The calculation of the acute toxicity of mixtures of poisons to rainbow trout. Water Research (Pergamon Press) 2:723-733.

Brown, W. G., A. Singh, and E. N. Castle. 1965. Net economic value of the Oregon salmon-steelhead sport fishery. Journal of Wildlife Management 29:266-279.

Buchsbaum, R. 1948. Animals Without Backbones. 2nd ed., rev. University of Chicago Press, Chicago. xii + 405pp.

Bucksteeg, W., H. Thiele, and K. Stöltzel. 1955. Die Beeinflussung von Fischen durch Giftstoffe aus Abwässern. Vom Wasser (Jahrbuch für Wasser-Chemie) 22:194-211.

Buffon, L. L. de. 1777. 4th Supplement to Histoire Naturelle générale et particulière. Paris.

Bull, H. O. 1957. Behavior: conditioned responses. Pages 211-228. *In* Margaret E. Brown (Editor), The Physiology of Fishes. Vol. 2, Behavior. Academic Press Inc., New York. xi + 526pp.

Bullock, T. H. 1955. Compensation for temperature in the metabolism and activity of poikilotherms. Biological Reviews 30:311-342.

Burdick, G. E. 1957. A graphical method for deriving threshold values of toxicity and the equation of the toxicity curve. New York Fish and Game Journal 4:102-108.

———, H. J. Dean, and E. J. Harris. 1958. Toxicity of cyanide to brown trout and smallmouth bass. New York Fish and Game Journal 5:133-163.

———, E. J. Harris, H. J. Dean, T. M. Walker, J. Skea, and D. Colby. 1964. The accumulation of DDT in lake trout and the effect on reproduction. Transactions of the American Fisheries Society 93:127-136.

Butcher, R. W. 1946. The biological detection of pollution. Journal of the Institute of Sewage Purification 1946(2):92-97.

Cannon, W. B. 1932. The Wisdom of the Body. W. W. Norton & Company, Inc., New York. xv + 312pp.

Carline, R. F. 1968. Laboratory studies on the food consumption, growth, and activity of juvenile coho salmon. M.S. thesis. Oregon State University, Corvallis. 75pp.

Carlyle, T. 1830. On history. Pages 83-95. *In* Critical and Miscellaneous Essays, Vol. II. The Works of Thomas Carlyle, Vol. 27. Centenary edition, 1899. Chapman and Hall, London. 507pp.

Carpenter, Kathleen E. 1924. A study of the fauna of rivers polluted by lead mining in the Aberystwyth district of Cardiganshire. Annals of Applied Biology 11:1-23.

———. 1925. On the biological factors involved in the destruction of river-fisheries by pollution due to lead-mining. Annals of Applied Biology 12:1-13.

———. 1927. The lethal action of soluble metallic salts on fishes. British Journal of Experimental Biology 4:378-390.

———. 1930. Further researches on the action of metallic salts on fishes. Journal of Experimental Zoology 56:407-422.

Carson, Rachel. 1962. Silent Spring. Houghton Mifflin Company, Boston. x + 368pp.

Chadwick, G. G., and R. W. Brocksen. 1969. Accumulation of dieldrin by fish and selected fish-food organisms. Journal of Wildlife Management 33:693-700.

Chamberlin, T. C. 1897. The method of multiple working hypotheses. Journal of Geology 5:837-848.

――――, and R. D. Salisbury. 1904. Geology. Vol. 1: Geologic Processes and Their Results. 2nd ed. Henry Holt and Company, New York. xix + 684pp.

Chapman, D. W. 1962. Aggressive behavior in juvenile coho salmon as a cause of emigration. Journal of the Fisheries Research Board of Canada 19:1047-1080.

――――. 1965. Net production of juvenile coho salmon in three Oregon streams. Transactions of the American Fisheries Society 94:40-52.

――――. 1966. Food and space as regulators of salmonid populations in streams. American Naturalist 100:345-357.

――――. 1967. Production in fish populations. Pages 3-29. *In* S. D. Gerking (Editor), The Biological Basis of Freshwater Fish Production. Blackwell Scientific Publications, Oxford. xiv + 495pp.

――――. 1968. Production. Pages 182-196. *In* W. E. Ricker (Editor), Methods for Assessment of Fish Production in Fresh Waters. International Biological Programme Handbook 3. Blackwell Scientific Publications, Oxford. xiii + 313pp.

Chapman, G. A. 1965. Effects of sub-lethal levels of pentachlorophenol on the growth and metabolism of a cichlid fish. M.S. thesis. Oregon State University, Corvallis. 77pp.

――――. 1969. Toxicity of pentachlorophenol to trout alevins. Ph.D. thesis. Oregon State University, Corvallis. 87pp.

Chapman, R. N. 1931. Animal Ecology. McGraw-Hill Book Company, Inc., New York. x + 464pp.

Chase, E. S. 1964a. Nine decades of sanitary engineering. Water Works and Wastes Engineering 1(4):57, 98.

――――. 1964b. Nine decades of sanitary engineering. Part 2 – The awakening. Water Works and Wastes Engineering 1(6):58-59, 86.

――――. 1964c. Nine decades of sanitary engineering. Part 3 – Back to the land. Water Works and Wastes Engineering 1(7):48-49.

――――. 1964d. Nine decades of sanitary engineering. Part 4 – Science enters. Water Works and Wastes Engineering 1(8):49-50, 78.

――――. 1964e. Nine decades of sanitary engineering. Part 5 – English pioneers. Water Works and Wastes Engineering 1(10):60-61.

――――. 1964f. Nine decades of sanitary engineering. Part 6 – The American pioneers. Water Works and Wastes Engineering 1(11):50-52.

――――. 1965a. Nine decades of sanitary engineering. Part 7 – Pioneers in wastes treatment practices (1880-1920). Water Works and Wastes Engineering 2(1):56-58.

――――. 1965b. Nine decades of sanitary engineering. Part 8 – More pioneers. Water Works and Wastes Engineering 2(3):48-51.

――――. 1965c. Nine decades of sanitary engineering. Part 9 – Controversies and litigation. Water Works and Wastes Engineering 2(6):132-134.

Clark, J. W., and W. Viessman, Jr. 1965. Water Supply and Pollution Control. International Textbook Company, Scranton, Pennsylvania. xiv + 575pp.

Clarke, F. W. 1924. The composition of the river and lake waters of the United States. U. S. Geological Survey Professional Paper 135. iv + 199pp.

Clarke, G. L. 1946. Dynamics of production in a marine area. Ecological Monographs 16:321-335.

――――, W. T. Edmondson, and W. E. Ricker. 1946. Mathematical formulation of biological productivity. Ecological Monographs 16:336-337.

Clawson, M. 1959. Methods of measuring the demand for and value of outdoor recreation. Reprint 10. Resources for the Future, Inc., Washington, D.C. 36pp.

Clements, F. E., and V. E. Shelford. 1939. Bio-ecology. John Wiley & Sons, Inc., New York. vi + 425pp.

Coble, D. W. 1961. Influence of water exchange and dissolved oxygen in redds on survival of steelhead trout embryos. Transactions of the American Fisheries Society 90:469-474.

Coe, W. R. 1932. Development of the gonads and the sequence of the sexual phases in the California oyster (Ostrea lurida). Bulletin of the Scripps Institution of Oceanography 3:119-144.

Cole, H. A. 1941. The fecundity of *Ostrea edulis*. Journal of the Marine Biological Association of the United Kingdom 25:243-260.

———. 1942. Primary sex-phases in *Ostrea edulis*. Quarterly Journal of Microscopical Science 83:317-356.

Cole, L. C. 1957. Sketches of general and comparative demography. Pages 1-15. *In* Population Studies: Animal Ecology and Demography. Cold Spring Harbor Symposia on Quantitative Biology, Vol. 22. xiv + 437pp.

Collins, G. B. 1952. Factors influencing the orientation of migrating anadromous fishes. Fishery Bulletin 73. (Woods Hole Oceanographic Institution Contribution 585.) Fishery Bulletin of the U. S. Fish and Wildlife Service 52:375-396.

Copeland, D. E. 1948. The cytological basis of chloride transfer in the gills of Fundulus heteroclitus. Journal of Morphology 82:201-227.

———. 1950. Adaptive behavior of the chloride cell in the gill of Fundulus heteroclitus. Journal of Morphology 87:369-379.

Coulson, H. J. W., and U. A. Forbes. 1952. The Law of Waters, Sea, Tidal, and Inland, and of Land Drainage, 6th ed. Edited by S. R. Hobday. Sweet & Maxwell, London. lx + 1320pp.

Craig, W. 1918. Appetites and aversions as constituents of instincts. Biological Bulletin 34:91-107.

Crandall, Catherine A., and C. J. Goodnight. 1963. The effects of sublethal concentrations of several toxicants to the common guppy, *Lebistes reticulatus*. Transactions of American Microscopical Society 82:59-73.

Crisp, G., and L. Lloyd. 1954. The community of insects in a patch of woodland mud. Transactions of the Royal Entomological Society of London 105:269-[314] + 3 plates.

Cummins, K. W., W. P. Coffman, and P. A. Roff. 1966. Trophic relationships in a small woodland stream. Verhandlung der Internationalen Vereinigung für Theoretische und Angewandte Limnologie 16:627-637.

Dam, L. van. 1938. On the utilization of oxygen and regulation of breathing in some aquatic animals. Dissertation, Univ. Gröningen. 143pp.

Darnell, R. M. 1958. Food habits of fishes and larger invertebrates of Lake Pontchartrain, Louisiana, an estuarine community. Publications of the Institute of Marine Science (University of Texas) 5:353-416.

———. 1961. Trophic spectrum of an estuarine community, based on studies of Lake Pontchartrain, Louisiana. Ecology 42:553-568.

———. 1968. Animal nutrition in relation to secondary production. American Zoologist 8:83-93.

Darwin, C. 1859. On the Origin of Species by Means of Natural Selection, or the Preservation of Favoured Races in the Struggle for Life. John Murray, London. ix + [502]pp. (A facsimile of the first edition. 1964. Introduction by E. Mayr. Harvard University Press, Cambridge, Massachusetts.)

Davis, C. C. 1964. Evidence for the eutrophication of Lake Erie from phytoplankton records. Limnology and Oceanography 9:275-283.

Davis, G. E., and C. E. Warren. 1965. Trophic relations of a sculpin in laboratory stream communities. Journal of Wildlife Management 29:846-871.

———, and ———. 1968. Estimation of food consumption rates. Pages 204-225. *In* W. E. Ricker (Editor), Methods for Assessment of Fish Production in Fresh Waters. International Biological Programme Handbook 3. Blackwell Scientific Publications, Oxford. xiii + 313pp.

Davis, H. C. 1958. Survival and growth of clam and oyster larvae at different salinities. Biological Bulletin 114:296-307.

———. 1961. Effects of some pesticides on eggs and larvae of oysters (*Crassostrea virginica*) and clams (*Venus mercenaria*). Commercial Fisheries Review 23(12):8-23.

———, and P. E. Chanley. 1956. Spawning and egg production of oysters and clams. Biological Bulletin 110:117-128.

Dawson, A. B. 1935. The hemopoietic response in the catfish, Ameiurus nebulosus, to chronic lead poisoning. Biological Bulletin 68:335-346.

Deevey, E. S., Jr. 1947. Life tables for natural populations of animals. Quarterly Review of Biology 22:283-314.

———. 1950. The probability of death. Scientific American 182(4):58-60.

Degens, P. N., H. van der Zee, J. D. Kommer, and A. H. Kamphuis. 1950. Synthetic detergents and sewage processing. Part 5. The effect of synthetic detergents on certain water fauna. Journal of the Institute of Sewage Purification 1950(1):63-68.

Dewitt, J. W., Jr. 1963. Effects of pollutional conditions on stream organisms with especial emphasis on stonefly naiads. Ph.D. thesis. Oregon State University, Corvallis. 216pp.

Dilling, W. J., C. W. Healey, and W. C. Smith. 1926. Experiments on the effects of lead on the growth of plaice (*Pleuronectes platessa*). Annals of Applied Biology 13:168-176.

Dimick, R. E., and W. P. Breese. 1965. Bay mussel embryo bioassay. Pacific Northwest Industrial Waste Conference 12:165-175.

Dittmar, W. 1884. Report on researches into the composition of ocean-water collected by H.M.S. Challenger during the years 1873-76. Report of the Voyage of H.M.S. Challenger, Physics and Chemistry Vol. 1:1-251 + 3 plates.

Dobzhansky, T. 1957. Mendelian populations as genetic systems. Pages 385-393. *In* Population Studies: Animal Ecology and Demography. Cold Spring Harbor Symposia on Quantitative Biology, Vol. 22. xiv + 437pp.

Doudoroff, P. 1938. Reactions of marine fishes to temperature gradients. Biological Bulletin 75:494-509.

———. 1942. The resistance and acclimatization of marine fishes to temperature changes. I. Experiments with Girella nigricans (Ayres). Biological Bulletin 83:219-244.

———. 1945. The resistance and acclimatization of marine fishes to temperature changes. II. Experiments with Fundulus and Atherinops. Biological Bulletin 88:194-206.

———. 1956. Some experiments on the toxicity of complex cyanides to fish. Sewage and Industrial Wastes 28:1020-1040.

———, B. G. Anderson, G. E. Burdick, P. S. Galtsoff, W. B. Hart, Ruth Patrick, E. R. Strong, E. W. Surber, and W. M. Van Horn. 1951. Bio-assay methods for the evaluation of acute toxicity of industrial wastes to fish. Sewage and Industrial Wastes 23:1381-1397.

———, and M. Katz. 1950. Critical review of literature on the toxicity of industrial wastes and their components to fish. 1. Alkalies, acids, and inorganic gases. Sewage and Industrial Wastes 22:1432-1458.

———, and ———. 1953. Critical review of literature on the toxicity of industrial wastes and their components to fish. II. The metals, as salts. Sewage and Industrial Wastes 25:802-839.

———, G. Leduc, and C. R. Schneider. 1966. Acute toxicity to fish of solutions containing complex metal cyanides, in relation to concentrations of molecular hydrocyanic acid. Transactions of the American Fisheries Society 95:6-22.

———, and D. L. Shumway. 1970. Dissolved Oxygen Requirements of Freshwater Fishes. European Inland Fisheries Advisory Commission, Food and Agricultural Organization of the United Nations. In press.

———, and C. E. Warren. 1957. Biological indices of water pollution, with special reference to fish populations. Pages 144-163. *In* C. M. Tarzwell (Editor), Biological Problems in Water Pollution. Transactions of the 1956 Seminar. R. A. Taft Engineering Center, U. S. Department of Health, Education and Welfare. 272pp.

Dubos, R. 1965. Man Adapting. Yale University Press, New Haven. xxii + 527pp.

Edmondson, W. T. 1968. Water-quality management and lake eutrophication: the Lake Washington case. Pages 139-178. *In* T. H. Campbell and R. O. Sylvester (Editors), Water Resources Management and Public Policy. University of Washington Press, Seattle. 253pp.

Edwards, R. W., and V. M. Brown. 1967. Pollution and fisheries: a progress report. Water Pollution Control (Journal of the Institute of Water Pollution Control) 66:63-78.

Ellis, M. M. 1937. Detection and measurement of stream pollution. Bulletin of the U. S. Bureau of Fisheries 48(22):365-437.

Ellis, R. H. 1968. Effects of kraft pulp mill effluent on the production and food relations of juvenile chinook salmon in laboratory streams. M.S. thesis. Oregon State University, Corvallis. 55pp.

Ellis, R. J., and H. Gowing. 1957. Relationship between food supply and condition of wild brown trout, *Salmo trutta* Linnaeus, in a Michigan stream. Limnology and Oceanography 2:299-308.

Elton, C. 1933. The Ecology of Animals. Methuen & Co. Ltd., London. vii + 97pp.

Emerson, A. E. 1939a. Social coordination and the superorganism. American Midland Naturalist 21:182-209.

_____. 1939b. Populations of social insects. Ecological Monographs 9:287-300.

_____. 1960. The evolution of adaptation in population systems. Pages 307-348. *In* S. Tax (Editor), The Evolution of Life. (Evolution after Darwin, Vol. 1.) The University of Chicago Press, Chicago. viii + 629pp.

Fabricius, E. 1950. Heterogeneous stimulus summation in the release of spawning activities in fish. Fishery Board of Sweden, Institute of Fresh-Water Research (Drottningholm) Report 31:57-99.

_____, and K.-J. Gustafson. 1954. Further aquarium observations on the spawning behaviour of the char, *Salmo alpinus* L. Fishery Board of Sweden, Institute of Fresh-Water Research (Drottningholm) 35:58-104.

_____, and _____. 1955. Observations on the spawning behaviour of the grayling, *Thymallus thymallus* (L.). Fishery Board of Sweden, Institute of Fresh-Water Research (Drottningholm) Report 36:75-103.

Fair, G. M., J. C. Geyer, and D. A. Okun. 1968. Water and Wastewater Engineering. Vol. 2. Water Purification and Wastewater Treatment and Disposal. John Wiley & Sons, Inc., New York.

Falconer, D. S. 1960. Introduction to Quantitative Genetics. Oliver and Boyd, London. ix + 365pp.

Federal Water Pollution Control Administration. 1966a. Delaware Estuary Comprehensive Study: Preliminary Report and Findings. U. S. Department of the Interior, Philadelphia. [xxvi] + 94pp. + appendices.

_____. 1966b. Guidelines for Establishing Water Quality Standards for Interstate Waters. U. S. Department of the Interior. 12pp.

Ferguson, D. E., and C. R. Bingham. 1966. Endrin resistance in the yellow bullhead, *Ictalurus natalis*. Transactions of the American Fisheries Society 95:325-326.

_____, D. D. Culley, W. D. Cotton, and R. P. Dodds. 1964. Resistance to chlorinated hydrocarbon insecticides in three species of freshwater fish. Bioscience 14(11):43-44.

Ferguson, R. G. 1958. The preferred temperature of fish and their midsummer distribution in temperate lakes and streams. Journal of the Fisheries Research Board of Canada 15:607-624.

Filice, F. P. 1954. A study of some factors affecting the bottom fauna of a portion of the San Francisco Bay estuary. Wasmann Journal of Biology 12:257-292.

_____. 1958. Invertebrates from the estuarine portion of San Francisco Bay and some factors influencing their distributions. Wasmann Journal of Biology 16:159-211.

_____. 1959. The effect of wastes on the distribution of bottom invertebrates in the San Francisco Bay estuary. Wasmann Journal of Biology 17:1-17.

Finer, S. E. 1952. The Life and Times of Sir Edwin Chadwick. Methuen and Company, Ltd., London. xi + 555pp.

Fingerman, M., and L. D. Fairbanks. 1956. Osmotic behavior and bleeding of the oyster *Crassostrea virginica*. Tulane Studies in Zoology 3:149-168.

Fingl, E., and D. M. Woodbury. 1965. General principles. Pages 1-36. *In* L. S. Goodman and A. Gilman (Editors), The Pharmacological Basis of Therapeutics. 3rd ed. The Macmillan Company, New York. xviii + 1785pp.

Fisher, K. C. 1958. An approach to the organ and cellular physiology of adaptation to temperature in fish and small mammals. Pages 3-49. *In* C. L. Prosser (Editor), Physiological Adaptation. American Physiological Society, Washington, D.C. 185pp.

Fisher, R. J. 1963. Influence of oxygen concentration and of its diurnal fluctuations on the growth of juvenile coho salmon. M.S. thesis. Oregon State University, Corvallis. 48pp.

Florey, E. 1966. An Introduction to General and Comparative Animal Physiology. W. B. Saunders Company, Philadelphia. xi + 713pp.

Florkin, M. 1966. Nitrogen metabolism. Pages 309-351. *In* K. M. Wilbur and C. M. Yonge (Editors), Physiology of Mollusca, Vol. 2. Academic Press Inc., New York. xiii + 645pp.

Forbes, S. A. 1887. The lake as a microcosm. Bulletin of the Peoria Scientific Association. (Reprinted 1925. Bulletin of the Illinois State Natural History Survey 15:537-550.)

_____, and R. E. Richardson. 1913. Studies on the biology of the Upper Illinois River. Bulletin of the Illinois State Laboratory of Natural History 9:[481]-574 + 21 plates.

_____, and _____. 1919. Some recent changes in Illlinois River biology. Bulletin of the Illinois Natural History Survey 13:[139]-156.

Ford, E. B. 1964. Ecological Genetics. Methuen & Co. Ltd., London. xv + 335pp.

Forsythe, R. A. 1961. The needs and obligations of federal agencies. Pages 250-269. *In* Proceedings The National Conference on Water Pollution, 1960. U. S. Department of Health, Education, and Welfare, Washington, D.C. x + 607pp.

Fox, C. J. J. 1909. On the coefficients of absorption of nitrogen and oxygen in distilled water and sea-water, and of atmospheric carbonic acid in sea-water. Transactions of the Faraday Society 5:68-87.

Fraenkel, G. S., and D. L. Gunn. 1940. The Orientation of Animals. Oxford, at the Clarendon Press. [viii] + 352pp.

Frank, P. W. 1960. Prediction of population growth form in *Daphnia pulex* cultures. American Naturalist 94:357-372.

_____, Catherine D. Boll, and R. W. Kelly. 1957. Vital statistics of laboratory cultures of Daphnia pulex De Geer as related to density. Physiological Zoology 30:287-305.

Freeman, P. R. 1963. Judicial expression. Pages 64-87. *In* J. E. McKee and H. W. Wolf (Editors), Water Quality Criteria. 2nd ed. California State Water Quality Control Board Publication 3-A. xiv + 548pp. + map.

Frey, D. G. (Editor). 1963. Limnology in North America. The University of Wisconsin Press, Madison. [xviii] + 734pp.

Fry, F. E. J. 1947. Effects of the environment on animal activity. University of Toronto Studies Biological Series 55. Ontario Fisheries Research Laboratory Publication 68. 62pp.

_____. 1957. The aquatic respiration of fish. Pages 1-63. *In* Margaret E. Brown (Editor), The Physiology of Fishes. Vol. 1, Metabolism. Academic Press Inc., New York. xiii + 447pp.

_____. 1964. Animals in aquatic environments: fishes. Pages 715-728. *In* D. B. Dill, E. F. Adolph, and C. G. Wilber (Editors), Adaptation to the Environment. (Handbook of Physiology, Section 4.) American Physiological Society, Washington, D.C. ix + 1056pp.

_____, J. R. Brett, and G. H. Clawson. 1942. Lethal limits of temperature for young gold fish. Revue Canadienne de Biologie 1:50–56.

_____, and J. S. Hart. 1948. Cruising speed of goldfish in relation to water temperature. Journal of the Fisheries Research Board of Canada 7:169-175.

_____, _____, and K. F. Walker. 1946. Lethal temperature relations for a sample of young speckled trout, *Salvelinus fontinalis*. University of Toronto Studies Biological Series 54. Ontario Fisheries Research Laboratory Publication 66:9-35.

Fujiya, M. 1961. Use of electrophoretic serum separation in fish studies. Journal Water Pollution Control Federation 33:250-257.

Fuller, G. W. 1912. Sewage Disposal. McGraw-Hill Book Company, New York. xv + 767pp.

_____, and J. R. McClintock. 1926. Solving Sewage Problems. McGraw-Hill Book Company, Inc., New York. x + 548pp.

Gaarder, T., and E. Eliassen. 1955. The energy-metabolism of *Ostrea edulis*. Universitetet i Bergen Årbok. Naturvitenskapelig Rekke 1954(3): 1–6.

_____, and H. H. Gran. 1927. Investigations of the production of plankton in the Oslo Fjord. Conseil Permanent International pour l'Exploration de la Mer. Rapports et Procès-Verbaux des Réunions 42. 48pp.

Gaddum, J. H. 1953. Pharmacology. 4th ed. Oxford University Press, London. xviii + 562pp.

Gaffney, M. M. 1966. Welfare economics and the environment. Pages 88-101. *In* H. Jarrett (Editor), Environmental Quality in a Growing Economy. Published for Resources for the Future, Inc., by The Johns Hopkins Press, Baltimore. xv + 173pp.

Galtsoff, P. S. 1964. The American Oyster (*Crassostrea virginica* Gmelin). Fishery Bulletin Vol. 64, U. S. Fish and Wildlife Service. iii + 480pp.

_____, W. A. Chipman, Jr., A. D. Hasler, and J. B. Engle. 1938. Preliminary report on the cause of the decline of the oyster industry of the York River, Va., and the effects of pulp-mill pollution on oysters. Bureau of Fisheries Investigational Report 37. 42pp.

_____, and Dorothy V. Whipple. 1930. Oxygen consumption of normal and green oysters. Bulletin of the Bureau of Fisheries 46:489-508. (Reprinted 1931. Bureau of Fisheries Document 1094.)

Garrey, W. E. 1916. The resistance of fresh water fish to changes of osmotic and chemical conditions. American Journal of Physiology 39:313-329.

Garside, E. T. 1966. Effects of oxygen in relation to temperature on the development of embryos of brook trout and rainbow trout. Journal of the Fisheries Research Board of Canada 23:1121-1134.

Gaufin, A. R., and C. M. Tarzwell. 1956. Aquatic macro-invertebrate communities as indicators of organic pollution in Lytle Creek. Sewage and Industrial Wastes 28:906-924.

Gause, G. F. 1934. The Struggle for Existence. The Williams and Wilkins Company, Baltimore. ix + 163pp.

Gehm, H. W. 1956. Activated sludge at high temperatures and high pH values. Pages 352-355. *In* J. McCabe and W. W. Eckenfelder, Jr. (Editors), Biological Treatment of Sewage and Industrial Wastes. Vol. 1. Aerobic Oxidation. Reinhold Publishing Corporation, New York. vii + 393pp.

Gerking, S. D. 1962. Production and food utilization in a population of bluegill sunfish. Ecological Monographs 32:[31]-78.

Glass, B. 1965. Nor any drop to drink. 1. Chicago: keeping it clean. Saturday Review, October 23, 1965. pp. 35-36.

Goodman, L. S., and A. Gilman (Editors). 1965. The Pharmacological Basis of Therapeutics. 3rd ed. The Macmillan Company, New York. xviii + 1785pp.

Gorlinski, J. S. 1957. Legal basis for water pollution control in California. Pages 61-63. *In* Waste Treatment and Disposal Aspects to Development of California's Pulp and Paper Resources. California State Water Pollution Control Board Publication 17. [102]pp.

Grassé, P.-P. 1958. Agnathes et poissons: Anatomie, Ethologie, Systématique. (Traité de Zoologie: Anatomie, Systématique, Biologie. Vol. 13.) Masson et Cie Éditeurs, Paris. Tome 13, Fascicule 2 [925]-1812.

Gray, H. F. 1940. Sewerage in ancient and mediaeval times. Sewage Works Journal 12:939-946.

Gray, J. 1928. The growth of fish. III. The effect of temperature on the development of the eggs of *Salmo fario.* British Journal of Experimental Biology 6:125-130.

Greenberg, B. 1947. Some relations between territory, social hierarchy, and leadership in the green sunfish (Lepomis cyanellus). Physiological Zoölogy 20:267-299.

Grinnell, J. 1904. The origin and distribution of the chestnut-backed chickadee. Auk 21:364-382.

———. 1917. The niche-relationships of the California thrasher. Auk 34:427-433.

———. 1928. Presence and absence of animals. University of California Chronicles 30:429-450. (Reprinted 1943. Joseph Grinnell's Philosophy of Nature. University of California Press, Berkeley.)

Happold, F. C., and A. Key. 1932. The bacterial purification of gas works liquors. —The action of the liquors on the bacterial flora of sewage. Journal of the Institute of Sewage Purification 1932(2):252-257.

Harden Jones, F. R. 1968. Fish Migration. St. Martin's Press, New York. viii + 325pp.

Hardin, G. 1960. The competitive exclusion principle. Science 131:1292-1297.

Hardy, A. 1959. The Open Sea: Its Natural History. Part 2, Fish and Fisheries. Houghton Mifflin Company, Boston. xiv + 322pp.

Harnly, M. H. 1941. Flight capacity in relation to phenotypic and genotypic variations in the wings of Drosophila melanogaster. Journal of Experimental Zoology 88:263-273.

Hart, J. S. 1947. Lethal temperature relations of certain fish of the Toronto region. Transactions of the Royal Society of Canada, Series 3, 41(5):57-71.

Hart, W. B., P. Doudoroff, and J. Greenbank. 1945. The Evaluation of the Toxicity of Industrial Wastes, Chemicals and Other Substances to Freshwater Fishes. Waste Control Laboratory, The Atlantic Refining Co., Philadelphia. 317 +[14]pp.

Hasler, A. D. 1947. Eutrophication of lakes by domestic drainage. Ecology 28:383-395.

———. 1957. The sense organs: olfactory and gustatory senses of fishes. Pages 187-209. *In* Margaret E. Brown (Editor), The Physiology of Fishes. Vol. 2, Behavior. Academic Press Inc., New York. xi + 526pp.

———. 1964. Experimental limnology. Bioscience 14(7):36-38.

———, R. H. Horrall, W. J. Wisby, and W. Braemer. 1958. Sun-orientation and homing in fishes. Limnology and Oceanography 3:353-361.

————, and W. J. Wisby. 1951. Discrimination of stream odors by fishes and its relation to parent stream behavior. American Naturalist 85:223-238.

Hatch, R. W., and D. A. Webster. 1961. Trout production in four central Adirondack mountain lakes. Cornell Agricultural Experiment Station Memoir 373:1-81.

Hathaway, E. S. 1927a. Quantitative study of the changes produced by acclimatization in the tolerance of high temperatures by fishes and amphibians. Bulletin of the U. S. Bureau of Fishes 43(2):169-192. (1928. Government Printing Office Document 1030.)

————. 1927b. The relation of temperature to the quantity of food consumed by fishes. Ecology 8:428-434.

Hawkes, H. A. 1952. The ecology of *Anisopus fenestralis* Scop. (Diptera) in sewage bacteria beds. Annals of Applied Biology 39:181-192.

————. 1961. An ecological approach to some bacteria bed problems. Journal of the Institute of Sewage Purification 1961(2):105-132.

————. 1962. Biological aspects of river pollution. Pages 311-432. *In* L. Klein, River Pollution. 2. Causes and Effects. Butterworth, London. xiv + 456pp.

————. 1963. The Ecology of Waste Water Treatment. Pergamon Press Ltd., Oxford. ix + 203pp.

————, and S. H. Jenkins. 1951. Biological principles in sewage purification. Journal of the Institute of Sewage Purification 1951(3):300-317.

Hebb, D. O. 1953. Heredity and environment in mammalian behaviour. British Journal of Animal Behaviour 1:43-47.

Heinroth, O. 1911. Beiträge zur Biologie, namentlich Ethologie und Physiologie der Anatiden. Verhandlungen des Internationalen Ornithologen-Kongresses 5:589-702.

Henderson, C. 1957. Application factors to be applied to bioassays for the safe disposal of toxic wastes. Pages 31-37. *In* C. M. Tarzwell (Editor), Biological Problems in Water Pollution. Transactions of the 1956 Seminar. R. A. Taft Sanitary Engineering Center, U. S. Department of Health, Education, and Welfare. 272pp.

————, and C. M. Tarzwell. 1957. Bio-assays for control of industrial effluents. Sewage and Industrial Wastes 29:1002-1017.

Henderson, L. J. 1913. The Fitness of the Environment. The Macmillan Company, New York. xv + 317pp.

Herbert, D. W. M. 1965. Pollution and fisheries. Pages 173-195. *In* G. T. Goodman, R. W. Edwards, and J. M. Lambert (Editors), Ecology and the Industrial Society. British Ecological Society Symposium 5, 1964. viii + 395pp.

————, Dorothy H. M. Jordan, and R. Lloyd. 1965. A study of some fishless rivers in the industrial midlands. Journal of the Institute of Sewage Purification 1965(6):569-579.

————, and J. C. Merkens. 1952. The toxicity of potassium cyanide to trout. Journal of Experimental Biology 29:632-649.

Herrick, F. H. 1909. Natural history of the American lobster. Bulletin of the Bureau of Fisheries 24:149-408 + 15 plates. (Reprinted 1911. Government Printing Office Document 747.)

Herrington, W. C. 1948. Limiting factors for fish populations. Some theories and an example. Bulletin of the Bingham Oceanographic Collection 11:229-283.

Hinde, R. A. 1959. Some recent trends in ethology. Pages 561-610. *In* S. Koch (Editor), Psychology, A Study of a Science. Study 1, Conceptual and Systematic. Vol. 2, General Systematic Formulations, Learning, and Special Processes. McGraw-Hill Book Company, Inc., New York. x + 706pp.

————. 1966. Animal Behaviour: A Synthesis of Ethology and Comparative Psychology. McGraw-Hill Book Company, New York. x + 534pp.

Hoar, W. S. 1951. The behaviour of chum, pink and coho salmon in relation to their seaward migration. Journal of the Fisheries Research Board of Canada 8:241-263.

————. 1957a. Endocrine organs. Pages 245-285. *In* Margaret E. Brown (Editor), The Physiology of Fishes. Vol. 1, Metabolism. Academic Press Inc., New York. xiii + 447pp.

————. 1957b. The gonads and reproduction. Pages 287-321. *In* Margaret E. Brown (Editor), The Physiology of Fishes. Vol. 1, Metabolism. Academic Press Inc., New York. xiii + 447pp.

————. 1958. The evolution of migratory behaviour among juvenile salmon of the genus *Oncorhynchus*. Journal of the Fisheries Research Board of Canada 15:391-428.

Hogetsu, K., and S. Ichimura. 1954. Studies on the biological production of Lake Suwa. 6. The ecological studies on the production of phytoplankton. Japanese Journal of Botany 14:280-303.

Höglund, L. B. 1961. The reactions of fish in concentration gradients. Fishery Board of Sweden, Institute of Fresh-Water Research (Drottningholm) Report 43. 147pp.

Holeton, G. F., and D. J. Randall. 1967. The effect of hypoxia upon the partial pressure of gases in the blood and water afferent and efferent to the gills of rainbow trout. Journal of Experimental Biology 46:317-327.

Holst, E. von. 1934. Studien über Reflexe and Rythmen beim Goldfisch (Carassius auratus). Zeitschrift für Ergleichende Physiologie 20:582-599.

Hopkins, A. E., P. S. Galtsoff, and H. C. McMillin. 1931. Effects of pulp mill pollution on oysters. Bulletin of the U. S. Bureau of Fisheries 47(6):125-186.

Howard, L. O., and W. F. Fiske. 1911. The importation into the United States of the parasites of the gipsy moth and the brown-tail moth. U. S. Department of Agriculture, Bureau of Entomology Bulletin 91. 344pp. + errata.

Hughes, G. M. 1964. Fish respiratory homeostasis. Pages 81-107. *In* Homeostasis and Feedback Mechanisms. (Symposia of the Society for Experimental Biology 18.) Academic Press Inc., New York. viii + 460pp.

Hunt, R. L. 1966. Production and angler harvest of wild brook trout in Lawrence Creek, Wisconsin. Wisconsin Conservation Department Bulletin 35. 52pp.

Huntsman, A. G. 1948. Method in ecology-biapocrisis. Ecology 29:30-42.

Hutchinson, G. E. 1957a. A Treatise on Limnology. Vol. 1: Geography, Physics, and Chemistry. John Wiley and Sons, Inc., New York. xiv + 1015pp.

———. 1957b. Concluding remarks. Pages 415-427. *In* Population Studies: Animal Ecology and Demography. Cold Spring Harbor Symposia on Quantitative Biology, Vol. 22. xiv + 437pp.

———. 1959. Homage to Santa Rosalia or why are there so many kinds of animals? American Naturalist 93:145-159.

———. 1963. The prospect before us. Pages 683-690. *In* D. G. Frey (Editor), Limnology in North America. The University of Wisconsin Press, Madison. [xviii] + 734pp.

———. 1967. A Treatise on Limnology. Vol. 2: Introduction to Lake Biology and the Limnoplankton. John Wiley and Sons, Inc., New York. xi + 1115pp.

Huxley, T. H. 1884. The Crayfish: An Introduction to the Study of Zoology. 4th ed. Kegan Paul, Trench & Co., London. xiv + 371pp.

Hynes, H. B. N. 1960. The Biology of Polluted Waters. Liverpool University Press, Liverpool. xiv + 202pp.

Ide, F. P. 1954. Pollution in relation to stream life. Pages 86-108. *In* First Ontario Industrial Waste Conference. Pollution Control Board of Ontario, Toronto. 121pp.

Imhoff, K., and G. M. Fair. 1956. Sewage Treatment. 2nd ed. John Wiley and Sons, Inc., New York. vi + 338pp.

Ivlev, V. S. 1939a. Energy balance of the growing larva of *Silurus glanis*. Comptes Rendus (Doklady) de l'Académie des Sciences de l'URSS, N.S. 25:87-89.

———. 1939b. Effect of starvation on energy transformation during the growth of fish. Comptes Rendus (Doklady) de l'Académie des Sciences de l'URSS, N.S. 25:90-92.

———. 1939c. Energy balance in the carp. Zoologicheskii Zhurnal 18:303-318.

———. 1945. Biologicheskaya produktivnost' vodoemov. Uspekhi Sovremennoi Biologii 19:98-120. (Translated by W. E. Ricker, 1966. The biological productivity of waters. Journal of the Fisheries Research Board of Canada 23:1707-1759.)

———. 1947. [Effect of density of planting on the growth of carp.] Byulleten' Moskovskogo Obshchestva Ispytatelei̇. Prirody 52:29-38.

———. 1961a. Experimental Ecology of the Feeding of Fishes. (Translated by D. Scott.) Yale University Press, New Haven. viii + 302pp.

———. 1961b. Ob utilizatsii pishchi rybami-planktofagami. Trudy Sevastopoloskoi Biologicheskoi Stantsii 14:188-201. (Translated by W. E. Ricker, 1963. On the utilization of food by plankton-eating fishes. Fisheries Research Board of Canada Translation Series 447. 15pp. + [ii].)

Jarrett, H. (Editor). 1966. Environmental Quality in a Growing Economy. Published for Resources for the Future, Inc., by The Johns Hopkins Press, Baltimore. xv + 173pp.

Jensen, L. D., and A. R. Gaufin. 1966. Acute and long-term effects of organic insecticides on two species of stonefly naiads. Journal Water Pollution Control Federation 38:1273-1286.

Johnson, J. W. H. 1914a. A contribution to the biology of sewage disposal. Journal of Economic Biology 9:[105]-124.

––––––. 1914b. A contribution to the biology of sewage disposal, Part II. Journal of Economic Biology 9:[127]-164.

Johnson, W. E. 1961. Aspects of the ecology of a pelagic, zooplankton-eating fish. Verhandlungen Internationale Vereinigung für Theoretische und Angewandte Limnologie 14:727-731.

Jones, J. R. E. 1938. The relative toxicity of salts of lead, zinc and copper to the stickleback (*Gasterosteus aculeatus* L.) and the effect of calcium on the toxicity of lead and zinc salts. Journal of Experimental Biology 15:394-407.

––––––. 1939. The relation between the electrolytic solution pressures of the metals and their toxicity to the stickleback (*Gasterosteus aculeatus* L.) Journal of Experimental Biology 16:425-437.

––––––. 1947. The reactions of *Pygosteus pungitius* L. to toxic solutions. Journal of Experimental Biology 24:110-122.

––––––. 1952. The reactions of fish to water of low oxygen concentration. Journal of Experimental Biology 29:403-415.

––––––. 1964. Fish and River Pollution, Butterworth, London. viii + 203pp.

Jones, J. W., and J. N. Ball. 1954. The spawning behaviour of brown trout and salmon. British Journal of Animal Behaviour 2:103-114.

Jordan, Dorothy H. M., and R. Lloyd. 1964. The resistance of rainbow trout (*Salmo gairdnerii* Richardson) and roach (*Rutilus rutilus* (L.)) to alkaline solutions. International Journal Air and Water Pollution 8:405-409.

Juday, C. 1940. The annual energy budget of an inland lake. Ecology 21:438-450.

Kalleberg, H. 1958. Observations in a stream tank of territoriality and competition in juvenile salmon and trout (*Salmo salar* L. and *S. trutta* L.). Fishery Board of Sweden, Institute of Fresh-Water Research (Drottningholm) Report 39:55-98.

Katz, M., and W. C. Howard. 1955. The length and growth of 0-year class creek chubs in relation to domestic pollution. Transactions of the American Fisheries Society 84:228-238.

Kawamoto, N. 1929. Physiological studies on the eel. 2. The influence of temperature and of the relative volumes of the red corpuscles and plasma upon the haemoglobin dissociation curve. Tohoku Imperial University Science Report, 4th Series (Biology) 4:643-659.

Keenleyside, M. H. A., and F. T. Yamamoto. 1962. Territorial behaviour of juvenile Atlantic salmon (*Salmo salar* L.). Behaviour 19:139-169.

Kennedy, J. S. 1954. Is modern ethology objective? British Journal of Animal Behaviour 2:12-19.

Kerswill, C. J., and P. F. Elson. 1955. Preliminary observations on effects of 1954 DDT spraying on Miramichi salmon stocks. Fisheries Research Board of Canada, Progress Reports of the Atlantic Coast Stations 62:17-23.

Keup, L. E., W. M. Ingram, and K. M. Mackenthun (Editors). 1967. Biology of Water Pollution: A Collection of Selected Papers on Stream Pollution, Waste Water, and Water Treatment. Federal Water Pollution Control Administration, U. S. Department of the Interior. iv + 290pp.

Keys, A. B. 1931. Chloride and water secretion and absorption by the gills of the eel. Zeitschrift für Vergleichende Physiologie 15:366-389.

Kirchshofer, R. 1954. Ökologie und Revierverhältnisse beim Schriftbarsch. Österreichische Zoologische Zeitschrift 5:329-349.

Kleiber, M. 1961a. Metabolic rate and food utilization as a function of body size. (Brody Memorial Lecture 1.) Missouri Agricultural Experiment Station Research Bulletin 767. 42pp.

––––––. 1961b. The Fire of Life: An Introduction to Animal Energetics. John Wiley and Sons, Inc., New York. xxii + 454pp.

Klein, L. 1962. River Pollution. 2. Causes and Effects. Butterworth, London. xiv + 456pp.

––––––. 1966. River Pollution. 3. Control. Butterworth, Washington. xv + 484pp.

Klopfer, P. H., and J. P. Hailman. 1967. An Introduction to Animal Behaviour: Ethology's First Century. Prentice-Hall, Inc., Englewood Cliffs, New Jersey. xiv + 297pp.

Koelz, W. 1927. Coregonid fishes of the Great Lakes. Bulletin of the U. S. Bureau of Fisheries 43(2):297-643. (1929. Government Printing Office Document 1048.)

Kohn, A. J. 1957. Primary organic productivity of a Hawaiian coral reef. Limnology and Oceanography 2:241-251.

Kolkwitz, R. 1911. Biologie des Trinkwassers, Abwassers und der Vorfluter. Rubner, Gruber, und Ficker's Handbuch der Hygiene II, 2. S. Herzel, Leipzig.

_____, and M. Marsson. 1908. Ökologie der pflanzlichen Saprobien. Berichte der Deutschen Botanischen Gesellschaft 26a:505-519. (Translated 1967. Ecology of plant saprobia. Pages 47-52. *In* L. E. Keup, W. M. Ingram, and K. M. Mackenthun [Editors], Biology of Water Pollution. Federal Water Pollution Control Administration, U. S. Department of the Interior. iv + 290pp.)

_____, and _____. 1909. Ökologie der tierischen Saprobien. Beiträge zur Lehre von der biologischen Gewässerbeurteilung. Internationale Revue der Gesamten Hydrobiologie und Hydrogeographie 2:126-152. (Translated 1967. Ecology of animal saprobia. Pages 85-95. *In* L. E. Keup, W. M. Ingram, and K. M. Mackenthun [Editors], Biology of Water Pollution. Federal Water Pollution Control Administration, U. S. Department of the Interior. iv + 290pp.)

Korringa, P. 1952. Recent advances in oyster biology. Quarterly Review of Biology 27(3):266-308, (4):339-365.

Koski, K. V. 1966. The survival of coho salmon (*Oncorhynchus kisutch*) from egg deposition to emergence in three Oregon coastal streams. M.S. thesis. Oregon State University, Corvallis. 84pp.

Kozlovsky, D. G. 1968. A critical evaluation of the trophic level concept. I. Ecological efficiencies. Ecology 49:48-60.

Krogh, A. 1937. Osmotic regulation in fresh water fishes by active absorption of chloride ions. Zeitschrift für Vergleichende Physiologie 24:656-666.

_____. 1939. Osmotic Regulation in Aquatic Animals. Cambridge at the University Press. [ix] + 242pp.

_____, and I. Leitch. 1919. The respiratory function of the blood in fishes. Journal of Physiology (London) 52:288-300.

Krogius, F. V. 1961. O sviaziakh tempa rosta i chislennosti krasnoi. Trudy Soveschanii Ikhtiologicheskoi Komissii Akademii Nauk SSSR 13:132-146. (Translated by R. E. Foerster, 1962. On the relation between rate of growth and population density in sockeye salmon. Fisheries Research Board of Canada Translation Series 411. 17 + [6]pp.)

_____, and E. M. Krokhin. 1948. Ob urozhainosti molodi krasnoi (*Oncorhynchus nerka* Walb.). Izvestiia Tikhookeanskovo Nauchno-Issledovatelskovo Instituta Rybnovo Khoziaistva i Okeanografii 28:3-27. (Translated by R. E. Foerster, 1958. On the production of young sockeye salmon (*Oncorhynchus nerka* Walb.). Fisheries Research Board of Canada Translation Series 109. 27 + [3]pp.)

Krumholz, L. A. 1956. Observations on the fish population of a lake contaminated by radioactive wastes. Bulletin of the American Museum of Natural History 110:277-368.

Lack, D. 1954. The Natural Regulation of Animal Numbers. Clarendon Press, Oxford. viii + 343pp.

_____. 1966. Population Studies of Birds. Clarendon Press, Oxford. v + 341pp.

Lackey, J. B. 1945. Plankton productivity of certain southeastern Wisconsin lakes as related to fertilization. II. Productivity. Sewage Works Journal 17:795-802.

_____, and C. N. Sawyer. 1945. Plankton productivity of certain southeastern Wisconsin lakes as related to fertilization. I. Surveys. Sewage Works Journal 17:573-585.

Lagler, K. F., J. E. Bardock, and R. R. Miller. 1962. Ichthyology. John Wiley and Sons, Inc., New York. xiii + 545pp.

Lavoisier, A. L. 1777. Expériences sur la respiration des animaux et sur les changements qui arrivent a l'air en passant par leur poumon. Mémoires de l'Académie des Sciences, Paris. p. 185.

Leduc, G. 1966. Some physiological and biochemical responses of fish to chronic poisoning by cyanide. Ph.D. thesis. Oregon State University, Corvallis. 146pp.

Lee, R. A. 1969. Bioenergetics of feeding and growth of largemouth bass in aquaria and ponds. M.S. thesis. Oregon State University, Corvallis. 63pp.

Léger, L. 1912. Études sur l'action nocive des produits de déversements industriels chimiques dans les eaux douces. 2e série. Eaux de décapage des métaux. Annales de l'Université de Grenoble 24:41-122.

Lehrman, D. S. 1953. A critique of Konrad Lorenz's theory of instinctive behavior. Quarterly Review of Biology 28:337-363.

Lemke, A. E., and D. I. Mount. 1963. Some effects of alkyl benzene sulfonate on the bluegill, *Lepomis macrochirus*. Transactions of the American Fisheries Society 92:372-378.

Leslie, P. H. 1948. Some further notes on the use of matrices in population mathematics. Biometrika 35:213-245.

Lewis, G. N. 1926. The Anatomy of Science. Yale University Press, New Haven. [xi] + 219pp.

Lewontin, R. C. 1957. The adaptations of populations to varying environments. Pages 395-408. *In* Population Studies: Animal Ecology and Demography. Cold Spring Harbor Symposia on Quantitative Biology, Vol. 22. xiv + 437pp.

Lindeman, R. L. 1941a. Seasonal food-cycle dynamics in a senescent lake. American Midland Naturalist 26:636-673.

——. 1941b. The developmental history of Cedar Creek Bog, Minnesota. American Midland Naturalist 25:101-112.

——. 1942. The trophic-dynamic aspect of ecology. Ecology 23:399-418.

Lloyd, D., and Lydia D. Orr. 1969. The diuretic response by rainbow trout to sub-lethal concentrations of ammonia. Water Research (Pergamon Press) 3:335-344.

Lloyd, L. 1943. Materials for a study in animal competition. Part III. The seasonal rhythm of *Psychoda alterata* Say and an effect of intraspecific competition. Annals of Applied Biology 30:358-364.

——. 1944. Sewage bacteria bed fauna in its natural setting. Nature 154:397.

——. 1945. Animal life in sewage purification processes. Journal of the Institute of Sewage Purification 1945(2):119-139.

——, J. F. Graham, and T. B. Reynoldson. 1940. Materials for a study in animal competition. The fauna of the sewage bacteria beds. Annals of Applied Biology 27:122-150.

Lloyd, R. 1960. The toxicity of zinc sulfate to rainbow trout. Annals of Applied Biology 48:84-94.

——. 1961a. The toxicity of ammonia to rainbow trout (*Salmo gairdnerii* Richardson). Water Waste Treatment Journal 8:278-279.

——. 1961b. Effect of dissolved oxygen concentrations on the toxicity of several poisons to rainbow trout (*Salmo gairdnerii* Richardson). Journal of Experimental Biology 38:447-455.

——. 1961c. The toxicity of mixtures of zinc and copper sulphates to rainbow trout (*Salmo gairdnerii* Richardson). Annals of Applied Biology 49:535-538.

——, and Dorothy H. M. Jordon. 1964. Some factors affecting the resistance of rainbow trout (*Salmo gairdnerii* Richardson) to acid waters. International Journal Air and Water Pollution 8:393-403.

Locke, J. 1689. Essay Concerning Human Understanding. Reprinted 1952. Pages 93-395. *In* Great Books of the Western World. Vol. 35. Encyclopaedia Brittanica, Inc., Chicago. x + 509pp.

Loeb, J., and H. Wasteneys. 1912. On the adaptation of fish (Fundulus) to higher temperatures. Journal of Experimental Zoology 12:543-557.

Longwell, J., and F. T. K. Pentelow. 1935. The effect of sewage on brown trout (*Salmo trutta* L.). Journal of Experimental Biology 12:1-12.

Loosanoff, V. L. 1952. Behavior of oysters in water of low salinities. Proceedings National Shellfish Association, Atlantic City, New Jersey. 1952:135-151.

——. 1961. Effects of turbidity on some larval and adult bivalves. Proceedings of the Gulf and Caribbean Fisheries Institute 14:80-95.

——, and H. C. Davis. 1963. Rearing of bivalve mollusks. Pages 1-136. *In* F. S. Russell (Editor), Advances in Marine Biology, Vol. 1. Academic Press Inc., New York. xiii + 410pp.

Lorenz, K. 1935. Der Kumpan in der Umwelt des Vögels. Journal für Ornithologie 83:137-213, 289-413.

——. 1937. Uber den Begriff der Instinkthandlung. Folia Biotheoretica 2:18-50.

——. 1961. Phylogenetische Anpassung und adaptive Modifikation des Verhaltens. Zeitschrift für Tierpsychologie 18:139-187.

Lorz, H. W., and T. G. Northcote. 1965. Factors affecting stream location, and timing and intensity of entry by spawning kokanee (*Oncorhynchus nerka*) into an inlet of Nicola Lake, British Columbia. Journal of the Fisheries Research Board of Canada 22:665-687.

Lotka, A. J. 1925. Elements of Physical Biology. Williams & Wilkins Company, Baltimore. xxx + 460pp.

————. 1932. The growth of mixed populations: Two species competing for a common food supply. Journal of the Washington Academy of Sciences 22:461-469.

————. 1945. Population analysis as a chapter in the mathematical theory of evolution. Pages 355-385. *In* W. E. L. Clark and P. B. Medawar (Editors), Essays on Growth and Form: Presented to D'Arcy Wentworth Thompson. Clarendon Press, Oxford. viii + 408pp.

Lowenstein, O. 1957. The sense organs: the acoustico-lateralis system. Pages 155-186. *In* Margaret E. Brown (Editor), The Physiology of Fishes. Vol. 2, Behavior. Academic Press Inc., New York. xi + 526pp.

McCabe, J., and W. W. Eckenfelder, Jr. (Editors). 1956. Biological Treatment of Sewage and Industrial Wastes. Vol. 1. Aerobic Oxidation. Reinhold Publishing Corporation, New York. vii + 393pp.

————, and ———— (Editors). 1958. Biological Treatment of Sewage and Industrial Wastes. Vol. 2, Anaerobic Digestion and Solids-Liquid Separation. Reinhold Publishing Corporation, New York. vi + 330pp.

McDougall, W. 1923. Outline of Psychology. Charles Scribner's Sons, New York. xvi + 456pp.

McGauhey, P. H. 1968. Engineering Management of Water Quality. McGraw-Hill Book Company, New York. [vii] + 295pp.

McIntire, C. D., R. L. Garrison, H. K. Phinney, and C. E. Warren. 1964. Primary production in laboratory streams. Limnology and Oceanography 9:92-102.

————, and H. K. Phinney. 1965. Laboratory studies of periphyton production and community metabolism in lotic environments. Ecological Monographs 35:[237]-258.

McKee, J. E. (Editor). 1952. Water Quality Criteria. California State Water Pollution Control Board Publication 3. 512pp. + map.

————. 1960. The need for water quality criteria. Pages 15-26. *In* H. A. Faber and Lena J. Bryson (Editors), Proceedings of the Conference on Physiological Aspects of Water Quality. U. S. Public Health Service, Washington, D.C. [xi] + 244pp.

————. 1967. Parameters of marine pollution—an overall evaluation. Pages 259-266. *In* T. A. Olson and F. J. Burgess (Editors), Pollution and Marine Ecology. Interscience Publishers (Wiley), New York. xvi + 364pp.

————, and H. W. Wolf (Editors). 1963. Water Quality Criteria. 2nd ed. California State Water Quality Control Board Publication 3-A. xiv + 548pp. + map.

McKinney, R. E. 1956. Biological flocculation. Pages 88-100. *In* J. McCabe and W. W. Eckenfelder, Jr. (Editors), Treatment of Sewage and Industrial Wastes. Vol. 1, Aerobic Oxidation. Reinhold Publishing Corporation, New York. vii + 393pp.

————. 1957. Activity of microorganisms in organic waste disposal. Applied Microbiology 5:167-174.

————, and A. Gram. 1956. Protozoa and activated sludge. Sewage and Industrial Wastes 28:1219-1231.

McLeese, D. W. 1956. Effects of temperature, salinity and oxygen on the survival of the American lobster. Journal of the Fisheries Research Board of Canada 13:247-272.

Mackenthun, K. M., and W. M. Ingram. 1967. Biological Associated Problems in Freshwater Environments. Federal Water Pollution Control Administration, U. S. Department of the Interior. x + 287pp.

Maloeuf, N. S. R. 1937. Studies on the respiration (and osmoregulation) of animals, I. Aquatic animals without an oxygen transporter in their internal medium. Zeitschrift für Vergleichende Physiologie 25:1-28.

Malthus, T. R. 1798. First Essay on Population. Reprinted 1965. Augustus M. Kelley, Reprints of Economic Classics, New York. ix + 396pp.

Margalef, R. 1951. Diversidad de especies en las comunidades naturales. [Diversity of species in natural communities.] Publicaciones Instituto de Biologia Aplicada 9:5-28.

————. 1956. Información y diversidad específica en las comunidades de organismos. [Information and specific diversity in communities of organismos.] Investigacion Pesquera 3:99-106.

———. 1960. Ideas for a synthetic approach to the ecology of running waters. Internationale Revue der Gesamten Hydrobiologie 45:133-153.

———. 1968. Perspectives in Ecological Theory. University of Chicago Press, Chicago. viii + 111pp.

Marler, P. 1961. The filtering of external stimuli during instinctive behaviour. Pages 150-166. *In* W. H. Thorpe and O. L. Zangwill (Editors), Current Problems in Animal Behaviour. University Press, Cambridge. xiv + 424pp.

———, and W. J. Hamilton III. 1966. Mechanisms of Animal Behavior. John Wiley & Sons, Inc., New York. xi + 771pp.

Marshall, S. M., and A. P. Orr. 1960. Feeding and nutrition. Pages 227-258. *In* T. H. Waterman (Editor), The Physiology of Crustacea. Vol. 1, Metabolism and Growth. Academic Press Inc., New York. xvii + 670pp.

Martin, A. W., and Florence M. Harrison. 1966. Excretion. Pages 353-386. *In* K. M. Wilbur and C. M. Yonge (Editors), Physiology of Mollusca, Vol. 2. Academic Press Inc., New York. xiii + 645pp.

Mason, J. C. 1969. Hypoxial stress prior to emergence and competition among coho salmon fry. Journal of the Fisheries Research Board of Canada 26:63-91.

———, and D. W. Chapman. 1965. Significance of early emergence, environmental rearing capacity, and behavioral ecology of juvenile coho salmon in stream channels. Journal of the Fisheries Research Board of Canada 22:173-190.

Maucha, R. 1924. Upon the influence of temperature and intensity of light on the photosynthetic production of nannoplankton. Verhandlungen der Internationalen Vereinigung für Theoretische und Angewandte Limnologie 2:381-401.

Maynard, D. M. 1960. Circulation and heart function. Pages 161-226. *In* T. H. Waterman (Editor), The Physiology of Crustacea. Vol. 1, Metabolism and Growth. Academic Press Inc., New York. xvii + 670pp.

Mayr, E. 1963. Animal Species and Evolution. The Belknap Press of Harvard University Press, Cambridge, Massachusetts. xiv + 797pp.

Mendel, G. 1866. Versuche über Pflanzen-Hybriden. Verhandlungen Naturforschender Verein in Brunn 4:3-47. Reprinted 1951. Journal of Heredity 42:1-47. (English translation 1938. Experiments in Plant-hybridisation. Harvard University Press, Cambridge, Massachusetts. 41pp.)

Metcalf, L., and H. P. Eddy. 1930. Sewerage and Sewage Disposal. 2nd ed. McGraw-Hill Book Company, Inc., New York and London. xvi + 783pp.

Mills, E. V. 1945. The treatment of settled sewage in percolating filters in series with periodic change in the order of the filters. — Results of operation of the experimental plant at Minworth, Birmingham, 1940-1944. Journal of the Institute of Sewage Purification 1945(1):35-49.

Milne, A. 1957. Theories of natural control of insect populations. Pages 253-271. *In* Population Studies: Animal Ecology and Demography. Cold Spring Harbor Symposia on Quantitative Biology, Vol. 22. xiv + 437pp.

Minder, L. 1938. Der Zürichsee als Eutrophierungsphänomen. Summarische Ergebnisse aus fünfzig Jahren Zürichseeforschung. Geologie der Meere und Binnengewässer 2:284-299.

Minot, C. S. 1908. The Problem of Age, Growth, and Death: A Study of Cytomorphosis. G. P. Putnam's Sons, New York. xxii + 280pp.

Möbius, von K. 1877. Die Auster und die Austernwirthschaft. Wiegandt, Hempel & Parey, Berlin. 69pp. (Translated by H. J. Rice, 1880. The oyster and oyster-culture. Report of the Commissioner, U. S. Commission of Fish and Fisheries Appendix H 27:683-751.)

Monod, J. 1949. The growth of bacterial cultures. Annual Review of Microbiology 3:371-394.

Moore, G. T., and K. F. Kellerman. 1905. Copper as an algicide and disinfectant in water supplies. U.S.D.A., Bureau of Plant Industry Bulletin 76. 55pp.

Morgan, H. R. 1952. The Salmonella. Pages 420-436. *In* R. J. Dubos (Editor), Bacterial and Mycotic Infections of Man. 2nd ed. J. B. Lippincott Company, Philadelphia. xiv + 886pp.

Mott, J. C. 1957. The cardiovascular system. Pages 81-108. *In* Margaret E. Brown (Editor), The Physiology of Fishes. Vol. 1, Metabolism. Academic Press Inc., New York. xiii + 447pp.

Mount, D. I. 1968. Chronic toxicity of copper to fathead minnows (*Pimephales promelas*, Rafinesque). Water Research (Pergamon Press) 2:215-223.

———, and C. E. Stephan. 1967. A method for establishing acceptable toxicant limits for fish—malathion and the butoxyethanol ester of 2,4-D. Transactions of the American Fisheries Society 96:185-193.

National Research Council. Committee on Pollution. 1966. Waste Management and Control. National Academy of Sciences, Washington, D.C. [xii] + 257pp. (NAS Publication 1400.)

National Technical Advisory Committee. 1968. Water Quality Criteria. Report to the Secretary of the Interior. Federal Water Pollution Control Administration, Washington, D.C. x + 234pp.

Needham, J. G., and J. T. Lloyd. 1937. The Life of Inland Waters. Comstock Publishing Company, Inc., Ithaca, New York. 438pp.

Needham, P. R., and A. C. Taft. 1934. Observations on the spawning of steelhead trout. Transactions of the American Fisheries Society 64:332-338 + 1 plate.

Neess, J., and R. C. Dugdale. 1959. Computation of production for populations of aquatic midge larvae. Ecology 40:425-430.

Neil, J. H. 1957. Some effects of potassium cyanide on speckled trout (*Salvelinus fontinalis*). Ontario Industrial Waste Conference 4:74-96.

Nelson, P. R. 1958. Relationship between rate of photosynthesis and growth of juvenile red salmon. Science 128:205-206.

Nelson, T. C. 1938. The feeding mechanism of the oyster. I. On the pallium and the branchial chambers of Ostrea virginica, O. edulis and O. angulata, with comparisons with other species of the genus. Journal of Morphology 63:1-61.

Newman, M. A. 1956. Social behavior and interspecific competition in two trout species. Physiological Zoölogy 29:64-81.

Nicholson, A. J. 1933. The balance of animal populations. Journal of Animal Ecology 2:132-178.

————. 1954. An outline of the dynamics of animal populations. Australian Journal of Zoology 2:9-65.

————. 1957. The self-adjustment of populations to change. Pages 153-173. *In* Population Studies: Animal Ecology and Demography. Cold Spring Harbor Symposia on Quantitative Biology, Vol. 22. xiv + 437pp.

Nilsson, N.-A. 1955. Studies on the feeding habits of trout and char in North Swedish lakes. Fishery Board of Sweden, Institute of Fresh-Water Research (Drottningholm) Report 36:163-225.

Norris, K. S. 1963. The functions of temperature in the ecology of the percoid fish *Girella nigricans* (Ayres). Ecological Monographs 33:23-62.

Odum, E. P. 1959. Fundamentals of Ecology. 2nd ed., in collaboration with H. T. Odum. W. B. Saunders Company, Philadelphia. xvii + 546pp.

Odum, H. T. 1956. Primary production in flowing waters. Limnology and Oceanography 1:102-117.

————. 1957a. Trophic structure and productivity of Silver Springs, Florida. Ecological Monographs 27:[55]-112.

————. 1957b. Primary production measurements in eleven Florida springs and a marine turtle-grass community. Limnology and Oceanography 2:85-97.

Ogilvie, D. M., and J. M. Anderson. 1965. Effect of DDT on temperature selection by young Atlantic salmon, *Salmo salar*. Journal of the Fisheries Research Board of Canada 22:503-512.

Orton, J. H. 1937. Oyster Biology and Oyster-Culture. Edward Arnold & Company, London. 211pp.

Owen, G. 1966. Digestion. Pages 53-96. *In* K. M. Wilbur and C. M. Yonge (Editors), Physiology of Mollusca, Vol. 2. Academic Press Inc., New York. xiii + 645pp.

Packard, W. H. 1905. On resistance to lack of oxygen and on a method of increasing this resistance. American Journal of Physiology 15:30-41.

Paloheimo, J. E., and L. M. Dickie. 1965. Food and growth of fishes. I. A growth curve derived from experimental data. Journal of the Fisheries Research Board of Canada 22:521-542.

————, and ————. 1966a. Food and growth of fishes. II. Effects of food and temperature on the relation between metabolism and body weight. Journal of the Fisheries Research Board of Canada 23:869-908.

————, and ————. 1966b. Food and growth of fishes. III. Relations among food, body size, and growth efficiency. Journal of the Fisheries Research Board of Canada 23:1209-1248.

Parker, R. R., and P. A. Larkin. 1959. A concept of growth in fishes. Journal of the Fisheries Research Board of Canada 16:721-745.

Parry, G. 1960. Excretion. Pages 341-366. *In* T. H. Waterman (Editor), The Physiology of Crustacea. Vol. 1, Metabolism and Growth. Academic Press Inc., New York. xvii + 670pp.

Patrick, Ruth. 1949. A proposed biological measure of stream conditions, based on a survey of the Conestoga Basin, Lancaster County, Pennsylvania. (With appendices by J. B. Graham, J. M. Ward, and D. G. Reihard, Jr.) Proceedings of the Academy of Natural Sciences of Philadelphia 101:277-341.

———. 1950. Biological measure of stream conditions. Sewage and Industrial Wastes 22:926-938.

———. 1953. Biological phases of stream pollution. Proceedings of the Pennsylvania Academy of Science 27:33-36.

———, M. H. Hohn, and J. H. Wallace. 1954. A new method for determining the pattern of the diatom flora. The Academy of Natural Sciences of Philadelphia, Notulae Naturae 259. 12pp.

Patten, B. C. 1962. Species diversity in net phytoplankton of Raritan Bay. Journal of Marine Research 20:57-75.

Pearl, R., and L. J. Reed. 1920. On the rate of growth of the population of the United States since 1790 and its mathematical representation. Proceedings of the National Academy of Sciences 6:275-288.

Pedersen, E. 1947. Østersens respirasjon. Undersøkelser utført ved Statens Utklekningsanstalt Flødevigen. Fiskeridir. Skr. Havundersøk. 8:1-51.

Penny, C., and C. Adams. 1863. Fourth Report, Royal Commission on Pollution of Rivers in Scotland. London. Vol. 2, Evidence:377-391.

Petersen, C. G. J. 1918. The sea bottom and its production of fish-food. Danish Biological Station Report 25. 62pp. + appendices.

———, and P. B. Jensen. 1911. Valuation of the sea. I. Animal life of the sea-bottom, its food and quantity. Danish Biological Station Report 20. 76pp. + appendices.

Phelps, E. B. 1944. Stream Sanitation. John Wiley and Sons, Inc., New York. xi + 276pp.

Phillips, J. 1935. Succession, development, the climax, and the complex organism: an analysis of concepts. Part III. The complex organism: conclusion. Journal of Ecology 23:488-508.

Phillips, R. W., and H. J. Campbell. 1961. The embryonic survival of coho salmon as influenced by some environmental conditions in gravel beds. Pacific Marine Fisheries Commission 14:60–73.

———, ———, W. L. Hug, and E. W. Claire. 1966. A study of the effect of logging on aquatic resources, 1960-66. Progress Memorandum, Fisheries 3. Research Division, Oregon Game Commission. 28pp.

Pianka, E. R. 1966. Latitudinal gradients in species diversity: a review of concepts. American Naturalist 100:33-46.

Pilgrim, R. L. C. 1953. Osmotic relations in molluscan contractile tissues, I. Isolated ventricle-strip preparations from lamellibranchs (*Mytilus edulis* L., *Ostrea edulis* L., *Anodonta cygnea* L.). Journal of Experimental Biology 30:297-317.

Planck, M. 1949. Scientific Autobiography and Other Papers. (Translated from German by Frank Gaynor.) Philosophical Library, New York. 192pp.

Poincaré, H. 1913. The Foundations of Science: Science and Hypothesis, The Value of Science, Science and Method. (Translated by G. B. Halsted.) The Science Press, New York. xi + 553pp.

Powers, E. B. 1917. The goldfish (*Carassius carassius*) as a test animal in the study of toxicity. Illinois Biological Monographs 4:[121]-193.

Precht, H. 1958. Concepts of the temperature adaptation of unchanging reaction systems of cold-blooded animals. Pages 50-78. *In* C. L. Prosser (Editor), Physiological Adaptation. American Physiological Society, Washington, D.C. 185pp.

Prosser, C. L. 1950. Temperature: metabolic aspects and perception. Pages 341-380. *In* C. L. Prosser (Editor), Comparative Animal Physiology. W. B. Saunders Company, Philadelphia. ix + 888pp.

———. 1958a. General summary: the nature of physiological adaptation. Pages 167-180. *In* C. L. Prosser (Editor), Physiological Adaptation. American Physiological Society, Washington, D.C. 185pp.

———. (Editor). 1958b. Physiological Adaptation. American Physiological Society, Washington, D.C. 185pp.

———. 1964. Perspectives of adaptation: theoretical aspects. Pages 11-25. *In* D. B. Dill, E. F. Adolph, and C. G. Wilber (Editors), Adaptation to the Environment. (Handbook of Physiology, Section 4.) American Physiological Society, Washington, D.C. ix + 1056pp.

———, and F. A. Brown, Jr. 1961. Comparative Animal Physiology. 2nd ed. W. B. Saunders Company, Philadelphia. ix + 688pp.

Prytherch, H. F. 1924. Experiments in the artificial propagation of oysters. U. S. Bureau of Fisheries Document 961. 14pp. (Appendix 11 to the Report of the U. S. Commissioner of Fisheries, 1923.)

Putman, G. B. 1967. The influence of metabolites on the growth and development of salmonid embryos and sac fry. M.S. thesis. Oregon State University, Corvallis. 40pp.

Pütter, A. 1924. Der Umfang der Kohlensäurereduktion durch die Planktonalgen. Pflüger's Archiv für die Gesamte Physiologie des Menschen und der Tiere 205:293–312.

Quetelet, A. 1835. Sur l'Homme et le Développement de ses Facultés, ou Essai de Physique Sociale. Louis Hauman et Comp., Bruxelles. xvi + 343pp. + plates.

Reish, D. J., and H. A. Winter. 1954. The ecology of Alamitos Bay, California, with special reference to pollution. California Fish and Game 40:105–121.

Renn, C. E. 1960. Biological aspects of industrial waste problems. Pages 156–159. *In* C. M. Tarzwell (Editor), Biological Problems in Water Pollution. Transactions of the Second Seminar, 1959. R. A. Taft Sanitary Engineering Center, U. S. Department of Health, Education, and Welfare. 285pp.

Reynolds, R. 1946. Cleanliness and Godliness. Doubleday and Company, Inc., Garden City, New York. viii + 326pp.

Reynoldson, T. B. 1947a. An ecological study of the enchytraeid worm population of sewage bacteria beds. Field Investigations. Journal of Animal Ecology 16:26-[37].

———. 1947b. An ecological study of the enchytraeid worm population in sewage bacteria beds. Laboratory experiments. Annals of Applied Biology 34:331–345.

———. 1948. An ecological study of the enchytraeid worm population of sewage bacteria beds: synthesis of field and laboratory data. Journal of Animal Ecology 17:27–38.

Richardson, R. E. 1921. Changes in the bottom and shore fauna of the middle Illinois River and its connecting lakes since 1913–1915 as a result of the increase, southward, of sewage pollution. Bulletin of the Illinois State Natural History Survey 14 (Article 4):33–75.

———. 1925. Changes in the small bottom fauna of Peoria Lake, 1920 to 1922. Bulletin of the Illinois State Natural History Survey 15 (Article 5): 327–388.

———. 1928. The bottom fauna of the middle Illinois River, 1913–1925: its distribution, abundance, valuation, and index value in the study of stream pollution. Bulletin of the Illinois State Natural History Survey 17:387-475.

Richman, S. 1958. The transformation of energy by *Daphnia pulex*. Ecological Monographs 28:[273]–291.

Ricker, W. E. 1946. Production and utilization of fish populations. Ecological Monographs 16:373–391.

———. 1954. Stock and recruitment. Journal of the Fisheries Research Board of Canada 11:559-623.

———. 1958. Handbook of Computations for Biological Statistics of Fish Populations. Fisheries Research Board of Canada Bulletin 119. 300pp.

———, and R. E. Foerster. 1948. Computation of fish production. Bingham Oceanographic Collection Bulletin 11:173–211.

Riley, G. A. 1956. Oceanography of Long Island Sound, 1952-1954. IX. Production and utilization of organic matter. Bulletin of the Bingham Oceanographic Collection 15:324-344.

———. 1957. Phytoplankton of the north central Sargasso Sea. Limnology and Oceanography 2:252-270.

Robertson, J. D. 1960. Osmotic and ionic regulation. Pages 317-339. *In* T. H. Waterman (Editor), The Physiology of Crustacea. Vol. 1, Metabolism and Growth. Academic Press Inc., New York. xvii + 670 pp.

Root, R. W. 1931. The respiratory function of the blood of marine fishes. Biological Bulletin 61:427-456.

Roslansky, J. D. (Editor). 1967. The Control of Environment. (1966 Nobel Conference discussion.) North-Holland Publishing Company, Amsterdam. xi + 112pp.

Rubner, M. 1885. Calorimetrische Untersuchungen. Zeitschrift für Biologie 21:250-334, 337-410.

———. 1894. Die Quelle der thierischen Wärme. Zeitschrift für Biologie 30:73-142.

Rudolfs, W., G. E. Barnes, G. P. Edwards, H. Heukelekian, E. Hurwitz, C. E. Renn, S. Steinberg, and W. F. Vaughan. 1950. Review of literature on toxic materials affecting sewage treatment processes, streams, and B.O.D. determinations. Sewage and Industrial Wastes 22:1157-1191.

Ruggles, C. P. 1965. Juvenile sockeye studies in Owikeno Lake, British Columbia. Canadian Fish Culturist 36:3-21.

Ruttner, F. 1953. Fundamentals of limnology. (English translation by D. G. Frey and F. E. J. Fry.) University of Toronto Press, Toronto. xi + 242pp.

Sauberer, F. 1939. Beiträge zur Kenntnis des Lichtklimas einiger Alpenseen. Internationale Revue der Gesamten Hydrobiologie und Hydrographie 39:20-55.

Savvaitova, K. A., and Ia. S. Reshetnikov. 1961. Pitanie razlichnykh biologicheskikh goltsa—*Salvelinus malma* (Walb.) v nekotorykh vodoemakh Kamchatki. Voprosy Ikhtiologii 1:127-135. (Translated by R. E. Foerster, 1962. The food of different biological forms of the Dolly Varden char, *Salvelinus malma* (Walb.), in certain Kamchatka waters. Fisheries Research Board of Canada Translation Series 373. 12pp.)

Schiffman, R. H., and P. O. Fromm. 1959. Chromium-induced changes in the blood of rainbow trout, *Salmo gairdnerii*. Sewage and Industrial Wastes 31:205-211.

Schneirla, T. C. 1951. A consideration of some problems in the ontogeny of family life and social adjustment in various infrahuman animals. Pages 81-106. *In* M. J. E. Senn (Editor), Problems of Infancy and Childhood, Transaction of 4th Conference. Josiah Macy, Jr., Foundation, New York. 181pp.

Schröter, C. 1902. Die Vegetation des Bodensee, II. Teil. *In* Der Bodensee-Forschungen neunter Abschnitt. Lindau. 86pp.

_____, and O. Kirchner. 1896. Die Vegetation des Bodensee, I. Teil. *In* Der Bodensee-Forschunger neunter Abschnitt. Lindau. 122pp.

Sedgwick, W. T. 1889. Recent progress in biological water-analysis. Journal of the New England Water Works Association 4:50-55.

Seltzer, L. B. 1965. Cleveland: saving Lake Erie. Saturday Review, October 23, 1965. pp.36, 41.

Shaw, P. A., and J. A. Maga. 1943. The effect of mining silt on yield of fry from salmon spawning beds. California Fish and Game 29:29-41.

Shelford, V. E. 1917. An experimental study of the effects of gas waste upon fishes, with especial reference to stream pollution. Bulletin of the Illinois State Laboratory of Natural History 11:[381]-412 + figs.

_____. 1929. Laboratory and Field Ecology. The Williams & Wilkins Company, Baltimore. xii + 608pp.

_____, and W. C. Allee. 1913. The reactions of fishes to gradients of dissolved atmospheric gases. Journal of Experimental Zoology 14:207-266.

Shepard, M. P. 1955. Resistance and tolerance of young speckled trout (*Salvelinus fontinalis*) to oxygen lack, with special reference to low oxygen acclimation. Journal of the Fisheries Research Board of Canada 12:387-446.

Sherrington, C. 1906. The Integrative Action of the Nervous System. Yale University Press, New Haven, Connecticut. xvi + 411pp.

Shumway, D. L., C. E. Warren, and P. Doudoroff. 1964. Influence of oxygen concentration and water movement on the growth of steelhead trout and coho salmon embryos. Transactions of the American Fisheries Society 93:342-356.

Silliman, R. P. 1943. Studies on the Pacific pochard or sardine (*Sardinops caerulea*). 5. A method of computing mortalities and replacements. U.S. Fish and Wildlife Service, Special Scientific Report 24. 10pp

Silver, S. J., C. E. Warren, and P. Doudoroff. 1963. Dissolved oxygen requirements of developing steelhead trout and chinook salmon embryos at different water velocities. Transactions of the American Fisheries Society 92:327-343.

Slobodkin, L. B. 1959. Energetics in *Daphnia pulex* populations. Ecology 40:232-243.

_____. 1960. Ecological energy relationships at the population level. American Naturalist 94:213-236.

_____. 1961. Growth and Regulation of Animal Populations. Holt, Rinehart, and Winston, New York. viii + 184pp.

_____. 1962. Energy in animal ecology. Pages 69-101. *In* J. B. Cragg (Editor), Advances in Ecological Research. Vol. 1. Academic Press Inc., New York. xi + 203pp.

Sluckin, W. 1965. Imprinting and Early Learning. Aldine Publishing Company, Chicago. x + 147pp.

Smith, F. E. 1954. Quantitative aspects of population growth. Pages 277–294. *In* E. J. Boell (Editor), Dynamics of Growth Processes. Growth Symposium 11. Princeton University Press, Princeton, New Jersey. vii + 304pp.

Smith, H. S. 1935. The role of biotic factors in the determination of population densities. Journal of Economic Entomology 28:873–898.

Smith, H. W. 1930. The absorption and excretion of water and salts by marine teleosts. American Journal of Physiology 93:480–505.

——. 1932. Water regulation and its evolution in the fishes. Quarterly Review of Biology 7:1–26.

Smith, S. 1957. Early development and hatching. Pages 323–359. *In* Margaret E. Brown (Editor), The Physiology of Fishes. Vol. 1, Metabolism. Academic Press Inc., New York. xiii + 447pp.

Smoker, W. A. 1954. A preliminary review of salmon fishing trends on inner Puget Sound. Washington Department of Fisheries Research Bulletin 2. 55pp.

Solomon, M. E. 1949. The natural control of animal populations. Journal of Animal Ecology 18:1–35.

Spalding, D. A. 1873. Instinct: with original observations on young animals. Macmillan's Magazine 27:282–293. (Reprinted 1954. British Journal of Animal Behaviour 2:2–11.)

Sprague, J. B. 1964a. Avoidance of copper-zinc solutions by young salmon in the laboratory. Journal of the Water Pollution Control Federation 36:990–1004.

——. 1964b. Lethal concentrations of copper and zinc for young Atlantic salmon. Journal of the Fisheries Research Board of Canada 21:17–26.

——. 1969. Review paper: Measurement of pollutant toxicity to fish—1. Bioassay methods for acute toxicity. Water Research (Pergamon Press) 3:793–821.

——. 1970a. Review paper: Measure of pollutant toxicity to fish—2. Utilizing and applying bioassay results. Water Research (Pergamon Press) 4: In press.

——. 1970b. Review paper: Measurement of pollutant toxicity to fish—3. Sublethal effects and "safe" concentrations. Water Research (Pergamon Press) 4: In press.

——, and D. E. Drury. 1969. Avoidance reactions of salmonid fish to representative pollutants. *In* Advances in Water Pollution Research, Vol. 1. Proceedings Fourth International Conference. Pergamon Press, Oxford. In press.

——, P. F. Elson, and R. L. Saunders. 1965. Sublethal copper-zinc pollution in a salmon river—a field and laboratory study. International Journal Air and Water Pollution (Pergamon Press) 9:531–543.

——, and B. Ann Ramsay. 1965. Lethal levels of mixed cooper-zinc solutions for juvenile salmon. Journal of the Fisheries Research Board of Canada 22:425–432.

Steemann Nielsen, E. 1955. The production of organic matter by the phytoplankton in a Danish lake receiving extraordinarily great amounts of nutrient salts. Hydrobiologia 7:68–74.

Strickland, J. D. H. 1960. Measuring the production of marine phytoplankton. Fisheries Research Board of Canada Bulletin 122. viii + 172pp.

Sumner, F. B. 1905. The physiological effects upon fishes of changes in the density and salinity of water. Bulletin of the Bureau of Fisheries 25:53–108.

——, and P. Doudoroff. 1938. Some experiments upon temperature acclimatization and respiratory metabolism in fishes. Biological Bulletin 74:403–429.

Suter, R., and Emmeline Moore. 1922. Stream Pollution Studies. State of New York Conservation Commission. J. B. Lyon Company, Printers, Albany. pp. 3–27.

Sverdrup, H. U., M. W. Johnson, and R. H. Fleming. 1942. The Oceans: Their Physics, Chemistry, and General Biology. Prentice-Hall, Inc., New York. x + 1087pp.

Sykes, J. E., and B. A. Lehman. 1957. Past and present Delaware River shad fishery and considerations for its future. U.S. Fish and Wildlife Research Report 46. 25pp.

Tamiya, H. 1957. Mass culture of algae. Annual Review of Plant Physiology 8:309–334.

Tansley, A. G. 1935. The use and abuse of vegetational concepts and terms. Ecology 16:284–307.

Tarzwell, C. M. 1963. Sanitational limonology. Pages 653–666. *In* D. G. Frey (Editor), Limnology in North America. University of Wisconsin Press, Madison. xvii + 734pp.

———, and A. R. Gaufin. 1953. Some important biological effects of pollution often disregarded in stream surveys. Pages 295–316. *In* Proceedings 8th Industrial Waste Conference. Purdue University Engineering Bulletin, Extension Series 83. viii + 562pp.

———, and C. Henderson. 1957. Toxicity of dieldrin to fish. Transactions of the American Fisheries Society 86:245–257.

Tchernavin, V. 1939. The origin of salmon. Salmon and Trout Magazine 95:105–140.

Teal, J. M. 1957. Community metabolism in a temperate cold spring. Ecological Monographs 27:[283]–302.

Terroine, E. F., and R. Wurmser. 1922. L'énergie de croissance. I. Le développement de l'*Aspergillus niger*. Bulletin de la Société de Chimie Biologique 4:519–567.

Thomas, E. A. 1957. Der Zürichsee, sein Wasser und sein Boden. Jahrbuch vom Zürichsee 17:173–208.

Thomas, H. J. 1954. The oxygen uptake of the lobster (*Homarus vulgaris* Edw.). Journal of Experimental Biology 31:228–251.

Thompson, D. H. 1925. Some observations on the oxygen requirements of fishes in the Illinois River. Bulletin of the Illinois State Natural History Survey 15:423–437.

———, and F. D. Hunt. 1930. The fishes of Champaign County: a study of the distribution and abundance of fishes in small streams. Bulletin of the Illinois State Natural History Survey 19:1–101.

Thompson, D. W. 1942. On Growth and Form. The Macmillan Company, New York. 1116pp.

Thompson, W. R. 1956. The fundamental theory of natural and biological control. Annual Review of Entomology 1:379–402.

Thorpe, W. H. 1951. The definition of some terms used in animal behaviour studies. Bulletin of Animal Behaviour 9:34–40.

———. 1956. Learning and Instinct in Animals. Methuen and Co. Ltd., London. viii + 493pp.

———. 1963a. Ethology and the coding problem in germ cell and brain. Zeitschrift für Tierpsychologie 20:529–551.

———. 1963b. Learning and Instinct in Animals. 2nd ed. Methuen and Co. Ltd., London. x + 558pp.

Thucydides. c. 431 B.C. The History of the Peloponnesian War. Reprinted 1952. (Translated by R. Crawley.) Pages 343–616. *In* Great Books of the Western World, Vol. 6. Encyclopaedia Brittanica, Inc., Chicago. xi + 616pp.

Tinbergen, N. 1942. An objectivistic study of the innate behaviour of animals. Bibliotheca Biotheoretica (Leiden series) 1942:37–98.

———. 1951. The Study of Instinct. Oxford, at the Clarendon Press. xii + 228pp.

———. 1963. On aims and methods of ethology. Zeitschrift für Tierpsychologie 20:410–433.

———. 1965. Behavior and natural selection. *In* J. A. Moore (Editor), Ideas in Modern Biology. Proceedings 16th International Congress of Zoology 6:519–542.

Tokar, E. M. 1968. Some chronic effects of exposure to kraft mill effluent on juvenile chinook salmon. M.S. thesis. Oregon State University, Corvallis. 59pp.

Trama, F. B. 1954. The acute toxicity of copper to the common bluegill (*Lepomis macrochirus* Rafinesque). The Academy of Natural Sciences of Philadelphia, Notulae Naturae 257. 13pp.

Udall, S. L. 1964. A message for biologists. Bioscience 14(11):17–18.

Uexküll, J. von. 1909. Umwelt und Innenwelt der Tiere. Springer-Verlag, Berlin.

U.S. Federal Water Pollution Control Administration. 1968. Water Quality Criteria. Report of the National Technical Advisory Committee. U.S. Department of the Interior. x + 234pp.

U.S. President's Science Advisory Committee. 1965. Restoring the Quality of Our Environment. Environmental Pollution Panel Report. U.S. Government Printing Office. xii + 317pp.

U.S. Public Health Service. 1962. Drinking Water Standards. Publication 956. U.S. Department of Health, Education, and Welfare, Washington, D.C. viii + 61pp.

Vasnetsov, V. V. 1953. O zakonomernostyakh rosta ryb. [Patterns of fish growth.] *In* Ocherki po Obshchim Voprosam Biologii. Akademiya Nauk Press, Moscow.

Vaux, W. G. 1962. Interchange of stream and intragravel water in a salmon spawning riffle. U.S. Fish and Wildlife Service, Special Scientific Report—Fisheries 405. iii + 11pp.

Verduin, J. 1956. Primary production in lakes. Limnology and Oceanography 1:85-91.

Verhulst, P. F. 1838. Notice sur la loi que la population suit dans son accroissement. Correspondance Mathématique et Physique 10:113-121.

Virgil (Publius Vergilius Maro) 30 B.C. Georgic II. (Translated by J. Rhoades.) Reprinted 1952. Pages 52–66. *In* The Poems of Virgil. Great Books of the Western World, Vol. 13. Encyclopaedia Brittanica, Inc., Chicago. vii + 379pp.

Volterra, V. 1927. Variazioni e fluttuazion del numero d'individui in specie animali conviventi. Atti dell'Accademia Nazionale dei Lincei. Memorie Series 6(2):31–113.

_____. 1931. Leçons sur la Théorie Mathématique de al Lutte pour la Vie. (Cahiers Scientifiques Fasc. 7.) Gauthier-Villars, Paris. vi + 214pp.

Vonk, H. J. 1960. Digestion and metabolism. Pages 291–316. *In* T. H. Waterman (Editor), The Physiology of Crustacea. Vol. 1, Metabolism and Growth. Academic Press Inc., New York. xvii + 670pp.

Waddington, C. H. 1957. The Strategy of the Genes. The Macmillan Company, New York. ix + 262pp.

_____. 1959. Canalization of development and genetic assimilation of acquired characters. Nature 183:1654-1655.

_____. 1960. Evolutionary adaptation. Pages 381–402. *In* S. Tax (Editor), The Evolution of Life. (Evolution after Darwin, Vol. 1.) The University of Chicago Press, Chicago. viii + 629pp.

Warner, R. E. 1967. Bio-assays for microchemical environmental contaminants: with special reference to water supplies. Bulletin of the World Health Organization 36:181-207.

_____, Karen K. Peterson, and L. Borgman. 1966. Behavioural pathology in fish: a quantitative study of sublethal pesticide toxication. Journal of Applied Ecology 3(Supplement):223-247.

Warren, C. E., and G. E. Davis. 1967. Laboratory studies on the feeding, bioenergetics, and growth of fishes. Pages 175-214. *In* S. D. Gerking (Editor), The Biological Basis of Freshwater Fish Production. Blackwell Scientific Publications, Oxford. xiv + 495pp.

_____, and P. Doudoroff. 1958. The development of methods for using bioassays in the control of pulp mill waste disposal. Tappi 41(8):211A-216A.

_____, J. H. Wales, G. E. Davis, and P. Doudoroff. 1964. Trout production in an experimental stream enriched with sucrose. Journal of Wildlife Management 28:617-660.

Water Pollution Research Board. 1956. Efficiency of micro-organisms in sewage treatment. Water Pollution Research (Department of Scientific and Industrial Research, London) 1955:55-57.

Water Quality Act of 1956. Public Law 89-235. U.S. Statutes at Large 79:903-910.

Waterman, T. H. (Editor). 1960. The Physiology of Crustacea. Vol. 1, Metabolism and Growth. Academic Press Inc., New York. xvii + 670pp.

_____ (Editor). 1961. The Physiology of Crustacea. Vol. 2, Sense Organs, Integration, and Behavior. Academic Press Inc., New York. xiv + 681pp.

Weigelt, C., O. Saare, and L. Schwab. 1885. Die Schädigung von Fischerei und Fischzucht durch Industrie und Haus Abwasser. Archiv für Hygiene 3:39-117.

Weiss, C. M. 1961. Physiological effect of organic phosphorus insecticides on several species of fish. Transactions of the American Fisheries Society 90:143-152.

Welch, P. S. 1952. Limnology. McGraw-Hill Book Company, Inc., New York. xi + 538pp.

Wells, W. F. 1927. Report of the experimental shellfish station. New York State Conservation Department Report 16. 22pp.

Whipple, G. C. 1927. The Microscopy of Drinking Water. 4th ed. Revised by G. M. Fair and M. C. Whipple. John Wiley & Sons, Inc., New York. xix + 586pp. + 19 plates.

_____, and M. C. Whipple. 1911. Solubility of oxygen in sea water. Journal of the American Chemical Society 33:362-365.

White, G. F. 1966. Formation and role of public attitudes. Pages 105-127. *In* H. Jarrett (Editor), Environmental Quality in a Growing Economy. Published for Resources for the Future, Inc., by The Johns Hopkins Press, Baltimore. xv + 173pp.

Whitehead, A. N. 1925. Science and the Modern World. Lowell lectures, 1925. The Macmillan Company, New York. xi + 296pp.

Whitman, C. O. 1899. Animal behavior. 1898 Biological Lectures, Woods Hole 6:285–338.

Whitmore, C. M., C. E. Warren, and P. Doudoroff. 1960. Avoidance reactions of salmonid and centrarchid fishes to low oxygen concentration. Transactions of the American Fisheries Society 89:17-26.

Wickett, W. P. 1954. The oxygen supply to salmon eggs in spawning beds. Journal of the Fisheries Research Board of Canada 11:933-953.

Wilber, C. G. 1965. The biology of water toxicants in sublethal concentrations. Pages 326-331. *In* Biological Problems in Water Pollution. Third Seminar, 1962. R. A. Taft Sanitary Engineering Center, U.S. Department of Health, Education, and Welfare. 424pp.

Wilbur, K. M., and C. M. Yonge. 1964a. Physiology of Mollusca. Vol. 2. Academic Press Inc., New York. xii + 645pp.

_____, and _____. 1964b. Physiology of Mollusca. Vol. 1. Academic Press Inc., New York. xiii + 473pp.

Wilhm, J. L., and T. C. Dorris. 1968. Biological paramenters for water quality criteria. Bioscience 18:477-481.

Williams, R. T. 1962. Detoxication mechanisms *in vivo*. Pages 1-12. *In* B. B. Brodie and E. G. Erdös (Editors), Metabolic Factors Controlling Duration of Drug Action. (First International Pharmacological Meeting, Vol. 6.) The Macmillan Company, New York. xviii + 330pp.

Wilson, A. T. 1952. The cholera vibrios. Pages 507-514. *In* R. J. Dubos (Editor), Bacterial and Mycotic Infections of Man. 2nd ed. J. B. Lippincott Company, Philadelphia. xiv + 886pp.

Winberg, G. G. 1956. Intensivnost obmena i pishchevye potrebnosti ryb. Nauchnye Trudy Belorusskova Gosudarstvennovo Universiteta imeni V. I. Lenina, Minsk. 253pp. (Translated 1960. Rate of metabolism and food requirements of fishes. Fisheries Research Board of Canada Translation Series 194. 202pp + 32 tables.)

Wisby, W. J., and A. D. Hasler. 1954. Effect of olfactory occlusion on migrating silver salmon (*O. kisutch*). Journal of the Fisheries Research Board of Canada 11:472-478.

Woelke, C. E. 1960. Preliminary report of laboratory studies on the relationship between fresh sulfite waste liquor and the reproductive cycle of the Olympic oyster, *Ostrea lurida*. Pages 107-148. *In* Reports on Sulfite Waste Liquor in a Marine Environment and Its Effects on Oyster Larvae. Washington Department of Fisheries Research Bulletin 6. 161pp.

_____. 1965. Environmental requirements of marine invertebrates. Pages 67–77. *In* C. M. Tarzwell (Editor), Biological Problems in Water Pollution. Transactions of the Third Seminar, 1962. R. A. Taft Sanitary Engineering Center, U.S. Department of Health, Education, and Welfare. 424pp.

_____. 1968. Application of shellfish bioassay results to the Puget Sound pulp mill pollution problem. Northwest Science 42:125-133.

Wolvekamp, H. P., and T. H. Waterman. 1960. Respiration. Pages 35-100. *In* T. H. Waterman (Editor), The Physiology of Crustacea. Vol. 1, Metabolism and Growth. Academic Press Inc., New York. xvii + 670pp.

Wood, E. M. 1960. Definitive diagnosis of fish mortalities, Journal Water Pollution Control Federation 32:994–999.

Wright, J. 1966. The Coming Water Famine. Coward-McCann, Inc., New York. 255pp.

Wright, S., and W. M. Tidd. 1963. Summary of limnological investigations in western Lake Erie in 1929 and 1930. Transactions of the American Fisheries Society 63:271-281.

Wuhrmann, K. 1952. Sur quelques principes de la toxicologie du poisson. (The protection of rivers against pollution.) Bulletin Centre Belge Étude et Document. Eaux 15:77-85.

Wurtz, C. B. 1955. Stream biota and stream pollution. Sewage and Industrial Wastes 27:1270-1278.

Wynne-Edwards, V. C. 1962. Animal Dispersion in Relation to Social Behavior. Oliver and Boyd, Edinburgh. xi + 653pp.

Yonge, C. M. 1960. Oysters. Collins, London. xiv + 209pp.

INDEX

Note: Page numbers in *italics* refer to illustrations. Page numbers followed by (t) refer to tables.